VOLUME ONE HUNDRED AND TWENTY NINE

PROGRESS IN MOLECULAR BIOLOGY AND TRANSLATIONAL SCIENCE

The Molecular Basis of Viral Infection

VOLUME ONE HUNDRED AND TWENTY NINE

Progress in MOLECULAR BIOLOGY AND TRANSLATIONAL SCIENCE

The Molecular Basis of Viral Infection

Edited by

P.J. KLASSE

Department of Microbiology and Immunology, Weill Cornell Medical College, Cornell University, New York, USA

ELSEVIER

AMSTERDAM • BOSTON • HEIDELBERG • LONDON
NEW YORK • OXFORD • PARIS • SAN DIEGO
SAN FRANCISCO • SINGAPORE • SYDNEY • TOKYO

Academic Press is an imprint of Elsevier

Academic Press is an imprint of Elsevier
225 Wyman Street, Waltham, MA 02451, USA
525 B Street, Suite 1800, San Diego, CA 92101-4495, USA
125 London Wall, London, EC2Y 5AS, UK
The Boulevard, Langford Lane, Kidlington, Oxford OX5 1GB, UK

First edition 2015

ISBN: 978-0-12-802461-4
ISSN: 1877-1173

For information on all Academic Press publications
visit our website at store.elsevier.com

CONTENTS

CONTRIBUTORS

Kristine Niss Arfelt
Faculty of Health and Medical Sciences, Department of Neuroscience and Pharmacology, Laboratory for Molecular Pharmacology, University of Copenhagen, Copenhagen, Denmark

Marie Borggren
Virus Research and Development Laboratory, Department of Microbiological Diagnostics and Virology, Statens Serum Institut, Copenhagen, Denmark

Dennis Brown
Department of Molecular and Structural Biochemistry, North Carolina State University, Raleigh, North Carolina, USA

François-Loïc Cosset
CIRI, International Center for Infectiology Research, Team EVIR, Université de Lyon; Inserm U1111; Ecole Normale Supérieure de Lyon; Centre International de Recherche en Infectiologie, Université Claude Bernard Lyon 1; CNRS, UMR 5308, and LabEx Ecofect, Université de Lyon, Lyon, France

Emilia Cristiana Cuccurullo
Centre for Integrative Biology, University of Trento, Trento, Italy

Nick Davis-Poynter
Queensland Children's Medical Research Institute, Sir Albert Sakzewski Virus Research Centre, The University of Queensland & Royal Children's Hospital, Brisbane, Queensland, Australia

Michael S. Diamond
Departments of Medicine, Molecular Microbiology, Pathology & Immunology, Center for Human Immunology and Immunotherapy Programs, Washington University School of Medicine, St. Louis, Missouri, USA

Florian Douam
CIRI, International Center for Infectiology Research, Team EVIR, Université de Lyon; Inserm U1111; Ecole Normale Supérieure de Lyon; Centre International de Recherche en Infectiologie, Université Claude Bernard Lyon 1; CNRS, UMR 5308; LabEx Ecofect, Université de Lyon, Lyon, and CNRS, UMR 5557 Ecologie Microbienne, Microbial Dynamics and Viral Transmission Team, Université Claude Bernard Lyon 1, Villeurbanne, France

Suzan Fares
Faculty of Health and Medical Sciences, Department of Neuroscience and Pharmacology, Laboratory for Molecular Pharmacology, University of Copenhagen, Copenhagen, Denmark

Helen Elizabeth Farrell
Queensland Children's Medical Research Institute, Sir Albert Sakzewski Virus Research Centre, The University of Queensland & Royal Children's Hospital, Brisbane, Queensland, Australia

Eric O. Freed
Virus-Cell Interaction Section, HIV Drug Resistance Program, Center for Cancer Research, National Cancer Institute, Frederick, Maryland, USA

Alexander L. Greninger
School of Medicine, University of California, San Francisco, California, USA

Raquel Hernandez
Department of Molecular and Structural Biochemistry, North Carolina State University, Raleigh, North Carolina, USA

Leo C. James
Protein and Nucleic Acid Chemistry Division, Medical Research Council Laboratory of Molecular Biology, Cambridge, United Kingdom

Marianne Jansson
Department of Laboratory Medicine, Lund University, Lund, Sweden, and Department of Microbiology, Tumor and Cell biology, Karolinska Institute, Stockholm, Sweden

Eric M. Jurgens
Department of Pediatrics, Weill Cornell Medical College, Cornell University, New York, USA

P.J. Klasse
Department of Microbiology and Immunology, Weill Cornell Medical College, Cornell University, New York, USA

Dimitri Lavillette
CIRI, International Center for Infectiology Research, Team EVIR, Université de Lyon; Inserm U1111; Ecole Normale Supérieure de Lyon; Centre International de Recherche en Infectiologie, Université Claude Bernard Lyon 1; CNRS, UMR 5308, Lyon, and CNRS, UMR 5557 Ecologie Microbienne, Microbial Dynamics and Viral Transmission Team, Université Claude Bernard Lyon 1, Villeurbanne, France

Carsten Magnus
Institute of Medical Virology, University of Zurich, Zurich, Switzerland

William A. McEwan
Protein and Nucleic Acid Chemistry Division, Medical Research Council Laboratory of Molecular Biology, Cambridge, United Kingdom

Ann-Sofie Mølleskov-Jensen
Department of Neuroscience and Pharmacology, Laboratory for Molecular Pharmacology, University of Copenhagen, Copenhagen, Denmark

Anne Moscona
Department of Pediatrics, and Department of Microbiology and Immunology, Weill Cornell Medical College, Cornell University, New York, USA

Martha Trindade Oliveira
Queensland Children's Medical Research Institute, Sir Albert Sakzewski Virus Research Centre, The University of Queensland & Royal Children's Hospital, Brisbane, Queensland, Australia

Laura M. Palermo
Department of Pediatrics, and Department of Microbiology and Immunology, Weill Cornell Medical College, Cornell University, New York, USA

Jean-Louis Palgen
Department of Pediatrics, Weill Cornell Medical College, Cornell University, New York, USA, and Department of Biology, Ecole Normale Supérieure, Lyon, France

Theodore C. Pierson
Viral Pathogenesis Section, National Institute of Allergy and Infectious Diseases, National Institutes of Health, Bethesda, Maryland, USA

Massimo Pizzato
Centre for Integrative Biology, University of Trento, Trento, Italy

Matteo Porotto
Department of Pediatrics, Weill Cornell Medical College, Cornell University, New York, USA

Roland R. Regoes
Institute of Integrative Biology, ETH Zurich, Zurich, Switzerland

Mette M. Rosenkilde
Faculty of Health and Medical Sciences, Department of Neuroscience and Pharmacology, Laboratory for Molecular Pharmacology, University of Copenhagen, Copenhagen, Denmark

Philip R. Tedbury
Virus-Cell Interaction Section, HIV Drug Resistance Program, Center for Cancer Research, National Cancer Institute, Frederick, Maryland, USA

Chiara Valentini
Centre for Integrative Biology, University of Trento, Trento, Italy

Ricardo Vancini
Department of Molecular and Structural Biochemistry, North Carolina State University, Raleigh, North Carolina, USA

PREFACE

> *True, in a single conversation with someone we can discern particular traits. But it is only through repeated encounters in varied circumstances that we can recognize these traits as characteristic and essential. For a writer, for a musician, or for a painter, this variation of circumstances that enables us to discern, by a sort of experimentation, the permanent features of character is found in the variety of the works themselves.*
>
> ***From Marcel Proust's preface to John Ruskin's* The Bible of Amiens.**

The Ebola River, a tributary to the Congo, flows north of the village of Yambuku. There, in 1976, hundreds of people rapidly succumbed to a lethal hemorrhagic fever. The cause, Ebola virus, is a member of the genus Filoviridae, comprising single-stranded negative-RNA viruses with the idiosyncratic filamentous or worm-like morphology that has given them their name.[1] As a tragic Ebola epidemic now rages in West Africa, killing thousands, efforts to find a cure and a vaccine will intensify. It is already striking how the advancing field of filovirus studies shares questions and problems with the investigations—some old and established, some rapidly evolving—of other viruses, as exemplified in this book. Thus, knowledge is developing of how filoviruses enter cells,[2] the identity of the receptors for the virus on susceptible cells,[3,4] which cellular genes these viruses activate, how that activation affects the innate immune responses and pathogenesis,[5,6] how the virus is neutralized by antibodies, and which antibodies protect against infection.[7–10]

Thomas Milton Rivers, working at The Rockefeller Institute, which I see through the window when composing this Preface, established virology as a discipline separate from bacteriology.[11] He perspicaciously stated: "Viruses appear to be obligate parasites in the sense that their reproduction is dependent on living cells." His anthology *Filterable Viruses* (Baltimore: Williams and Wilkins, 1928) covered everything worth knowing about viruses at the time. Today, when the number of PubMed entries in virology is around a million, an anthology in general virology must be considerably less comprehensive. The current collection encompasses a number of topical forays into molecular aspects of viral replication and coexistence with host organisms. The chapters in this anthology offer rich opportunities to compare how specific questions are answered for different viruses. As with the example of Ebola virus above, certain themes recur and the emerging patterns of similarities and differences may provoke new questions and stimulate collaborations among virologists with distinct specialties.

Their eclectic diversity notwithstanding, the chapters form a narrative of sorts, first adhering kairologically to the replicative cycle that viruses largely share, and then broadening to depict wider aspects of virus–host interactions. Thus, the first three chapters depict entry into susceptible cells by different viruses: paramyxoviruses (Chapter "Unity in Diversity: Shared Mechanism of Entry Among Paramyxoviruses," Palgen *et al.*), alphavirus (Chapter "Alphavirus Entry into Host Cells," Vancini *et al.*), and hepatitis C virus (Chapter "The Mechanism of HCV Entry into Host Cells," Douam *et al.*). Entry requires viral interactions with specific receptors, as delineated in these chapters. Enveloped viruses can potentially enter either by fusing at the cell surface or by first following one of several distinct endocytic routes and then fusing with the endocytic vesicle. The exact mechanisms have been hotly debated for many viruses and these chapters bring new clarity and perhaps some surprises.

Then we shift the scope somewhat and consider the evolution of the entry mediator of HIV, viz., its envelope glycoprotein, Env. Now Env is extremely variable and capable of modulating its interactions with various host molecules: with mannose C-type lectins, which are possibly involved in attachment and transmission, and with the main receptor for the virus, CD4, as well as with the obligate coreceptors, which the virus fastidiously picks among a subset of the seven-transmembrane chemokine receptors. The strengths of the receptor interactions evolve concomitantly with the selection pressure that waxes and wanes as the virus escapes from the coevolving specificities of neutralizing antibodies and gets transmitted to immunologically naïve host organisms (Chapter "The Evolution of HIV Interactions with Coreceptors and Mannose C-Type Lectin Receptors," Borggren and Jansson).

Having obliquely touched on neutralization, we then narrow the focus to what is probably the quantitatively best understood example of how antibodies block viral infectivity, i.e., neutralization of flaviviruses: in Chapter "A Game of Numbers: The Stoichiometry of Antibody-Mediated Neutralization of Flavivirus Infection," Pierson and Diamond analyze the fine stoichiometric details of neutralizing antibody binding to flavivirions and explain why the same antibodies can either neutralize or enhance infectivity depending on what numbers bind to the virion.

We continue the theme of neutralization but switch to the naked adenoviruses, common causes of gastroenteritis, conjunctivitis, otitis, and respiratory tract infections. In Chapter "TRIM21-Dependent Intracellular

Antibody Neutralization of Virus Infection," McEwan and James describe the groundbreaking discovery that the cytoplasmic factor TRIM21 joins antibodies to effect cytoplasmic neutralization of adenovirus. TRIM21 might also augment the antibody-mediated neutralization of other naked viruses. That cytoplasmic neutralization occurs has long been suggested, even for enveloped viruses, but without decisive evidence; such claims have sometimes been erroneously linked to the kinetics and stoichiometry of neutralization.[12] But the newly discovered definitive mechanism, which depends on the traversal of antibody–capsid complexes into the cytoplasm, has its own distinct quantitative implications.

We then extend the consideration of postentry events to later steps in the replicative cycle, including viral assembly and release. The first example is how picornaviruses, although they as naked viruses lack membranes in their virions, interact with intracellular membranes and highjack components of the secretory pathway for their replication (Chapter "Picornavirus–Host Interactions to Construct Viral Secretory Membranes," Greninger). The story then returns to enveloped viruses in the form of retroviruses and the extensive cast of auxiliary factors they have evolved to counteract cellular barriers to their replication (Chapter "Retroviral Factors Promoting Infectivity," Cuccurullo *et al.*). Thereafter, the tale turns to the cytoplasmic domains of the retroviral Env proteins (Chapter "The Cytoplasmic Tail of Retroviral Envelope Glycoproteins," Tedbury and Freed). These cytoplasmic and intravirional tails are particularly long among the lentiviruses, to which HIV belongs. They contain motifs for endocytosis and trafficking of the Env proteins; they even exert transmembraneous conformational effects on the outer Env, the target for neutralizing antibodies. Toward the end of the replicative cycle, when Env gets incorporated into the viral envelope, these tails juxtapose the internal Gag precursor that drives the budding of virions from the cell surface. Furthermore, when retroviruses and other enveloped viruses assemble and egress, they usurp multiple cellular factors, evincing quintessential parasitism.

The scene is then set for some analyses of the free virus particles themselves. First, the classic virological measurement of inert-to-infective particle ratio is examined in general and for particular viruses (Chapter "Molecular determinants of the ratio of inert to infectious virus particles," Klasse). Then, taking the primate lentiviruses, which include HIV, as examples, Regoes and Magnus quantitatively dissect the contributions of individual Env subunits to the function of Env trimers, and of trimers to virion infectivity. These insights segue into analyses of the probabilities that inocula containing certain infectious doses establish infection in the host organism (Chapter "The Role of Chance in Primate Lentiviral Infectivity: From Protomer to Host Organism").

The ascent from the molecular determinants of individual virion infectivity up to the establishment of infection at the level of a host organism, thus crowning the accounts of the progression through the viral replicative cycle at the cellular level, finally ushers in the topic of virus–host coexistence. Viruses often cause disease. Their interactions with the innate and adaptive immune systems modulate their pathogenesis. Host and virus have evolved together, sometimes for a long time. Herpesviruses may have diverged into the three families alpha-, beta-, and gammaherpesvirinae 180–220 million years ago, cospeciations among mammals having continued during the past 80 million years.[13] In spite of those time lapses, herpesviruses can still get on our nerves (as when herpes simplex virus survives in the ganglion Gasseri or Varicella-Zoster virus gives facial palsy). Although far from perfect, the host's adaptation to these longtime companions is sophisticated. Thus, the vast majority of humans carry latent infections with Epstein–Barr virus but are symptom-free. And the herpesviruses have developed intricate interactions with the host-immune system in long-running evolutionary games with continually tied outcomes. As described in Chapters "Virus-Encoded 7 Transmembrane Receptors" by Mølleskov-Jensen *et al.* and "EBV, the Human Host, and the 7TM Receptors: Defense or Offense?" by Arfelt *et al.*, herpesviruses not only encode seven-transmembrane receptors—with similarities to the chemokine receptors usurped by HIV—but also modulate the expression of the host-cell genes for such receptors.

From the oldest to the newest, the molecular interactions underlying viral propagation are biologically fascinating. Many are also medically consequential. Better knowledge of those interactions can save lives. And structural knowledge of viruses guides rational vaccine and drug design, generating paradigms of translational science.

I am deeply grateful not only to all the contributing authors of this volume for their splendid work but also for the patient assistance by Helene Kabes and Mary-Ann Zimmerman at Elsevier and last but not least to Dr. Michael Conn, the Editor of *PMBTS*, who gave me the opportunity to take on this rewarding project.

P.J. Klasse
Department of Microbiology and Immunology, Weill Cornell Medical College, Cornell University, New York, USA

REFERENCES

1. Burton DR, Parren PW. Fighting the Ebola virus. *Nature*. 2000;408:527–528.
2. Bhattacharyya S, Warfield KL, Ruthel G, Bavari S, Aman MJ, Hope TJ. Ebola virus uses clathrin-mediated endocytosis as an entry pathway. *Virology*. 2010;401:18–28.
3. Bhattacharyya S, Hope TJ. Cellular factors implicated in filovirus entry. *Adv Virol*. 2013;2013:487585.
4. Kondratowicz AS, Lennemann NJ, Sinn PL, et al. T-cell immunoglobulin and mucin domain 1 (TIM-1) is a receptor for Zaire Ebolavirus and Lake Victoria Marburgvirus. *Proc Natl Acad Sci USA*. 2011;108:8426–8431.
5. Wahl-Jensen V, Kurz S, Feldmann F, et al. Ebola virion attachment and entry into human macrophages profoundly effects early cellular gene expression. *PLoS Negl Trop Dis*. 2011;5:e1359.
6. Xu W, Edwards MR, Borek DM, et al. Ebola virus VP24 targets a unique NLS binding site on karyopherin alpha 5 to selectively compete with nuclear import of phosphorylated STAT1. *Cell Host Microbe*. 2014;16:187–200.
7. Lee JE, Fusco ML, Hessell AJ, Oswald WB, Burton DR, Saphire EO. Structure of the Ebola virus glycoprotein bound to an antibody from a human survivor. *Nature*. 2008;454:177–182.
8. Oswald WB, Geisbert TW, Davis KJ, et al. Neutralizing antibody fails to impact the course of Ebola virus infection in monkeys. *PLoS Pathog*. 2007;3:e9.
9. Parren PW, Geisbert TW, Maruyama T, Jahrling PB, Burton DR. Pre- and post-exposure prophylaxis of Ebola virus infection in an animal model by passive transfer of a neutralizing human antibody. *J Virol*. 2002;76:6408–6412.
10. Shedlock DJ, Bailey MA, Popernack PM, Cunningham JM, Burton DR, Sullivan NJ. Antibody-mediated neutralization of Ebola virus can occur by two distinct mechanisms. *Virology*. 2010;401:228–235.
11. Rivers TM. Filterable viruses a critical review. *J Bacteriol*. 1927;14:217–258.
12. Klasse PJ. Neutralization of virus infectivity by antibodies: old problems in new perspectives. *Adv Biol*. 2014;2014:1–24, Article ID 157895.
13. McGeoch DJ, Cook S, Dolan A, Jamieson FE, Telford EA. Molecular phylogeny and evolutionary timescale for the family of mammalian herpesviruses. *J Mol Biol*. 1995;247:443–458.

CHAPTER ONE

Unity in Diversity: Shared Mechanism of Entry Among Paramyxoviruses

Jean-Louis Palgen*,†, Eric M. Jurgens*, Anne Moscona*,‡, Matteo Porotto*,1, Laura M. Palermo*,‡

*Department of Pediatrics, Weill Cornell Medical College, Cornell University, New York, USA
†Department of Biology, Ecole Normale Supérieure, Lyon, France
‡Department of Microbiology and Immunology, Weill Cornell Medical College, Cornell University, New York, USA
[1]Corresponding author: e-mail address: map2028@med.cornell.edu

Contents

Abstract

The *Paramyxoviridae* family includes many viruses that are pathogenic in humans, including parainfluenza viruses, measles virus, respiratory syncytial virus, and the emerging zoonotic Henipaviruses. No effective treatments are currently available for these

Progress in Molecular Biology and Translational Science, Volume 129
ISSN 1877-1173
http://dx.doi.org/10.1016/bs.pmbts.2014.10.001

viruses, and there is a need for efficient antiviral therapies. Paramyxoviruses enter the target cell by binding to a cell surface receptor and then fusing the viral envelope with the target cell membrane, allowing the release of the viral genome into the cytoplasm. Blockage of these crucial steps prevents infection and disease. Binding and fusion are driven by two virus-encoded glycoproteins, the receptor-binding protein and the fusion protein, that together form the viral "fusion machinery." The development of efficient antiviral drugs requires a deeper understanding of the mechanism of action of the *Paramyxoviridae* fusion machinery, which is still controversial. Here, we review recent structural and functional data on these proteins and the current understanding of the mechanism of the paramyxovirus cell entry process.

1. INTRODUCTION TO PARAMYXOVIRUSES

1.1. Classification and medical significance

The *Paramyxoviridae* family, among the *Mononegavirales* order, is composed of enveloped viruses containing nonsegmented negative-strand RNA (reviewed in Refs. 1–3). Its members are found worldwide (Fig. 1) and

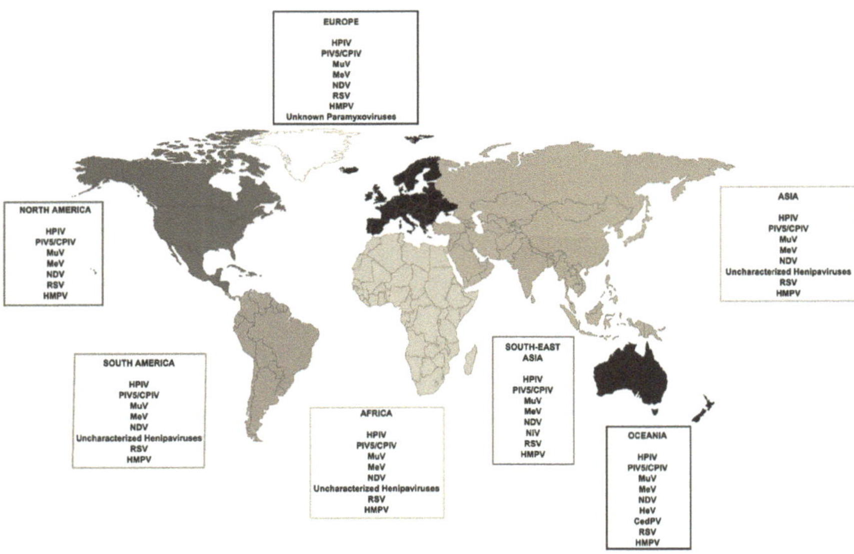

Figure 1 World distribution of major paramyxoviruses. Paramyxoviruses are found on every continent. Henipa- and Henipa-like viruses have been found in Oceania, Asia, Africa, and South America, but human infections have only been reported in Oceania and South-East Asia. Abbreviations are as in Table 1. *Data gathered from Enders,*[2] *Ganar* et al.,[6] *Croser and Marsh,*[7] *the World Health Organization, the World Organization for Animal Health, and recent studies.*[8]

infect a broad range of host species including humans, pigs, horses, and birds. Several paramyxoviruses such as measles virus (MeV), mumps virus (MuV), human parainfluenza virus (HPIV), and respiratory syncytial virus (RSV) continue to have a major impact on global health. These viruses cause severe infections mainly affecting the respiratory tract of children and immunocompromised patients (Table 1).

The *Paramyxoviridae* family is divided into two subfamilies: *Paramyxovirinae* and *Pneumovirinae*. The *Paramyxovirinae* subfamily consists of seven genera, *Respirovirus* (which includes human parainfluenza virus type 3; HPIV3), *Rubulavirus* (which includes MuV), *Morbillivirus* (which includes MeV), *Avulavirus* (which includes Newcastle disease virus; NDV), *Aquaparamyxovirus* (which only includes Atlantic salmon paramyxovirus; ASPV[4]), *Ferlavirus* (which only includes Fer-de-Lance virus; FDLV[5]), and *Henipavirus* (which includes Nipah virus [NiV] and Hendra virus [HeV] as well as Cedar virus [CedPV] recently discovered in bats in Australia[9]) (Table 1). Other paramyxoviruses such as J-virus (JPV) and Beilong virus (BeiPV), as well as some recently discovered bat paramyxoviruses, are closely related, but remain unassigned to any subfamily.[8,10] The *Pneumovirinae* subfamily consists of two genera: *Pneumovirus* (which includes RSV) and *Metapneumovirus* (which includes human metapneumovirus; HMPV) (Table 1).

Epidemiological studies have shown that HPIV is responsible for around 7% of hospitalizations for fever and/or respiratory diseases in children under 5.[11] RSV alone is responsible for at least 3–9% (66,000–199,000) of deaths caused by acute lower respiratory tract infection worldwide, mainly in children under the age of 5.[12] HMPV also causes acute respiratory infections. Studies led on hospitalized patients in Virginia revealed that HMPV is involved in 90% of wheezing cases requiring hospitalization.[13] In terms of the pathogens that do not infect humans but cause problems to society, NDV infects poultry and is associated with a high mortality rate due to respiratory tract infections, generally occurring in developing countries where this disease has a negative economic impact (reviewed in Ref. 6).

The emerging Henipaviruses NiV and HeV are associated with high mortality and/or lethal outbreaks. In the first outbreak of NiV in Malaysia in 1999, 265 people were infected, and 105 patients died of fatal encephalitis. HeV first emerged in Australia in 1994 primarily affecting horses; however, seven people have been infected resulting in four deaths. All Henipaviruses are classified as Biosafety level 4 agents, due to the high lethality of infection and the lack of established treatment. The main reservoir for Henipaviruses is fruit bats, notably the *Pteropus* genus. These bats are mainly present in

Table 1 Paramyxoviruses classification and associated pathologies

Order	Family	Subfamily	Genus	Species	Associated diseases
Mononegavirales	*Paramyxoviridae*	*Paramyxovirinae*	*Respirovirus*	HPIV1, HPIV3	Rhinitis, pharyngitis, pneumonia
			Rubulavirus	MuV, HPIV2, HPIV4, PIV5/CPIV	Mumps, orchitis, meningoencephalitis, tracheobronchitis
			Morbillivirus	MeV	Measles, encephalitis
			Avulavirus	NDV	Fatal respiratory tract infection of poultry
			Aquaparamyxovirus	ASPV	Salmonid gill disease?
			Ferlavirus	FDLV	Unknown
			Henipavirus	HeV, NiV	Fatal encephalitis
		Pneumovirinae	*Pneumovirus*	RSV	Rhinitis, pneumonia
			Metapneumovirus	HMPV	Rhinitis, fever, bronchiolitis

The *Paramyxoviridae* family is divided into two subfamilies and seven genera. Most paramyxoviruses cause a wide range of pathology. Data gathered from ICTV, Enders[2] and recent articles.[4,5] PIV, parainfluenza virus; HPIV, human parainfluenza virus; CPIV, canine parainfluenza virus; MuV, mumps virus; MeV, measles virus; NDV, Newcastle disease virus; ASPV, Atlantic salmon paramyxovirus; FDLV, Fer-de-Lance virus; HeV, Hendra virus; NiV, Nipah virus; RSV, respiratory syncytial virus; HMPV, human metapneumovirus.

Africa, South-East Asia, and Oceania (reviewed in Refs. 7,14). Recent studies have identified new paramyxoviruses in European insectivorous bats[8] which produce symptoms resembling HeV infection. While these viruses remain unassigned to any genus, they are more closely related to *Paramyxovirinae* than to *Pneumovirinae*.[8]

The only preventive vaccines currently available for members of the *Paramyxoviridae* family are those against MeV, MuV, and NDV (for poultry). Even these viruses are still a major health concern. According to the World Health Organization (WHO), each year, around 200,000 deaths are associated with MeV infection mainly in developing countries. The Centers for Disease Control and Prevention declares that more people have been infected with measles in the United States during the first 4 months of 2014 than have been infected in the first 4 months of the past 18 years.

In 2012, 687,000 cases of MuV infection were reported across the world (WHO). Usually, MuV infection is not lethal but it can lead to complications such as meningitis, encephalitis, and orchitis, with possible permanent sequelae (WHO). Furthermore, there are currently no therapies to treat patients infected by any paramyxovirus (reviewed in Ref. 2), making these viruses a significant public health issue.

1.2. Structure

Paramyxoviruses are 150–300 nm in diameter with envelopes composed of host cell lipids and viral glycoproteins (reviewed in Refs. 1,2). The genome is a nonsegmented RNA strand of negative polarity, between 15,210 (RSV type 2, GU591759.1, Kumaria *et al.*[15]) and 15,894 nucleotides (MeV, NC_001498.1, Takeuchi *et al.*[16]) with the exception of Henipaviruses which contain a longer genome (NC_001906, Wang *et al.*,[17] Yu *et al.*[18]; NC_007454.1, Jack *et al.*[19]; NC_007803.1, Li *et al.*[20]). The length of the genome of each paramyxovirus is always a multiple of six nucleotides, an organization required for efficient replication by the viral polymerase.[21,22] The genomic RNA strand is encapsidated by the helical nucleocapsid protein, N or NP (Fig. 2). The large protein (L) and the phosphoprotein (P) constitute the viral RNA-dependent transcriptase/replicase complex. In the virion, L and P are associated with the RNA–nucleocapsid complex (Fig. 2). The N or NP protein also interacts with the matrix protein (M), a nonstructural protein that lines the envelope of the viral particle (Fig. 2). The lipid bilayer envelope of the virus is derived from the host cell membrane, formed when the virus buds from a region of membrane expressing the viral

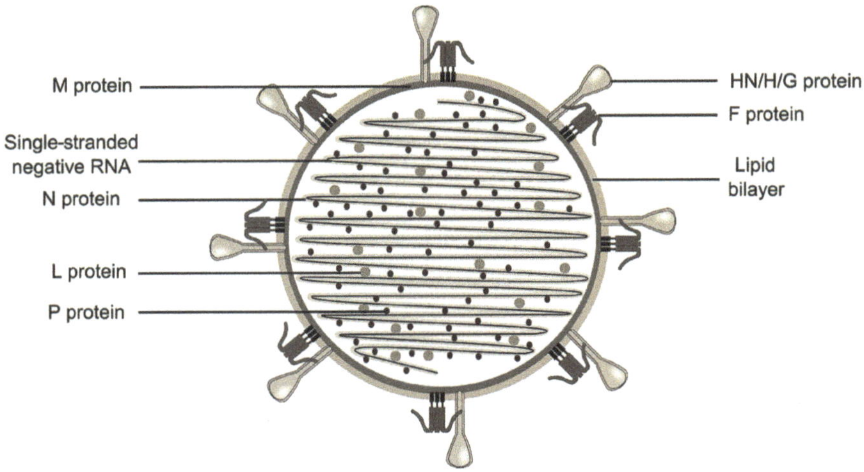

Figure 2 Schematic representation of the common structure of Paramyxoviruses. *Paramyxoviridae* are enveloped viruses. They contain single-stranded negative RNA coated with nucleocapsid (N) protein as well as a large (L) protein and a phosphoprotein (P) that carries out polymerase activity. The matrix (M) protein lines the viral lipid bilayer. The two viral glycoproteins—hemagglutinin–neuraminidase (HN)/hemagglutinin (H)/glycoprotein (G) and fusion (F)—protrude from the viral membrane.

receptor-binding protein (hemagglutinin–neuraminidase (HN)/hemagglutinin (H)/glycoprotein (G)) and the fusion protein (F). As for many other enveloped viruses, the virions are labile and can be easily inactivated *ex vivo* by heat, organic solvents such as ethanol, or detergents.

The six proteins N/NP, L, P, M, HN/H/G, and F are conserved among the *Paramyxoviridae* family. In addition, some structural proteins are restricted to specific viruses and their roles may be less clear, for example, the small hydrophobic proteins and the transmembrane (TM) proteins.[10,23–26] *Paramyxoviridae* also encode for nonstructural proteins that are involved in the inhibition of the interferon response.[27] In addition, the alternative splicing of the P-gene leads to the expression of C, V, and W proteins, whose role is to counteract host innate immunity (reviewed in Ref. 28).

1.3. Viral entry and life cycle

Paramyxovirus fusion is mediated by two different viral proteins that in most cases must work in concert to accomplish viral entry. The receptor-binding protein first engages the cellular receptor, then in most cases activates the fusion protein, and the fusion protein inserts itself into the target cellular membrane, allowing the viral envelope and target cellular membrane to merge. Upon fusion of the viral envelope with the target cell membrane, the genetic material is released into the cytoplasm. The negative sense RNA, which is present in the form of a nucleocapsid or RNA/protein complex, is converted into positive-sense message-length RNAs by the RNA-dependent RNA polymerase that is provided by the virus. This step allows for translation of virally encoded proteins. Replication of the viral genome occurs via transcription of a full-length positive-sense strand which is then copied into a full-length negative-sense new genome and encapsidated by the viral nucleocapsid protein. The matrix protein binds to the nucleocapsid and interacts with the cytosolic tails of the membrane-bound HN/H/G and F proteins, facilitating the process of budding of progeny virions. Release of new viral particles from the cell surface, in some paramyxoviruses requiring a receptor-cleaving enzymatic function carried out by the receptor-binding protein,[29] permits infection of new target cells and spread of infection.

Most *paramyxovirus* fusion events occur in a pH-independent manner, at the cell surface; however some viruses enter the cell via endocytosis (reviewed in Refs. 1,3). How the receptor-binding protein and the fusion protein (together called "fusion machinery") work together to promote fusion has been an area of active investigation since it was first shown that the paramyxovirus receptor-binding protein plays an active role in the fusion process during entry.[30–32] Several models have been proposed, and the molecular details of the fusion process mediated by the paramyxovirus fusion machinery remain controversial. Previous models postulate a duality among *Paramyxoviridae* (reviewed in Refs. 1,3). It has been proposed that for paramyxoviruses that bind a proteinaceous receptor, the role of the receptor-binding protein is mainly a repressive one[33–38] (reviewed in Refs. 39–41) and that upon receptor binding, the fusion protein is released and proceeds to fusion; on the other hand, the receptor-binding proteins of sialic acid-binding viruses have been thought to interact with F only upon receptor engagement[33,34,42,43] (reviewed in Refs. 39,40,44,45). Our data suggest that a common mechanism applies to all paramyxoviruses that use a receptor-binding protein to activate a fusion protein, including those that bind a proteinaceous receptor.[46,47] The debate will be detailed in the sections below. In this chapter, we review recent advances in the field of paramyxovirus entry. We first summarize structural data about *Paramyxoviridae* virions and specifically HN/H/G and F. The focus then turns to receptor engagement and its effects on HN/H/G. We detail the interaction between these two surface glycoproteins before, during, and after receptor engagement, as well as the membrane fusion process mediated by F, and propose a potential unifying model for *Paramyxoviridae* fusion.

2. STRUCTURE AND FUNCTION OF THE PARAMYXOVIRUS GLYCOPROTEINS

2.1. The receptor-binding protein

The paramyxovirus receptor-binding proteins present on different members of the virus family are known as HN, H, or G. These proteins are distinguished by the type of receptor they engage, their ability to cleave sialic acid (neuraminidase activity), and their ability to agglutinate red blood cells (reviewed in Ref. 1). The HN protein carried by the *Respirovirus*, *Rubulavirus*, and *Avulavirus* genera (Table 1) possesses both sialic acid-binding (hemagglutinating) and sialic acid-cleaving (neuraminidase) activities. Sialic acid binding is active during viral entry while neuraminidase activity is involved in viral budding and prevents the virus from self-aggregating. The H protein carried by the *Morbillivirus* genus (Table 1) does not bind to sialic acid during viral entry. Both HN and H proteins have the ability to agglutinate red blood cells, but the H binds proteinaceous receptors during MV entry. The H protein lacks neuraminidase activity, suggesting that following viral release, self-aggregation mediated by sialic acid binding does not occur. The G protein carried by the *Pneumovirinae* and *Henipaviruses* genera (Table 1) does not bind sialic acid and does not possess neuraminidase activity. Like H, G proteins bind proteinaceous receptors.

The three types of receptor-binding proteins differ in the type of receptor they bind, but share the same general architecture. HN, H, and G are type II TM proteins, with N-termini inside the viral particle (Fig. 3). Each is present on the viral membrane as a tetramer composed of two dimers, an arrangement known as a dimer of dimers. A dimer consists of an association of two monomers (Fig. 3), each of which monomer contains a cytoplasmic tail domain, a TM domain, a stalk domain, and a globular head domain (reviewed in Refs. 1,3).

Dimers of the receptor-binding proteins are formed by disulfide bridges between the stalk domains of two monomers[48–59] and are also linked via the stalk domain of the proteins, as described for PIV5-HN,[49] although the TM domain may stabilize the tetramer, as described for NDV-HN.[48] These interdimer links mainly involve noncovalent bonds which are weaker than the intradimer disulfide linkages. Tetramers of HN/H/G are more suitable for crystallization than dimers, as described for HeV-G.[59] Alteration of

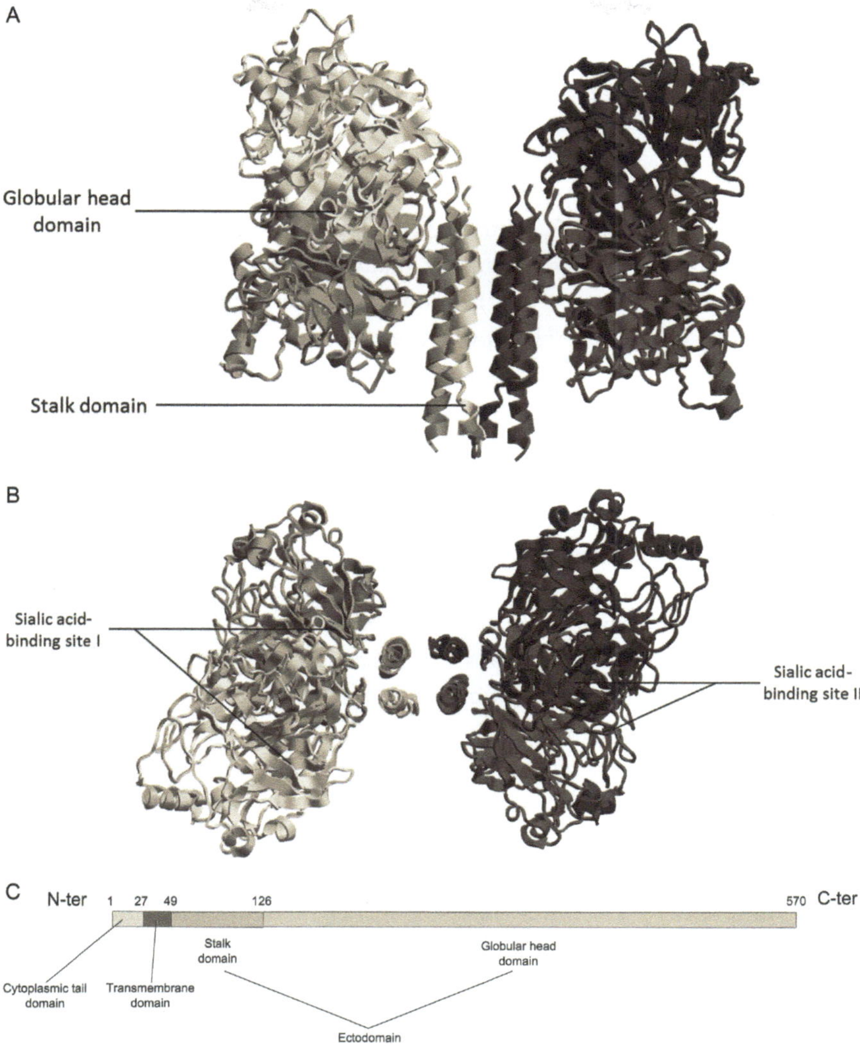

Figure 3 Structure of the Newcastle disease virus hemagglutinin–neuraminidase protein. (A) Side view of the crystal structure of the tetramerized NDV-HN ectodomain showing the stalk and the globular domains of each monomer. Each color represents one monomer of the receptor-binding protein. One dimer is composed of green and yellow monomers, and the other of red and blue monomers. (PDB ID: 3T1E; Yuan *et al.*[48]). (B) Top view of the crystal structure of the tetramerized NDV-HN ectodomain, showing sialic acid-binding sites I and II. Each monomer bears a site I and a site II (PDB ID: 3T1E; Yuan *et al.*[48]). (C) Schematic representation of the domains of NDV-HN. (See the color plate.)

disulfide bridges via *in vitro* mutagenesis alters dimer and tetramer stability.[57,58] Interestingly, a recent study from Navaratnarajah *et al.*[58] reported that the complete stalk domain of MeV-H is not directly involved in the tetrameric structure, and the extent of involvement of the stalk in formation of the tetramer for other members of the *Paramyxoviridae* family is unclear.

Crystallographic studies of the tetrameric globular heads show each monomer carrying an N-terminal six-blade β-propeller, characteristic of neuraminidase enzymes.[48,50–61] Interestingly, HN, H, and G share this structure, although only HN possesses neuraminidase activity. H and G carry a structural vestigial neuraminidase site,[50–54] consistent with the hypothesis of a common evolutionary origin for these three receptor-binding proteins.

In the case of HN, the sialic acid-binding site of each monomer, known as sialic acid-binding site I, is located at the top of the globular head domain, in the center of the β-propeller. In addition, a second sialic acid-binding site (known as site II) was identified crystallographically at the dimer interface of NDV-HN.[62] This site II is involved in receptor binding.[63,64] Functional analysis has suggested a second sialic acid site on HPIV1[65,66] and has identified a second sialic acid-binding site on HPIV3 that is also important for activating F,[67] although these sites have not been demonstrated crystallographically. The receptor-binding site of G shares the same location as site I,[53,54] whereas the H-binding sites are located on the side of the β-propeller.[52,61,68] The structure of the tetrameric receptor-binding protein ectodomain, comprised of the head and stalk domains, has been solved for PIV5[55] and NDV.[48] The stalk domain adopts a 4-helix bundle conformation (Fig. 3) with a hydrophobic core located at the upper part of the stalk domain.[48,55]

2.2. The fusion protein

The paramyxovirus fusion protein, F, is a type I TM protein, with its N-terminus outside the viral particle. It is synthesized as an inactive F_0 precursor (reviewed in Refs. 1,3). F_0 is then cleaved into its active form, F, which is composed of two subunits, F_1 and F_2. The two subunits are linked by a disulfide bridge between the heptad repeat N-terminal domain (HRN) of F_1 and F_2 (Fig. 4). The cleavage creates the hydrophobic fusion peptide, which is inserted into the target membrane during the fusion process, once an activation step exposes the peptide at the surface of the molecule.

F acts as a homotrimer in which each monomer is linked to each other via a TM domain[71,72] and contains a cytoplasmic tail, a TM domain, a heptad repeat C-terminal domain (HRC), a HRN, and the fusion peptide (Fig. 4). The heptad repeat domains are regions of 7-mer repeats in which every seventh residue is either a leucine, isoleucine or valine, and the whole structure is an amphipathic α-helix.

The F cleavage step is crucial for the viruses, as uncleaved F proteins are unable to promote fusion. For most *Paramyxoviridae* (with the exception of Henipaviruses), furin proteases within the *trans*-Golgi network cleave F_0 at an R–X–K/R–R consensus motif (reviewed in Ref. 1). Unlike most *Paramyxoviridae*, RSV F possesses two cleavage sites which are required for efficient fusion.[73] For Henipaviruses, F_0 is cleaved in the endosomal

A

Figure 4 Structure of the paramyxovirus fusion protein. (A) Crystal structure of the prefusion state of the trimeric fusion protein of PIV5 showing the fusion peptide (purple) in the hydrophobic pocket formed by a hydrophobic domain (deep green), the HRN domain (deep blue), and the F_2 subunit (yellow) (PDB ID: 4GIP; Welch *et al.*[69]).
(Continued)

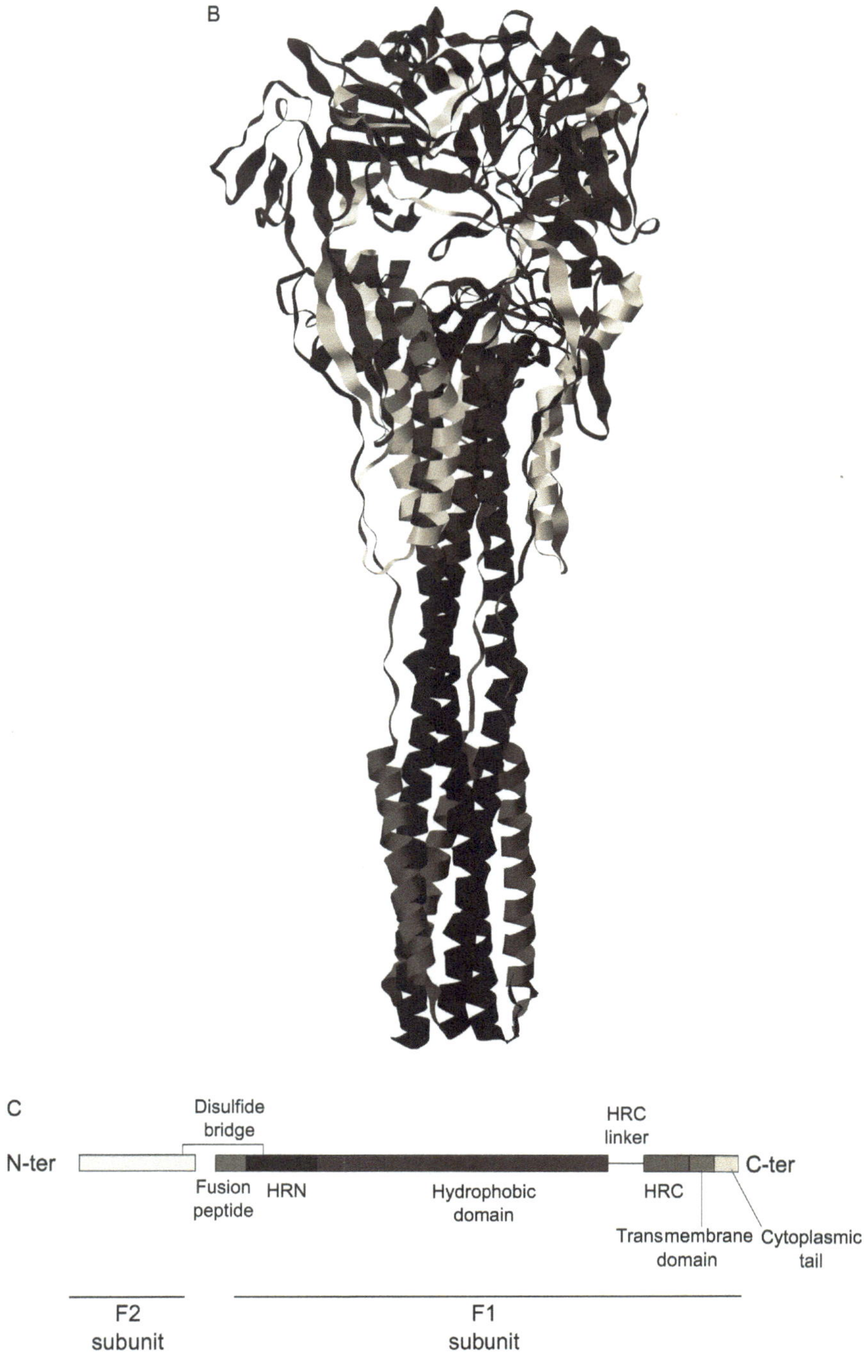

Figure 4—Cont'd (B) Crystal structure of the postfusion state of the fusion protein of HPIV3 showing the 6-helix bundle structure formed by the HRN (deep blue) and HRC (red) domains interacting together (PDB ID: 1ZTM; Yin *et al.*[70]). (C) Schematic representation of the main domains of a monomer of the cleaved paramyxovirus fusion protein. HRC/HRN: heptad repeat C-/N-terminal domain. (See the color plate.)

compartment by cathepsins L and B at a VGDVR/K consensus motif.[74–77] Henipavirus F_0 is expressed at the plasma membrane, reinternalized, and then cleaved before associating with the rest of the viral particle. The cleavage also seems highly dependent on the valine content of the fusion peptide, as reported for HeV.[78]

The structure of the cleaved PIV5-F in its prefusion state has been solved.[69] The fusion peptide is initially buried in a hydrophobic pocket, preventing premature exposure.[69] This pocket is composed of the HRN (Fig. 4). In this prefusion state, HRC forms an α-helix close to the viral membrane.[79] The crystallographic data suggested that few conformational changes occurred after cleavage, when compared to uncleaved forms of PIV5-F and HPIV3-F.[70,80] However, uncleaved F is fusion incompetent. Established models describe the state of prefusion F as being metastable, and destabilized following activation by the receptor-binding protein.

In the postfusion state, the fusion peptide is exposed in an open α-helical domain, and the heptad repeat domains associate, forming a highly stable 6-helix bundle. The formation of this stable structure is a significant driver of the process of membrane fusion[81,82] (reviewed in Ref. 83) (Fig. 4). Like the receptor-binding proteins, F is highly glycosylated. For NiV-F, it seems that some of these glycosylation sites decrease the fusogenicity of the virus.[84] *In vivo*, these additional carbohydrates may protect the virus from recognition by the host immune system.[84]

3. PROPOSED MECHANISMS OF RECEPTOR-BINDING PROTEIN AND FUSION PROTEIN INTERACTIONS

3.1. The globular heads of the receptor-binding protein selectively engage specific cellular receptors

The fusion process begins when the receptor-binding protein engages its receptor. HN recognizes sialic acid-bearing membrane proteins, whereas H and G bind proteinaceous receptors. H binds different proteinaceous receptors for each virus. For example, MeV-H engages CD46, CD150/SLAM (signaling lymphocyte-activation molecule), and Nectin 4.[52,61,85–89] CD46 binding seems to be unique to laboratory-adapted strains. CD150 is expressed on the cell surface of macrophages and dendritic cells, and MeV engages this receptor to infect the host immune system.[90] Nectin-4 is expressed on the basal surface of the epithelium cells, allowing MeV to be spread from macrophages to epithelium and then into the lung lumen.[88,89] The neurotropic Henipavirus G engages ephrin B2 and B3 on cell surfaces[9,53,54,91]; these molecules are expressed in neurons in the brain. Ephrin B2 and B3 are also found in other cell types and are conserved among many species, allowing Henipaviruses to infect a range of species including humans, pigs, horses, and bats. Henipaviruses can spread within the host by binding lymphocytes and using them as transporters.[92] The G protein of *Pneumovirinae* binds heparan sulfate proteoglycans.[93–96] RSV-G has been shown to interact with the chemokine receptor CXC3CR1, through a CX3C motif.[97] While it is unlikely that this interaction would promote fusion, this interaction strongly inhibits the host immune response.[98] The diversity in receptor usage confers paramyxoviruses the ability to adapt, gain access, and infect new tissues and new hosts.

3.2. The stalk domain of the receptor-binding protein interacts with and activates F

HN/H/G is the driving force for fusion initiation and then for sustaining F's role in mediating viral entry[46] (reviewed in Refs. 1,99). Under a variety of *in vitro* experimental situations, F can fuse alone,[46,64,100,101] or a "headless" HN/H/G may be sufficient to mediate F activation[47,101,102] (see specific examples below). However, as discussed in Section 3.6 the function of specific residues in the globular head of HN is essential for infection in the host, and any subtle change at the dimer interface of the globular domain can affect HN dimer association, impact the HN/F fusion machinery, and markedly alter host infection.

The globular heads of the HN/H/G proteins bind the cellular receptor. The stalk domains of HN/H/G proteins are responsible for specific interaction with the homologous F proteins and are critical for F activation once they receive the signal from the receptor-bound globular head.[33,43,48,49,56–58,103–106] After initial identification of the importance of the stalk of the receptor-binding protein for activating F, this stalk function has been assessed using a variety of approaches including the use of the "headless" receptor-binding proteins mentioned above. A construct consisting of the PIV5-HN stalk domain (residues 1–117) lacking the globular binding domain was sufficient to activate F. This activation seemed to be specific; the PIV5-HN stalk could not activate heterotypic Fs and required direct interaction with F.[101] This set of experiments was used to postulate that for PIV5-HN, activation of F requires that the stalk domain be "freed." Receptor engagement would drive the movement of the heads that would free the stalk. Similar experiments have been performed using different "headless" stalk proteins with varied results. However, only very specific MeV-H, NiV-G, and PIV5 stalk lengths can activate the F protein[101,102,104] and, for MeV-H, the stalk must be partially stabilized in order to be functional.[104] Only one out of several different headless stalk constructs of mumps HN,[107] NDV-HN,[107] and NiV-G[102] can activate F suggesting that the specific sequence of the receptor-binding protein stalk and the F protein are crucial for this activity. For HPIV3, a headless HN does not seem to be capable of activating HPIV3-F. Thus, how the stalk domain of paramyxovirus HN/H/G activates F remains to be further characterized, and as described below, we contend that the interaction of the globular head of the receptor-binding protein with its receptor provides a critical signal to the stalk in the process of F activation.

Chimeric proteins bearing the globular domain from NDV and the stalk domain of HPIV3-HN, NiV-G, or MeV-H revealed that receptor engagement by the NDV-HN globular head is sufficient for transmitting the activating signal through the stalk domain of these other paramyxoviruses and triggering the homologous fusion protein.[64,108] These chimeric receptor-binding proteins are only capable of triggering an F protein that is homologous to the stalk domain of the chimeric protein. Thus, a chimeric protein with an NDV-HN globular head and an HPIV3-HN stalk can only activate HPIV3-F.[64] The only exceptions are Henipaviruses NiV-G and HeV-G whose stalks demonstrate enough sequence similarity to activate both F proteins.[109] Closely related Henipaviruses, such as the recently discovered Cedar virus,[9] may share the same property. The chimeric receptor-binding proteins reveal one of the ways in which HN, H, and G protein function is conserved at least among the *Paramyxovirinae* subfamily, and support the hypothesis of a unified model for the paramyxovirus fusion machinery in which the globular head domain of the receptor-binding protein acts as a receiving unit that is independent of the rest of the protein. The receptor-binding protein engages its receptor and transmits a signal to the stalk domain. The stalk domain likely undergoes conformational changes allowing it to activate its homologous fusion protein[57,110] (reviewed in Ref. 45).

3.3. The role of the receptor-binding protein before receptor engagement

Several distinct models describing the interaction between the HN/H/G protein and the F protein have been proposed (reviewed in Refs. 1,3). One model, the dissociation or clamp model, postulates that the HN/H/G and F proteins interact prior to receptor engagement and that receptor engagement abrogates this interaction. Another model, the association or provocateur model, suggests that HN/H/G only interacts with the homologous F protein following receptor engagement. Recent studies from our group have uncovered elements in support of a unified model among paramyxoviruses.[47,111] We used a Bimolecular Fluorescence Complementation (BiFC) strategy where HPIV3-HN and HPIV3-F were, respectively, fused with the N-terminus of YFP and C-terminus of CFP. Only if HN and F proteins interact, the fluorescent protein is reconstituted and fluorescence is emitted upon excitation. We observed that HPIV3-HN interacts with F in the absence of receptor engagement. Upon receptor engagement, HN and F continue to interact and cluster at the point where the fusion pore will form.[47] Whether clustering occurs before or after activation of the F protein has not been firmly established, but recent data are more consistent with clustering occurring first and F activation occurring in the cluster (unpublished). HN and F continue to interact throughout the fusion process[46] and dissociate only once fusion is complete (unpublished).

For HPIV3, it appears that nonreceptor-engaged HN protein stabilizes F, maintaining it in the prefusion state. When HPIV3-F alone is exposed to high temperatures, it enters the postfusion state (as assessed by acquisition of sensitivity to proteinase K digestion); however, in the presence of nonreceptor-engaged HN, F remains in its prefusion state, resistant to proteinase digestion.[111] These data support the idea that HPIV3-HN serves a "protective" role for the fusion protein.[111] Prior to receptor engagement, the receptor-binding protein stabilizes HPIV3-F and prevents it from premature activation. Ader *et al.*[110] recently showed that the F protein of some morbilliviruses is highly stable and suggested that it is unlikely that the H stabilizes F; however it cannot be excluded that *in vivo*, stabilization may be required since many parameters could prematurely trigger fusion.

An intriguing role of pH in the NDV cell entry process has recently emerged.[112] NDV entry is reduced when caveolin-associated traffic is inhibited. Cholesterol seems to be important in the process since the drug methyl-β-cyclodextrin, which inhibits cholesterol trafficking, also diminishes NDV-HN binding. Moreover, NDV particles were shown to colocalize with EEA1, a marker of early endosome formation suggesting that NDV could enter the cell through caveolin-mediated endocytosis. Past work showed that HMPV, NiV, and RSV can use the endosomal pathway to enter cells.[113–115] Low pH exposure increases NDV fusion and subsequent syncytia formation while, reciprocally, fusion decreases in the presence of pH-acidification inhibitors.[116] Consistent with these results, the stability of NDV-F (assessed through fusion assays) decreases after exposure to low pH; F is more easily activated under these conditions.[116] There may be an accessory pH-dependent pathway through the caveolin-mediated endocytosis for NDV.

For some paramyxoviruses, the F protein can mediate fusion in the absence of the receptor-binding protein, in some cases permitting viral infection.[117–123] However, very few viruses are infectious when lacking their receptor-binding protein. Infectious virions that lack a receptor-binding protein have been studied for RSV, HMPV, and laboratory-adapted strains of NDV; however in these cases, infectivity is enhanced when the receptor-binding protein is present.[96,118,124,125] In the case of RSV, the F protein itself can bind nucleolin,[126] the HMPV F can bind integrins,[127] and both can engage heparan sulfate,[128–130] potentially permitting G-independent entry. It will be important to determine whether in these cases the receptor-binding protein serves a role *in vivo* in stabilization of F prior to receptor engagement; an F-stabilizing role for HN/H/G protein may be a conserved feature at least among the *Paramyxovirinae* subfamily.

3.4. The receptor-binding protein transmits a triggering signal to the fusion protein upon receptor engagement

Upon receptor engagement, the HN/H/G protein activates the F protein to undergo its final fusion-readiness structural changes. The mechanism whereby this activation occurs—where the signal for activation originates, and how it is transmitted from the receptor-binding protein to the fusion protein—is a topic of significance. The stalk domain of HN/H/G protein is critical to this activation process.[58,103–107,131,132] Recent studies from our group indicate that for NiV the domain that connects the globular head to the stalk domain is required for transmission of the triggering signal to F.[131] Chimeric receptor-binding proteins containing the globular heads of NDV and the stalk domain of NiV can efficiently activate NiV-F only if specific residues are present in the head–stalk junction.[131]

A proposed model deriving from PIV5 postulated that the heads of HN/H/G could change their position upon receptor engagement, shifting from a "heads-down" conformation, in which the stalk domain is masked, to a "heads-up" conformation exposing the stalk domain. A masked stalk domain would prevent HN/H/G and F interaction, and upon receptor engagement the heads would move aside, allowing the stalk to activate the F protein.[48,49,55,101] However, recent analyses of the MV fusion process do not support this model, suggesting that at least this mechanism may not apply to other paramyxoviruses (reviewed in Ref. 45). MeV-H and MeV-F interact prior to receptor engagement,[133] indicating that an H head must be up to allow H/F interaction during transit to the cell membrane. However, F is not triggered prematurely (i.e., before receptor engagement during infection), suggesting that for MeV the exposure of the stalk is not the crucial requirement for F activation. It has been suggested that MeV receptor engagement with a "pulling" of the H molecule could induce a conformational change in the stalk that triggers F.[33] Brindley *et al.* noted that MeV-H proteins with truncated globular heads promoted fusion *in vitro*,[104] implying that the globular heads may mask a portion of the stalk domain responsible for activating the F protein, consistent with a "heads-down, heads-up" model. However, the truncated H proteins that were studied are highly specific, with truncation at precise sites on H being required for fusion complementation. Headless Hs, to promote F-mediated fusion, require stabilization at the C-terminus, e.g., by a yeast-derived GCN4 motif.[104] Thus, it seems likely that the specific truncated H proteins do not represent a general mechanism, except when they are stabilized or otherwise modified to adopt a structure similar to the receptor-engaged conformation that activates F. Finally, key residues on the MeV-H that are involved in F protein activation are located at the membrane-proximal part of the stalk domain, where the globular heads of MeV-H do not reach even when the protein is in a "heads-down" conformation.[58,103,107,131]

How does receptor engagement affect the receptor-binding protein and modulate the F activation step? The first studies to analyze how receptor engagement modifies the receptor-binding protein structure showed that receptor engagement does not appear to affect the monomer–monomer interface.[134] Stabilization of NDV-HN monomers by disulfide bridges restricted monomer movement but enhanced HN's fusion promotion activity, suggesting that large movements between monomers are not needed for successful fusion promotion. Crystallographic data of MeV-H bound to SLAM receptors revealed H present in two tetrameric conformations. In the pre-F triggering conformation, H is in a planar form and upon receptor engagement, a "sliding" movement between the dimers occurs that gives rise to the post F-triggering form.[52] One recently proposed model suggests that upon receptor engagement, after exposure of the stalk of the receptor-binding protein, a specific conformational change occurs in the complex formed by HN/H/G and F, to reach an "induced-fit" state that leads to F activation.[107] In this case, the nature of the signal-activating F is a modulation of its interaction with HN/H/G rather than a loss or a gain of interaction (except perhaps for PIV5).

Even after F activation and during the fusion process itself, the receptor-binding protein continues to regulate the fusion process. For HPIV3, the presence of HN is crucial until merger of the viral and cellular membranes.[46] Using a mutant HN that lacks neuraminidase activity and thus constitutively engages its sialic acid receptor, we specifically disrupted receptor engagement at precise times using the small-molecule zanamivir. Only when HN continuously engages its receptor can F proceed through the fusion process. Even after insertion of the fusion peptide into the target membrane, F still requires the activating signal from HN/H/G to complete fusion.[46]

3.5. The fusion protein inserts its hydrophobic fusion peptide into the target membrane leading to the formation of the fusion pore

After activation, F undergoes several conformational changes (Fig. 4). Fusion can be prevented by introducing disulfide bridges to stabilize F.[135] F acts as an indivisible unit. The cytoplasmic tail of F protein must bear conserved Ser/Thr residues for fusion to properly occur, and the cytoplasmic tail is likely to be virus-specific; it cannot be interchanged with that of another virus.[136] The current model for F action is based on several available F protein structures (Fig. 4) and predicts that F inserts its fusion peptide into the target membrane by extending its α-helical domains. The amphipathic HRC and HRN domain then interact with each other, driving fusion. Supporting this hypothesis, Donald *et al.* found that the α-helical domains on PIV5-F interact avidly with each other.[137] The model postulates that once the fusion peptide is inserted, the F protein is in a prehairpin intermediate state (Fig. 5). This transient intermediate was recently observed for PIV5.[79] Kim *et al.* used nanobeads bearing a lipid bilayer as targets for viral particles, and HRC-derived peptides that interact with the HRN domain to prevent F from refolding after fusion peptide insertion. They then measured the distance between the viral bilayer and the nanobead bilayer using EM and compared it to *in silico* predictions based on the hypothesis of the prehairpin fusion intermediate. The experimental results fit the computational model, supporting the existence of the intermediate state.[78] F proteins blocked in the transient state were identified by electronic microscopy by gold coupling of HRC-derived peptides and were observed to be present in clusters. This observation is consistent with our model for HPIV3, where receptor-binding and fusion protein clustering is required in

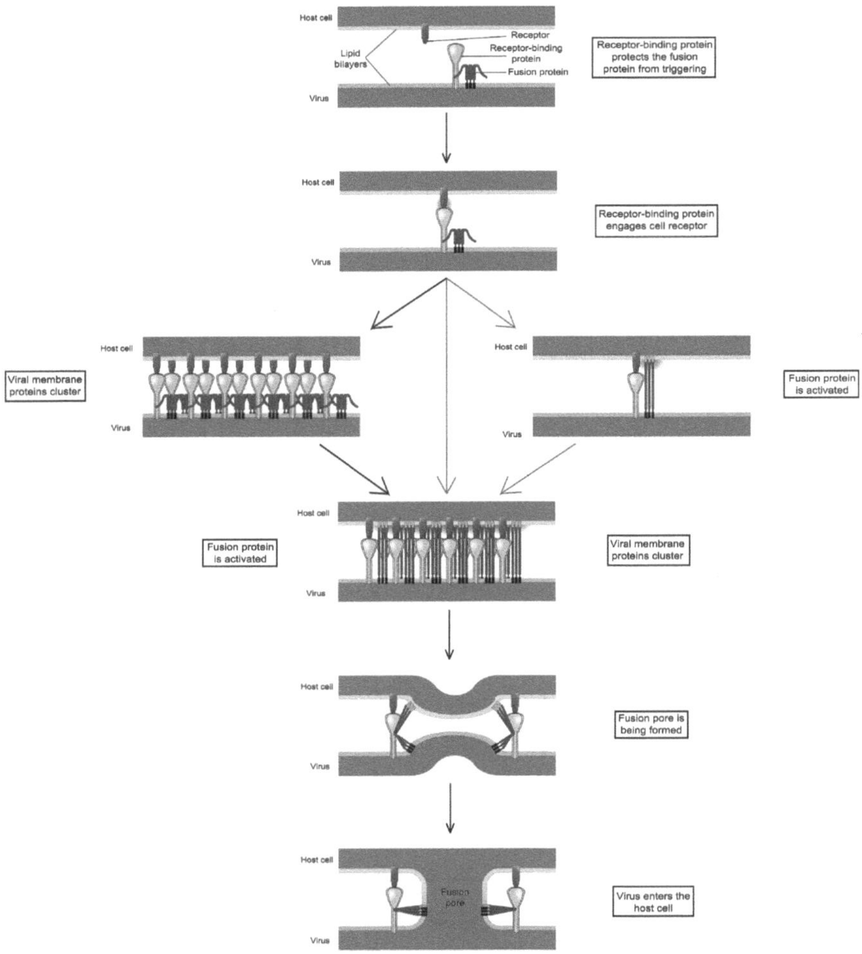

Figure 5 Unified model for *Paramyxoviridae* fusion process. See text for description. Pink (gray in the print version): HRN domain and fusion peptide. HRC/HRN: heptad repeat C-/N-terminal domain.

order for fusion to occur.[47] Attaining this unstable fusion intermediate is likely to be the event that is prevented by the nonreceptor-engaged HN/H/G protein.[111]

After the unstable intermediate state, F adopts a hairpin intermediate state (Fig. 5). In this form, the two heptad repeat regions (HRN and HRC) interact with each other, leading to a 6-helix bundle conformation, which mechanically forces the two membranes together (Fig. 5). In this process, the length of the HRC-linking region separating the HRC domain from the membrane (Fig. 4) is crucial. Linkers that are too short or too long were associated with fusion-incompetent F proteins, as described for NDV.[82] This is consistent with the hypothesis that the HRC linker contributes to the stabilization of the HRC domain and thus can modulate its function, as previously described for HeV-F.[72]

A critical element of this refolding is that the HRN and HRC domains form a close interaction. Inhibitors of this interaction, an attractive strategy for blocking fusion and viral entry, and peptides derived from the HRC domain of paramyxovirus F molecules, have recently been shown to inhibit the fusion process *in vivo*.[138,139] The peptide inhibitors are far more efficient when conjugated with lipids. This conjugation allows for peptide insertion into target cell membranes, placing the antifusion peptides in close proximity to the viral glycoproteins. These conjugated peptides can reach the brain and may be efficient against neurotropic viruses such as NiV, or neurotropic variants of MeV.[138–140]

At the end of the F refolding process, F is in its stable postfusion state. Experimentally expressed, soluble F protein naturally adopts this postfusion state.[70,141] The stability of this final state implies that the fusion process is irreversible, and supports the importance of the "protective role"[111] of the receptor-binding protein in preventing premature initiation of the activation process. Indeed, virions whose F proteins have been prematurely activated cannot fuse and are noninfectious.[142,143]

3.6. The interaction between HN/H/G and F modulates infection in the natural host

A series of recent experiments showed that the communication between the HPIV3-HN and -F directly impacts infection in the natural host.[108,142] An HPIV3 virus bearing HN molecules that trigger F rapidly or interact with F avidly is effective at fusing in cell culture, but fares poorly in natural tissues or *in vivo*.[32,47,67,142] Enhanced F-triggering by HN, while advantageous *in vitro*, results in noninfectious viral particles in airway epithelium,[144] in which the F protein may have been prematurely activated before contacting target cells.[142,143] Recent structural analysis of the HN molecules with specific biological effects of revealed properties that are critical for infection *in vivo*. The second sialic acid-binding site that was identified by functional assays as a key operative site on HPIV3-HN (site II)[67] has a clear structural correlate and modulates viral growth *in vivo*.[108] Specific structural changes at the HN dimer interface, where site II exists, modulate the interaction between HN and F, impact fusion triggering, and directly impact viral infection. For example, the HN from a virus that is well adapted to growth in lung tissue has a wider separation across the dimer interface near the regions of site II than a variant that is restricted in growth in the lung. These conformational changes are propagated to nearby loops at the dimer interface and reduce dimer association in the lung-adapted virus's HN, as measured by buried interface areas, indicating that the dimer interface (and its modulation of HN/F interaction) is critical to infection in the host.[108] The viruses that grow well in lung tissue and *in vivo* bear a less active HN/F fusion machinery, with lower receptor avidity and less efficient fusion triggering.[144] One may speculate that during natural infection in the host, receptors may be widely available, and an overly fusogenic virus may prematurely trigger its F before reaching the target membrane, a notion consistent with the evolution of laboratory-adapted HPIV3 variants *in vivo* to become less fusogenic than reference strains.[144] In support of this concept, several highly fusogenic glycosylation mutants from NiV have not been identified *in vivo*.[84] It seems apparent that most paramyxoviruses require a specific balance between the various properties of the fusion machinery for viability *in vivo*.

4. CONCLUSIONS

A variety of models that reflect the diversity of the paramyxoviruses have been proposed for the steps in entry. However, recent data may be taken together to support a common fusion mechanism. Reviewing recent advances in the field of *Paramyxoviridae* entry, we propose a unified model for the fusion process (Fig. 5). Before receptor engagement, the receptor-binding protein interacts with the fusion protein and, if necessary, prevents its untimely activation. After the globular heads of the receptor-binding protein engage the target receptor, an activating signal is transmitted via the stalk domain to the fusion protein. This signal appears likely to be in the form of a conformational change in the stalk domain and—in an "induced-fit" model—induces a structural change in the fusion protein. The fusion protein is destabilized and inserts its hydrophobic fusion peptide into the target membrane. The two heptad repeat domains of the fusion protein interact with each other as the molecule progresses to its stable postfusion state. This process drives the formation of the fusion pore (Fig. 5).

Many open questions remain about the process of fusion activation. While the importance of the HN/H/G stalk domain is well-established,[49,103,104] the mechanism by which the signal is transmitted to the homologous F protein is just beginning to be understood. Structural analysis of the complete HN/H/G molecule, both engaged to its receptor and free, has been challenging in the face of the important hydrophobic domains.[48,49,55,59] Measuring the precise structural changes that trigger F protein activation will require novel approaches including cryoelectron microscopic analysis of specific stages in the fusion activation process.

Understanding the fusion and entry process of paramyxoviruses is key for the design of new therapies. The unifying model we suggest for the *Paramyxoviridae* fusion machinery can support the development of new antiviral strategies that may have broad-spectrum potential for inhibiting entry. Potentially promising therapies that target the entry step include small molecules that prematurely trigger the fusion protein,[143] or lipid-conjugated HRC-derived peptides.[138,139] Together with other antiviral approaches that target common mechanisms, these strategies hold promise for wide applicability for preventing and treating this important group of pathogens.

ACKNOWLEDGMENTS

We are grateful to Ashton Kutcher for his support for microscopy, to Dan and Nancy Paduano for their essential support of innovative research projects, and to the Friedman Family Foundation for our laboratories at Weill Cornell Medical College. M. P. is a Friedman Family Research Scholar in Pediatric Infectious Diseases. The work was supported by NIH R01 AI31971 and NIH 3R01 AI031971-19S1 to A. M., NIH Region II Center of Excellence for Bio-Defense and Emerging Infectious Disease Research U54AI057158 Research Grant to A. M., NIH R21 AI100292, NIH R21 EB011707, and NIH R21 NS073781 to M. P.

REFERENCES

1. Chang A, Dutch RE. Paramyxovirus fusion and entry: multiple paths to a common end. *Viruses*. 2012;4(4):613–636. http://dx.doi.org/10.3390/v4040613.
2. Enders G. Paramyxoviruses. In: Baron S, ed. *Medical Microbiology*. 4th ed. Galveston, TX: University of Texas Medical Branch at Galveston; 1996 Available at, http://www.ncbi.nlm.nih.gov/books/NBK8461/, Accessed April 11, 2014.
3. Bossart KN, Fusco DL, Broder CC. Paramyxovirus entry. *Adv Exp Med Biol*. 2013;790:95–127. http://dx.doi.org/10.1007/978-1-4614-7651-1_6.
4. Falk K, Batts WN, Kvellestad A, Kurath G, Wiik-Nielsen J, Winton JR. Molecular characterisation of Atlantic salmon paramyxovirus (ASPV): a novel paramyxovirus associated with proliferative gill inflammation. *Virus Res*. 2008;133(2):218–227. http://dx.doi.org/10.1016/j.virusres.2008.01.006.
5. Kurath G, Batts WN, Ahne W, Winton JR. Complete genome sequence of Fer-de-Lance virus reveals a novel gene in reptilian paramyxoviruses. *J Virol*. 2004;78(4):2045–2056.
6. Ganar K, Das M, Sinha S, Kumar S. Newcastle disease virus: current status and our understanding. *Virus Res*. 2014;184C:71–81. http://dx.doi.org/10.1016/j.virusres.2014.02.016.
7. Croser EL, Marsh GA. The changing face of the henipaviruses. *Vet Microbiol*. 2013;167(1-2):151–158. http://dx.doi.org/10.1016/j.vetmic.2013.08.002.

8. Kurth A, Kohl C, Brinkmann A, et al. Novel paramyxoviruses in free-ranging European bats. *PLoS One*. 2012;7(6):e38688. http://dx.doi.org/10.1371/journal.pone.0038688.
9. Marsh GA, de Jong C, Barr JA, et al. Cedar virus: a novel Henipavirus isolated from Australian bats. *PLoS Pathog*. 2012;8(8):e1002836. http://dx.doi.org/10.1371/journal.ppat.1002836.
10. Magoffin DE, Mackenzie JS, Wang L-F. Genetic analysis of J-virus and Beilong virus using minireplicons. *Virology*. 2007;364(1):103–111. http://dx.doi.org/10.1016/j.virol.2007.01.045.
11. Weinberg GA, Hall CB, Iwane MK, et al. Parainfluenza virus infection of young children: estimates of the population-based burden of hospitalization. *J Pediatr*. 2009;154(5):694–699. http://dx.doi.org/10.1016/j.jpeds.2008.11.034.
12. Nair H, Nokes DJ, Gessner BD, et al. Global burden of acute lower respiratory infections due to respiratory syncytial virus in young children: a systematic review and meta-analysis. *Lancet*. 2010;375(9725):1545–1555. http://dx.doi.org/10.1016/S0140-6736(10)60206-1.
13. Williams JV, Tollefson SJ, Heymann PW, Carper HT, Patrie J, Crowe JE. Human metapneumovirus infection in children hospitalized for wheezing. *J Allergy Clin Immunol*. 2005;115(6):1311–1312. http://dx.doi.org/10.1016/j.jaci.2005.02.001.
14. Clayton BA, Wang LF, Marsh GA. Henipaviruses: an updated review focusing on the pteropid reservoir and features of transmission. *Zoonoses Public Health*. 2013;60(1):69–83. http://dx.doi.org/10.1111/j.1863-2378.2012.01501.x.
15. Kumaria R, Iyer LR, Hibberd ML, Simões EAF, Sugrue RJ. Whole genome characterization of non-tissue culture adapted HRSV strains in severely infected children. *Virol J* 2011;8:372. http://dx.doi.org/10.1186/1743-422X-8-372.
16. Takeuchi K, Miyajima N, Kobune F, Tashiro M. Comparative nucleotide sequence analyses of the entire genomes of B95a cell-isolated and vero cell-isolated measles viruses from the same patient. *Virus Genes*. 2000;20(3):253–257.
17. Lo MK, Harcourt BH, Mungall BA, et al. Determination of the henipavirus phosphoprotein gene mRNA editing frequencies and detection of the C, V and W proteins of Nipah virus in virus-infected cells. *J Gen Virol*. 2009;90(Pt 2):398–404. http://dx.doi.org/10.1099/vir.0.007294-0.
18. Yu M, Hansson E, Langedijk JP, Eaton BT, Wang LF. The attachment protein of Hendra virus has high structural similarity but limited primary sequence homology compared with viruses in the genus Paramyxovirus. *Virology*. 1998;251(2):227–233. http://dx.doi.org/10.1006/viro.1998.9302.
19. Jack PJM, Boyle DB, Eaton BT, Wang L-F. The complete genome sequence of J virus reveals a unique genome structure in the family Paramyxoviridae. *J Virol*. 2005;79(16):10690–10700. http://dx.doi.org/10.1128/JVI.79.16.10690-10700.2005.
20. Li Z, Yu M, Zhang H, et al. Beilong virus, a novel paramyxovirus with the largest genome of non-segmented negative-stranded RNA viruses. *Virology*. 2006;346(1):219–228. http://dx.doi.org/10.1016/j.virol.2005.10.039.
21. Hausmann S, Jacques JP, Kolakofsky D. Paramyxovirus RNA editing and the requirement for hexamer genome length. *RNA*. 1996;2(10):1033–1045.
22. Halpin K, Bankamp B, Harcourt BH, Bellini WJ, Rota PA. Nipah virus conforms to the rule of six in a minigenome replication assay. *J Gen Virol*. 2004;85(pt 3):701–707.
23. Jin H, Zhou H, Cheng X, Tang R, Munoz M, Nguyen N. Recombinant respiratory syncytial viruses with deletions in the NS1, NS2, SH, and M2-2 genes are attenuated in vitro and in vivo. *Virology*. 2000;273(1):210–218. http://dx.doi.org/10.1006/viro.2000.0393.
24. Takeuchi K, Tanabayashi K, Hishiyama M, Yamada A. The mumps virus SH protein is a membrane protein and not essential for virus growth. *Virology*. 1996;225(1):156–162. http://dx.doi.org/10.1006/viro.1996.0583.

25. He B, Leser GP, Paterson RG, Lamb RA. The paramyxovirus SV5 small hydrophobic (SH) protein is not essential for virus growth in tissue culture cells. *Virology*. 1998;250(1):30–40. http://dx.doi.org/10.1006/viro.1998.9354.
26. Biacchesi S, Pham QN, Skiadopoulos MH, Murphy BR, Collins PL, Buchholz UJ. Infection of nonhuman primates with recombinant human metapneumovirus lacking the SH, G, or M2-2 protein categorizes each as a nonessential accessory protein and identifies vaccine candidates. *J Virol*. 2005;79(19):12608–12613. http://dx.doi.org/10.1128/JVI.79.19.12608-12613.2005.
27. Atreya PL, Peeples ME, Collins PL. The NS1 protein of human respiratory syncytial virus is a potent inhibitor of minigenome transcription and RNA replication. *J Virol*. 1998;72(2):1452–1461.
28. Audsley MD, Moseley GW. Paramyxovirus evasion of innate immunity: diverse strategies for common targets. *World J Virol*. 2013;2(2):57–70. http://dx.doi.org/10.5501/wjv.v2.i2.57.
29. Huberman K, Peluso RW, Moscona A. Hemagglutinin-neuraminidase of human parainfluenza 3: role of the neuraminidase in the viral life cycle. *Virology*. 1995;214(1):294–300. http://dx.doi.org/10.1006/viro.1995.9925.
30. Moscona A, Peluso RW. Fusion properties of cells persistently infected with human parainfluenza virus type 3: participation of hemagglutinin-neuraminidase in membrane fusion. *J Virol*. 1991;65(6):2773–2777.
31. Morrison T, McQuain C, McGinnes L. Complementation between avirulent Newcastle disease virus and a fusion protein gene expressed from a retrovirus vector: requirements for membrane fusion. *J Virol*. 1991;65(2):813–822.
32. Moscona A, Peluso RW. Relative affinity of the human parainfluenza virus type 3 hemagglutinin-neuraminidase for sialic acid correlates with virus-induced fusion activity. *J Virol*. 1993;67(11):6463–6468.
33. Navaratnarajah CK, Oezguen N, Rupp L, et al. The heads of the measles virus attachment protein move to transmit the fusion-triggering signal. *Nat Struct Mol Biol*. 2011;18(2):128–134. http://dx.doi.org/10.1038/nsmb.1967.
34. Mirza AM, Aguilar HC, Zhu Q, et al. Triggering of the newcastle disease virus fusion protein by a chimeric attachment protein that binds to Nipah virus receptors. *J Biol Chem*. 2011;286(20):17851–17860. http://dx.doi.org/10.1074/jbc.M111.233965.
35. Corey EA, Iorio RM. Mutations in the stalk of the measles virus hemagglutinin protein decrease fusion but do not interfere with virus-specific interaction with the homologous fusion protein. *J Virol*. 2007;81(18):9900–9910. http://dx.doi.org/10.1128/JVI.00909-07.
36. Aguilar HC, Matreyek KA, Filone CM, et al. N-glycans on Nipah virus fusion protein protect against neutralization but reduce membrane fusion and viral entry. *J Virol*. 2006;80(10):4878–4889. http://dx.doi.org/10.1128/JVI.80.10.4878-4889.2006.
37. Bishop KA, Hickey AC, Khetawat D, et al. Residues in the stalk domain of the hendra virus g glycoprotein modulate conformational changes associated with receptor binding. *J Virol*. 2008;82(22):11398–11409. http://dx.doi.org/10.1128/JVI.02654-07.
38. Plemper RK, Hammond AL, Gerlier D, Fielding AK, Cattaneo R. Strength of envelope protein interaction modulates cytopathicity of measles virus. *J Virol*. 2002;76(10):5051–5061.
39. Iorio RM, Melanson VR, Mahon PJ. Glycoprotein interactions in paramyxovirus fusion. *Future Virol*. 2009;4(4):335–351. http://dx.doi.org/10.2217/fvl.09.17.
40. Lee B, Ataman ZA. Modes of paramyxovirus fusion: a Henipavirus perspective. *Trends Microbiol*. 2011;19(8):389–399. http://dx.doi.org/10.1016/j.tim.2011.03.005.
41. Iorio RM, Mahon PJ. Paramyxoviruses: different receptors—different mechanisms of fusion. *Trends Microbiol*. 2008;16(4):135–137. http://dx.doi.org/10.1016/j.tim.2008.01.006.

42. Porotto M, Murrell M, Greengard O, Doctor L, Moscona A. Influence of the human parainfluenza virus 3 attachment protein's neuraminidase activity on its capacity to activate the fusion protein. *J Virol.* 2005;79(4):2383–2392. http://dx.doi.org/10.1128/JVI.79.4.2383-2392.2005.
43. Porotto M, Murrell M, Greengard O, Moscona A. Triggering of human parainfluenza virus 3 fusion protein (F) by the hemagglutinin-neuraminidase (HN) protein: an HN mutation diminishes the rate of F activation and fusion. *J Virol.* 2003;77(6):3647–3654.
44. Dutch RE. Entry and fusion of emerging paramyxoviruses. *PLoS Pathog.* 2010;6(6):e1000881. http://dx.doi.org/10.1371/journal.ppat.1000881.
45. Plattet P, Plemper RK. Envelope protein dynamics in paramyxovirus entry. *MBio.* 2013;4(4):e00413. http://dx.doi.org/10.1128/mBio.00413-13.
46. Porotto M, Devito I, Palmer SG, et al. Spring-loaded model revisited: paramyxovirus fusion requires engagement of a receptor binding protein beyond initial triggering of the fusion protein. *J Virol.* 2011;85(24):12867–12880. http://dx.doi.org/10.1128/JVI.05873-11.
47. Porotto M, Palmer SG, Palermo LM, Moscona A. Mechanism of fusion triggering by human parainfluenza virus type III: communication between viral glycoproteins during entry. *J Biol Chem.* 2012;287(1):778–793. http://dx.doi.org/10.1074/jbc.M111.298059.
48. Yuan P, Swanson KA, Leser GP, Paterson RG, Lamb RA, Jardetzky TS. Structure of the Newcastle disease virus hemagglutinin-neuraminidase (HN) ectodomain reveals a four-helix bundle stalk. *Proc Natl Acad Sci USA.* 2011;108(36):14920–14925. http://dx.doi.org/10.1073/pnas.1111691108.
49. Bose S, Welch BD, Kors CA, Yuan P, Jardetzky TS, Lamb RA. Structure and mutagenesis of the parainfluenza virus 5 hemagglutinin-neuraminidase stalk domain reveals a four-helix bundle and the role of the stalk in fusion promotion. *J Virol.* 2011;85(24):12855–12866. http://dx.doi.org/10.1128/JVI.06350-11.
50. Colf LA, Juo ZS, Garcia KC. Structure of the measles virus hemagglutinin. *Nat Struct Mol Biol.* 2007;14(12):1227–1228. http://dx.doi.org/10.1038/nsmb1342.
51. Hashiguchi T, Kajikawa M, Maita N, et al. Crystal structure of measles virus hemagglutinin provides insight into effective vaccines. *Proc Natl Acad Sci USA.* 2007;104(49):19535–19540. http://dx.doi.org/10.1073/pnas.0707830104.
52. Hashiguchi T, Ose T, Kubota M, et al. Structure of the measles virus hemagglutinin bound to its cellular receptor SLAM. *Nat Struct Mol Biol.* 2011;18(2):135–141. http://dx.doi.org/10.1038/nsmb.1969.
53. Bowden TA, Aricescu AR, Gilbert RJC, Grimes JM, Jones EY, Stuart DI. Structural basis of Nipah and Hendra virus attachment to their cell-surface receptor ephrin-B2. *Nat Struct Mol Biol.* 2008;15(6):567–572. http://dx.doi.org/10.1038/nsmb.1435.
54. Xu K, Rajashankar KR, Chan Y-P, Himanen JP, Broder CC, Nikolov DB. Host cell recognition by the henipaviruses: crystal structures of the Nipah G attachment glycoprotein and its complex with ephrin-B3. *Proc Natl Acad Sci USA.* 2008;105(29):9953–9958. http://dx.doi.org/10.1073/pnas.0804797105.
55. Welch BD, Yuan P, Bose S, Kors CA, Lamb RA, Jardetzky TS. Structure of the parainfluenza virus 5 (PIV5) hemagglutinin-neuraminidase (HN) ectodomain. *PLoS Pathog* 2013;9(8):e1003534. http://dx.doi.org/10.1371/journal.ppat.1003534.
56. Lawrence MC, Borg NA, Streltsov VA, et al. Structure of the haemagglutinin-neuraminidase from human parainfluenza virus type III. *J Mol Biol.* 2004;335(5):1343–1357.
57. Navaratnarajah CK, Negi S, Braun W, Cattaneo R. Membrane fusion triggering: three modules with different structure and function in the upper half of the measles virus attachment protein stalk. *J Biol Chem.* 2012;287(46):38543–38551. http://dx.doi.org/10.1074/jbc.M112.410563.

58. Navaratnarajah CK, Kumar S, Generous A, Apte-Sengupta S, Mateo M, Cattaneo R. The measles virus hemagglutinin stalk: structure and function of the central fusion-activation and membrane-proximal segments. *J Virol.* 2014;88(11):6158–6167. http://dx.doi.org/10.1128/JVI.02846-13.
59. Bowden TA, Crispin M, Harvey DJ, Jones EY, Stuart DI. Dimeric architecture of the Hendra virus attachment glycoprotein: evidence for a conserved mode of assembly. *J Virol.* 2010;84(12):6208–6217. http://dx.doi.org/10.1128/JVI.00317-10.
60. Crennell S, Takimoto T, Portner A, Taylor G. Crystal structure of the multifunctional paramyxovirus hemagglutinin-neuraminidase. *Nat Struct Biol.* 2000;7(11):1068–1074. http://dx.doi.org/10.1038/81002.
61. Santiago C, Celma ML, Stehle T, Casasnovas JM. Structure of the measles virus hemagglutinin bound to the CD46 receptor. *Nat Struct Mol Biol.* 2010;17(1):124–129. http://dx.doi.org/10.1038/nsmb.1726.
62. Zaitsev V, von Itzstein M, Groves D, et al. Second sialic acid binding site in Newcastle disease virus hemagglutinin-neuraminidase: implications for fusion. *J Virol.* 2004;78(7):3733–3741.
63. Bousse TL, Taylor G, Krishnamurthy S, Portner A, Samal SK, Takimoto T. Biological significance of the second receptor binding site of Newcastle disease virus hemagglutinin-neuraminidase protein. *J Virol.* 2004;78(23):13351–13355. http://dx.doi.org/10.1128/JVI.78.23.13351-13355.2004.
64. Porotto M, Salah Z, DeVito I, et al. The second receptor binding site of the globular head of the Newcastle disease virus hemagglutinin-neuraminidase activates the stalk of multiple paramyxovirus receptor binding proteins to trigger fusion. *J Virol.* 2012;86(10):5730–5741. http://dx.doi.org/10.1128/JVI.06793-11.
65. Bousse T, Takimoto T. Mutation at residue 523 creates a second receptor binding site on human parainfluenza virus type 1 hemagglutinin-neuraminidase protein. *J Virol.* 2006;80(18):9009–9016. http://dx.doi.org/10.1128/JVI.00969-06.
66. Alymova IV, Portner A, Mishin VP, McCullers JA, Freiden P, Taylor GL. Receptor-binding specificity of the human parainfluenza virus type 1 hemagglutinin-neuraminidase glycoprotein. *Glycobiology.* 2012;22(2):174–180. http://dx.doi.org/10.1093/glycob/cwr112.
67. Porotto M, Fornabaio M, Kellogg GE, Moscona A. A second receptor binding site on human parainfluenza virus type 3 hemagglutinin-neuraminidase contributes to activation of the fusion mechanism. *J Virol.* 2007;81(7):3216–3228. http://dx.doi.org/10.1128/JVI.02617-06.
68. Leonard VH, Sinn PL, Hodge G, et al. Measles virus blind to its epithelial cell receptor remains virulent in rhesus monkeys but cannot cross the airway epithelium and is not shed. *J Clin Invest.* 2008;118:2448–2458.
69. Welch BD, Liu Y, Kors CA, Leser GP, Jardetzky TS, Lamb RA. Structure of the cleavage-activated prefusion form of the parainfluenza virus 5 fusion protein. *Proc Natl Acad Sci USA.* 2012;109(41):16672–16677. http://dx.doi.org/10.1073/pnas.1213802109.
70. Yin H-S, Paterson RG, Wen X, Lamb RA, Jardetzky TS. Structure of the uncleaved ectodomain of the paramyxovirus (hPIV3) fusion protein. *Proc Natl Acad Sci USA.* 2005;102(26):9288–9293. http://dx.doi.org/10.1073/pnas.0503989102.
71. Smith EC, Smith SE, Carter JR, et al. Trimeric transmembrane domain interactions in paramyxovirus fusion proteins: roles in protein folding, stability, and function. *J Biol Chem.* 2013;288(50):35726–35735. http://dx.doi.org/10.1074/jbc.M113.514554.
72. Smith EC, Culler MR, Hellman LM, Fried MG, Creamer TP, Dutch RE. Beyond anchoring: the expanding role of the hendra virus fusion protein transmembrane domain in protein folding, stability, and function. *J Virol.* 2012;86(6):3003–3013. http://dx.doi.org/10.1128/JVI.05762-11.

73. González-Reyes L, Ruiz-Argüello MB, García-Barreno B, et al. Cleavage of the human respiratory syncytial virus fusion protein at two distinct sites is required for activation of membrane fusion. *Proc Natl Acad Sci USA*. 2001;98(17):9859–9864. http://dx.doi.org/10.1073/pnas.151098198.
74. Diederich S, Moll M, Klenk H-D, Maisner A. The nipah virus fusion protein is cleaved within the endosomal compartment. *J Biol Chem*. 2005;280(33):29899–29903. http://dx.doi.org/10.1074/jbc.M504598200.
75. Pager CT, Dutch RE. Cathepsin L is involved in proteolytic processing of the Hendra virus fusion protein. *J Virol*. 2005;79(20):12714–12720. http://dx.doi.org/10.1128/JVI.79.20.12714-12720.2005.
76. Pager CT, Craft Jr WW, Patch J, Dutch RE. A mature and fusogenic form of the Nipah virus fusion protein requires proteolytic processing by cathepsin L. *Virology*. 2006;346(2):251–257. http://dx.doi.org/10.1016/j.virol.2006.01.007.
77. Diederich S, Sauerhering L, Weis M, et al. Activation of the Nipah virus fusion protein in MDCK cells is mediated by cathepsin B within the endosome-recycling compartment. *J Virol*. 2012;86(7):3736–3745. http://dx.doi.org/10.1128/JVI.06628-11.
78. Smith EC, Gregory SM, Tamm LK, Creamer TP, Dutch RE. Role of sequence and structure of the Hendra fusion protein fusion peptide in membrane fusion. *J Biol Chem*. 2012;287(35):30035–30048. http://dx.doi.org/10.1074/jbc.M112.367862.
79. Kim YH, Donald JE, Grigoryan G, et al. Capture and imaging of a prehairpin fusion intermediate of the paramyxovirus PIV5. *Proc Natl Acad Sci USA*. 2011; 108(52):20992–20997. http://dx.doi.org/10.1073/pnas.1116034108.
80. Yin H-S, Wen X, Paterson RG, Lamb RA, Jardetzky TS. Structure of the parainfluenza virus 5 F protein in its metastable, prefusion conformation. *Nature*. 2006;439(7072):38–44. http://dx.doi.org/10.1038/nature04322.
81. McLellan JS, Yang Y, Graham BS, Kwong PD. Structure of respiratory syncytial virus fusion glycoprotein in the postfusion conformation reveals preservation of neutralizing epitopes. *J Virol*. 2011;85(15):7788–7796. http://dx.doi.org/10.1128/JVI.00555-11.
82. Swanson K, Wen X, Leser GP, Paterson RG, Lamb RA, Jardetzky TS. Structure of the Newcastle disease virus F protein in the post-fusion conformation. *Virology*. 2010;402(2):372–379. http://dx.doi.org/10.1016/j.virol.2010.03.050.
83. Harrison SC. Viral membrane fusion. *Nat Struct Mol Biol*. 2008;15(7):690–698. http://dx.doi.org/10.1038/nsmb.1456.
84. Biering SB, Huang A, Vu AT, et al. N-Glycans on the Nipah virus attachment glycoprotein modulate fusion and viral entry as they protect against antibody neutralization. *J Virol*. 2012;86(22):11991–12002. http://dx.doi.org/10.1128/JVI.01304-12.
85. Dörig RE, Marcil A, Chopra A, Richardson CD. The human CD46 molecule is a receptor for measles virus (Edmonston strain). *Cell*. 1993;75(2):295–305.
86. Tatsuo H, Ono N, Tanaka K, Yanagi Y. SLAM (CDw150) is a cellular receptor for measles virus. *Nature*. 2000;406(6798):893–897. http://dx.doi.org/10.1038/35022579.
87. Erlenhoefer C, Wurzer WJ, Löffler S, Schneider-Schaulies S, ter Meulen V, Schneider-Schaulies J. CD150 (SLAM) is a receptor for measles virus but is not involved in viral contact-mediated proliferation inhibition. *J Virol*. 2001;75(10):4499–4505. http://dx.doi.org/10.1128/JVI.75.10.4499-4505.2001.
88. Noyce RS, Bondre DG, Ha MN, et al. Tumor cell marker PVRL4 (nectin 4) is an epithelial cell receptor for measles virus. *PLoS Pathog*. 2011;7(8):e1002240. http://dx.doi.org/10.1371/journal.ppat.1002240.
89. Mühlebach MD, Mateo M, Sinn PL, et al. Adherens junction protein nectin-4 is the epithelial receptor for measles virus. *Nature*. 2011;480(7378):530–533. http://dx.doi.org/10.1038/nature10639.
90. Kato S-I, Nagata K, Takeuchi K. Cell tropism and pathogenesis of measles virus in monkeys. *Front Microbiol*. 2012;3:14. http://dx.doi.org/10.3389/fmicb.2012.00014.

91. Bonaparte MI, Dimitrov AS, Bossart KN, et al. Ephrin-B2 ligand is a functional receptor for Hendra virus and Nipah virus. *Proc Natl Acad Sci USA*. 2005;102(30):10652–10657. http://dx.doi.org/10.1073/pnas.0504887102.
92. Mathieu C, Pohl C, Szecsi J, et al. Nipah virus uses leukocytes for efficient dissemination within a host. *J Virol*. 2011;85(15):7863–7871. http://dx.doi.org/10.1128/JVI.00549-11.
93. Krusat T, Streckert HJ. Heparin-dependent attachment of respiratory syncytial virus (RSV) to host cells. *Arch Virol*. 1997;142(6):1247–1254.
94. Bourgeois C, Bour JB, Lidholt K, Gauthray C, Pothier P. Heparin-like structures on respiratory syncytial virus are involved in its infectivity in vitro. *J Virol*. 1998;72(9):7221–7227.
95. Feldman SA, Hendry RM, Beeler JA. Identification of a linear heparin binding domain for human respiratory syncytial virus attachment glycoprotein G. *J Virol*. 1999;73(8):6610–6617.
96. Thammawat S, Sadlon TA, Hallsworth PG, Gordon DL. Role of cellular glycosaminoglycans and charged regions of viral G protein in human metapneumovirus infection. *J Virol*. 2008;82(23):11767–11774. http://dx.doi.org/10.1128/JVI.01208-08.
97. Harcourt JL, Karron RA, Tripp RA. Anti-G protein antibody responses to respiratory syncytial virus infection or vaccination are associated with inhibition of G protein CX3C-CX3CR1 binding and leukocyte chemotaxis. *J Infect Dis*. 2004; 190(11):1936–1940. http://dx.doi.org/10.1086/425516.
98. Chirkova T, Boyoglu-Barnum S, Gaston KA, et al. Respiratory syncytial virus G protein CX3C motif impairs human airway epithelial and immune cell responses. *J Virol*. 2013;87(24):13466–13479. http://dx.doi.org/10.1128/JVI.01741-13.
99. Steffen DL, Xu K, Nikolov DB, Broder CC. Henipavirus mediated membrane fusion, virus entry and targeted therapeutics. *Viruses*. 2012;4(2):280–308. http://dx.doi.org/10.3390/v4020280.
100. Russell CJ, Jardetzky TS, Lamb RA. Membrane fusion machines of paramyxoviruses: capture of intermediates of fusion. *EMBO J*. 2001;20(15):4024–4034. http://dx.doi.org/10.1093/emboj/20.15.4024.
101. Bose S, Zokarkar A, Welch BD, Leser GP, Jardetzky TS, Lamb RA. Fusion activation by a headless parainfluenza virus 5 hemagglutinin-neuraminidase stalk suggests a modular mechanism for triggering. *Proc Natl Acad Sci USA*. 2012;109(39):E2625–E2634. http://dx.doi.org/10.1073/pnas.1213813109.
102. Liu Q, Stone JA, Bradel-Tretheway B, et al. Unraveling a three-step spatiotemporal mechanism of triggering of receptor-induced Nipah virus fusion and cell entry. *PLoS Pathog*. 2013;9(11):e1003770. http://dx.doi.org/10.1371/journal.ppat.1003770.
103. Ader N, Brindley MA, Avila M, et al. Structural rearrangements of the central region of the morbillivirus attachment protein stalk domain trigger F protein refolding for membrane fusion. *J Biol Chem*. 2012;287(20):16324–16334. http://dx.doi.org/10.1074/jbc.M112.342493.
104. Brindley MA, Suter R, Schestak I, Kiss G, Wright ER, Plemper RK. A stabilized headless measles virus attachment protein stalk efficiently triggers membrane fusion. *J Virol*. 2013;87(21):11693–11703. http://dx.doi.org/10.1128/JVI.01945-13.
105. Talekar A, Moscona A, Porotto M. Measles fusion machinery activated by sialic acid binding globular domain. *J Virol*. 2013;87(24):13619–13627. http://dx.doi.org/10.1128/JVI.02256-13.
106. Melanson VR, Iorio RM. Amino acid substitutions in the F-specific domain in the stalk of the newcastle disease virus HN protein modulate fusion and interfere with its interaction with the F protein. *J Virol*. 2004;78(23):13053–13061. http://dx.doi.org/10.1128/JVI.78.23.13053-13061.2004.

107. Bose S, Song AS, Jardetzky TS, Lamb RA. Fusion activation through attachment protein stalk domains indicates a conserved core mechanism of paramyxovirus entry into cells. *J Virol.* 2014;88(8):3925–3941. http://dx.doi.org/10.1128/JVI.03741-13.
108. Xu R, Palmer SG, Porotto M, et al. Interaction between the Hemagglutinin-neuraminidase and fusion glycoproteins of human parainfluenza virus type III regulates viral growth in vivo. *MBio.* 2013;4(5):e00803–e00813. http://dx.doi.org/10.1128/mBio.00803-13.
109. Bossart KN, Wang L-F, Flora MN, et al. Membrane fusion tropism and heterotypic functional activities of the Nipah virus and Hendra virus envelope glycoproteins. *J Virol.* 2002;76(22):11186–11198.
110. Ader N, Brindley M, Avila M, et al. Mechanism for active membrane fusion triggering by morbillivirus attachment protein. *J Virol.* 2013;87(1):314–326. http://dx.doi.org/10.1128/JVI.01826-12.
111. Porotto M, Salah ZW, Gui L, et al. Regulation of paramyxovirus fusion activation: the hemagglutinin-neuraminidase protein stabilizes the fusion protein in a pretriggered state. *J Virol.* 2012;86(23):12838–12848. http://dx.doi.org/10.1128/JVI.01965-12.
112. Cantín C, Holguera J, Ferreira L, Villar E, Muñoz-Barroso I. Newcastle disease virus may enter cells by caveolae-mediated endocytosis. *J Gen Virol.* 2007;88(pt 2):559–569. http://dx.doi.org/10.1099/vir.0.82150-0.
113. Schowalter RM, Chang A, Robach JG, Buchholz UJ, Dutch RE. Low-pH triggering of human metapneumovirus fusion: essential residues and importance in entry. *J Virol.* 2009;83(3):1511–1522. http://dx.doi.org/10.1128/JVI.01381-08.
114. Kolokoltsov AA, Deniger D, Fleming EH, Roberts Jr NJ, Karpilow JM, Davey RA. Small interfering RNA profiling reveals key role of clathrin-mediated endocytosis and early endosome formation for infection by respiratory syncytial virus. *J Virol.* 2007;81(14):7786–7800. http://dx.doi.org/10.1128/JVI.02780-06.
115. Diederich S, Thiel L, Maisner A. Role of endocytosis and cathepsin-mediated activation in Nipah virus entry. *Virology.* 2008;375(2):391–400. http://dx.doi.org/10.1016/j.virol.2008.02.019.
116. Sánchez-Felipe L, Villar E, Muñoz-Barroso I. Entry of Newcastle disease virus into the host cell: role of acidic pH and endocytosis. *Biochim Biophys Acta.* 2013;1838(1 pt B): 300–309. http://dx.doi.org/10.1016/j.bbamem.2013.08.008.
117. Kahn JS, Schnell MJ, Buonocore L, Rose JK. Recombinant vesicular stomatitis virus expressing respiratory syncytial virus (RSV) glycoproteins: RSV fusion protein can mediate infection and cell fusion. *Virology.* 1999;254(1):81–91. http://dx.doi.org/10.1006/viro.1998.9535.
118. Techaarpornkul S, Barretto N, Peeples ME. Functional analysis of recombinant respiratory syncytial virus deletion mutants lacking the small hydrophobic and/or attachment glycoprotein gene. *J Virol.* 2001;75(15):6825–6834. http://dx.doi.org/10.1128/JVI.75.15.6825-6834.2001.
119. Schowalter RM, Smith SE, Dutch RE. Characterization of human metapneumovirus F protein-promoted membrane fusion: critical roles for proteolytic processing and low pH. *J Virol.* 2006;80(22):10931–10941. http://dx.doi.org/10.1128/JVI.01287-06.
120. Sergel TA, McGinnes LW, Morrison TG. A single amino acid change in the Newcastle disease virus fusion protein alters the requirement for HN protein in fusion. *J Virol.* 2000;74(11):5101–5107.
121. Seth S, Vincent A, Compans RW. Mutations in the cytoplasmic domain of a paramyxovirus fusion glycoprotein rescue syncytium formation and eliminate the hemagglutinin-neuraminidase protein requirement for membrane fusion. *J Virol.* 2003;77(1):167–178.
122. Ito M, Nishio M, Kawano M, Komada H, Ito Y, Tsurudome M. Effects of multiple amino acids of the parainfluenza virus 5 fusion protein on its haemagglutinin-

neuraminidase-independent fusion activity. *J Gen Virol.* 2009;90(pt 2):405–413. http://dx.doi.org/10.1099/vir.0.006437-0.
123. Karron RA, Buonagurio DA, Georgiu AF, et al. Respiratory syncytial virus (RSV) SH and G proteins are not essential for viral replication in vitro: clinical evaluation and molecular characterization of a cold-passaged, attenuated RSV subgroup B mutant. *Proc Natl Acad Sci USA.* 1997;94(25):13961–13966.
124. Dutch RE, Joshi SB, Lamb RA. Membrane fusion promoted by increasing surface densities of the paramyxovirus F and HN proteins: comparison of fusion reactions mediated by simian virus 5 F, human parainfluenza virus type 3 F, and influenza virus HA. *J Virol.* 1998;72(10):7745–7753.
125. Herfst S, Mas V, Ver LS, et al. Low-pH-induced membrane fusion mediated by human metapneumovirus F protein is a rare, strain-dependent phenomenon. *J Virol.* 2008;82(17):8891–8895. http://dx.doi.org/10.1128/JVI.00472-08.
126. Tayyari F, Marchant D, Moraes TJ, Duan W, Mastrangelo P, Hegele RG. Identification of nucleolin as a cellular receptor for human respiratory syncytial virus. *Nat Med.* 2011;17(9):1132–1135. http://dx.doi.org/10.1038/nm.2444.
127. Wei Y, Zhang Y, Cai H, et al. Roles of the putative integrin-binding motif of the human metapneumovirus fusion (f) protein in cell-cell fusion, viral infectivity, and pathogenesis. *J Virol.* 2014;88(8):4338–4352. http://dx.doi.org/10.1128/JVI.03491-13.
128. Feldman SA, Audet S, Beeler JA. The fusion glycoprotein of human respiratory syncytial virus facilitates virus attachment and infectivity via an interaction with cellular heparan sulfate. *J Virol.* 2000;74(14):6442–6447.
129. Barretto N, Hallak LK, Peeples ME. Neuraminidase treatment of respiratory syncytial virus-infected cells or virions, but not target cells, enhances cell-cell fusion and infection. *Virology.* 2003;313(1):33–43.
130. Chang A, Masante C, Buchholz UJ, Dutch RE. Human metapneumovirus (HMPV) binding and infection are mediated by interactions between the HMPV fusion protein and heparan sulfate. *J Virol.* 2012;86(6):3230–3243. http://dx.doi.org/10.1128/JVI.06706-11.
131. Talekar A, DeVito I, Salah Z, et al. Identification of a region in the stalk domain of the nipah virus receptor binding protein that is critical for fusion activation. *J Virol.* 2013;87(20):10980–10996. http://dx.doi.org/10.1128/JVI.01646-13.
132. Bose S, Heath CM, Shah PA, Alayyoubi M, Jardetzky TS, Lamb RA. Mutations in the parainfluenza virus 5 fusion (F) protein reveal domains important for fusion triggering and metastability. *J Virol.* 2013;87(24):13520–13531. http://dx.doi.org/10.1128/JVI.02123-13.
133. Brindley MA, Takeda M, Plattet P, Plemper RK. Triggering the measles virus membrane fusion machinery. *Proc Natl Acad Sci USA.* 2012;109(44):E3018–E3027. http://dx.doi.org/10.1073/pnas.1210925109.
134. Mahon PJ, Mirza AM, Musich TA, Iorio RM. Engineered intermonomeric disulfide bonds in the globular domain of Newcastle disease virus hemagglutinin-neuraminidase protein: implications for the mechanism of fusion promotion. *J Virol.* 2008;82(21):10386–10396. http://dx.doi.org/10.1128/JVI.00581-08.
135. Zokarkar A, Connolly SA, Jardetzky TS, Lamb RA. Reversible inhibition of fusion activity of a paramyxovirus fusion protein by an engineered disulfide bond in the membrane-proximal external region. *J Virol.* 2012;86(22):12397–12401. http://dx.doi.org/10.1128/JVI.02006-12.
136. Zokarkar A, Lamb RA. The paramyxovirus fusion protein C-terminal region: mutagenesis indicates an indivisible protein unit. *J Virol.* 2012;86(5):2600–2609. http://dx.doi.org/10.1128/JVI.06546-11.
137. Donald JE, Zhang Y, Fiorin G, et al. Transmembrane orientation and possible role of the fusogenic peptide from parainfluenza virus 5 (PIV5) in promoting fusion. *Proc*

Natl Acad Sci USA. 2011;108(10):3958–3963. http://dx.doi.org/10.1073/pnas.1019668108.
138. Porotto M, Rockx B, Yokoyama CC, et al. Inhibition of Nipah virus infection in vivo: targeting an early stage of paramyxovirus fusion activation during viral entry. *PLoS Pathog*. 2010;6(10):e1001168. http://dx.doi.org/10.1371/journal.ppat.1001168.
139. Welsch JC, Talekar A, Mathieu C, et al. Fatal measles infection prevented by brain-penetrant fusion inhibitors. *J Virol*. 2013;87(24):13785–13794. http://dx.doi.org/10.1128/JVI.02436-13.
140. Pessi A, Langella A, Capitò E, et al. A general strategy to endow natural fusion-protein-derived peptides with potent antiviral activity. *PLoS One*. 2012;7(5):e36833. http://dx.doi.org/10.1371/journal.pone.0036833.
141. Chen L, Gorman JJ, McKimm-Breschkin J, et al. The structure of the fusion glycoprotein of Newcastle disease virus suggests a novel paradigm for the molecular mechanism of membrane fusion. *Structure*. 2001;9(3):255–266.
142. Palermo LM, Porotto M, Yokoyama CC, et al. Human parainfluenza virus infection of the airway epithelium: viral hemagglutinin-neuraminidase regulates fusion protein activation and modulates infectivity. *J Virol*. 2009;83(13):6900–6908. http://dx.doi.org/10.1128/JVI.00475-09.
143. Farzan SF, Palermo LM, Yokoyama CC, et al. Premature activation of the paramyxovirus fusion protein before target cell attachment with corruption of the viral fusion machinery. *J Biol Chem*. 2011;286(44):37945–37954. http://dx.doi.org/10.1074/jbc.M111.256248.
144. Palmer SG, Porotto M, Palermo LM, Cunha LF, Greengard O, Moscona A. Adaptation of human parainfluenza virus to airway epithelium reveals fusion properties required for growth in host tissue. *MBio*. 2012;3(3):e00137. http://dx.doi.org/10.1128/mBio.00137-12.

CHAPTER TWO

Alphavirus Entry into Host Cells

Ricardo Vancini[1], Raquel Hernandez, Dennis Brown
Department of Molecular and Structural Biochemistry, North Carolina State University, Raleigh, North Carolina, USA
[1]Corresponding author: e-mail address: rgvancin@ncsu.edu

Contents

Abstract

Viruses have evolved to exploit the vast complexity of cellular processes for their success within the host cell. The entry mechanisms of enveloped viruses (viruses with a surrounding outer lipid bilayer membrane) are usually classified as being either endocytotic or fusogenic. Different mechanisms have been proposed for *Alphavirus* entry

Progress in Molecular Biology and Translational Science, Volume 129
ISSN 1877-1173
http://dx.doi.org/10.1016/bs.pmbts.2014.10.002

and genome delivery. Indirect observations led to a general belief that enveloped viruses can infect cells either by protein-assisted fusion with the plasma membrane in a pH-independent manner or by endocytosis and fusion with the endocytic vacuole in a low-pH environment. The mechanism of Alphavirus penetration has been recently revisited using direct observation of the processes by electron microscopy under conditions of different temperatures and time progression. Under conditions non-permissive for endocytosis or any vesicular transport, events occur which allow the entry of the virus genome into the cells. When drug inhibitors of cellular functions are used to prevent entry, only ionophores are found to significantly inhibit RNA delivery. Arboviruses are agents of significant human and animal disease; therefore, strategies to control infections are needed and include development of compounds which will block critical steps in the early infection events. It appears that current evidence points to an entry mechanism, in which alphaviruses infect cells by direct penetration of cell plasma membranes through a pore structure formed by virus and, possibly, host proteins.

1. INTRODUCTION

1.1. Alphaviruses

The *Alphavirus* genus currently comprises 26 identified members, among the commonly found are Eastern Equine Encephalitis (EEE), Venezuelan Equine Encephalitis (VEE), Semliki Forest (SF), Chikungunya, Ross River, Western Equine Encephalitis (WEE), and Sindbis Virus (SINV).[1] Among the genomes sequenced, they share about 45% sequence identity in the structural proteins and about 60% identity in the nonstructural proteins. SINV is recognized as an insect virus that has adapted to infect vertebrates, allowing it to thrive in both vertebrate and invertebrate cells. Sindbis infection in humans is rather innocuous and results in mild fever, rash, and diarrhea, with these symptoms often misdiagnosed and attributed to other enteric and dermal virus infections. In contrast, all members of the *Alphavirus* genus: the VEE virus, WEE virus, and EEE virus are known to cause fatal encephalitis, thus posing a great veterinary risk in certain regions of the world.[1,2]

1.2. Alphavirus life cycle

The alphaviruses are included in the arbovirus group, among the arboviruses there are over 700 currently known and they are considered a major source of animal and human disease worldwide.[3] The alphaviruses have a wide host range and are transmitted in nature among birds, rodents, and domestic animals.[4] These viruses are transmitted in nature by blood-sucking insects, mostly mosquitoes. The blood-sucking insects carry viruses obtained from a blood meal and infect other host animals when feeding again. The transmission can be enzootic within the same animal population and have small mammals and birds as their reservoir. In the epizootic cycle, they infect domestic animals and humans as a dead-end infection.

1.3. Alphavirus structure

Alphaviruses are structurally distinct from other membrane-containing viruses in that they have a precise size, shape, and composition. SINV is very simple in its composition and its structure has been extensively studied.[5–8] The mature Sindbis virion contains three structural proteins (E1, E2, and capsid) in a 1:1:1 stoichiometric ratio. The virion is 70 nm in diameter, containing two protein shells with a host-derived membrane bilayer sandwiched in between the two shells. The particle is organized into two geometrically identical $T=4$ icosahedral shells. The inner nucleocapsid shell is built from 240 copies of the capsid protein. Within this inner shell is a densely packed, (+polarity) single-stranded, 117-kb viral genomic RNA. The outer shell is composed of 240 copies each of two viral glycoproteins, E1 and E2, which form protruding spikes on the surface which have threefold rotational symmetry.[7] They are composed of 439 and 423 amino acids, respectively, and both proteins have a molecular mass of about 50 kDa. E1 and E2 form heterodimers in the endoplasmic reticulum of infected cells. These dimers trimerize prior to delivery to the plasma membrane.[9,10] The host-derived membrane is positioned between the inner and outer shell of Sindbis and is penetrated by the transmembrane domain anchors of E1 and E2 proteins.[11,12] Assembly occurs as the endo domains of the E2 proteins interact with the core protein in a one-to-one ratio; thus, 80 E1–E2 heterotrimers make up the outer surface of the Sindbis virion representing a $T=4$ icosahedral structure equivalent to the geometry of the inner capsid (Fig. 1).[6,7] Thus, the geometry of the capsid determines the structure of the outer glycoprotein shell.[13] The interaction of the E2 endodomain with a hydrophobic cleft in the capsid protein locks the shells together and gives stability to the overall structure of the virus.[14] The level of lateral and transmembrane interactions between the two protein shells results in a precisely organized particle, unlike the structure of other membrane-containing viruses.[15,16] The E1 and E2 proteins have been assigned different functions in the virion. E2 has been implicated in receptor recognition and binding to host cells,[7]

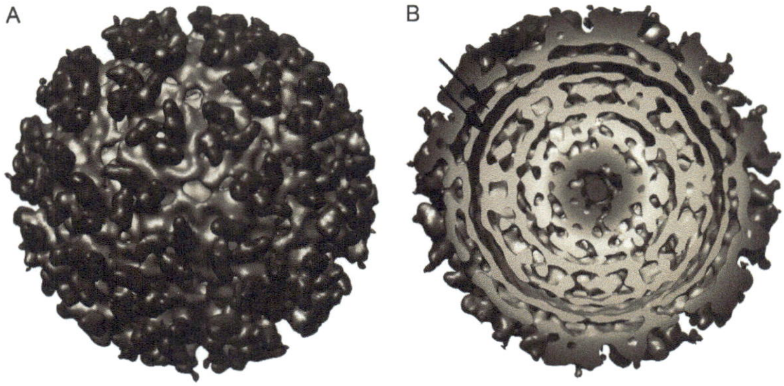

Figure 1 Sindbis virus structure. (A) Cryo-EM reconstruction of Sindbis virus, view from the fivefold axis. The outer most envelope of E1/E2 trimers is shown in blue. (B) Cutaway view of the structure, the inner most nucleocapsid shell of the virus particle, is shown in green/yellow/red. The host-derived lipid bilayer is sandwiched between the layers (arrows). (See the color plate.)

while E1 is implicated in maintaining the overall icosahedral structure of the particle, and for mediating the process of infection.

Recently, small-angle neutron-scattering studies have shown that particles grown in mammalian and insect cells are not structurally identical.[16] This method has the advantage of observing structure in that lipid and RNA densities are easily detected through the use of contrast variation technique.[17] Particles obtained from insect cells showed to be more compact, lacking some degree of RNA intercalation in the capsid shell, as seen in mammalian grown virus.[18] In addition, the outer protein shell seems to be more extended in the mammalian grown virus, which suggests the protein organization for each virus might be significantly different. The biological consequences of the structural differences uncovered by this technique are not known. It may be a result of different temperatures of assembly and different biochemical environments.

2. ALPHAVIRUS INTERACTION WITH HOST CELLS

As obligate intracellular pathogens, viruses must have the ability to cross cellular barriers. This primary step involves receptor binding, capsid destabilization, and genome delivery, resulting in an infection event. Many of these changes result from conformational modifications in metastable viral protein structures.[19]

2.1. Role of attachment factors and receptors

It is believed that Alphavirus infection is started by the attachment of a virion to a cell receptor as occurs with most viruses. These generalized attachment factors concentrate virus particles on the host cell and create favorable conditions for receptor binding.[20] This first critical step results in the penetration event which delivers the viral genome to its site of replication. The binding of E2 envelope protein to a receptor(s) at the plasma membrane of the host cells is considered the primary event in Sindbis infection.[21] However, neither the cellular receptor(s) nor the precise E2 amino acids involved in the binding step are known. Despite many efforts, identification of the specific *Alphavirus* receptor(s) has remained elusive. Since SINV can reproduce in many laboratory cell lines derived from different organs of both insects and mammals which provide widely divergent biochemical environments, the cell receptor must be a ubiquitous class of molecules present in all of these cells. It is also possible that alphaviruses can utilize different or multiple receptors. Furthermore, this process is complicated by the ability of Sindbis and related viruses to evolve into lab-adapted strains as well as interact transiently with host factors at the plasma membrane.[22]

2.1.1 Putative receptors

Unlike attachment factors, receptors not only bind viruses but also promote their entry into cells. Receptor binding on the plasma membrane is one of the first steps in the infection process and usually highly specific.[23] There are many candidate molecules reported in the literature, including heparan sulfate, a cell surface glycosaminoglycan to which Sindbis demonstrated affinity for binding.[22] However, this affinity seems to be acquired after a series of passages in cell culture. In addition, DC-sign and L-sign have been shown to bind to Sindbis in dendritic cells.[24] The major histocompatibility complex,[25] heat-shock protein 70, and Laminin proteins have been suggested as receptors utilized by Sindbis.[21,25–27] Recently, the natural resistance-associated macrophage protein has also been implicated in SINV entry using a *Drosophila* system.[28] However, none of them have been proven to have specific interactions with E2. The length of the candidate list suggests that there are multiple proteins that alphaviruses may use, and if there is a specific receptor, it might be both cell and virus specific. If there is one widely used receptor, it would have to be one conserved through evolution.[29] It is possible that the virus associates nonspecifically with various proteins before finding and binding to its specific receptor or receptor complex.

Thus, the susceptibility of different cell types to Sindbis in the absence of a characterized receptor might indicate that alternative receptors may be used.

2.2. Routes for enveloped virus internalization

In animal cells, efficient viral infection depends on multiple events, including receptor binding, membrane penetration, genome internalization, and replication.[30] In order to be successful, viruses employ different strategies to mediate a productive infection as a result of their interaction with cells. To avoid lysosomal degradation, alphaviruses are believed to fuse with endosomal membranes at low pH, whereas nonenveloped viruses lyse endosomes or form pores in membranes to release their genome into the cells.[31] The mechanism of delivery of the alphaviruses genome remains one of the most debated aspects of the *Togaviridae* lifecycle.[29,32,33] There are two proposed mechanisms for this process to occur. One pathway proposes that it might happen as with Sendai virus which engages a receptor on the surface of the cell causing the viral proteins to undergo a conformational change which fuses the virus membrane with the cell membrane. This is referred to as the pH-independent pathway.[34] A second pathway proposes that infection of alphaviruses requires fusion of the viral envelope with a cell membrane in a pH-dependent manner, after receptor-mediated internalization.[35] This infection pathway was first described in influenza viruses and later proposed to be the entry mechanism for all enveloped viruses and some nonenveloped viruses.[36]

2.2.1 Endocytic routes

There are several endocytic routes by which viruses are internalized into cells: clathrin-mediated endocytosis, caveolar endocytosis, and also clathrin-independent and caveola-independent mechanisms.[23] Infection by membranous viruses such as influenza is well described, the virus binds to a receptor on the cell surface, and it is taken into an endosome. Once in the interior of the endosome, the pH is then lowered by H^+ ion pumps, and this change in pH triggers conformational rearrangements of the structural proteins resulting in HA-catalyzed fusion (Fig. 2A).[37] This allows the penetration process to occur while escaping the route to lysosomes and viral degradation. This pH-dependent pathway has been proposed for the alphaviruses and other arboviruses such as SFV and TBE.[38] It is mediated by the formation of clathrin-coated pits and transport to early endosomes, where the low-pH environment triggers the fusion event.[39] Viruses such as influenza, VSV, and SFV have been observed in clathrin-coated vesicles

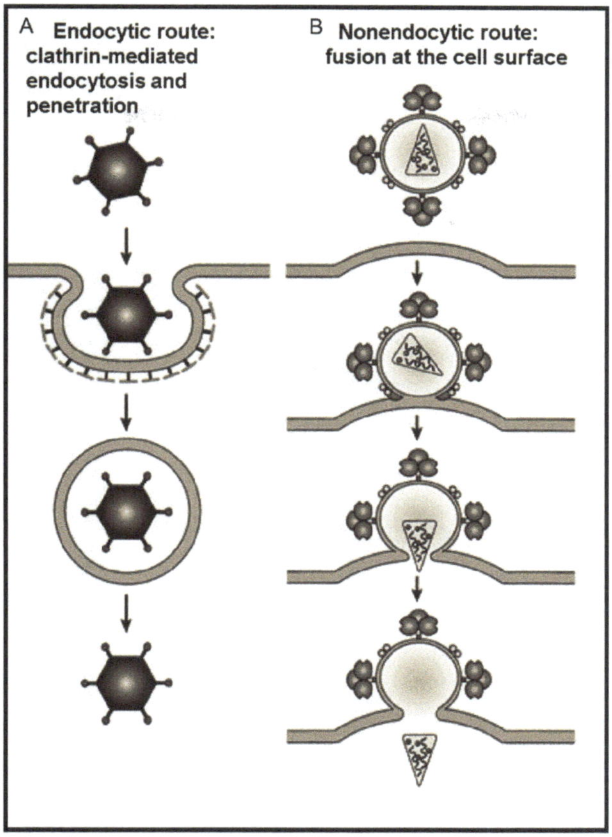

Figure 2 Virus entry pathways. (A) Clathrin-mediated endocytosis, for example, adenovirus. Endocytosis by caveolae can also occur, for example, SV40. (B) Fusion at the cell membrane, for example, Sendai virus. Fusion can also occur from inside an endosome, for example, influenza. *Reprinted by permission from Macmillan Publishers Ltd: Ref. 31 (copyright 2004).*

by electron microscopy.[36] However, the relative infectivity of these viruses was not determined and the particles seen may not be infectious.

The number of viruses that enter cell via caveolar endocytosis is reduced, most of these viruses are nonenveloped and are less than 55 nm.[23] It is a well-characterized entry portal for the nonenveloped simian virus 40 (SV40).[40] This endocytic route is dynamin and cholesterol dependent, but compared with CME, it is much slower and resulting vesicles do not become acidified.[41] In some particular cases, such as HIV, the entry point remains controversial. HIV fusion has long been thought to occur at the cell surface, but cumulative evidence supports the endocytic entry of HIV leading to a productive infection[42] and that this route could be the predominant HIV entry pathway into different cells.[43]

2.2.2 Nonendocytic routes

An example of a classic nonendocytic mechanism is the direct membrane fusion, where viruses enter cells via fusion with the host cell plasma membrane at neutral pH.[31] This route, used by enveloped viruses, has been considered the most efficient way to deliver the viral genome to the cytosol.[23] Fusion occurs after a fusion-promoting viral protein is recognized by a specific cell receptor (Fig. 2B). This process is limited to some species of retroviridae such as murine leukemia virus,[20] paramyxoviridae such as Sendai virus,[44] and herpesviridae such as HSV-1,[45] even though a similar process has been suggested to occur in arboviruses such as the flaviviruses.[46]

2.3. Conformational changes during entry

A discussion of the Alphavirus structure is particularly important regarding the entry process. While many membrane-containing viruses are structurally pleomorphic, such as influenza, which consists of a membrane bilayer with embedded surface glycoproteins. The Alphavirus SINV has a precise and rigid structure (Fig. 3A) in which the proteins with their lateral associations are not free to move in the fluid lipid as may happen with influenza.[47] It is well established that the influenza glycoproteins HA1 and HA2 drive the

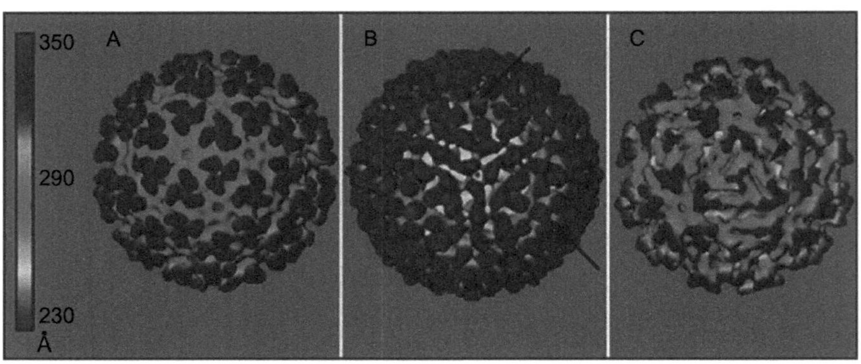

Figure 3 The three-dimensional structures of Sindbis virus surface at 28 Å resolution viewed along icosahedral threefold axis. (A) Sindbis virus structure at pH 7.2. The conformational changes in Sindbis virus after exposure to pH 5.3 (B) and back to neutral pH (C) are shown. The arrows point to the protruding structure at the fivefold vertices. Arrowhead shows the fissure that appears at the twofold axis upon return to neutral pH. *Reprinted with permission from Elsevier from Ref. 32.* (See the color plate.)

fusion of viral and host membranes in a low-pH environment.[48] In the case of Sindbis, cryoelectron microscopy studies of the virus at pH 5.3 revealed the presence of a novel protruding structure at one of the fivefold vertices of the particle that seems to be induced by the low-pH treatment (Fig. 3B).[32] This proteinaceous structure is suggested to be comprised primarily of E1 proteins and is proposed to form a pore structure in the host membrane bilayer through which the viral genome is delivered. In the same set of experiments, upon return to neutral pH, the virus undergoes a second set of conformational changes in which the protruding structure at the fivefold axis is withdrawn into the virus structure and large fissures appear on the virus surface along the twofold axis (Fig. 3C).[32] These changes may compromise the structural integrity of the virions and allow the fusion of virus and cell membranes to take place. The fusion of virus with living cells has been shown to be a two-step event requiring exposure to low pH followed by return to neutral pH, conditions which are not met in the endocytic pathway. In immuno-EM studies, where Sindbis virions are allowed to adsorb to cells at 4 °C and then briefly incubated at room temperature prior to fixation, it is possible to observe electron-dense particles and also particles that apparently have lost their electron-dense core (genome) at the surface.[32,49] These phenomena suggest that the virus RNA has been transferred to the host cell at the surface in the absence of endocytosis and at neutral pH. This observation forms the basis for the hypothesis that suggests that Alphavirus fusion capabilities are not involved in the process of infection.

3. MEASURING VIRAL ENTRY

Virus cell entry is a complex area of research in which many controversial observations have been reported. This is not only a result of difficulties associated with the direct visualization of the viral entry and its gene delivery but also the poorly characterized downstream effects of many inhibitors that have been used in previous studies of the role of endocytosis and acidic pH.[50,51] The techniques used for virus entry studies can generally fall into two categories, direct and indirect observation. Previous studies employing electron microscopy to directly analyze the entry enveloped virus have provided valuable data. However, these data came from experiments designed to investigate membrane fusion,[52] rather than focusing on genome delivery or virus uncoating. In addition, direct observation of virus–cell interaction by electron microscopy can be challenging when the infectivity ratio and physical integrity of these particles are not taken into

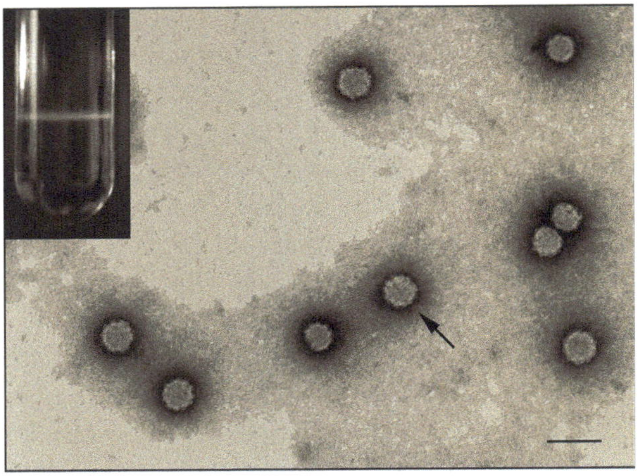

Figure 4 Electron microscopy of purified Sindbis virus. Micrograph of negatively stained Sindbis particles, showing the preservation, homogeneous size (700 Å), and icosahedral symmetry (arrow). The symmetry is emphasized by the ability to detect the particles in different orientations (bar = 100 nm).

consideration (Fig. 4). If the infectivity ratio of these preparations is poor, the observation of particles in endosomes, for example, may lead to an inaccurate interpretation of the infection process.[53] Thus, the sensitivity of some enveloped viruses to laboratory purification and storage procedures needs to be considered when designing any penetration study with membrane-containing viruses. The ability of SINV to grow to very high titers in laboratory cell cultures, its stable structure, and its nonpathogenicity to humans make Sindbis an ideal prototype virus to study alphaviruses life cycle and infection mechanism. It is also an excellent model to study membrane proteins in their native structural form. Purified Sindbis, in contrast to other membrane-containing viruses, can have a particle to plaque-forming unit (pfu) ratio that approaches unity.[54] These features of SINV are optimal for the study of entry involving direct observation by electron microscopy.[47] The term virus entry refers specifically to the mechanism by which the virus binds to the host cell receptor, penetrates the cell membrane barrier, and releases the infectious RNA into the cell initiating the infection.

3.1. Direct observations by electron microscopy

Much of the general model of penetration for enveloped viruses has originated from early electron microscopy studies with Sendai virus and SFV, which suggested the pathway of endocytosis[55] and membrane fusion.[56]

Alphaviruses were found in endosomes and clathrin-coated vesicles; however, a convincing image of an Alphavirus fusing with the endosomal membrane has yet to be presented. Early studies proposed that virus entry for Sindbis and VSV involved a mechanism which did not require fusion.[57] While direct observation holds great potential for visualizing the process of cell penetration, it is incapable of distinguishing a virus which is engaged in a productive infection from a virus involved in a dead-end interaction with the cell. When direct observation has proven difficult to perform or cannot properly answer questions, indirect or secondary reports have been used. Some of the indirect reporters that have been used to measure entry include viral RNA production,[58] virus protein synthesis,[59] and virus production/titer.[60]

Recent studies using direct EM observation in conjunction with immunogold antibody labeling have shown that infectious Sindbis particles are found as electron-dense or translucent particles while engaging the plasma membrane[32] at neutral pH. It has also been shown using this approach that the process of converting full particles into empty ones is time and temperature dependent.[49] These studies did not detect any evidence of membrane fusion or involvement of an endocytic process.

3.2. Role of membrane models in studies of virus entry

The evidence supporting Alphavirus entry by low-pH-dependent fusion comes from data produced in virus–liposome model systems, where it has been shown that alphaviruses can induce membrane fusion on exposure to acidic pH.[35,61,62] This fusion event is proposed to be an event identical to the interaction between the viral and endosomal membranes, when endosome acidification occurs. However, it was shown that this fusion process could only occur when a high amount (up to 50 mol%) of cholesterol was present in the liposome membrane.[63,64] There are important issues associated with this model. Liposomes contain no proteins and thus no receptor, have no metabolic activity, and have no ionic or osmotic gradient across their membranes. Thus, liposomes are very different from the biochemical context in which a virus–cell membrane interaction occurs. Moreover, the insects, which are vectors of these viruses, do not have the capacity to produce cholesterol,[65,66] and it is not known if cholesterol obtained by dietary routes is incorporated in any significant amount into insect cell membranes. It was later shown that this fusion of living cell membranes is actually a two-step event requiring a return to neutral pH,[67] rather than an event that occurs only by lowering the pH as takes place in liposome fusion and as would occur in the endosome environment. Furthermore, the suggested process of fusion of the virus membrane with liposomes has been shown to be nonleaky, in that the fusion event seals the mixed membrane product.[68] In contrast, infection of cells with alphaviruses has been demonstrated to produce pores in the membrane,[69] which allows entry of small molecules,[69] leaks ions, and could be plugged with transitional metal ions.[70,71]

3.3. Role of inhibitors in studies of virus entry

The treatment of cells with chemicals that inhibit the acidification of endosomes has been used as an assay for entry of alphaviruses. The initial evidence that endocytosis and fusion upon exposure to low pH were required for Alphavirus infection came from the use of lysosomotropic weak bases.[50] Compounds such as ammonium chloride and chloroquine have been used in such experiments by treating cells prior to infection.[72,73] The ability of these drugs to block endosome acidification and prevent viral RNA or protein synthesis has been used as evidence to support the receptor-mediated endocytosis model for genome delivery.[59,74] There are important aspects to consider in these approaches. First, the assays employed to analyze the effects of the inhibitor drugs used in these experiments are based on indirect observation of viral entry. Viral RNA or structural protein synthesis occurs downstream of the initial events of entry, after expression of the nonstructural proteins, negative strand synthesis, and subgenomic RNA production have taken place.[47] These latter events likely require the normal function of the infected cell, and these drugs may directly or indirectly impair other functions related to viral RNA and structural protein synthesis. Indirect, secondary effects like these have been demonstrated in mosquito cells in which it was shown that some weak bases such as chloroquine cannot inhibit virus infection (though it blocks diphtheria intoxication which requires exposure to acid pH)[75] and others interfere with virus production by blocking proteolytic processing of the nonstructural precursor protein, rather than preventing viral entry.[75] In a recent report, a specific inhibitor of the V-ATPase, Bafilomycin A1, which has been demonstrated to specifically prevent the acidification of endosomes, was used to assess its effect on Alphavirus infection.[76] The effects were analyzed using a SINV construct containing a GFP reporter gene. The results showed that BafA1 did not prevent virus entry; instead, it inhibited only protein synthesis and folding as demonstrated by transfecting cells with the Sindbis RNA, thus bypassing the process of entry. This research suggested that BafA1 was inhibiting the proper folding of newly synthesized proteins and not the entry of virus RNA.

4. ALPHAVIRUS GENOME DELIVERY

4.1. Role of membrane fusion

The ability of *Alphavirus* to fuse to liposomes and host cell membranes in cell cultures under laboratory conditions are the primary evidence that led to the proposal that membrane fusion is involved in the Alphavirus infection process. Although valid for some membrane-containing viruses, there are some problems with this model for alphaviruses. Unlike influenza, which can be described as a pleomorphic membranous structure with embedded proteins, alphaviruses such as Sindbis are rigid protein shells with an associated membrane.[7] The two icosahedral protein shells with many transmembrane and lateral interactions stabilizing the particle result in a precise structure in which the membrane bilayer is occluded by the outer protein shell (Fig. 1, arrows).[7,32] The lateral interactions which stabilize the outer protein shell are further stabilized by intermolecular disulfide bridges.[9] Therefore, the membrane is not the form-determining factor as with influenza, and because it is not exposed on the surface, it is not readily available to engage the host cell membrane. In order for Sindbis and other alphaviruses to use the influenza model, the inner and outer $T=4$ icosahedral shells formed by the viral structural proteins would have to disassemble and reassemble, resulting in a novel set of lateral associations between the structural proteins. Early studies proposed that the E1 of the alphaviruses referred to as group II fusion protein, to be the one responsible for fusion.[77,78] Interestingly, removal of the E2 protein by digestion suggested that E1 protein alone might be sufficient for the fusion event.[79,80] It is proposed that within endosomal vesicles the E1–E2 heterodimer would undergo irreversible conformational changes upon exposure to low pH.[81] It is further suggested (though not directly demonstrated) that these proteins insert a small fusion loop into the host endosomal membrane after endocytosis and a shift to low pH.[82] While elegant models have been presented to mechanistically illustrate how these fusion events may occur, the mechanism by which the repulsive forces of the opposing phosphate groups on the lipid bilayer are overcome remains elusive.[47] Furthermore, the highly curved and constrained radius of the liposome combined with the high concentration of cholesterol used may contribute to the fusion/lipid mixing properties demonstrated. The ability of alpha and flaviviruses to fuse membranes is not disputed; however, that this is the mechanism of virus penetration and infection remains to be proved.

4.2. Role of low pH

The Alphavirus infection process has been characterized as an event dependent on the acidic environment of endosomes. Once inside the endosome, the activity of vacuolar ATPase pumps lowers the pH of the vesicles and at some critical pH value the viral membrane proteins undergo dramatic conformational rearrangements.[83]

The studies in support of this model are extensive and rely on the observations that alphaviruses have membrane fusion capabilities.[29] In original experiments, treatment of cultured cells with adsorbed virus with low pH resulted in cell–cell fusion.[84] However in these studies, cells were always returned to neutral pH before fusion was seen. This two-step process was later shown to be required for fusion of viral and cell membranes to occur[67] and endosomes do not return to neutrality after pH is lowered. Other studies have used the virus–liposome system in order to demonstrate, quantify, and determine the requirements of the fusion capability of alphaviruses. Indeed, alpha virus fusion with artificial membranes is clearly demonstrated. However, biologic membranes such as virus membranes carry a net negative pK, and when a net negatively charged liposome is added to net positively charged liposomes, they may fuse.[85] Therefore, in previous studies where the pH of the virus–cell interaction was lowered, one passes a point where there is ionic bonding between virus and cell. Thus, the low pH may add tight binding and may also destabilize a metastable protein.

A low-pH study of SINV structure was undertaken with the idea that low pH may trigger the same or similar conformational rearrangements as does the contact of the virus with the cell receptor.[32] Infectious SINV at pH 7.4 was exposed to pH 5.3, returned to neutral pH, and analyzed by cryo-EM. The results revealed that low-pH treatment triggered a substantial rearrangement of both E1 and E2 proteins and the formation of knobs of E1 protruding from each of the fivefold axes. Returning to neutral pH did not restore the native structure and resulted in noninfectious virus. Interpretation of these data in the context of the prevailing model of fusion would predict that the virus and cell membranes fuse and the nucleocapsid is delivered to the cytosol. However, it is evident from the cryo-EM reconstructions that the lipid membrane is not exposed at low pH or upon return to neutrality.[29,32] Therefore, the hypothesis raised here is that there is no *a priori* reason to conclude that because a virus has a membrane it must fuse to enter host cells.

4.3. Role of pores in the cell membrane

It has been reported that during the entry of animal viruses into host cells, not only is the virus able to permeabilize the membrane to its genome but also to other macromolecules.[70,86] Several animal viruses demonstrate this phenomenon such as SFV, VSV, adenoviruses, and polioviruses.[86–88] With SINV, it has been shown that infected cells become permeable to alpha-sarcin[69] and develop pores in the cell plasma membrane.[70] These pores have been shown to be of different sizes and can be effectively plugged with transitional metal ions.[71] Evidence supporting the formation of such pores permeabilizing the plasma membrane during viral infections has also been provided by others. Were the alphaviruses simply to bind to its receptor and be taken into the cell, as in the endocytosis model, there would be no loss of membrane integrity, as demonstrated in the virus–liposome system.[68]

5. ALPHAVIRUS ENTRY IN THE ABSENCE OF MEMBRANE FUSION

5.1. A direct assay for entry at the plasma membrane

SINV is known for not producing empty particles because they need the nucleocapsid scaffold on which the envelope proteins will be assembled and the nucleocapsid cannot assemble without the RNA.[89,90] To observe the early interaction of a 70-nm particle with a host cell that is on average 15 μm in diameter, using thin section electron microscopy (~220 sections across a single cell), there are a few obstacles that need to be overcome. The protocol requires a high multiplicity of infection in order to increase the probability of observing a significant number of particles in the same field and it requires a low particle/pfu ratio in order to observe infectious particles. Identifying the virus proteins with gold-conjugated virus-specific antibody is also required to allow the identification of the virus in any structural reconfiguration that may occur during virus–cell interactions.[32] In 2004, Paredes *et al.* were able to detect full (electron-dense) and empty particles on the surface of mammalian cells using this approach.

The experimental approach for the process is straightforward; virus particles are allowed to adsorb to the cell surface at 4 °C, which would slow entry, synchronizing infection until temperature is raised to physiological conditions. The temperature is then raised and the sample fixed concurrently. The virus–cell complexes are then labeled with virus-specific antiserum and gold bead-conjugated secondary antibodies. This allows not only intact viruses but also virus membranes that have lost their electron-dense RNA content to be identified.[49] By using this approach, it is possible to determine if the virus' infectious path follows one of the three different outcomes. First, if the virus fuses with the plasma membrane, it would be possible to see labeled viral proteins, dispersing on the cell surface.[32] Second, if the virus triggered receptor-mediated endocytosis, it should be possible to see particles labeled on the surface of the cell over a crescent clathrin-coated pit or nonlabeled particles already in endosomes close to the cell surface.[55] Finally, if the virus uses a pathway of direct penetration at the plasma membrane, it should be possible to detect labeled particles at the cell surface at different stages of the entry process. A population of virus particles undergoing the process of infection would be expected to include particles that are electron-dense with their genomic material still undelivered (Fig. 5A),

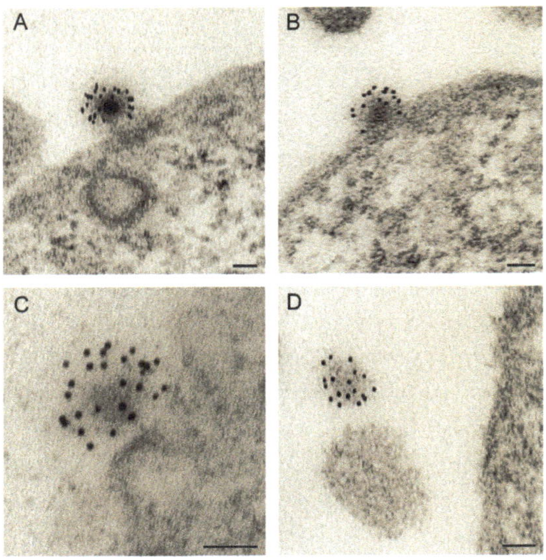

Figure 5 Thin-section electron microscopy of Sindbis virus–cell complexes at pH 7.2. Panel of TEM observations illustrating the representative populations of particles at different incubation temperatures. High magnification showing examples of (A) electron-dense full particles at the cell surface at 4 °C; (B) intermediate penetration stage particles with loss of electron density at 15–22 °C; and (C and D) empty particles with loss of RNA electron density mostly at 37 °C, with eventual breakage of the virions. Bars = 50 nm.

electron-dense particles connected with the cell surface (stalk or pore) (Fig. 5B and C), and finally, empty (electron-transparent) antibody-labeled particles that have undergone RNA transfer into the cell (Fig. 5C and D).[49]

5.2. The role of temperature in the process of infection

Alphavirus entry has been previously shown to occur at temperatures that did not allow fusion or endocytosis.[49,91] Wang *et al.* successfully used low temperature (15 °C) to analyze Sindbis infection using infectivity assays and found that significant numbers of viral particles could establish infection. These experiments demonstrated that while the low temperature did not block infection, it efficiently blocked low-pH-induced fusion, showing that the two processes are unrelated.

To characterize the initial steps of Sindbis interaction with mammalian cells and determine the kinetics of the cell penetration process, previous studies took advantage of early cell biology observations in which temperatures below 18 °C are shown to arrest endocytosis in mammalian cells and at 15 °C all cellular vesicular transport is halted.[92] This method has proven to be a nonlethal and nontoxic procedure to inhibit endocytosis and analyze virus penetration. Previous work have also used immunogold-labeling procedures,[32] in order to clearly identify and confirm the presence of viral particles without the electron-dense RNA core. These particles could not be determined to be virus related without immunolabeling and could possibly have been overlooked in previous studies (Fig. 6). When the process of virus entry (virus–cell complexes) was analyzed at several different temperatures 4, 15, 22, and 37 °C at neutral pH, it was shown that after initial attachment at 4 °C, the viral particles were found to become transformed from full to empty by a process that was a function of increasing temperature.[49] Full particles were defined by electron density at the core, equivalent to that seen in a native virus structure. Empty viruses lost the electron-dense core and were transparent and could only be identified by immuno-EM (Fig. 6).

This method was used to give a percentage of virions at different stages of penetration at different temperatures.[49] There was a striking difference in the relative numbers of all of the penetration stages at every temperature examined. At 4 °C about 84% of the particles are full, and as warmer temperatures are employed (from 4 to 37 °C) they all became empty (Fig. 7A). The percentage of particles that were found to be empty started at 16% at 4 °C and increased to 100% at 37 °C. The temperature data could be fit to an Arrhenius plot showing that entry is not by a physical force (injection)

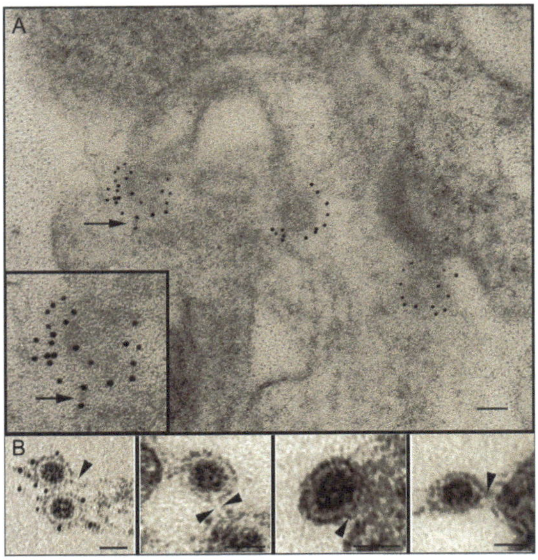

Figure 6 Thin-section electron micrographs of the cell membrane-associated SINV with pore stalks. Sindbis virions were attached to the cells, fixed for electron microscopy at neutral pH, and immunolabeled with anti-Sindbis antibody conjugated with gold beads. (A) At 37 °C, most virions have lost their electron-dense RNA containing core at and it is possible to observe a reminiscent connecting structure labeled (arrow—inset). (B) Panel of Sindbis virus with an electron-dense core attached to the cell surface by a pore structure at 4 °C (arrowheads). Bars = 50 nm. *Copyright © American Society for Microbiology, Ref. 49.*

mechanism. The energy of activation for Sindbis entry was calculated to be 27 kcal/mol (Fig. 7C).

These results support a mechanism of entry in which the viral particles attach to the cell surface, followed by a reorganization of the RNA, and the subsequent passage of the viral RNA through a pore-like structure that connects the virus to and penetrates the plasma membrane, as the particles become empty and eventually break up and disperse due to the loss of the core (RNA) structure (Fig. 5D).[32,49] The nature of the pore-like structure that is formed as the virus contacts the cell is of major interest; the structure that develops at the fivefold axis of the free virus upon exposure to acid pH is 52 Å wide and 62 Å.[47] In support that major structural changes in the virion structure take place during the penetration process, there is the observation that attachment to cell surface results in exposure of otherwise buried virus epitopes.[93]

In SINV–cell complexes both at 4 and 37 °C, the formation of a stalk connecting the virus to the cell surface could be seen.[49] Interestingly, the

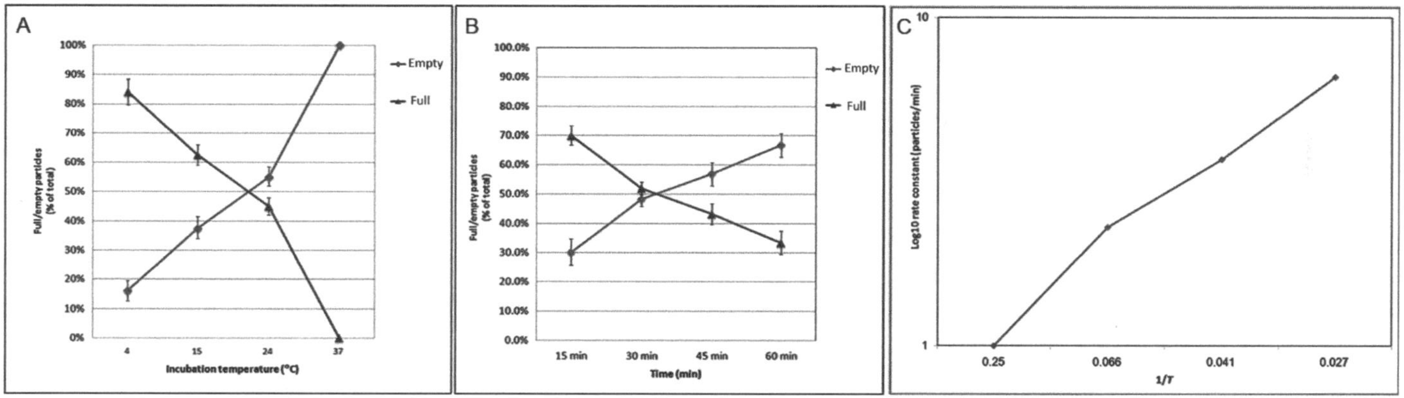

Figure 7 Temperature and time dependence of genome delivery of Sindbis virus into BHK cells. (A) Sindbis virus–cell complexes incubated and fixed at several temperatures were directly analyzed by electron microscopy. The graph shows that as temperature rises from 4 to 37 °C there is a linear rise in the population of empty particles at the cell surface (blue line (gray in the print version)) reaching the total population, and a linear decrease in the population of full particles (red (dark gray in the print version)) reaching zero at 37 °C. (B) Interaction of Sindbis with BHK cells at 4 °C at 15–60 min. As it occurs with the temperature progression, the empty particles population increases with time from 30.1% to 66.6% (blue (gray in the print version)) and the population of full particles decreases significantly from 48.1% to 17.6% at 60 min (red (dark gray in the print version)). (C) The number of particles which became empty in 15 min at each temperature was converted to apparent rate constants, providing an $E_a = 27.7$ kcal/mol. Data are shown as the mean of triplicate counts of two independent samples. Error bars are SEM ($n = 4$). *Copyright American Society for Microbiology, Ref. 49.*

connecting stalk/pore could also be labeled with the anti-Sindbis-conjugated gold beads, suggesting the penetration pore is composed of virus proteins (Fig. 6A, inset). In virus–cell complexes fixed at 4 °C, the stalk is also clearly observed (Fig. 6B). Additional evidence supporting Alphavirus entry via a pore at the plasma membrane was obtained using freeze-fracture immunolabeling,[94] suggesting it is indeed a proteinaceous structure that traverses the cell membrane, and is often seen associated with Sindbis virions. The formation of the pore-like stalk that precedes genome delivery should not to be confused with the hemifusion intermediate states which occur in membrane viruses such as influenza which do enter by a process of membrane fusion.[95] Sindbis membrane fusion has been extensively shown to be a two-step process requiring the return from low pH back to neutrality to occur and recently described to be formed under nonpermissive conditions for membrane fusion.[49]

5.3. The role time in the process of infection

In a second set of experiments, Sindbis particles were incubated with BHK cells as described earlier but the infection temperature was held at 4 °C and particle quantitation was made at different times postvirus addition to determine if the penetration process was also time dependent. Infections were done at 4 °C as it was the only temperature which slowed the rate of entry enough for satisfactory data collection. SINV exhibited a similar pattern of entry progression to that described for experiments using an increase in temperature (Fig. 7B). At early times (15 min), most particles were observed as full (~70%). At 30 min of cell–virus interaction, there was a significant increase in the number of empty particles and a concomitant decrease in the number of full virions. At 45 min, the difference is more pronounced (~43%) full and (57%) empty, reaching a maximum (34% full and 66% empty) as the time progressed to 60 min. Incubation at 4 °C significantly slowed the entry process as 60 min was not long enough to complete the transition to all empty particles, while 15 min at 37 °C completed the transition. Previous work has suggested that the events leading to infection of living cells by SINV can occur in the absence of endocytosis and membrane fusion.[32,91] The progression of virus entry with time at 4 °C was an additional and compelling evidence that the entry process may occur at the cell surface independently of endocytosis and membrane fusion.

In addition, the effect of time on the rate of entry on mammalian cells (BHK) was recently compared to entry rate into insect (C7-10) cells

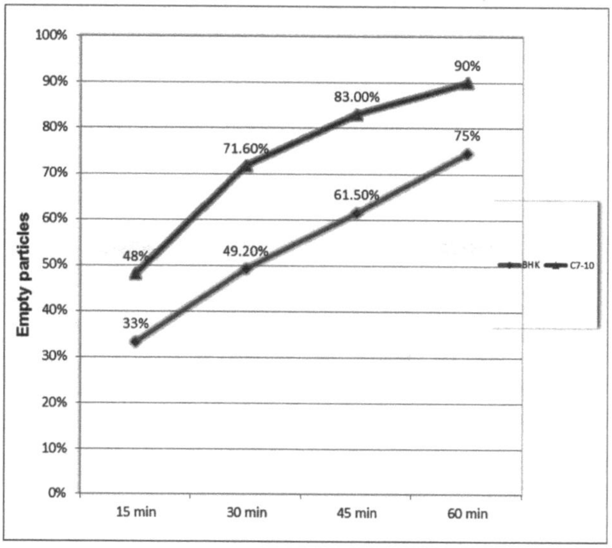

Figure 8 Comparison of the effect of time on entry of Sindbis virus into BHK or C7-10 cells. By 60 min postinfection at 4 °C, 90% of the virus particles were empty as compared to the 75% seen in BHK cells at the same time point.

(unpublished observations). The experiment was conducted as described for Fig. 7B and it demonstrates (Fig. 8) that entry into insect cells is faster possibly due to the difference in membrane composition and the lack of cholesterol which increases membrane viscosity.

5.4. The role of membrane potential

Current models describing virus entry are based on mechanistic insights from crystal structures of envelope proteins and their theoretical interaction with cell membranes and indirect reporters of viral entry (replication)[29,47]; however, no energy requirements are envisaged. As mentioned previously, the fact that low pH is required for fusion of some viruses has been taken as evidence that an acidic pH is necessary for alphaviruses genome uncoating.

It is well known that during infection of cells by animal viruses, membrane permeability can be altered.[96] In addition, a number of different-sized molecules have been reported to be able to coenter the cytoplasm during viral infection.[71,86] Such changes in membrane permeability cause the diffusion of cations and protons, thereby resulting in a drop in membrane potential.[97] Once viral proteins are inserted into the membrane, they might be able to open pores, through which protons, other ions, and possibly RNA can pass.[94] With the Alphavirus, SFV entry does not occur when the membrane potential is abolished by modifying the concentration of monovalent cations even under acidic conditions.[98] Even though viruses can attach to dead cells or isolated membranes, they only effectively fuse with and enter into cells that possess an energized membrane.[99] In addition, Poliovirus is one of the viruses that efficiently promotes entry of protein toxins such as alpha-sarcin,[88] which has also been reported for Sindbis.[69]

The effects of metabolic inhibitors on the formation of empty particles is currently being investigated, in a 30-min pretreatment at 4 °C followed by virus addition at the same temperature. After 30 min of incubation, the cells were processed for electron microscopy as in Fig. 5. The results have shown that the only significant reductions were seen when cells were treated with the ionophores monensin and carbonyl cyanide *m*-chlorophenyl hydrazone. The results indicate that a proton gradient may facilitate the infection process of Sindbis and possibly other alphaviruses.

5.5. Similarities with other viruses

Early reports on arbovirus entry have shown that flaviviruses could be detected on the cell surface with no evidence of fusion or endocytosis.[46,100] Recent work on flavivirus entry has shown that when conducted with infectious virus obtained directly from the mosquito vector (Fig. 9), preserving the infectivity and preventing purification processes, West Nile, and Dengue virus presented similar results.[101]

In the case of flaviviruses, measurement of permeability by patch-clamp techniques has shown that these viruses can form ion-permeable pores on

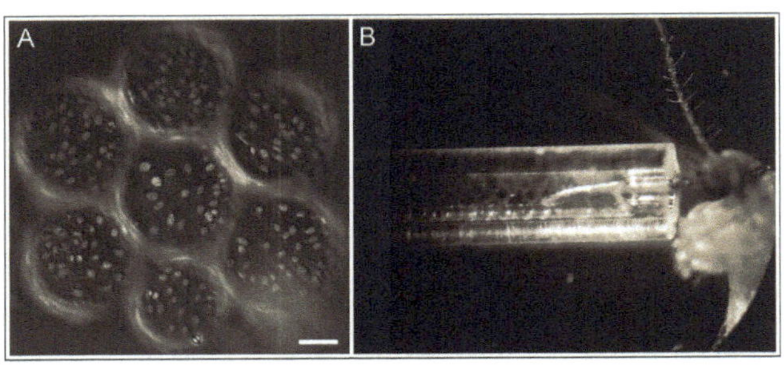

Figure 9 Morphology of Vero cells grown on microcarriers and *in vitro* capillary tube infection assay. (A) Phase contrast and fluorescence microscopy of Vero cells grown on cytodex-3 beads for 3 days the same cells labeled with DAPI. Bar = 50 μm. (B) Individual *Culex pipiens* mosquitoes were allowed to probe and feed from capillary tubes containing Vero cells grown on cytodex-3 beads. *Partially reprinted with permission from Elsevier from Ref. 101.*

cell membranes at neutral pH.[70] For viruses such as Poliovirus or Adenovirus, that are devoid of a membrane, suggested mechanism for genome entry also involves pore formation in membranes through which the genome would pass.[102] Hemolytic viruses such as Sendai or Influenza viruses are also able to induce permeability changes, rendering the host cell membrane leaky toward ions and small metabolites.[103,104] In the case of Sendai virus, they are also able to fuse their lipid envelope directly at the plasma membrane in a pH-independent manner; however, this mechanism has been recently revisited.[53]

Interestingly, the process by which alphaviruses have been proposed to infect cells recently[29,32,49] is very similar to the entry mechanism proposed for nonenveloped viruses such as polioviruses.[105] In this case, it was hypothesized that polioviruses interacts with cellular membranes through exposure of hydrophobic helices at the fivefold vertex of the virion upon interaction with a cell receptor.

5.6. Implications of a new model for entry

The data compiled here support a new model for the cell entry of alphaviruses and is a significant departure from the existing paradigm regarding the entry of membrane-containing viruses. In this model, the interaction of the virus with a putative receptor would set the essential conditions to allow conformational changes that produce a new structure in the center of the fivefold axis of the virion.[32] These events would then lead to the formation of a protein pore-like structure formed by viral and possibly host proteins that would penetrate the plasma membrane. The passage of RNA through the pore would then initiate infection (Fig. 10). The slow rate of entry seen at nonphysiological cold temperatures may reflect an impairment of cellular functions that may be required as driving forces to favor the genome delivery process. This process is very similar in many aspects to the entry mechanism proposed for nonenveloped viruses such as polioviruses[105] and later suggested to occur with Sendai virus.[53]

Regardless of whether the virus genome is delivered at the plasma membrane or through an endosome, it has to cross a membrane. How such a big macromolecule traverses a barrier such as a lipid membrane remains unknown. So far, the entry process seems to require a membrane potential and is affected by the chemistry of the host cell membrane. The obvious implication of these data is that studies in which virus has been allowed to attach on ice for 1 h during the infection phase may not have

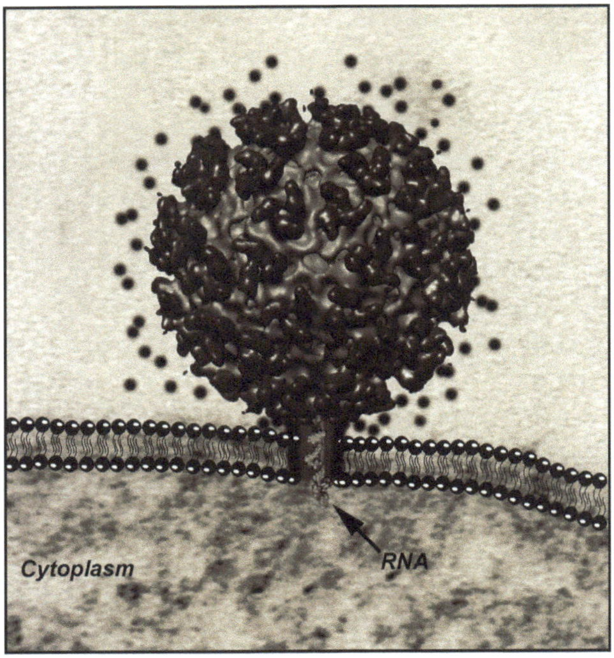

Figure 10 Schematic view of Sindbis virus genome delivery. Illustration of immunolabeled cross-linked virus-attached cell surface. After attachment and interaction with a receptor, conformational changes at the fivefold axis of the virion would allow the formation of a protein pore-like structure that would penetrate the cell membrane and allow the passage of virus RNA initiating the infection. The cell membrane and pore-like structure are emphasized in the cartoon.

synchronized the infection, as proposed, but rather allowed infectious RNA to be internalized. It is possible that synchronization does occur at low temperature by arresting some early event after entry. It is significant that in the process of infection by alphaviruses, the viral membranes remain outside the cell. If the viral membrane disengages from the host cell instead of merging with it, then not only would the cell not have viral proteins on its surface but also the release of virus membrane pieces with embedded proteins could serve as immunologic decoys.[49,53] The alphaviruses are clearly capable of fusing with artificial membranes at low pH; they are also capable of fusing with membranes of living cells if briefly exposed to low pH and then returned to neutral pH.[47,67] Therefore, the quest for an entry inhibitor based on a fusion trigger mechanism may be in vain if the fusion capabilities of these viruses have little to do with the process of infection.

6. CHALLENGES AND PERSPECTIVES

The characterization of cell entry pathway for alphaviruses has been challenging. Even though some of these viruses are now reemerging and account for thousands of human and animal infections each year, there are no effective therapies or vaccines available. The difficulties that have been encountered to understand their entry mechanism resulted from experiments examining entry that relied on indirect analysis, indicating virus entry has taken place, such as viral RNA or protein synthesis, which occurs after entry. The poor infectivity ratio typical of animal virus preparations and the inability to identify virus structures which have altered morphology have made conclusions from experiments involving direct observation difficult to interpret. By the use of new approaches that overcome these obstacles, it has been shown that these membrane-containing viruses may actually use a different mechanism to infect cells.

Arthropod-borne viruses such as the alphaviruses are significant source of disease to man and animals. A precise understanding of the entry pathway and the identification of cell receptors may not only lead to new therapies against viral infection but also provide new insights into cellular functions.

ACKNOWLEDGMENTS

The authors are supported by a grant from the Foundation for Research (Carson City, NV) and by the North Carolina Agricultural Research Service.

REFERENCES

1. Strauss JH, Strauss EG. The alphaviruses: gene expression, replication, and evolution. *Microbiol Rev*. 1994;58(3):491–562.
2. Gould EA, Coutard B, Malet H, et al. Understanding the alphaviruses: recent research on important emerging pathogens and progress towards their control. *Antiviral Res*. 2010;87(2):111–124.
3. Morens DM, Fauci AS. Dengue and hemorrhagic fever: a potential threat to public health in the United States. *JAMA*. 2008;299(2):214–216.
4. Schlesinger RW. New opportunities in biological research offered by arthropod cell cultures. I. Some speculations on the possible role of arthropods in the evolution of arboviruses. *Curr Top Microbiol Immunol*. 1971;55:241–245.
5. Enzmann PJ, Weiland F. Studies on the morphology of alphaviruses. *Virology*. 1979;95(2):501–510.
6. Anthony RP, Brown DT. Protein-protein interactions in an alphavirus membrane. *J Virol*. 1991;65(3):1187–1194.
7. Paredes AM, Brown DT, Rothnagel R, et al. Three-dimensional structure of a membrane-containing virus. *Proc Natl Acad Sci USA*. 1993;90(19):9095–9099.

8. Choi HK, Lu G, Lee S, Wengler G, Rossmann MG. Structure of Semliki Forest virus core protein. *Proteins*. 1997;27(3):345–359.
9. Mulvey M, Brown DT. Formation and rearrangement of disulfide bonds during maturation of the Sindbis virus E1 glycoprotein. *J Virol*. 1994;68(2):805–812.
10. Carleton M, Lee H, Mulvey M, Brown DT. Role of glycoprotein PE2 in formation and maturation of the Sindbis virus spike. *J Virol*. 1997;71(2):1558–1566.
11. Rice CM, Strauss JH. Association of Sindbis virion glycoproteins and their precursors. *J Mol Biol*. 1982;154(2):325–348.
12. Strauss EG, Lenches EM, Strauss JH. Molecular genetic evidence that the hydrophobic anchors of glycoproteins E2 and E1 interact during assembly of alphaviruses. *J Virol*. 2002;76(20):10188–10194.
13. Ferreira D, Hernandez R, Horton M, Brown DT. Morphological variants of Sindbis virus produced by a mutation in the capsid protein. *Virology*. 2003;307(1):54–66.
14. Lee H, Ricker PD, Brown DT. The configuration of Sindbis virus envelope proteins is stabilized by the nucleocapsid protein. *Virology*. 1994;204(1):471–474.
15. Hernandez R, Lee H, Nelson C, Brown DT. A single deletion in the membrane-proximal region of the Sindbis virus glycoprotein E2 endodomain blocks virus assembly. *J Virol*. 2000;74(9):4220–4228.
16. Whitehurst CB, Willis JH, Sinodis CN, Hernandez R, Brown DT. Single and multiple deletions in the transmembrane domain of the Sindbis virus E2 glycoprotein identify a region critical for normal virus growth. *Virology*. 2006;347(1):199–207.
17. Heller WT. Small-angle neutron scattering and contrast variation: a powerful combination for studying biological structures. *Acta Crystallogr D Biol Crystallogr*. 2010;66(pt 11):1213–1217.
18. He L, Piper A, Meilleur F, et al. The structure of Sindbis virus produced from vertebrate and invertebrate hosts as determined by small-angle neutron scattering. *J Virol*. 2010;84(10):5270–5276.
19. Greber UF. Signalling in viral entry. *Cell Mol Life Sci*. 2002;59(4):608–626.
20. Kalia M, Jameel S. Virus entry paradigms. *Amino Acids*. 2011;41(5):1147–1157.
21. Byrnes AP, Griffin DE. Binding of Sindbis virus to cell surface heparan sulfate. *J Virol*. 1998;72(9):7349–7356.
22. Klimstra WB, Ryman KD, Johnston RE. Adaptation of Sindbis virus to BHK cells selects for use of heparan sulfate as an attachment receptor. *J Virol*. 1998;72(9):7357–7366.
23. Mudhakir D, Harashima H. Learning from the viral journey: how to enter cells and how to overcome intracellular barriers to reach the nucleus. *AAPS J*. 2009;11(1):65–77.
24. Klimstra WB, Nangle EM, Smith MS, Yurochko AD, Ryman KD. DC-SIGN and L-SIGN can act as attachment receptors for alphaviruses and distinguish between mosquito cell- and mammalian cell-derived viruses. *J Virol*. 2003;77(22):12022–12032.
25. Helenius A, Morein B, Fries E, et al. Human (HLA-A and HLA-B) and murine (H-2K and H-2D) histocompatibility antigens are cell surface receptors for Semliki Forest virus. *Proc Natl Acad Sci USA*. 1978;75(8):3846–3850.
26. Smith AL, Tignor GH. Host cell receptors for two strains of Sindbis virus. *Arch Virol*. 1980;66(1):11–26.
27. Strauss JH, Wang KS, Schmaljohn AL, Kuhn RJ, Strauss EG. Host-cell receptors for Sindbis virus. *Arch Virol Suppl*. 1994;9:473–484.
28. Rose PP, Hanna SL, Spiridigliozzi A, et al. Natural resistance-associated macrophage protein is a cellular receptor for Sindbis virus in both insect and mammalian hosts. *Cell Host Microbe*. 2011;10(2):97–104.
29. Kononchik Jr JP, Hernandez R, Brown DT. An alternative pathway for alphavirus entry. *Virol J*. 2011;8:304.

30. Condit RC. Principles of virology. In: Knipe DMH, Peter M, eds. *Fields Virology*. 4th ed. Philadelphia, PA: Lippincott Williams & Wilkins; 2001:19–54.
31. Dimitrov DS. Virus entry: molecular mechanisms and biomedical applications. *Nat Rev Microbiol*. 2004;2(2):109–122.
32. Paredes AM, Ferreira D, Horton M, et al. Conformational changes in Sindbis virions resulting from exposure to low pH and interactions with cells suggest that cell penetration may occur at the cell surface in the absence of membrane fusion. *Virology*. 2004;324(2):373–386.
33. Ubol S, Griffin DE. Identification of a putative alphavirus receptor on mouse neural cells. *J Virol*. 1991;65(12):6913–6921.
34. Blumenthal R, Clague MJ, Durell SR, Epand RM. Membrane fusion. *Chem Rev*. 2003;103(1):53–69.
35. Kielian M. Membrane fusion and the alphavirus life cycle. *Adv Virus Res*. 1995;45:113–151.
36. Marsh M, Helenius A. Virus entry: open sesame. *Cell*. 2006;124(4):729–740.
37. Lakadamyali M, Rust MJ, Zhuang X. Endocytosis of influenza viruses. *Microbes Infect*. 2004;6(10):929–936.
38. Lescar J, Roussel A, Wien MW, et al. The fusion glycoprotein shell of Semliki Forest virus: an icosahedral assembly primed for fusogenic activation at endosomal pH. *Cell*. 2001;105(1):137–148.
39. Mayor S, Pagano RE. Pathways of clathrin-independent endocytosis. *Nat Rev Mol Cell Biol*. 2007;8(8):603–612.
40. Pelkmans L, Puntener D, Helenius A. Local actin polymerization and dynamin recruitment in SV40-induced internalization of caveolae. *Science*. 2002;296(5567):535–539.
41. Pelkmans L, Burli T, Zerial M, Helenius A. Caveolin-stabilized membrane domains as multifunctional transport and sorting devices in endocytic membrane traffic. *Cell*. 2004;118(6):767–780.
42. Klasse PJ. The molecular basis of HIV entry. *Cell Microbiol*. 2012;14(8):1183–1192.
43. Melikyan GB. HIV entry: a game of hide-and-fuse? *Curr Opin Virol*. 2014;4:1–7.
44. Ghosh JK, Peisajovich SG, Ovadia M, Shai Y. Structure–function study of a heptad repeat positioned near the transmembrane domain of Sendai virus fusion protein which blocks virus–cell fusion. *J Biol Chem*. 1998;273(42):27182–27190.
45. Spear PG, Eisenberg RJ, Cohen GH. Three classes of cell surface receptors for alphaherpesvirus entry. *Virology*. 2000;275(1):1–8.
46. Hase T, Summers PL, Eckels KH, Baze WB. An electron and immunoelectron microscopic study of dengue-2 virus infection of cultured mosquito cells: maturation events. *Arch Virol*. 1987;92(3–4):273–291.
47. Brown DT, Hernandez R. Infection of cells by alphaviruses. *Adv Exp Med Biol*. 2012;726:181–199.
48. Skehel JJ, Wiley DC. Receptor binding and membrane fusion in virus entry: the influenza hemagglutinin. *Annu Rev Biochem*. 2000;69:531–569.
49. Vancini R, Wang G, Ferreira D, Hernandez R, Brown DT. Alphavirus genome delivery occurs directly at the plasma membrane in a time- and temperature-dependent process. *J Virol*. 2013;87(8):4352–4359. http://dx.doi.org/10.11.28/JVI.03412-12.
50. Helenius A, Marsh M, White J. Inhibition of Semliki Forest virus penetration by lysosomotropic weak bases. *J Gen Virol*. 1982;58(pt 1):47–61.
51. Marsh M, Wellsteed J, Kern H, Harms E, Helenius A. Monensin inhibits Semliki Forest virus penetration into culture cells. *Proc Natl Acad Sci USA*. 1982;79(17):5297–5301.
52. Duzgunes N, Pedroso de Lima MC, Stamatatos L, et al. Fusion activity and inactivation of influenza virus: kinetics of low pH-induced fusion with cultured cells. *J Gen Virol*. 1992;73(pt 1):27–37.

53. Haywood AM. Membrane uncoating of intact enveloped viruses. *J Virol.* 2010;84(21):10946–10955.
54. Hernandez R, Sinodis C, Brown DT. Sindbis virus: propagation, quantification, and storage. *Curr Protoc Microbiol.* 2005, [chapter 15], Unit 15B.1.
55. Helenius A, Kartenbeck J, Simons K, Fries E. On the entry of Semliki Forest virus into BHK-21 cells. *J Cell Biol.* 1980;84:404–420.
56. Lin X, Derdeyn CA, Blumenthal R, West J, Hunter E. Progressive truncations C terminal to the membrane-spanning domain of simian immunodeficiency virus Env reduce fusogenicity and increase concentration dependence of Env for fusion. *J Virol.* 2003;77(12):7067–7077.
57. Fan DP, Sefton BM. The entry into host cells of Sindbis virus, vesicular stomatitis virus and Sendai virus. *Cell.* 1978;15(3):985–992.
58. Glomb-Reinmund S, Kielian M. The role of low pH and disulfide shuffling in the entry and fusion of Semliki Forest virus and Sindbis virus. *Virology.* 1998;248(2):372–381.
59. DeTulleo L, Kirchhausen T. The clathrin endocytic pathway in viral infection. *EMBO J.* 1998;17(16):4585–4593.
60. Perez L, Carrasco L. Involvement of the vacuolar H(+)-ATPase in animal virus entry. *J Gen Virol.* 1994;75(pt 10):2595–2606.
61. Kielian M, Jungerwirth S. Mechanisms of enveloped virus entry into cells. *Mol Biol Med.* 1990;7(1):17–31.
62. Kielian M, Chatterjee PK, Gibbons DL, Lu YE. Specific roles for lipids in virus fusion and exit. Examples from the alphaviruses. *Subcell Biochem.* 2000;34:409–455.
63. Kielian MC, Helenius A. Role of cholesterol in fusion of Semliki Forest virus with membranes. *J Virol.* 1984;52(1):281–283.
64. Lu YE, Cassese T, Kielian M. The cholesterol requirement for Sindbis virus entry and exit and characterization of a spike protein region involved in cholesterol dependence. *J Virol.* 1999;73(5):4272–4278.
65. Cleverley D, Geller H, Lenard J. Characterization of cholesterol-free insect cells infectible by baculoviruses: effects of cholesterol on VSV fusion and infectivity and on cytotoxicity induced by influenza M2 protein. *Exp Cell Res.* 1997;233(2):288–296.
66. Mitsuhashi J, Nakasone S, Horie Y. Sterol-free eukaryotic cells from continuous cell lines of insects. *Cell Biol Int Rep.* 1983;7(12):1057–1062.
67. Edwards J, Brown DT. Sindbis virus-mediated cell fusion from without is a two-step event. *J Gen Virol.* 1986;67(pt 2):377–380.
68. Smit JM, Li G, Schoen P, et al. Fusion of alphaviruses with liposomes is a non-leaky process. *FEBS Lett.* 2002;521(1–3):62–66.
69. Madan V, Sanz MA, Carrasco L. Requirement of the vesicular system for membrane permeabilization by Sindbis virus. *Virology.* 2005;332(1):307–315.
70. Koschinski A, Wengler G, Repp H. The membrane proteins of flaviviruses form ion-permeable pores in the target membrane after fusion: identification of the pores and analysis of their possible role in virus infection. *J Gen Virol.* 2003;84(pt 7):1711–1721.
71. Koschinski A, Wengler G, Repp H. Rare earth ions block the ion pores generated by the class II fusion proteins of alphaviruses and allow analysis of the biological functions of these pores. *J Gen Virol.* 2005;86(pt 12):3311–3320.
72. Talbot PJ, Vance DE. Biochemical studies on the entry of Sindbis virus into BHK-21 cells and the effect of NH4Cl. *Virology.* 1982;118(2):451–455.
73. Cassell S, Edwards J, Brown DT. Effects of lysosomotropic weak bases on infection of BHK-21 cells by Sindbis virus. *J Virol.* 1984;52(3):857–864.
74. Helenius A, Marsh M. Endocytosis of enveloped animal viruses. *Ciba Found Symp.* 1982;92:59–76.
75. Hernandez R, Luo T, Brown DT. Exposure to low pH is not required for penetration of mosquito cells by Sindbis virus. *J Virol.* 2001;75(4):2010–2013.

76. Hunt SR, Hernandez R, Brown DT. Role of the vacuolar-ATPase in Sindbis virus infection. *J Virol.* 2011;85(3):1257–1266.
77. Kielian M, Helenius A. pH-induced alterations in the fusogenic spike protein of Semliki Forest virus. *J Cell Biol.* 1985;101(6):2284–2291.
78. Justman J, Klimjack MR, Kielian M. Role of spike protein conformational changes in fusion of Semliki Forest virus. *J Virol.* 1993;67(12):7597–7607.
79. Omar A, Koblet H. Semliki Forest virus particles containing only the E1 envelope glycoprotein are infectious and can induce cell-cell fusion. *Virology.* 1988;166(1):17–23.
80. Sanz MA, Rejas MT, Carrasco L. Individual expression of Sindbis virus glycoproteins. E1 alone promotes cell fusion. *Virology.* 2003;305(2):463–472.
81. Bron R, Wahlberg JM, Garoff H, Wilschut J. Membrane fusion of Semliki Forest virus in a model system: correlation between fusion kinetics and structural changes in the envelope glycoprotein. *EMBO J.* 1993;12(2):693–701.
82. Levy-Mintz P, Kielian M. Mutagenesis of the putative fusion domain of the Semliki Forest virus spike protein. *J Virol.* 1991;65(8):4292–4300.
83. Kielian M, Rey FA. Virus membrane-fusion proteins: more than one way to make a hairpin. *Nat Rev Microbiol.* 2006;4(1):67–76.
84. White J, Kartenbeck J, Helenius A. Fusion of Semliki Forest virus with the plasma membrane can be induced by low pH. *J Cell Biol.* 1980;87(1):264–272.
85. Jan Wilschut DH, ed. *Membrane Fusion.* New York: Dekker; 1991.
86. Carrasco L, Otero MJ, Castrillo JL. Modification of membrane permeability by animal viruses. *Pharmacol Ther.* 1989;40(2):171–212.
87. FitzGerald DJ, Padmanabhan R, Pastan I, Willingham MC. Adenovirus-induced release of epidermal growth factor and pseudomonas toxin into the cytosol of KB cells during receptor-mediated endocytosis. *Cell.* 1983;32(2):607–617.
88. Almela MJ, Gonzalez ME, Carrasco L. Inhibitors of poliovirus uncoating efficiently block the early membrane permeabilization induced by virus particles. *J Virol.* 1991;65(5):2572–2577.
89. Owen KE, Kuhn RJ. Identification of a region in the Sindbis virus nucleocapsid protein that is involved in specificity of RNA encapsidation. *J Virol.* 1996;70(5):2757–2763.
90. Wengler G, Boege U, Bischoff H, Wahn K. The core protein of the alphavirus Sindbis virus assembles into core-like nucleoproteins with the viral genome RNA and with other single-stranded nucleic acids in vitro. *Virology.* 1982;118(2):401–410.
91. Wang G, Hernandez R, Weninger K, Brown DT. Infection of cells by Sindbis virus at low temperature. *Virology.* 2007;362(2):461–467.
92. Lippincott-Schwartz J, Roberts TH, Hirschberg K. Secretory protein trafficking and organelle dynamics in living cells. *Annu Rev Cell Dev Biol.* 2000;16:557–589.
93. Flynn DC, Meyer WJ, Mackenzie Jr JM, Johnston RE. A conformational change in Sindbis virus glycoproteins E1 and E2 is detected at the plasma membrane as a consequence of early virus-cell interaction. *J Virol.* 1990;64(8):3643–3653.
94. Kononchik JP, Vancini R, Brown DT. Alphavirus adsorption to mosquito cells as viewed by freeze fracture immunolabeling. *Virology.* 2011;415(2):132–140.
95. Melikyan GB, Niles WD, Ratinov VA, Karhanek M, Zimmerberg J, Cohen FS. Comparison of transient and successful fusion pores connecting influenza hemagglutinin expressing cells to planar membranes. *J Gen Physiol.* 1995;106(5):803–819.
96. Carrasco L. Modification of membrane permeability by animal viruses. *Adv Virus Res.* 1995;45:61–112.
97. Carrasco L. Entry of animal viruses and macromolecules into cells. *FEBS Lett.* 1994;350(2–3):151–154.
98. Helenius A, Kielian M, Wellsteed J, Mellman I, Rudnick G. Effects of monovalent cations on Semliki Forest virus entry into BHK-21 cells. *J Biol Chem.* 1985;260(9):5691–5697.

99. Fuchs P, Kohn A. Changes induced in cell membranes adsorbing animal viruses, bacteriophages, and colicins. *Curr Top Microbiol Immunol.* 1983;102:57–99.
100. Hase T, Summers PL, Eckels KH. Flavivirus entry into cultured mosquito cells and human peripheral blood monocytes. *Arch Virol.* 1989;104(1–2):129–143.
101. Vancini R, Kramer LD, Ribeiro M, Hernandez R, Brown D. Flavivirus infection from mosquitoes in vitro reveals cell entry at the plasma membrane. *Virology.* 2013;435(2):406–414.
102. Fricks CE, Hogle JM. Cell-induced conformational change in poliovirus: externalization of the amino terminus of VP1 is responsible for liposome binding. *J Virol.* 1990;64(5):1934–1945.
103. Gray MA, Micklem KJ, Pasternak CA. Protein synthesis in cells infected with Semliki Forest virus is not controlled by intracellular cation changes. *Eur J Biochem.* 1983;135(2):299–302.
104. Patel K, Pasternak CA. Ca^{2+}-sensitive permeability changes caused by influenza virus. *Biosci Rep.* 1983;3(8):749–755.
105. Belnap DM, Filman DJ, Trus BL, et al. Molecular tectonic model of virus structural transitions: the putative cell entry states of poliovirus. *J Virol.* 2000;74(3):1342–1354.

CHAPTER THREE

The Mechanism of HCV Entry into Host Cells

Florian Douam[*,†,‡,§,¶,||,#], **Dimitri Lavillette**[*,†,‡,§,¶,#,1], **François-Loïc Cosset**[*,†,‡,§,¶,||,1,2]

*CIRI, International Center for Infectiology Research, Team EVIR, Université de Lyon, Lyon, France
†Inserm U1111, Lyon, France
‡Ecole Normale Supérieure de Lyon, Lyon, France
§Centre International de Recherche en Infectiologie, Université Claude Bernard Lyon 1, Lyon, France
¶CNRS, UMR5308, Lyon, France
||LabEx Ecofect, Université de Lyon, Lyon, France
#CNRS, UMR 5557 Ecologie Microbienne, Microbial Dynamics and Viral Transmission Team, Université Claude Bernard Lyon 1, Villeurbanne, France
[2]Corresponding author: e-mail address: flcosset@ens-lyon.fr

Contents

Abstract

Hepatitis C virus (HCV) is an enveloped, positive strand RNA virus classified within the *Flaviviridae* family and is a major cause of liver disease worldwide. HCV life cycle and propagation are tightly linked to several aspects of lipid metabolism. HCV propagation depends on and also shapes several aspects of lipid metabolism such as cholesterol

[1] These authors contributed equally to the work.

Progress in Molecular Biology and Translational Science, Volume 129
ISSN 1877-1173
http://dx.doi.org/10.1016/bs.pmbts.2014.10.003

uptake and efflux through different lipoprotein receptors during its entry into cells, lipid metabolism modulating HCV genome replication, lipid droplets acting as a platform for recruitment of viral components, and very low density lipoprotein assembly pathway resulting in incorporation of neutral lipids and apolipoproteins into viral particles.

During the first steps of infection, HCV enters hepatocytes through a multistep and slow process. The initial capture of HCV particles by glycosaminoglycans and/or lipoprotein receptors is followed by coordinated interactions with the scavenger receptor class B type I, a major receptor of high-density lipoprotein, the CD81 tetraspanin, and the tight junction proteins Claudin-1 and Occludin. This tight concert of receptor interactions ultimately leads to uptake and cellular internalization of HCV through a process of clathrin-dependent endocytosis. Over the years, the identification of the HCV entry receptors and cofactors has led to a better understanding of HCV entry and of the narrow tropism of HCV for the liver. Yet, the role of the two HCV envelope glycoproteins, E1 and E2, remains ill-defined, particularly concerning their involvement in the membrane fusion process. Here, we review the current knowledge and advances addressing the mechanism of HCV cell entry within hepatocytes and we highlight the challenges that remain to be addressed.

1. INTRODUCTION

Hepatitis C virus (HCV) is an enveloped, positive strand RNA virus classified within the *Flaviviridae* family. The *Flaviviridae* family is composed of four virus genera: the flaviviruses, the pestiviruses, the hepaciviruses, and the pegiviruses that all replicate within the cell cytosol. Their RNA genome encodes for a large polyprotein cleaved into several structural and non-structural proteins and then involved in the biogenesis of new viral particles. However, HCV appears as an unusual *Flaviviridae*. Viral particles are structurally heterogeneous and harbor a buoyant density that is lower than most of the known viruses. This unique property is probably linked to the ability of HCV to hijack the host lipid metabolism. The specific association with some lipids and apolipoproteins has critical consequences for its entry and assembly into hepatocytes. This association makes complex the relationship between the virus and its host and, thus, hampers the development of relevant experimental systems. Although HCV presents a high degree of genetic and functional similarities with some well-studied flaviviruses, many steps of the HCV life cycle remain to be precisely understood.

HCV is a major cause of liver disease worldwide. Acute infection often leads to the setting up of a chronic liver disease to the development of hepatocellular carcinoma. HCV's high ability to generate neutralizing antibody-escape viral mutants and to restrict host innate immune response likely contributes to the installation of chronic infection. Although no vaccine is still available, the recent development of several new direct-acting agents (DAAs) targeting viral nonstructural proteins has led to serious progress in patient treatments. However, these treatments remain today extremely expensive and the risk of emerging viruses carrying resistances needs to be taken into account. Thus, the constant improvement of our understanding of the HCV life cycle may be crucial to address future challenges of patient treatments.

Entry is a critical step in the life cycle of enveloped viruses. Through attachment, internalization, and fusion of the viral particles with cellular membranes, this process ensures the release of the viral genome into the host cell cytosol and the initiation of virus replication. Over the years, the identification of a large number of HCV entry factors has led to a better understanding of HCV entry and of the narrow tropism of HCV for the liver. However, by involving lipoprotein components, viral structural mediators as well as an important numbers of cellular receptors or factors, HCV entry is a complex process whose characterization remains highly challenging in a spatiotemporal context. Moreover, the roles of the two HCV envelope glycoproteins E1 and E2 are still not well understood. Indeed, despite the fact that E2 has been identified to be the major HCV binding protein and recently characterized structurally, the precise role of E1 during entry and its interrelation with E2 are still not fully elucidated.

Here, we review the current knowledge and advances addressing the mechanism of HCV cell entry within hepatocytes, in an attempt to draw a relevant overview of this process. In parallel, we will point out some challenges that remain to be addressed.

2. THE VIRAL PARTICLE ORGANIZATION AND COMPOSITION: A FUNDAMENTAL KEY TO DECRYPT VIRUS ENTRY

The structure of the HCV particle remains poorly characterized despite substantial progress in cell culture systems enabling viral particle production as well as several biochemical and morphological studies. This contrasts with the well-characterized flavivirus viral particles. A striking and unique feature of HCV biology is the association with lipoproteins of HCV particles, which exhibit an unusually low buoyant density.[1–3] HCV particles are 50–80 nm in diameter[4] and contain a single-stranded RNA genome, core, and the envelope glycoproteins, E1 and E2.[5] HCV genome interacts with the core protein to form the nucleocapsid that is surrounded by a lipid membrane, called the viral envelope, in which are anchored two envelope glycoproteins, E1 and E2.

E1 and E2 glycoproteins are type I transmembrane proteins which form a noncovalent heterodimer within infected cells, whereas they assemble as large covalent complexes stabilized by disulfide bonds on the viral particle.[6] E1 and E2 are major viral determinants of HCV entry. They indeed play a role in receptor binding and ultimately mediate the fusion process between the viral envelope and an endosomal host cell membrane.

Importantly, due to virion association with lipoproteins, apolipoproteins such as apoE, apoB, apoA1, apoC1, apoC2 and apoC3 can also be found in association with HCV particles.[1,4,7–9] Furthermore, a characterization of cell culture-produced particles indicates that their lipid composition resembles the one of the very low-density lipoproteins (VLDLs) and low-density lipoproteins (LDLs) with cholesteryl esters (CEs) accounting for almost half of the total HCV lipids.[10] Electron microscopy analyses of purified infectious virions confirm the pleomorphic nature of HCV virions and show viral particles with a rather smooth and even surface.[4]

The nature of the interactions involved between HCV virion components and the lipoprotein remains undetermined. It has been suggested that HCV particles could be a lipo-viro-particle (LVP), i.e., an hybrid structure composed of a virion moiety and a lipoprotein moiety.[11] However, alternative models have also been suggested, with lipoproteins peripherally associated with canonical viral particles via interaction between apolipoproteins and HCV envelope lipids or proteins.[12] In both particle types, the interaction with lipoproteins could contribute to the shielding of HCV glycoproteins from the host antibody-neutralizing response and explain the poor detection or availability of HCV glycoproteins at the virion surface.[4,10,13] Importantly, apolipoprotein(s) associated with HCV particles play a critical role in HCV entry (see below).

The investigation of natural HCV, or LVP, and of its different lipoprotein- or immunoglobulin-associated forms has been laborious because the infectivity of HCV-containing sera *ex vivo*, in either primary hepatocytes or hepatoma cell lines, is very low.[14,15] Furthermore, the mechanisms by which LVPs mediate cell penetration, leading to release of HCV genetic material and replication, have not yet been characterized functionally.

To overcome these limitations toward the molecular characterization of HCV infection, several surrogate assays or tools have been developed. Two relevant and complementary infection assays consist of cell culture-grown genuine HCV (HCVcc) derived from a fulminant hepatitis C (Fig. 1),

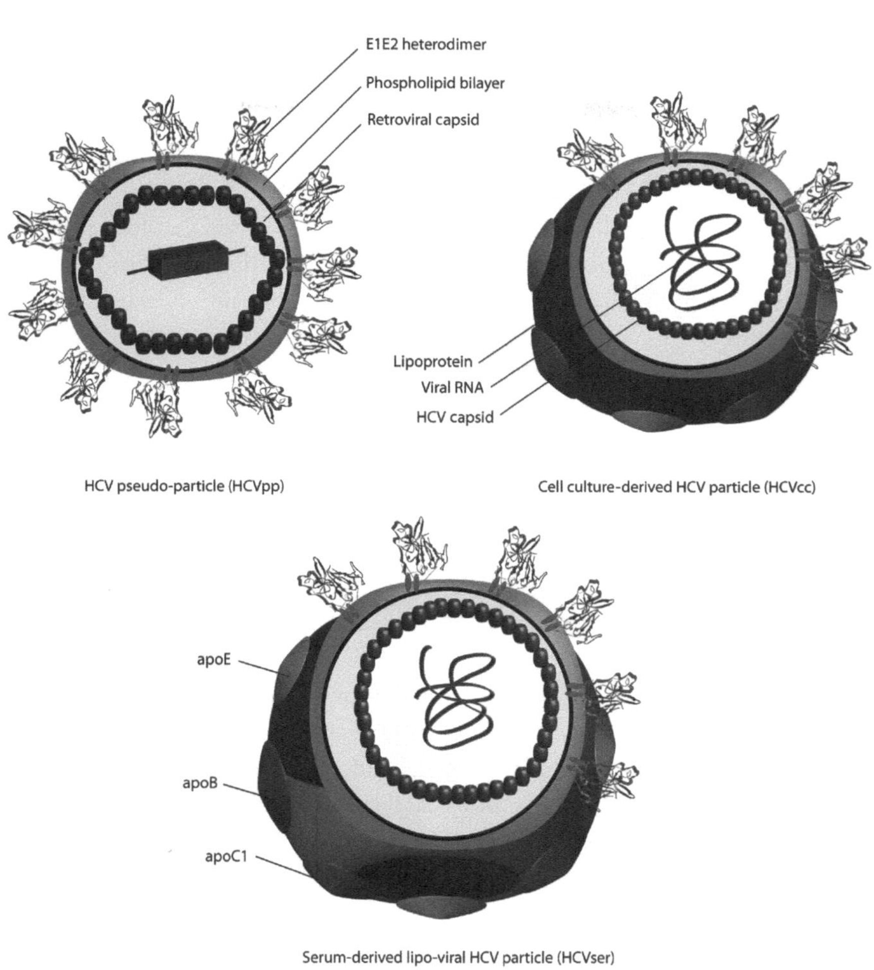

Figure 1 Experimental systems for the study of HCV. The different components for each type of viral particles are indicated. apo, apolipoprotein.

JFH-1 and JFH-1-derived recombinant genomes,[16–18] and of HCV pseudo-particles (HCVpp) harboring authentic E1E2 glycoproteins (Fig. 1), which are particularly amenable to mutagenesis analysis.[19–21] Similar to HCV derived from patients' plasmas, HCVcc RNA displays a broad density profile with high specific infectivity associated to low-density fractions, which might reproduce some features of HCV association to lipoprotein compounds.[16,22–27] In a recent study, the lipid composition of HCVcc particles was determined and found to resemble that of VLDL and LDL.[10] Highly purified HCVcc particles contain several apoE molecules on their surface as well as apoC-I and, to some extent, apoB, in line with the finding that HCV formation and secretion depend on VLDL assembly and secretion pathway[8,28,7,29–34] (Fig. 1). However, there has been much recent interest in propagating and studying HCV recovered in the serum from human liver mice,[35–37] i.e., HCVser, owing to the fact that hepatoma cell lines used to produce HCVcc have impaired production of VLDL[38] and that the resulting HCV particles have biophysical and biochemical characteristics that are likely different from those of the virus recovered *in vivo*[27] (Calattini *et al.*, 2014, submitted for publication). The three different experimental HCV systems are resumed in Fig. 1.

While HCVcc, which are produced in human hepato-carcinoma cells—typically Huh-7.5 cells, further permit investigation of the late steps of infection, HCVpp offer a particularly flexible platform to study the structure/function relationship of HCV glycoproteins both in cell culture and *in vitro*, in liposome membrane fusion assays.[39,40] Of note, since they are produced from 293T kidney cells, HCVpp are not associated to lipoproteins,[41] which, hence, allows to investigate the cell entry events specifically linked to the functions of the E1E2 glycoproteins and to reconstitute some of the interactions of HCV with lipoproteins or apolipoproteins.[42–46] Thus, both HCVcc and HCVpp (as well as HCVser) infection assays reproduce several cell entry characteristics of native HCV and, combined together, allow a precise dissection of the cellular and viral factors involved in the early events of HCV infection.

3. EARLY STEPS OF VIRUS ENTRY

3.1. Viral particle capture

Upon infection, HCV particles are transported via the blood stream and enter in contact with hepatocytes after crossing the fenestrated endothelium of the liver sinusoids. Once present in the space of Disse where basolateral membranes of hepatocyte are exposed, HCV entry is first engaged by initial capture of the viral particles by attachment factors and receptors on the surface of these cells in a spatiotemporally regulated manner. Initial attachment of HCV particles to hepatocytes is mediated by heparan sulfate proteoglycan (HSPG) syndecan-1 or syndecan-4[47,48] or by the scavenger receptor class B type I (SR-BI),[13] depending on virion density. It was initially thought that HCV glycoproteins are responsible for virion binding to HSPGs[49] or SR-BI.[50] However, more recent data suggest that apoE, rather than HCV glycoproteins themselves, could be involved in this initial contact.[13,51]

Due to HCV particle interaction with lipoproteins, the LDL receptor (LDL-r) has also been proposed to play a role in the early phase of HCV entry.[52] However, HCV–LDL-r interaction seems to involve a nonproductive entry pathway that can potentially lead to viral particle degradation.[53]

3.1.1 The heparan sulfate proteoglycans

Glycosaminoglycans (GAGs) are large polysaccharides expressed on the surface of most mammalian cells and serve as attachment molecules for a large number of viruses.[54] Interestingly, several *Flaviviridae* such as Dengue or CSFV use a particular GAG, the HSPG as an initial attachment factor before binding to more specific receptors.[54,55] Consistently, it has been shown that soluble HCV E2 (sE2) attachment to cell surface was abrogated through competition by heparin, a heparan sulfate homolog, as well as by a heparinase cell treatment, an enzyme able to degrade heparan sulfate at the cell surface.[49,56,57] Moreover, both intracellular and secreted HCVcc viral particles were also able to bind heparin.[56] Kinetic studies with the HCVcc system have demonstrated that HSPG plays a role in the very early cell entry step as a nonspecific attachment HCV factor.[56] However, the molecular determinants of this interaction remain difficult to determine. No precise GAG binding motifs have been identified so far on HCV E2, but the deletion of the hypervariable region 1 (HVR1) E2 or antibodies targeting this domain impaired sE2 binding to heparin,[58] indicating that HVR1 may play a role in this attachment. This assumption has also been confirmed by the high affinity of a HVR1-synthetic peptide for heparin. However, surprisingly, HCVpp binding is not inhibited by heparin,[59] suggesting that HSPG attachment of viral particles is independent of functional E1E2 envelope glycoprotein complexes. apoE is known to be harbored by LVPs but is also considered to be a GAG ligand. Consistently, suppression of the endogenous expression of apoE in Huh7.5 hepatoma cell lines and complementation with apoE mutants' deficient for HSPG binding resulted in the production of poorly infectious viral particles with a reduced attachment phenotype.[57] Moreover, a synthetic peptide derived from the apoE receptor binding domain (RBD) to HSPG was also able to inhibit HCVcc infection. In conclusion, GAGs appear to be important initial attachment factors for HCV particles during entry, through the mediation of apoE. More recently, a study screened a large number of HSPG core molecule and pointed out Syndecan-1 (SDC1) as the major HSPG attachment molecule involved in HCV entry.[48] Downregulation of the molecule impaired virus attachment, which was restored following restoration of

SDC1 expression. Alternatively, via mutagenesis and construction of soluble peptide, another study recently identified apoE key residues that are critical for SDC interaction and so far, HCV entry.[47] However, authors also showed that SDC4, and not SDC1, seems to be the main mediator of viral particle attachment through binding to apoE.

In conclusion, HSPGs appear as important capture components for HCV particles during early steps of entry. This role seems to be mainly mediated by several members of the SDC family, through the attachment of apoE molecules associated with lipo-viral particles.

3.1.2 The LDL-r

The LDL-r is expressed in a large variety of tissues.[60] However, its major function, which consists to capture and clear from the blood circulation the LDL, is restricted to the liver.[60,61] The LDL-r internalizes the LDLs through a clathrin-dependent endocytosis mechanism before delivery into the early endosome. The LDL-r is composed of three major extracellular domains.[62] Its N-terminal cysteine-rich ligand binding domain interacts with the apolipoprotein B-100 which is the main component of the LDL[60,62] but LDL-r also exhibits a high affinity for other types of lipoprotein such as the VLDL through interaction with the apolipoprotein E.[60,62] The heterogeneous buoyant density of the HCV particles and their association with lipoproteins[16,2] constituted strong arguments to hypothesize that HCV particles from low-density fraction bind to LDL-r and allow internalization. Consistently, a study demonstrated, using patient-derived HCV particles, that only low-density viral particles could bind to LDL-r-expressing cell lines, whereas soluble E2 did not.[63] This interaction was correlated with the level of expression of LDL-r and could be inhibited with soluble human-LDL peptide.[63,52] As no direct interaction was observed between sE2 and LDL-r,[63,64] the HCVpp system appeared as a noncompetent system to study the molecular determinants of this interaction. In contrast, the HCVcc system contributed to answer these questions. HCVcc infection was inhibited in assays using competition with VLDL, anti-apoE antibodies, or LDL-r downregulation indicating that HCV attachment is dependent on LDL-r and apoE.[32,33] Moreover, the exogenous expression of apoE2 after downregulation of endogenous apoE in HCV-infected cells did not rescue HCVcc infectivity consistently with the fact that apoE2 presents a low affinity for LDL-r.[32]

To conclude, it is now clear that apoE constitutes a mediator of HCV particles attachment on LDL-r as well as on GAGs. However, the role of

LDL-r as an entry factor is challenged as it was demonstrated recently that attachment to LDL-r may lead to nonproductive internalization of viral particles.[53] Further studies are needed to validate the role of LDL-r in HCV entry, particularly *in vivo*.

3.1.3 The scavenger receptor B-I

The SR-BI is a cell surface receptor found on many cell types but that is highly expressed within the liver and the steroidogenic tissues. This protein contains two C- and N-terminal cytoplasmic domains and one extracellular domain forming a large loop involved in the physiological functions of the receptor.[65] SR-BI is involved in the lipid metabolism and it binds different classes of lipoproteins such as the VLDL, LDL, or high-density lipoprotein (HDL).[66] More particularly, after binding to SR-BI, the CE is delivered from the HDL to the plasma membrane.[67] The HDL is not degraded in contrast to LDL after internalization by the LDL-r and the cholesterol mass of the plasma membrane is increased.[68]

SR-BI has been identified as a potential HCV receptor through binding assays with soluble E2 protein and neutralization assays using anti-SR-BI antibodies.[50,69,70] Moreover, SR-BI downregulation was shown to decrease both HCVpp and HCVcc infection, confirming the critical role of this receptor for HCV entry.[71,72] The HVR1 of E2 was shown to mediate SR-BI binding and specific SR-BI residues involved in this interaction have been identified.[25,69,73,74] Interestingly, it has been also observed that purified VLDL and antibodies targeting β-lipoproteins were able to inhibit SR-BI binding of patient-derived HCV particles.[75] Consistently, studies have reported that binding of viral particles is not completely inhibited despite deletion of HVRI or use of anti-HVR1 antibodies,[25,74] suggesting that HCV can probably be captured by SR-BI, albeit not in an E1E2-dependent manner but rather via HCV-associated lipoprotein components. A recent study[74] showed that SR-BI mediates primary attachment to cell surface of HCV particles of intermediate density and that these initial interactions involve apolipoproteins, such as apoE. Consequently, this may highlight the role of GAGs or LDL-r in the capture of viral particles from low- and high-density fractions. Altogether, lipoprotein components in the virion act as important host-derived ligands for HCV particle capture.

3.2. Early particle rearrangements

The capture of viral particles to the cell surface is not sufficient to initiate the virus entry process. Indeed, several molecular mechanisms are thought to trigger the transition between virus capture and entry. Although the full understanding of these mechanisms remains challenging, SR-BI has been proposed to be an important mediator in these processes. This SR-BI "access" function has been identified after the observation that blocking of the lipid transfer activity of SR-BI blocks entry of HCVcc particles from all density.[74] This function of SR-BI does not depend on E2/SR-BI interaction as murine SR-BI mediates HCV infection without being able to bind soluble E2.[73,74] Thus, SR-BI seems to represent a critical link between virus capture and subsequent binding to other cellular receptors. The SR-BI "access" function could modify the lipoprotein profile of viral particles and, at the same time, enrich the cellular membrane with cholesterol. These lipoprotein rearrangements may modify the lipid domains of the plasma membrane, the localization, and movement of different cell surface molecules used as HCV receptor. In addition, the lipoprotein rearrangements of HCV particle may allow the presentation of new E1E2 epitopes and the E1E2 binding to other receptors, leading to virus internalization.[74]

4. RECEPTOR BINDING AND CLUSTERING

4.1. E1E2 glycoproteins: Viral mediator of particle binding

HCV E1 and E2 glycoproteins are two highly N-glycosylated type I transmembrane proteins of respectively 190 aa (31 kDa) and 365 aa (70 kDa), each one harboring a single C-terminal transmembrane domain.[76] Interestingly, HCV E1 and E2 harbor an amino acid length similar to that of pestiviruses E1 and E2, whereas the main flavivirus envelope glycoprotein E remains larger with 500 aa.[77] After polyprotein translation, E1 and E2 are processed by the ER signal peptidase. E1 is the first protein to be completely processed through cleavage in the C-terminal domain of core and in the transmembrane domain of E1. E2 is then released through processing of the E2/p7 junction via cleavage in the transmembrane domain of E2. The processing of the E1/E2 junction has been shown to be particularly important for a stable insertion of the E2 transmembrane domain into the ER membrane.[78,79] After polyprotein cleavage, E1 and E2 are inserted into the ER membrane and their ectodomains are exposed in the ER lumen.[78,80–82] Cleavage of the E1/E2 and E2/p7 junctions also induces E1 and E2 transmembrane rearrangements that allow E1 and E2 to associate as heterodimer complex, in a slow process stabilized by noncovalent interactions.

4.1.1 Heterodimerization

An important common point between pestiviruses and HCV envelope glycoproteins is that E1 or E2 does not undergo proteolytic cleavage during its maturation in contrast to flaviviruses' envelope glycoprotein M.[83] E1 and E2 are therefore expressed directly as functional proteins that only require proper heterodimerization and glycosylation. It has been shown in several studies that the functional conformation of both E1 and E2 is dependent on the presence of chaperones or other cell proteins.[84–88] E1 is thought to be the first protein to obtain a functional conformation, which would be then required for E2 correct conformation and heterodimerization suggesting that E1 acts therefore as a chaperone for E2. However, other studies also indicate that E2 may also act as a chaperone for E1. During heterodimer formation, intramolecular disulfide bridges are formed within E1 and E2.[87] E1 and E2 harbor 8 and 18 cysteines that form 4 and 9 disulfide bridges, respectively. E2 disulfide bridges have been shown to be essential for CD81 binding and for antibodies recognition.[89] Interestingly, it has also been shown that E1 and E2 are linked at the surface of HCVcc particles by important intermolecular disulfide bridges, which do not seem to be the case at the HCVpp surface.[6] These differences may be explained by assembly in different compartments as HCVpp assembled in the multivesicular bodies,[90] whereas HCVcc mainly follow the VLDL assembly pathway.

4.1.2 Glycosylation

HCV E1 and E2 glycoproteins are heavily glycosylated proteins as their glycan mass accounts for nearly half of the total protein mass, which contrasts with the low glycosylation level of flavivirus E that harbors one or two glycans.[83] HCV E1 and E2 possess, respectively, 5 and 11 N-glycosylation sites.[91,92] The majority of these sites are highly conserved among genotypes. ER-resident E1E2 is mainly glycosylated by high-mannose type glycans. However, both high-mannose type glycans and complex glycans have been found associated with HCVcc E2 glycoproteins, suggesting that HCVcc E1E2 complexes likely undergo maturation through the Golgi apparatus.[6] The HCV N-glycans have multifunctional roles. By using the HCVcc system, a study showed that an N-glycosylation site in E1 (residue 196) is critical for both E1E2 heterodimerization and viral particles secretion.[93] Moreover, several sites have been identified in E2 to be involved for E2 folding and virus entry.[94,95] N-Glycosylation sites also appear to be important for both viral particles binding to cellular receptors (such as CD81) and protection of conserved epitopes against Nt Abs.[93,94]

4.1.3 Envelope glycoproteins and virus morphogenesis

HCV morphogenesis requires the concentration and encountering of viral structural proteins and genomic RNA in a temporally and spatially organized manner.[12] Assembly, budding, or egress of viral particles are events difficult to detect within infected cells, suggesting that these processes are either rare or rapid. An important feature of HCV assembly shared with other members of the *Flaviviridae* family is the involvement of the viral nonstructural proteins in this process. Another striking particularity of HCV morphogenesis is its intricate connection with the lipid metabolism.

A major component of the viral particle is the core protein, which interacts with the genomic viral RNA to form the nucleocapsid. Another major component of the viral particle is the envelope glycoprotein complex. HCV glycoproteins E1 and E2 form a noncovalent heterodimer which is retained in the ER.[96] However, this glycoprotein complex needs to migrate in close proximity of the lipid droplet (LD)–ER interface where assembly could be initiated.[97] It has been shown that NS2 interacts with E1, E2 and p7, and that such interactions are essential for the migration of E1E2 heterodimer at the virion assembly site.[98–101] Moreover, it has recently been reported that the cellular factor signal peptidase complex subunit 1 is involved in HCV assembly by helping the formation of membrane-associated NS2–E2 complex.[102] It has been proposed that E1E2 heterodimer, NS2, and p7 form a functional unit that migrates close to the LDs.[100] The presence of disulfide bridges between HCV envelope glycoproteins at the surface of HCV particle suggests that lateral protein–protein interactions, assisted by disulfide bond formation, might play an active role in the budding process of the HCV particle.[6] Besides their role in helping in the transport of HCV envelope glycoproteins to the assembly site, p7 and NS2 might also play additional functions during the assembly process. p7 is indeed also necessary for the final steps of capsid assembly as well as for capsid envelopment.[103]

HCV virion biogenesis is closely related to the VLDL assembly pathway. Indeed, inhibitors of microsomal triglyceride transfer protein (MTP), a protein involved in VLDL biogenesis, block the production of viral particles.[28,29,104] Furthermore, apolipoproteins such as apoE, apoB, apoA1, apoC1, apoC2, and apoC3 can also be found in association with HCV particles[4,11] and the lipid composition of viral particles resembles that of VLDL and LDL with CEs accounting for almost half of the total HCV lipids.[10]

Among HCV particle-associated apolipoproteins, there is a consensus about the involvement of apoE in HCV morphogenesis,[30,105,106] whereas although initially reported as playing a role in HCV assembly and as being associated with HCV virion, apoB may not be essential for HCV particle biogenesis.[30,105–107]

After assembly and budding in the ER, HCV particles are released from cells by transit through the secretory pathway.[107] During this process, HCV virions acquire their characteristic low buoyant density.[24,28] Furthermore, glycans associated with the viral envelope glycoproteins are also modified.[6] Finally, during viral egress, it has also been suggested that HCV particles depend on p7 to neutralize acidic compartments within the secretory pathway.[108]

4.1.4 Structure

Deciphering how E1 and E2 are structured and organized at the particle surface is highly dependent on the structural resolution of the entire E1E2 complex. Indeed, the functional conformation of E1 and E2 has been shown to be interdependent[84–88] highlighting that the resolution of the E1 and E2 structure requires the analysis of the heterodimer as a unique protein entity. However, up to now, the structure of this complex remains unresolved. Beyond the fact that this finding would highlight new structural targets for antiviral drugs' design, this structure will provide precious information on how E1 and E2 may structurally behave during entry and fusion. Over the past years, the resolution of HCV envelope glycoprotein structure has been strongly hampered by several intrinsic properties of these proteins. Soluble E1 or E2 form aggregates in solution, they are heavily glycosylated and present hypervariable region (such as several flexible loops) which seriously hampered the production of analyzable crystal structure.

As an alternative, Krey *et al.* proposed in 2010 an E2 structure model (Fig. 2) based on the determination of the connectivity of its nine disulfide bonds, which are strictly conserved across HCV genotypes. The disulfide connectivity pattern, together with data on the CD81 binding site and reported E2 deletion mutants, enabled the threading of the E2 polypeptide chain onto the structural template of fusion proteins of related flavi- and alphaviruses.[109] Even if this model provided a useful structural basis for further studies on the role of E2 in entry,[110–112] the recent structure of a large part of the E2 ectodomain (called E2 core) did not fit with this previous E2 structure model (Fig. 2). Indeed, E2 core appears as a globular, nonextended protein constrained on itself by several disulfide bridges.[113,114] The structure

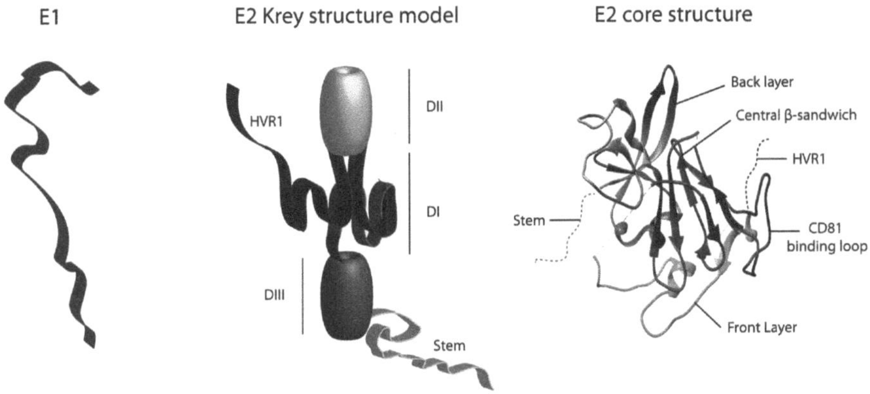

Figure 2 From the E2 Krey structure model to the E2 core structure. Tri-dimensional views of the E2 Krey structure model and of the E2 core structure are shown. The different domains for each structure are labeled. An E1 putative structure is suggested. D, domain; HVR1, hypervariable region 1. HCV E2 core (PDB 4MWF).

could be decomposed as two major sheets: a front sheet composed by a front layer and an Ig-fold domain forming a central β-sandwich region (which represents a common structure among fusion proteins) and a back sheet which harbors an original structure not previously described within envelope glycoproteins. In consequence, even though it is likely that the E2 core structure does not resume the E2 structure within incorporated E1E2 heterodimer complex, this protein likely harbors a very unique and unexpected structure that is clearly distinct from previously known envelope glycoproteins.

4.1.5 E2 functions during virus binding

Similar to flaviviruses E and pestiviruses E2, HCV E2 has been shown to be involved in receptors' recognition and virus binding at the cell surface. More precisely, E2 has been shown to specifically bind two particular cell surface molecules expressed at hepatocyte surface, the SR-BI and the tetraspanin CD81.

Assays using soluble E2 protein, mutagenesis, and anti-E2-neutralizing antibodies inhibiting CD81 binding permitted to precisely identify domains and residues in E2 that are involved in this interaction.[115–125] Although it is known that the proper E2 conformation is dependent on E1, soluble E2 has been shown to bind CD81, highlighting that the folding of the CD81 RBD in E2 does not require E1 to acquire its functionality.[116,120,126] Although several studies identified a large number of E2 residues critical for CD81 binding, the recent resolution of the E2 core structure allowed to precisely define the CD81 binding site within E2. Structural studies revealed that this binding is mediated by a conformational epitope mediated by a particular loop (CD81 binding loop) and few residues located within the front layer and the central β-sandwich.

E2 HVR1 has also been shown to be an important determinant of viral particle binding to SR-BI. Indeed, deletion of HVR1 and antibodies targeting this domain dramatically decrease the cell entry of viruses.[25,50,69,70,74] Despite its important genetic variability among genotypes, infectivity of viral particles has been shown to correlate with the quantity of basic residues present in HVR1.[59] Deletion of HVR1 renders anti-SR-BI antibodies inefficient against viral infection and abrogates SR-BI binding.[25] Moreover, HVR1 seems to be an important shielding domain as its deletion dramatically increases the accessibility of the CD81 binding domain for antibodies.[25] Accordingly, it has been shown that deletion of HVR1 increases sE2 binding to CD81.[50,120] This information suggests that SR-BI and CD81 binding are two tightly related events during virus entry.

4.1.6 From the role of E1 to the importance of E1E2 dialogs

In contrast to E2, E1 functions during virus binding have been poorly characterized. Indeed, there is actually no evidence that E1, nor soluble E1, directly binds to a cell surface molecule during virus entry. Although several residues have been identified in this protein to regulate heterodimer binding,[127] it is likely that these residues mediate heterodimer conformation and the accessibility of E2 RBDs. Accordingly, over the past decade, most of the studies focusing on E1 involvement in binding have been circumscribed to conserved amino acids that impact E1E2 conformation and E2 binding to CD81 or SR-BI. For instance, it has been shown that E1 cysteines influence E1E2 heterodimerization and the accessibility of the CD81 binding site within E2.[88] Besides the role of conserved amino acids, a study recently highlighted the importance of E1E2 dialogs in the binding phenotype of E1E2 complex. Indeed, it has been shown that a strong interplay between less conserved domains of E1 and E2 can regulate binding to both CD81 and SR-BI.[112] Moreover, although soluble E2 does not bind to Claudin-1, another cell surface molecule important for HCV cell entry, it has been shown that the association of E1 with E2 allows E1E2 binding to this molecule in an E1 sequence-dependent manner.[112] The maintenance of critical dialogs between E1 and E2 is tightly linked to the concept of domain coevolution, where a conserved protein function that involved different variable domains requires the coevolution of these domains to maintain the most optimal direct or indirect interaction (or dialog) between these domains.

Altogether, HCV binding seems more regulated by the entire E1E2 complex than by E1 or E2 individually. The E1E2 complex should therefore be more considered as a single protein entity rather than a combination of two proteins with complementary functions.

4.2. E2-CD81 binding engagement

Following attachment to SR-BI, particle early rearrangements trigger the exposure of receptor binding sites located within E2 glycoprotein, thus allowing binding of viral particles to the tetraspanin CD81. Such epitope uncovering could be due directly to particle composition modifications, lipoprotein rearrangements, or E1E2 conformational changes.

The tetraspanin CD81 is an adaptor cell surface molecule expressed on a very large panel of cell types.[128,129] CD81 harbors two extracellular domains, a small and a large extracellular loop (respectively, SEL and LEL) anchored to the cell membrane through four transmembrane domains. By performing binding assays with soluble E2 on a mouse cell line expressing a human cDNA library, the tetraspanin CD81 was the first cell entry receptor critical for HCV entry to be identified.[126] HCV–CD81 binding involves an interaction between the receptor LEL[130] and a conformational region of E2 (mainly composed by a loop) that has been structurally characterized recently.[113,114]

4.2.1 CD81: A critical HCV receptor

Several studies have reported that CD81 downregulation or the use of anti-CD81 antibodies blocks HCVpp, HCVcc, and serum-derived HCV particles' (HCVser) infection.[17,18,69,71,131,21,132,133] However, in contrast to HCVpp, no direct interaction has been observed between naïve HCVcc particles and CD81.[17,18,132] Although HCVpp are highly sensitive to inhibition by soluble CD81-LEL, HCVcc and HCVser have been shown to be, respectively, partially sensitive or totally nonsensitive to such inhibition.[133] It is likely that lipoprotein components associated with HCVcc and HCVser make the E2–CD81 binding site less accessible. This underlines the necessity for these particles to undergo several rearrangements during early attachment steps in order to uncover CD81 binding site. CD81 binding may therefore occur during later stage of viral entry, consistently with the fact that entry of prebound HCVpp has been shown to be inhibited by anti-CD81 antibodies.[134,135] However, E1E2 complex could be precipitated by CD81-LEL in the HCVcc context confirming a potential partial masking.[6] The accessibility can be increased by mutation or deletion in E1 and/or E2 which clearly indicates that the accessibility of CD81 binding site is also dependent on the general conformation of the E1E2 heterodimer or lipoprotein associations.[25,112]

4.2.2 CD81 as a major determinant for HCV-restricted species tropism

Human and chimpanzee liver cells appear to be the major HCV target both *in vitro* and *in vivo*. Important efforts have been done to understand the role of HCV receptors in the restricted tropism of HCV and CD81 has been identified to be an important determinant of this narrow tropism. It has been shown that the human, but not the rodent CD81 isoform, is able to bind soluble E2[134] and allow productive virus entry. Consistently, rodent soluble CD81-LEL does not inhibit HCVcc[134] or HCVser infection.[133] In hamster CHO or murine NIH3T3 cells, the expression of human isoform is crucial for infection.[136] However, expression of rodent CD81 in HepG2 cells (that does not express CD81) conferred susceptibility to HCVpp and HCVcc infection to various levels,[134] suggesting that the narrow tropism mediated by CD81 may be dependent on the cell type in which it is expressed. Moreover, another group was able to achieve viral adaptation of HCVcc Jc1 virus to murine CD81.[137] The three identified envelope glycoprotein mutations conferred to viral particle the ability to slightly infect murine hepatocytes in the absence of any human factor.

4.3. CD81-induced signaling and diffusion of receptor complexes

HCV binding to CD81 is known to trigger multiple signaling pathways and the diffusion of virus–receptor complexes toward site of viral particle internalization. Several reports have proposed that HCV binding to CD81 induces an actin-dependent relocalization of the virus/CD81 complexes toward apical membrane of hepatocytes, where viral particles are thought to be internalized.[138–141] This movement has been reported to be dependent of the CD81 cytoplasmic tail, which activates the MAPK and PI3K/AKT signaling pathways as well as the RhoGTPase family members.[138,142] Accordingly, inhibition of Rho GTPases is able to block HCV entry.

4.3.1 Epidermal growth factor-dependent signaling

The activation of the signaling pathways mentioned below has been shown to be also dependent on the phosphorylation of the Epidermal Growth Factor Receptor (EGFR).[138] Consistently, it has been hypothesized that

EGFR, and also EphA2, another member of the receptor tyrosine kinase (RTK) superfamily, could play an important role for HCV entry and CD81–virus complex lateral diffusion. Indeed, downregulation of these RTKs or use of specific protein kinase inhibitors impaired HCV entry both *in vitro* and *in vivo*[143] and use of RTK-specific ligands such as the epithelial growth factor (EGF) enhanced HCV infection through receptor dimerization and activation of intracellular signaling pathways. Consistently, another group reported in 2012 that HCV binding to CD81 induces the activation of EGFR by promoting its phosphorylation.[144] More recently, another report provided a direct link between EGFR-activated members and virus–receptor complexes by showing that HRas, a membrane-bound GTPase-activated downstream EGFR signaling, interact with CD81 molecules that are associated in plasma membrane with a tight junction protein, Claudin-1.[145] Hence, today, a current model would suggest that CD81 binding likely induces EGFR phosphorylation, then leading to the activation of the GTPase HRas (an enzyme also known as transforming protein p21). HRas would then promote CD81 lateral diffusion within plasma membrane and its stable clustering with Claudin-1 prior subsequent virus internalization.[145]

4.3.2 EWI-2wint

Interestingly, studies aiming to understand the dynamics of tetraspanins during HCV infection have also demonstrated the importance of CD81 lateral diffusion during HCV entry. Tetraspanins form at the cell surface dynamic clusters with several membrane proteins called tetraspanin-enriched areas (TEAs). It has been shown that HCV binding to CD81 is independent of the TEAs but that its interaction with EWI-2wint, a cleavage product of EWI-2, inhibits HCV entry by constraining CD81 into the TEAs.[146–148] EWI-2wint was shown to reduce the global diffusion of CD81 required for HCV infection, promotes its clustering into TEAs enriched in CD81 tetraspanin,[148] and thus impairs its cell surface colocalization with Claudin-1. In consequence, the expression of EWI-2 in most cell lines but not in hepatocytes could contribute to the restricted cell tropism of HCV.

4.4. A critical role of tight junction proteins

Subsequent to EGFR activation, CD81 lateral diffusion is thought to promote the association of virus–CD81 complexes with tight junction proteins.

Such receptor clustering has been shown to be critical for subsequent viral particle internalization.

The identification of CD81 and SR-BI as HCV receptors highlighted that several cell lines expressing ectopically these two molecules still remained nonpermissive for HCV entry, thus suggesting the existence of other critical entry factors. Through screening of a cDNA library derived from highly permissive Huh7.5 cells into nonpermissive HEK293t and NIH3T3 cells expressing CD81 and SR-BI, the tight junction proteins Claudin-1 and Occludin were identified as the last two cell entry factors allowing to recapitulate virus entry in nonpermissive cell lines.[136,149] In contrast to CD81 and SR-BI, these molecules are expressed into the hepatocyte apical membranes, as members of cellular tight junctions that are important mediators of the intercellular junctions between hepatocytes and essential for the maintenance of the large cellular plate constituting the liver tissue.

4.4.1 Claudin-1

Claudin-1 is composed of two extracellular domains, EL1 and EL2 (for extracellular loop 1 and 2) anchored to the cell membrane through four transmembrane domains.[150] Downregulation of Claudin-1 into Huh7.5 cells was reported to drastically inhibit HCV infection.[149] Expression of Claudin-1, -6, and -9, but not Claudin-7, into nonpermissive cell lines rendered these cells permissive for HCV infection,[149] through the mediation of a highly conserved motif located within Claudin-1 EL1.[149] Antibodies targeting EL1-flag tags were able to neutralize HCV entry at a postbinding step, suggesting that Claudin-1 likely plays a role subsequently to CD81 and SR-BI binding.[149] It was demonstrated that the localization of Claudin-1 into the tight junctions was critical for HCV infection,[151,152] suggesting that viral particles migration into these junctions may be important for virus entry. Indeed, CD81/Claudin-1 complexes colocalize in close proximity of intercellular junctions during HCV infection of Huh7.5 cells, likely through a Claudin-1 EL1–CD81 LEL molecular interface.[139–141,153] However, these receptor complexes seem to remain localized in basolateral intercellular junction and not in apically located tight junctions, which is consistent with a viral entry into hepatocyte cell layers via the sinusoidal blood.[140] Importantly, HCV has been shown to induce internalization of CD81–Claudin-1 complexes[141] and use of anti-Claudin-1 antibodies able to neutralize CD81–Claudin-1 interaction was shown to neutralize HCVcc infection, thus demonstrating the importance of such complex for HCV entry.[154] Altogether, the literature may support a model whereby the formation of CD81/Claudin-1 complexes following CD81 lateral diffusion could likely enhance HCV association with cell receptor complexes and, hence, promotes virus migration toward tight junctions and internalization.

In contrast to CD81, mouse Claudin-1 has been shown to render HEK293T cells susceptible to HCV entry,[149] suggesting that Claudin-1 may not be a major determinant of HCV-restricted species tropism. However, when expressed in human hepatoblastoma cell lines HuH6 (that express low levels of endogenous Claudin-1 but are able to replicated HCV RNA), this protein displayed moderate HCVpp and HCVcc entry although expression of human, rat, and hamster Claudin-1 allowed efficient virus entry.[155] Thus, similar to CD81, entry mediation ability of mouse Claudin-1 may be cell type dependent.

Regarding a potential interaction between viral particles and Claudin-1, it has been clearly shown that sE2 does not bind to Claudin-1, whereas anti-Claudin-1 antibodies are able to abrogate sE2 binding to Huh7.5 cells.[149,154] However, a recent report[112] showed that E1E2 complexes harbored by HCVpp, but not soluble E2, are indeed able to bind Claudin-1 in an E1-dependent manner. This suggests that E1 may structurally mediate, once in complex with E2, an E2 or E1E2 Claudin-1 conformational binding site. Such binding could then strengthen the association between viral particle and receptor complexes, thus enhancing viral particles' lateral diffusion and internalization.

4.4.2 Occludin

In contrast to Claudin-1, the critical function of Occludin during HCV entry remains to be precisely characterized. Occludin is also composed of two extracellular loops, EL1 and EL2, despite a larger size than Claudin-1. It has been suggested that Occludin EL2 is involved in HCVpp-mediated entry as well as in E2 binding. Indeed, an interaction between E2 and Occludin has been observed through microscopic confocal analyses and coimmunoprecipitation assays.[152,156,157] However, the relevance of this interaction is questionable so far as it has only been observed in the intracellular reticulum endoplasmic and not in the context of cell surface. A recent study also suggested a potential interaction between viral particles and Occludin, based on the observation that HCV types and subtypes display differences for Occludin utilization.[158] Interestingly, synchronized infection assays showed that Occludin acts subsequently to CD81 and Claudin-1 during HCV entry and that downregulation of Occludin decreases both HCV entry and glycoprotein-mediated cell fusion. These observations are in agreement with the notion that tight junction proteins likely act as late mediators of virus entry and are critical for virus internalization and, hence, membrane fusion.

In contrast to Claudin-1, Occludin has been shown to be a critical determinant for the restricted species tropism of HCV with CD81. Indeed, six residues located within EL2 have been shown to mediate HCV host tropism similar to the CD81 LEL.[136,159]

As discussed above, since Claudin-1 and Occludin are tight junction proteins, it was first believed that after binding to CD81, HCV particles would migrate to tight junction areas for internalization. However, Claudin-1 associates with CD81 at the basolateral membrane of polarized HepG2 cells, whereas pools of Claudin-1 in tight junctions show a minimal association with CD81.[139,140] Furthermore, disruption of tight junctions by calcium depletion of polarized HepG2 cells increases HCV entry.[160,161] Finally, live cell imaging experiments indicate that HCV particles do not migrate to cell–cell contact areas after binding, supporting the notion that tight junctions might not be required for HCV entry, at least in nonpolarized hepatoma cells.[162]

5. POSTBINDING STEPS AND VIRUS FUSION

5.1. Endocytosis and internalization

Viral particles' interactions with receptors, coreceptors, and cofactors lead to major molecular rearrangements at the plasma membrane. These rearrangements ultimately result in internalization of viral particle–receptor complexes through clathrin-dependent endocytosis. The fusion within the endosomal cell compartment between viral and host membranes, mediated by E1 and E2, is the final step of HCV entry that allows the release of the viral RNA into the cell host cytosol.

5.1.1 Endocytosis

Viruses have developed an important number of mechanisms to invade a cell and release their genetic material into the host cell cytosol. Among them, the hijacking of the cell endocytic pathway is commonly used by most of the known viruses. HCV is known to be internalized through clathrin-mediated endocytosis, a dynamin-dependent pinocytosis endocytic mechanism. During this process, particle attachment induces the inward budding of the plasma membrane and the formation of a clathrin pit that internalizes the particle–receptor complex. It has been shown that downregulation of the clathrin heavy chain inhibited HCVpp and HCVcc infection.[163] Moreover, internalization of CD81–Claudin-1 complexes through clathrin-dependent endocytosis has been observed during HCV infection, consistently with the fact that CD81–Claudin-1 association is critical for virus internalization and that virus internalization occurs prior fusion.[139–141,154] Following endocytosis, HCV-receptor complexes migrate toward RAB5A-containing endosomal compartments[107,141,164] where fusion is thought to occur, allowing release of viral RNA.

5.1.2 The transferrin receptor: A role to be defined

Transferrin receptor (TfR) 1 and 2 are transmembrane proteins expressed in the liver, involved in the iron uptake from blood to intracellular compartment. Interestingly, HCV infection in patients is often characterized by an iron overload. Consistently, a recent study found that HCV alters the expression of TfR1 within infected cells, thus deregulating iron homeostasis.[165] Moreover, TfR1 downregulation and antibodies-mediated neutralization against TfR1 inhibited both HCVcc and HCVpp infection, suggesting that TfR1 is important for glycoprotein-dependent viral entry.[165] However, the precise mechanism by which TfR1 participates to viral entry remains to identify. In a time-course study, the anti-TfR1 antibody lost its inhibitory activity after anti-CD81 blocking, which suggests that TfR1 interferes with HCV entry at a step after docking onto CD81.[165] Interestingly, the silencing of the TfR-1 trafficking protein, required for TfR1 internalization, was shown to inhibit HCV infection, which may suggest a potential role for TfR1 in virus internalization. A model where TfR1 would act in concert with tight junction proteins during receptor clustering and virus internalization remains to be defined.

5.2. Cell-to-cell transmission

It has been shown that HCV can propagate from infected cells to neighboring cells independent of the classical entry and internalization pathways used by circulating viral particles.[123,166–168] This cell-to-cell transfer of virus infection is thought to be an important route for efficient virus spread within liver tissue, as it constitutes a potential route to escape from the host neutralizing response.[167,168] SR-BI, Claudin-1, and Occludin-1 have been shown to be important for virus transmission between hepatocytes.[123,166–168] Consistently, a very recent study demonstrated that anti-SR-BI antibodies could bock cell-to-cell transmission. Moreover, authors showed that specific mutations in E1 and E2 were responsible of increased cell-to-cell spread ability, through reduction of SR-BI usage. However, this enhanced ability was presented as unlikely to occur *in vivo*, as mutations also increased virus sensitivity to neutralizing antibodies.[168]

In contrast with the other receptors, the role of CD81 in this process remains a matter of debates. Two studies previously reported that CD81 is dispensable for cell-to-cell transmission[123,169] although two others, more recent, reported that CD81 indeed plays a role in cell-to-cell transfer.[167,168] A recent study notably showed that expression of CD81 on producer cells modulates HCV cell-to-cell transmission.[168] In addition, HCV cell-to-cell spread has also been shown to be less dependent on TfR1.[165]

More recently, some studies have underlined the importance of target cell-to-cell spread for efficient virus clearance. Notably, a recent report showed that the Niemann–Pick C1-like 1 cholesterol absorption receptor (NPC1L1), recently shown to be critical for HCV cell entry (see below),[170] is essential for cell-to-cell spread.[171] Interestingly, this study reported that secretion of viral particles through the VLDL secretory pathway is not involved in cell-to-cell transmission. Importantly, this report showed that the combination of inhibitors targeting host factors involved in both cell-free and cell-to-cell transmission can lead to synergistic effects when combined with interferon, although no synergy was observed when only cell-free transmission was blocked.[171]

Additionally, another study showed that cell-to-cell transmission is the main route of escape for DAA-resistant HCV variants.[172] Thus, these two studies highlighted the importance of the cell-to-cell transmission route during virus infection and underlined that both cell-to-cell transmission (via HCV receptors–targeting inhibitors) and cell-free transmission (via DAAs or interferon) mechanisms should be inhibited to allow efficient and fast virus clearance.

5.3. Membrane fusion

Enveloped viruses release their genetic materials within cell cytosol by inducing fusion of their membrane with the host cell membrane. This mechanism can be generalized as a process including (i) conformational changes of the fusion protein(s), (ii) insertion of a fusion peptide into the host membrane, (iii) lipid mixing of external membrane leaflets (hemifusion), and (iv) full fusion of the two membranes, which ultimately unifies cell and virus membranes. Beyond such a general process, viruses harbor a large diversity of fusion mechanism, through different fusion protein structures, conformational changes, or cellular fusion sites.[83] Some viruses only require a prior attachment step and receptor-induced conformational changes of their envelope fusion protein(s) for fusion induction. This allows the merging of the virus membrane with the plasma membrane or an endosomal membrane. In contrast, some viruses require an acidification step to trigger the fusion protein(s) conformational changes and membrane merging. This results on membrane fusion inside endosome following a required step of virus internalization.

The use of bafilomycin A1, that affects endosomal acidic environment by preventing reacidification, has been shown to block HCV infection,[164,173] highlighting that the release of viral RNA through membrane fusion is dependent of endosomal acidification. The degree of acidification required for fusion is still not completely characterized. From experiments using the HCVpp model, it was proposed that HCV membrane fusion occurs in an early endosomal compartment as the use of transdominant-negative mutants indicate that it is dependent of RAB5 but not RAB7.[164] This is consistent with the idea that only a mild acidification is necessary to trigger fusion through action of envelope glycoproteins.[39] In cell–cell fusion assay, a weak fusion is even observed at neutral pH,[40,174] which questions the strict necessity of the pH drop and the possibility that cell-to-cell fusion may not require acid environment. However, HCVcc–liposome fusion assays[175] suggested that the pH range for HCVcc fusion was broad but shifted toward lower pH values (from pH 6.3 to 4.0). It is thus possible that HCV behaves in a slightly different manner, depending on the HCV fusion models. A plausible explanation would be that optimal HCVcc fusion requires additional factors or receptors that are lacking in some *in vitro* fusion assay. Indeed, in some experimental conditions, the secreted HCVcc virions are pH resistant and, therefore, not competent to mediate fusion suggesting that low pH-dependent fusion can be triggered in a time- and temperature-dependent manner only upon cell binding.[176] Recent evidence suggested that the interaction with CD81 may trigger a conformational change permitting low pH-induced membrane fusion.[173] This is reminiscent of avian leukosis retrovirus (ALV).[177] In contrast to typical pH-dependent virus, receptor interaction plays an essential role in priming ALV envelope glycoprotein for subsequent low-pH triggering. Thus, this priming step may be a way for HCV to control the fusogenicity of E1E2 glycoproteins and to prevent premature fusion within the producer cell upon virion secretion. However, one study argues that the pH sensitivity of E1E2 during virion egress may not be similar to that of E1E2 displayed onto the released viral particles.[108]

Altogether, the precise mechanism by which HCV mediates membrane fusion still remains poorly understood. Indeed, the lack of information regarding the structure of the E1E2 complex strongly limits our understanding of the membrane fusion mechanism. Importantly, the requirement of receptor-dependent priming steps, viral particle rearrangements, as well as low-pH endosomal environment potentially highlights that this process is modulated through the subsequent action of several host factors during HCV entry, hence suggesting that HCV fusion is a complex and perhaps unique mechanism among enveloped viruses.

5.3.1 Particle fusion-dependent rearrangements

From the early to the late steps of HCV entry, several molecular triggers are thought to prime viral particles for subsequent membrane fusion. Up to now, three main molecular triggers can be listed as priming elements of HCV membrane fusion: the SR-BI interaction with HVR1, the E2 binding to CD81, and the NPC1L1 lipid transfer activity (see below).

5.3.1.1 Early membrane fusion-dependent rearrangements

As SR-BI is able to bind virus-associated lipoprotein components during early steps of entry, several studies have attempted to understand the role of the interaction between HVR1 and SR-BI during entry. HDLs have been shown to enhance HCVpp infection by facilitating virus entry.[42,45] Interestingly, this enhancement was dependent on HVR1 although no direct interaction between HDL and HCVpp and no increase of HCV binding to SR-BI were observed.[42,45,178] Hence, it was proposed by other studies that the transfer of apolipoprotein C-1 (a major component of HDL that is also present on HCVcc particles) from HDL to HCV particles by SR-BI likely enhances viral fusion.[46,43] Consistently, a single mutation in HVR1, that abrogates HCVcc infection enhancement but not infectivity, induces a decrease of the infectivity of HCVcc from the lower intermediate density fraction,[74] thus underlining that such enhancement is dependent on both E2-SR-BI binding and SR-BI lipid transfer function. Moreover, as mouse/murine SR-BI does not support E2 binding and an enhancement function, the generation of chimeric human/murine SR-BI constructs made possible to identify the first domain of SR-BI as mediator of the enhancement function. Altogether, the E2–SR-BI interaction seems to represent an important mechanism that contributes to the high-infectious ability of the intermediate density fraction in contrast to other density fractions that do not undergo such particle rearrangements. A potential assumption would be that the apoCI transfer from HDL to HCV, by modifying the lipoprotein profile of viral particles, facilitates further fusogenic rearrangements of the viral particle.

5.3.1.2 Postattachment membrane fusion-dependent rearrangements

An interesting study in 2011 studied for the first time the role of CD81 in E1E2-mediated virus fusion.[173] The authors confirmed first that preexposure of E1E2 complex to low pH enhances virus entry and fusion and does not induce irreversible conformational changes. However, they were able to show that preincubation of viral particles with CD81 human LEL triggers irreversible refolding when exposed to low pH, thus inhibiting subsequent HCV infection. Interestingly, preincubation of viral particles with the soluble CD81 LEL alone was able to enhance virus entry and fusion more effectively than low-pH exposure. The induced E1E2 conformational changes were distinct from the refolding induced by low-pH exposure and were shown to form slowly and be transient in the absence of target membrane. CD81-primed E1E2 complex was stable in pH-neutral intracellular compartments through interaction with endosomal membrane, thus allowing their fusion competency to be maintained. Indeed, authors showed that human CD81 LEL promotes soluble E2 binding to liposomes and that the induced conformational changes were also able to perform low pH-dependent fusion.

Altogether, this study nicely demonstrated that CD81 binding likely triggers E1E2 conformational changes that are then stabilized after viral particle internalization, suggesting that E1 and/or E2 insertion within target membranes may occur quickly after internalization. These changes would then prime further E1E2 conformational changes during endosomal acidification, leading to membrane merging. Although this study highlights the critical role of E2 during fusion, this study does not bring major evidence that E2 would be the HCV fusion protein. However, authors showed that all the HCV entry steps are intimately related and highlight fusion as a complex mechanism which still needs to be better understood.

5.3.1.3 Late membrane fusion-dependent rearrangements

NPC1L1 is a large cell surface protein composed of 13 transmembrane domains mainly expressed on the apical surface of intestinal enterocytes and human hepatocytes. NPC1L1 acts as a cholesterol-sensing receptor and is responsible for cellular cholesterol absorption and cholesterol homeostasis. The involvement of this protein in the cholesterol uptake, its preferential expression on hepatocytes, highlighted a potential role for NPC1L1 in HCV entry. Indeed, very recently, a study demonstrated that

NPC1L1 is an active mediator of the HCV cell entry process. Authors observed that HCVcc-infected Huh7 presented a downregulation of NPC1L1 expression and that downregulation of this receptor significantly impaired HCVcc infection.[170] Use of anti-NPC1L1-specific antibodies significantly inhibited HCV infection and entry. Consistently, using a fluorescent-based HCVcc fusion assay, NPC1L1 has been shown to be critical for postbinding step events and for fusion.[170] The use of ezetimibe, a direct inhibitor of NPC1L1 internalization already used as a cholesterol-lowering medication, was able to impair propagation of HCV *in vitro* and *in vivo* suggesting that the biological function of NPC1L1 is the mediator of the NPC1L1-mediated cell entry.[170] Indeed, the authors demonstrated the existence of a correlation between the cholesterol content of viral particles and their dependence on NPC1L1 during entry.[170] However, how NPC1L1 precisely mediates virus entry remains unknown. A possible assumption would be that NPC1L1 lipid transfer activity could rearrange lipid-associated viral particles during postbinding step. Consistently, cholesterol of the target membranes has been found to have a strong promoting effect not only on the membrane fusion capacity of flaviviruses[179,180] but also on HCV fusion.[39] NPC1L1-dependent lipid rearrangements could occur subsequently to virus internalization as a potential priming step for virus fusion and contribute to fusion-related particle rearrangements initiated by E2–SR-BI interaction.

5.3.2 The mechanism of HCV fusion

5.3.2.1 The class of fusion protein

Based on their resolved structure, fusion proteins are commonly divided into three distinct classes. Class I proteins are mainly composed of α-helical structures and harbor a fusion peptide located at the N-terminal of the protein. They are assembled as trimeric structures onto virus surface, both in their pre- and postfusion conformational states,[83] but are concealed by an attachment protein in their prefusion conformation state. Fusogenic conformational changes can be initiated by binding of the attachment protein to a cellular receptor or by acidification of an endosomal compartment, leading to the exposure of the fusion peptide.

In contrast to class I fusion protein, class II fusion proteins are three-domain proteins mainly composed of β-sheet secondary structures[181,182] that harbor a fusion peptide located in an internal loop. In their prefusion state, class II fusion proteins are dimeric structures lying down parallel to the lipid bilayer, whereas in their post fusion state, they protrude as trimers from the viral particle surface.[181,182] Commonly, the class II fusion proteins require a low-pH induction to induce membrane merging.[181,182] For flaviviruses, acidification of endosomal compartment induces homodimer rotations that extend to form subsequent trimeric structures. Insertion of the fusion peptide and folding over of several trimeric structures is then required to induce membrane bridging and fusion.[181–183]

A third class of fusion protein has been recently identified in particular viruses such as the herpes simplex virus 1 (HSV-1) or the vesicular stomatitis virus (VSV). They harbor a prefusion trimeric structure with a central domain mainly composed of α-helices and their fusion peptide is divided into two domains, located in two fusion loops composed of β-sheets.[184] Acidification and/or attachment to viral receptors are thought to initiate conformational changes of class III fusion proteins.

5.3.2.2 The HCV fusion protein: An unusual suspect

An E2 model structure proposed by Krey *et al.*[109] strongly suggested that HCV E2 presented structural features of a class II fusion protein (Fig. 2), thus underlining HCV E2 as the HCV fusion protein. Indeed, similar to Dengue virus E surface protein, HCV E2 was the largest and the immunodominant envelope glycoprotein. However, HCV E2 harbors significant differences from E such as a high number of glycosylations and a putative domain II constituted of only one loop. HCV E1 was thus thought to harbor a similar function as Dengue PrM by chaperoning E2 functional conformation during heterodimerization. Subsequent to the publication of the Krey model, experimental assays have been conducted to challenge the proposed E2 structure. Importantly, some studies strengthened the relevance of the Krey model,[110] but no evidence confirmed the HCV putative domain II as carrying a fusion peptide despite the identification of other fusion-dependent domains within alternative domains of E2.[40] As mentioned below, several studies have also focused on the role of E1 during membrane fusion. Importantly, most of these studies pointed out important domains or residues within E1 that modulate virus fusion.[127,40,185] Thus, even though HCV E2 was (consensually) considered as the likely HCV fusion protein, these studies also suggested that E1 has an active role to play during virus entry and membrane fusion, or at least a role that is not restricted to chaperoning E2.

In 2013, a major breakthrough occurred with the resolution of the structure of the E2 envelope glycoprotein of BVDV-1,[186,187] a pestivirus member of the *Flaviviridae* Family (Fig. 3A). BVDV-1 harbors two envelope glycoproteins E1 and E2 of similar size as HCV E1 and E2. In consequence, their functions and structures were previously thought to be strongly similar. However, BVDV-1 E2 harbors a strongly extended structure mainly composed of β-strands that strongly contrast with class II fusion protein (Fig. 3B).[186,187] The protein, divided in four domains, does not harbor any fusion loop and is poorly responsive to low pH, undermining its ability to be a fusion protein. In contrast, E2 distal domains present Ig-like fold structures consistent with a potential attachment function.[186,187] This assumption is supported by the fact that a soluble form of E2 from another pestivirus, CSFV, has been shown to inhibit viral entry of both BVDV and CSFV.[188] In parallel, the most distal domain (Domain A or DA) appears to rearrange after low-pH induction, thus suggesting that this domain may conceal an E1 fusion peptide (Fig. 3B). During fusion, low-pH environment would disorder DA and unmask the E1 fusion peptide as for alphaviruses. Interestingly, the authors of these two studies extrapolated their results to HCV, challenging the relevance of the Krey model and suggesting that, as BVDV-1, HCV E1 could carry the HCV fusion peptide. Another striking similarity between HCV and pestivirus is the requirement for both viruses to undergo a postattachment priming step to ensure a correct fusion process.[173,189] Indeed, as flavivirus fusion does not require such priming, this strongly suggests that the fusion process of hepacivirus and pestivirus share similarities and is likely distinct from the one of flaviviruses.

The very recent resolution of the HCV E2 core structure has strengthened these assumptions. As mentioned below, despite harboring a central Ig-fold and β-strand sheets, which is common in class II fusion protein, HCV E2 core also harbors original domains such as its back layer (or back sheet domain) previously undescribed within a fusion protein (Fig. 2).[113,114] More importantly, its globular and nonextended structure makes unlikely a potential ability for E2 to rearrange and fuse membranes together. Consistently, a recent study confirmed that the E2 putative fusion peptide (located in domain II), previously highlighted by the Krey model, is indeed not involved in virus fusion but rather in modulating virus binding.[125]

Hence, similar to BVDV-1 E2, the apparent structure of HCV E2 undermines the potentiality for this protein to be a fusion protein and strongly suggests that the HCV and BVDV-1 E1 proteins are indeed fusion proteins (Fig. 3B). In contrast, the E2 proteins of HCV and BVDV-1 would thus represent their binding proteins. However, it is important to point out that the role of E2, for each virus, cannot be simply restricted to virus

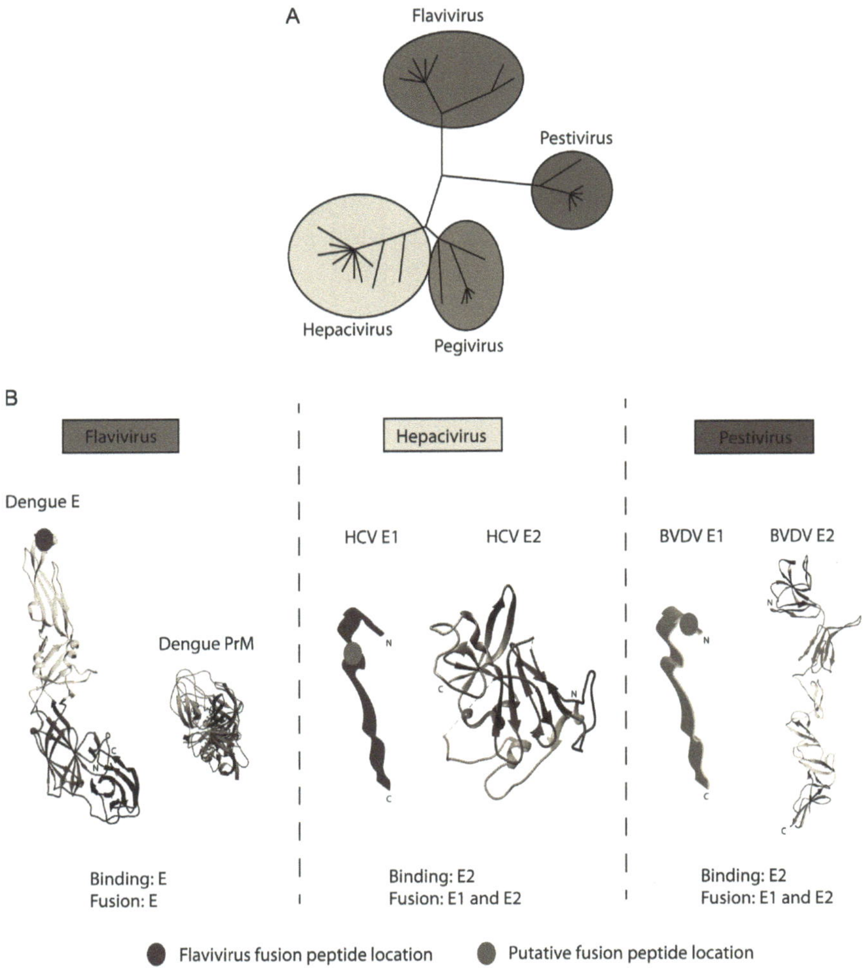

Figure 3 *Flaviviridae* envelope glycoproteins. (A) Schematic representation of the phylogenetic tree of the *Flaviviridae* family. Each genus is labeled with a particular color. Flavivirus, orange; pestivirus, blue; pegivirus, green; hepacivirus, yellow. (B) Envelope glycoproteins of flavivirus (Dengue E and PrM), hepacivirus (HCV E1 and E2), and pestivirus (BDVD-1 E1 and E2) are compared. Flavivirus fusion peptide location is highlighted as a large red dot within E domain II. HCV and BVDV E1 putative structures are suggested as well as the putative location of their fusion peptide (large blue dot). Known or putative functions for each protein are indicated below the protein structures. HCV, Hepatitis C Virus; BVDV, Bovine viral diarrhea virus. Dengue E (PDB 1K4R), Dengue PrM (PDB 3C6E), HCV E2 core (PDB 4MWF), and BVDV-1 E2 (PDB 4JNT). (See the color plate.)

binding. Indeed, structural resolution to BVDV-1 E2 has shown that rearrangement of the E2 distal domain following low-pH induction may be necessary to unmask a potential E1 fusion peptide,[186,187] thus making the BVDV-1 fusion a process mediated by the entire E1E2 heterodimer (Fig. 3B). Similarly, HCV E2 residues involved in fusion have been previously identified.[112,40] E1E2 dialogs, as well as the interaction between E2 and CD81, have also been shown to regulate virus fusion, thus highlighting a potential, active involvement of HCV E2 during the HCV fusion process.[112,173] It is however still unclear whether E2 can rearrange individually to low pH as BVDV-1 E2 can. Even though E2 core has been shown to not respond to acidification,[114] it is worth noting that E2 core behavior does not likely resume the behavior of the full E2 protein in a low-pH environment. Altogether, this may indicate that HCV fusion is a complex mechanism strongly dependent of the E1E2 interplays and rearrangements rather than being mediated by a single glycoprotein.

It has been suggested in recent reviews that HCV and BVDV-1 E1 may represent a new structural class of fusion protein,[77] according to their close genetic proximity and to the potential novel fusion mechanism they could mediate. However, there is actually no evidence that these two proteins share a similar structure and fusion mechanism. Indeed, as BVDV-1 and HCV E2 display dissimilar structural organizations, the potentiality that E1—from either virus—also harbors structural disparities and mediates a distinct fusion mechanism cannot be excluded. Moreover, even though BVDV-1 and HCV E1 would share a similar structure (Fig. 3B), the importance of E1E2 interplay during fusion for each virus, combined with the fact that their E2 proteins are structurally different, supports the assumption that HCV and BVDV-1 fusion are likely distinct mechanism. The structural resolution of the E1E2 complexes of both HCV and BVDV-1, in their pre- and postfusion conformational states, is needed to answer to these questions. Notably, how the E1 fusion peptide is concealed (by an E2, E1, or E1E2 conformational domain) and unmasked during HCV fusion remains a key element to understand this mechanism.

As a conclusion, even though a lot still needs to be understood regarding the molecular mechanisms governing the HCV fusion process, it is now certain that this phenomenon is not mediated by a single glycoprotein, but rather by an entire heterodimer complex which appears as a nondissociable, functional protein unit.

5.3.2.3 A model for HCV fusion

Following internalization, fusion-primed heterodimers are thought to have their conformations stabilized through insertion within endosomal membranes.[173] After having reached RAB5A-containing endosomal compartments, viral particles may then be exposed to an acidified environment, thus inducing additional, irreversible conformational changes to fusion-primed heterodimers. These induced rearrangements are thought to induce the exposure of a putative E1 fusion peptide. How this peptide is concealed and how structural rearrangements occur during acidification, however, remain to be discovered.

The insertion of the putative fusion peptide could then initiate the fusion process. Regarding recent findings, the merging of viral and endosomal membranes could be likely mediated by a refolding of the entire heterodimer complex (as inferred from results with BVDV-1) through interdependent, drastic structural rearrangements of both E1 and E2. Indeed, E1–E2 coevolution networks have been shown to be critical for E1E2-mediated fusion.[112] Inside these networks, the back layer domain of E2 has been shown to be an important effector of the global E1E2 refolding during membrane fusion.[112] Following insertion of the fusion peptide, this domain could be an important effector of the E2 structural rearrangements, which would then induce a drastic E1 fold back. Such fold back could finally allow membranes to get in close proximity and, subsequently, to undergo lipid mixing.

Up to now, no information is available regarding the number of heterodimer complexes that are required to ensure a complete membrane merging. However, this question also remains strongly linked to the lack of data regarding the structural organization of HCV heterodimer as displayed onto viral particles.

As a conclusion, although the HCV fusion process has started to reveal a small part of its secret via the resolution of the E2 core structure, this mechanism still remains to be fully understood, in both a structural and dynamic context using a full E1E2 heterodimer.

6. CONCLUDING REMARKS

HCV entry is a process with a rare degree of complexity (Fig. 4). This complexity finds its origin within the combination of several features, all interdependent, that makes HCV entry a unique mechanism within the viral world. One of the most important factors is likely the strong heterogeneity of viral particle composition. Due to an important, multifaceted interplay

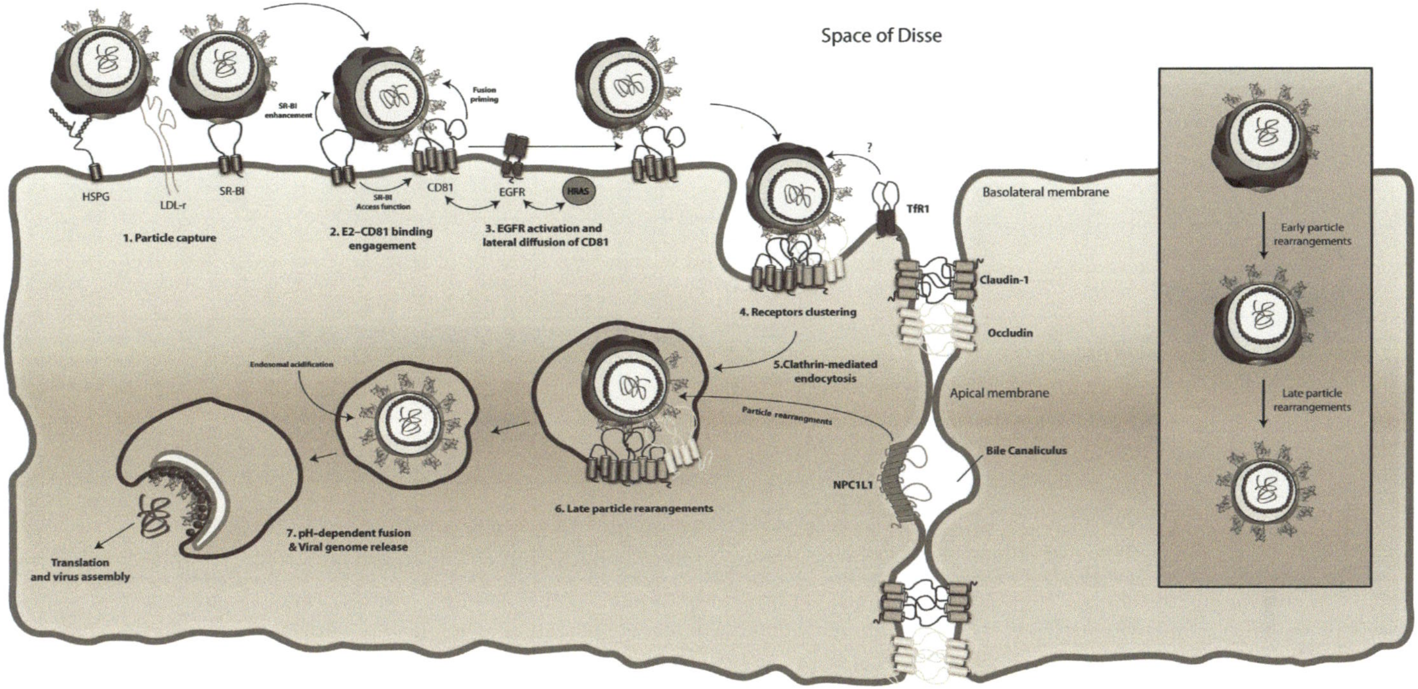

Figure 4 Proposed model of HCV entry. Hepatocytes (yellow-orange; light gray in the print version), their membrane (brown; gray in the print version), and the different HCV receptors, cofactors, and cell host components involved in HCV entry are indicated. The main steps of the HCV cell entry process are presented in bold (from 1 to 7). For each main step, substeps are mentioned when relevant. Arrows indicate the directional pathways used by lipo-viral particles during entry and may be associated with a step or with a substep. Lipo-viral particles attach to hepatocytes surface through interaction between apolipoproteins and HSPG, LDL-r, or SR-BI surface molecules (1). This attachment allows the engagement

(Continued)

between host lipid metabolism and HCV, the lipid composition of viral particles guides the early attachment steps of virus entry.

Another important feature is the great diversity of hepatocyte receptors and cofactors that mediate virus entry. The requirement for such a panel of molecules makes HCV entry dependent of the successive activity of several host proteins and thus sheds light on the origins of the restricted entry tropism of HCV to liver cells. Hence, HCV entry appears as a dynamic, interdependent multistep process. However, to be better understood, its spatiotemporal characterization remains today a considerable challenge.

Another complex feature of HCV entry is the nature of the E1E2 heterodimer. E1 and E2 appear now as two coevolving subunits of a single functional complex, which thus makes their individual analysis of limited relevance. This apparent complexity and originality, combined with the absence of structural resolution of the E1E2 complex and the lack of information regarding their organization onto the virus surface, strongly impact our understanding of HCV entry.

Lastly, the HCV fusion mechanism represents today one of the most elusive entry features of HCV. Indeed, its complete characterization is tightly linked to our understanding of the structural rearrangements of the E1E2 heterodimers. However, recent studies have started to pave the way toward a better elucidation of the HCV fusion mechanism. The current model suggests that E1, which would be the fusion protein, could mediate the merging

Figure 4—Cont'd of viral particles to CD81 and the activation of the EGFR receptor signaling (2). Lateral diffusion of the CD81–HCV complexes toward apical membrane (3) is mediated by EGFR and HRas signaling. This movement induces the association of CD81 with Claudin-1 (4), likely strengthened by a direct interaction between E1E2 and Claudin-1. Receptor clustering allows viral particle internalization (5) and migration toward RAB5-endosomal compartments through clathrin-mediated endocytosis (6). The role of the TfR1 is not known. Early (mediated by SR-BI and CD81 E2 interactions) and late (likely mediated by NPC1L1) fusion-dependent particle rearrangements allow particle priming for the subsequent fusion process. Endosomal acidification induces exposure of a putative E1 fusion peptide via E1 or E1E2 refolding. Dramatic E1–E2 global rearrangements allow heterodimers to fold over and the subsequent membrane merging (7), thus leading to the release of the viral RNA into the host cell cytosol. Evolution of particle rearrangements during entry is highlighted at the right of the cell entry model. For details regarding the different components of the lipo-viral particles, see Fig. 1. HSPG, heparan sulfate proteoglycan; LDL-r, low-density lipoprotein receptor; SR-BI, scavenger receptor B-I; EGFR, epidermal growth factor receptor; NPC1L1, Niemann–Pick C1-like 1 cholesterol absorption receptor; TfR1, transferrin receptor 1;?, unknown function.

of membranes via a strong interplay with E2. Thus, the potential exclusive characteristics of the HCV heterodimer may underline HCV fusion as a complex, but also unique, process among enveloped viruses.

In the future, the resolution of the structure of the E1E2 complex will allow us to better clarify the E1E2-mediated entry mechanism and hence, HCV fusion. Even though the strong coevolution networks between E1 and E2 likely confer a certain functional plasticity to the HCV envelope, these networks are also involved in the maintenance of critical E1E2 functional dialogs via purifying selection. Thus, the broad detection and analysis of the E1E2 coevolution networks may thus represent an interesting opportunity to improve our understanding, structurally and functionally, of the HCV heterodimer dynamic rearrangements during virus entry.

ACKNOWLEDGMENTS

We apologize to all colleagues whose work could not be referenced because of length restriction.

Work in our laboratory was supported by the French National Agency for Research on AIDS and Viral Hepatitis (ANRS), the European Research Council (ERC-2008-AdG-233130-HEPCENT), and the LabEx ECOFECT (ANR-11-LABX-0048) of Université de Lyon.

We are grateful to our co-workers, colleagues and members of the EVIR Team in the CIRI and UMR754 UCBL INRA ENVL EPHE for their encouragements and advices.

REFERENCES

1. Andre P, Komurian-Pradel F, Deforges S, et al. Characterization of low- and very-low-density hepatitis C virus RNA-containing particles. *J Virol.* 2002;76:6919–6928.
2. Thomssen R, Bonk S, Propfe C, Heermann KH, Kochel HG, Uy A. Association of hepatitis C virus in human sera with beta-lipoprotein. *Med Microbiol Immunol.* 1992;181:293–300.
3. Thomssen R, Bonk S, Thiele A. Density heterogeneities of hepatitis C virus in human sera due to the binding of beta-lipoproteins and immunoglobulins. *Med Microbiol Immunol.* 1993;182:329–334.
4. Catanese MT, Uryu K, Kopp M, et al. Ultrastructural analysis of hepatitis C virus particles. *Proc Natl Acad Sci USA.* 2013;110:9505–9510.
5. Vieyres G, Dubuisson J, Pietschmann T. Incorporation of hepatitis C virus e1 and e2 glycoproteins: the keystones on a peculiar virion. *Viruses.* 2014;6:1149–1187.
6. Vieyres G, Thomas X, Descamps V, Duverlie G, Patel AH, Dubuisson J. Characterization of the envelope glycoproteins associated with infectious hepatitis C virus. *J Virol.* 2010;84:10159–10168.
7. Chang KS, Jiang J, Cai Z, Luo G. Human apolipoprotein E is required for infectivity and production of hepatitis C virus in cell culture. *J Virol.* 2007;81:13783–13793.
8. Meunier JC, Russell RS, Engle RE, Faulk KN, Purcell RH, Emerson SU. Apolipoprotein c1 association with hepatitis C virus. *J Virol.* 2008;82:9647–9656.

9. Nielsen SU, Bassendine MF, Burt AD, Martin C, Pumeechockchai W, Toms GL. Association between hepatitis C virus and very-low-density lipoprotein (VLDL)/LDL analyzed in iodixanol density gradients. *J Virol.* 2006;80:2418–2428.
10. Merz A, Long G, Hiet MS, et al. Biochemical and morphological properties of hepatitis C virus particles and determination of their lipidome. *J Biol Chem.* 2011;286: 3018–3032.
11. Bartenschlager R, Penin F, Lohmann V, Andre P. Assembly of infectious hepatitis C virus particles. *Trends Microbiol.* 2011;19:95–103.
12. Lindenbach BD. Virion assembly and release. *Curr Top Microbiol Immunol.* 2013;369: 199–218.
13. Dao Thi VL, Granier C, Zeisel MB, et al. Characterization of hepatitis C virus particle sub-populations reveals multiple usage of the scavenger receptor BI for entry steps. *J Biol Chem.* 2012;287:31242–31257.
14. Rumin S, Berthillon P, Tanaka E, et al. Dynamic analysis of hepatitis C virus replication and quasispecies selection in long-term cultures of adult human hepatocytes infected in vitro. *J Gen Virol.* 1999;80(Pt 11):3007–3018.
15. Fournier C, Sureau C, Coste J, et al. In vitro infection of adult normal human hepatocytes in primary culture by hepatitis C virus. *J Gen Virol.* 1998;79(Pt 10):2367–2374.
16. Lindenbach BD, Evans MJ, Syder AJ, et al. Complete replication of hepatitis C virus in cell culture. *Science.* 2005;309:623–626.
17. Wakita T, Pietschmann T, Kato T, et al. Production of infectious hepatitis C virus in tissue culture from a cloned viral genome. *Nat Med.* 2005;11:791–796.
18. Zhong J, Gastaminza P, Cheng G, et al. Robust hepatitis C virus infection in vitro. *Proc Natl Acad Sci USA.* 2005;102:9294–9299.
19. Bartosch B, Dubuisson J, Cosset FL. Infectious hepatitis C virus pseudo-particles containing functional E1-E2 envelope protein complexes. *J Exp Med.* 2003;197:633–642.
20. Drummer HE, Maerz A, Poumbourios P. Cell surface expression of functional hepatitis C virus E1 and E2 glycoproteins. *FEBS Lett.* 2003;546:385–390.
21. Hsu M, Zhang J, Flint M, et al. Hepatitis C virus glycoproteins mediate pH-dependent cell entry of pseudotyped retroviral particles. *Proc Natl Acad Sci USA.* 2003;100: 7271–7276.
22. Podevin P, Carpentier A, Pene V, et al. Production of infectious hepatitis C virus in primary cultures of human adult hepatocytes. *Gastroenterology.* 2010;139:1355–1364.
23. Yi M, Ma Y, Yates J, Lemon SM. Compensatory mutations in E1, p7, NS2, and NS3 enhance yields of cell culture-infectious intergenotypic chimeric hepatitis C virus. *J Virol.* 2007;81:629–638.
24. Gastaminza P, Kapadia SB, Chisari FV. Differential biophysical properties of infectious intracellular and secreted hepatitis C virus particles. *J Virol.* 2006;80: 11074–11081.
25. Bankwitz D, Steinmann E, Bitzegeio J, et al. Hepatitis C virus hypervariable region 1 modulates receptor interactions, conceals the CD81 binding site, and protects conserved neutralizing epitopes. *J Virol.* 2010;84:5751–5763.
26. Grove J, Nielsen S, Zhong J, et al. Identification of a residue in hepatitis C virus E2 glycoprotein that determines scavenger receptor BI and CD81 receptor dependency and sensitivity to neutralizing antibodies. *J Virol.* 2008;82:12020–12029.
27. Lindenbach BD, Meuleman P, Ploss A, et al. Cell culture-grown hepatitis C virus is infectious in vivo and can be recultured in vitro. *Proc Natl Acad Sci USA.* 2006;103:3805–3809.
28. Gastaminza P, Cheng G, Wieland S, Zhong J, Liao W, Chisari FV. Cellular determinants of hepatitis C virus assembly, maturation, degradation, and secretion. *J Virol.* 2008;82:2120–2129.

29. Huang H, Sun F, Owen DM, et al. Hepatitis C virus production by human hepatocytes dependent on assembly and secretion of very low-density lipoproteins. *Proc Natl Acad Sci USA*. 2007;104:5848–5853.
30. Jiang J, Luo G. Apolipoprotein E but not B is required for the formation of infectious hepatitis C virus particles. *J Virol*. 2009;83:12680–12691.
31. Cun W, Jiang J, Luo G. The C-terminal alpha-helix domain of apolipoprotein E is required for interaction with nonstructural protein 5A and assembly of hepatitis C virus. *J Virol*. 2010;84:11532–11541.
32. Hishiki T, Shimizu Y, Tobita R, et al. Infectivity of hepatitis C virus is influenced by association with apolipoprotein E isoforms. *J Virol*. 2010;84:12048–12057.
33. Owen DM, Huang H, Ye J, Gale Jr M. Apolipoprotein E on hepatitis C virion facilitates infection through interaction with low-density lipoprotein receptor. *Virology*. 2009;394:99–108.
34. Benga WJ, Krieger SE, Dimitrova M, et al. Apolipoprotein E interacts with hepatitis C virus nonstructural protein 5A and determines assembly of infectious particles. *Hepatology*. 2010;51:43–53.
35. Mailly L, Robinet E, Meuleman P, Baumert TF, Zeisel MB. Hepatitis C virus infection and related liver disease: the quest for the best animal model. *Front Microbiol*. 2013;4:213.
36. von Schaewen M, Ploss A. Murine models of hepatitis C: what can we look forward to? *Antiviral Res*. 2014;104:15–22.
37. Ploss A. Mouse models for human infectious diseases. *J Immunol Methods*. 2014;410:1–2.
38. Meex SJ, Andreo U, Sparks JD, Fisher EA. Huh-7 or HepG2 cells: which is the better model for studying human apolipoprotein-B100 assembly and secretion? *J Lipid Res*. 2011;52:152–158.
39. Lavillette D, Bartosch B, Nourrisson D, et al. Hepatitis C virus glycoproteins mediate low pH-dependent membrane fusion with liposomes. *J Biol Chem*. 2006;281: 3909–3917.
40. Lavillette D, Pécheur E-I, Donot P, et al. Characterization of fusion determinants points to the involvement of three discrete regions of both E1 and E2 glycoproteins in the membrane fusion process of hepatitis C virus. *J Virol*. 2007;81:8752–8765.
41. Flint M, Logvinoff C, Rice CM, McKeating JA. Characterization of infectious retroviral pseudotype particles bearing hepatitis C virus glycoproteins. *J Virol*. 2004;78:6875–6882.
42. Bartosch B, Verney G, Dreux M, et al. An interplay between hypervariable region 1 of the hepatitis C virus E2 glycoprotein, the scavenger receptor BI, and high-density lipoprotein promotes both enhancement of infection and protection against neutralizing antibodies. *J Virol*. 2005;79:8217–8229.
43. Dreux M, Boson B, Ricard-Blum S, et al. The exchangeable apolipoprotein ApoC-I promotes membrane fusion of hepatitis C virus. *J Biol Chem*. 2007;282:32357–32369.
44. Dreux M, Pietschmann T, Granier C, et al. High density lipoprotein inhibits hepatitis C virus neutralising antibodies by stimulating cell entry via activation of the scavenger receptor BI. *J Biol Chem*. 2006;281:18285–18295.
45. Voisset C, Callens N, Blanchard E, Op De Beeck A, Dubuisson J, Vu-Dac N. High density lipoproteins facilitate hepatitis C virus entry through the scavenger receptor class B type I. *J Biol Chem*. 2005;280:7793–7799.
46. Meunier JC, Engle RE, Faulk K, et al. Evidence for cross-genotype neutralization of hepatitis C virus pseudo-particles and enhancement of infectivity by apolipoprotein C1. *Proc Natl Acad Sci USA*. 2005;102:4560–4565.
47. Lefevre M, Felmlee DJ, Parnot M, Baumert TF, Schuster C. Syndecan 4 is involved in mediating HCV entry through interaction with lipoviral particle-associated apolipoprotein E. *PLoS One*. 2014;9:e95550.

48. Shi Q, Jiang J, Luo G. Syndecan-1 serves as the major receptor for attachment of hepatitis C virus to the surfaces of hepatocytes. *J Virol*. 2013;87:6866–6875.
49. Barth H, Schafer C, Adah MI, et al. Cellular binding of hepatitis C virus envelope glycoprotein E2 requires cell surface heparan sulfate. *J Biol Chem*. 2003;278: 41003–41012.
50. Scarselli E, Ansuini H, Cerino R, et al. The human scavenger receptor class B type I is a novel candidate receptor for the hepatitis C virus. *EMBO J*. 2002;21:5017–5025.
51. Jiang J, Wu X, Tang H, Luo G. Apolipoprotein E mediates attachment of clinical hepatitis C virus to hepatocytes by binding to cell surface heparan sulfate proteoglycan receptors. *PLoS One*. 2013;8:e67982.
52. Agnello V, Abel G, Elfahal M, Knight GB, Zhang Q-X. Hepatitis C virus and other flaviviridae viruses enter cells via low density lipoprotein receptor. *Proc Natl Acad Sci USA*. 1999;96:12766–12771.
53. Albecka A, Belouzard S, de Beeck AO, et al. Role of low-density lipoprotein receptor in the hepatitis C virus life cycle. *Hepatology*. 2012;55:998–1007.
54. Lin LT, Chen TY, Lin SC, et al. Broad-spectrum antiviral activity of chebulagic acid and punicalagin against viruses that use glycosaminoglycans for entry. *BMC Microbiol*. 2013;13:187.
55. Hulst MM, van Gennip HG, Vlot AC, Schooten E, de Smit AJ, Moormann RJ. Interaction of classical swine fever virus with membrane-associated heparan sulfate: role for virus replication in vivo and virulence. *J Virol*. 2001;75:9585–9595.
56. Koutsoudakis G, Kaul A, Steinmann E, et al. Characterization of the early steps of hepatitis C virus infection by using luciferase reporter viruses. *J Virol*. 2006;80:5308–5320.
57. Jiang J, Cun W, Wu X, Shi Q, Tang H, Luo G. Hepatitis C virus attachment mediated by apolipoprotein E binding to cell surface heparan sulfate. *J Virol*. 2012;86:7256–7267.
58. Barth H, Schnober EK, Zhang F, et al. Viral and cellular determinants of the hepatitis C virus envelope-heparan sulfate interaction. *J Virol*. 2006;80:10579–10590.
59. Callens N, Ciczora Y, Bartosch B, et al. Basic residues in hypervariable region 1 of hepatitis C virus envelope glycoprotein e2 contribute to virus entry. *J Virol*. 2005; 79:15331–15341.
60. Hussain MM, Strickland DK, Bakillah A. The mammalian low-density lipoprotein receptor family. *Annu Rev Nutr*. 1999;19:141–172.
61. Rigotti A, Miettinen HE, Krieger M. The role of the high-density lipoprotein receptor SR-BI in the lipid metabolism of endocrine and other tissues. *Endocr Rev*. 2003;24:357–387.
62. Innerarity TL. Structural biology. LDL receptor's beta-propeller displaces LDL. *Science*. 2002;298:2337–2339.
63. Molina S, Castet V, Fournier-Wirth C, et al. The low-density lipoprotein receptor plays a role in the infection of primary human hepatocytes by hepatitis C virus. *J Hepatol*. 2007;46:411–419.
64. Wunschmann S, Medh JD, Klinzmann D, Schmidt WN, Stapleton JT. Characterization of hepatitis C virus (HCV) and HCV E2 interactions with CD81 and the low-density lipoprotein receptor. *J Virol*. 2000;74:10055–10062.
65. Dao Thi VL, Dreux M, Cosset FL. Scavenger receptor class B type I and the hypervariable region-1 of hepatitis C virus in cell entry and neutralisation. *Expert Rev Mol Med*. 2011;13:e13.
66. Van Eck M, Hoekstra M, Out R, et al. Scavenger receptor BI facilitates the metabolism of VLDL lipoproteins in vivo. *J Lipid Res*. 2008;49:136–146.
67. Krieger M. Charting the fate of the "good cholesterol": identification and characterization of the high-density lipoprotein receptor SR-BI. *Annu Rev Biochem*. 1999;68: 523–558.

68. Silver DL, Wang N, Xiao X, Tall AR. High density lipoprotein (HDL) particle uptake mediated by scavenger receptor class B type 1 results in selective sorting of HDL cholesterol from protein and polarized cholesterol secretion. *J Biol Chem.* 2001;276:25287–25293.
69. Bartosch B, Vitelli A, Granier C, et al. Cell entry of hepatitis C virus requires a set of co-receptors that include the CD81 tetraspanin and the SR-B1 scavenger receptor. *J Biol Chem.* 2003;278:41624–41630.
70. Catanese MT, Graziani R, von Hahn T, et al. High-avidity monoclonal antibodies against the human scavenger class B type I receptor efficiently block hepatitis C virus infection in the presence of high-density lipoprotein. *J Virol.* 2007;81: 8063–8071.
71. Lavillette D, Tarr AW, Voisset C, et al. Characterization of host-range and cell entry properties of the major genotypes and subtypes of hepatitis C virus. *Hepatology.* 2005;41:265–274.
72. Zeisel MB, Koutsoudakis G, Schnober EK, et al. Scavenger receptor class B type I is a key host factor for hepatitis C virus infection required for an entry step closely linked to CD81. *Hepatology.* 2007;46:1722–1731.
73. Catanese MT, Ansuini H, Graziani R, et al. Role of scavenger receptor class B type I in hepatitis C virus entry: kinetics and molecular determinants. *J Virol.* 2010;84:34–43.
74. Dao Thi VL, Granier C, Zeisel MB, et al. Characterization of hepatitis C virus particle subpopulations reveals multiple usage of the scavenger receptor BI for entry steps. *J Biol Chem.* 2012;287:31242–31257.
75. Maillard P, Huby T, Andreo U, Moreau M, Chapman J, Budkowska A. The interaction of natural hepatitis C virus with human scavenger receptor SR-BI/Cla1 is mediated by ApoB-containing lipoproteins. *FASEB J.* 2006;20:735–737.
76. Lavie M, Goffard A, Dubuisson J. Assembly of a functional HCV glycoprotein heterodimer. *Curr Issues Mol Biol.* 2007;9:71–86.
77. Li Y, Modis Y. A novel membrane fusion protein family in Flaviviridae? *Trends Microbiol.* 2014;22:176–182.
78. Op De Beeck A, Cocquerel L, Dubuisson J. Biogenesis of hepatitis C virus envelope glycoproteins. *J Gen Virol.* 2001;82:2589–2595.
79. Cocquerel L, Meunier JC, Op de Beeck A, Bonte D, Wychowski C, Dubuisson J. Coexpression of hepatitis C virus envelope proteins E1 and E2 in cis improves the stability of membrane insertion of E2. *J Gen Virol.* 2001;82:1629–1635.
80. Cocquerel L, Wychowski C, Minner F, Penin F, Dubuisson J. Charged residues in the transmembrane domains of hepatitis C virus glycoproteins play a major role in the processing, subcellular localization, and assembly of these envelope proteins. *J Virol.* 2000;74:3623–3633.
81. Cocquerel L, Op de Beeck A, Lambot M, et al. Topological changes in the transmembrane domains of hepatitis C virus envelope glycoproteins. *EMBO J.* 2002;21: 2893–2902.
82. Rouille Y, Helle F, Delgrange D, et al. Subcellular localization of hepatitis C virus structural proteins in a cell culture system that efficiently replicates the virus. *J Virol.* 2006;80:2832–2841.
83. Kielian M, Rey FA. Virus membrane-fusion proteins: more than one way to make a hairpin. *Nat Rev Microbiol.* 2006;4:67–76.
84. Michalak JP, Wychowski C, Choukhi A, et al. Characterization of truncated forms of hepatitis C virus glycoproteins. *J Gen Virol.* 1997;78(Pt 9):2299–2306.
85. Patel J, Patel AH, McLauchlan J. The transmembrane domain of the hepatitis C virus E2 glycoprotein is required for correct folding of the E1 glycoprotein and native complex formation. *Virology.* 2001;279:58–68.

86. Cocquerel L, Quinn ER, Flint M, Hadlock KG, Foung SK, Levy S. Recognition of native hepatitis C virus E1E2 heterodimers by a human monoclonal antibody. *J Virol.* 2003;77:1604–1609.
87. Brazzoli M, Helenius A, Foung SK, Houghton M, Abrignani S, Merola M. Folding and dimerization of hepatitis C virus E1 and E2 glycoproteins in stably transfected CHO cells. *Virology.* 2005;332:438–453.
88. Wahid A, Helle F, Descamps V, Duverlie G, Penin F, Dubuisson J. Disulfide bonds in hepatitis C virus glycoprotein E1 control the assembly and entry functions of E2 glycoprotein. *J Virol.* 2013;87:1605–1617.
89. Fenouillet E, Lavillette D, Loureiro S, et al. Contribution of redox status to hepatitis C virus E2 envelope protein function and antigenicity. *J Biol Chem.* 2008; 283:26340–26348.
90. Sandrin V, Boulanger P, Penin F, Granier C, Cosset FL, Bartosch B. Assembly of functional hepatitis C virus glycoproteins on infectious pseudoparticles occurs intracellularly and requires concomitant incorporation of E1 and E2 glycoproteins. *J Gen Virol.* 2005;86:3189–3199.
91. Goffard A, Dubuisson J. Glycosylation of hepatitis C virus envelope proteins. *Biochimie.* 2003;85:295–301.
92. Zhang M, Gaschen B, Blay W, et al. Tracking global patterns of N-linked glycosylation site variation in highly variable viral glycoproteins: HIV, SIV, and HCV envelopes and influenza hemagglutinin. *Glycobiology.* 2004;14:1229–1246.
93. Helle F, Vieyres G, Elkrief L, et al. Role of N-linked glycans in the functions of hepatitis C virus envelope proteins incorporated into infectious virions. *J Virol.* 2010;84:11905–11915.
94. Helle F, Goffard A, Morel V, et al. The neutralizing activity of anti-hepatitis C virus antibodies is modulated by specific glycans on the E2 envelope protein. *J Virol.* 2007;81:8101–8111.
95. Goffard A, Callens N, Bartosch B, et al. Role of N-linked glycans in the functions of hepatitis C virus envelope glycoproteins. *J Virol.* 2005;79:8400–8409.
96. Dubuisson J, Hsu HH, Cheung RC, Greenberg HB, Russell DG, Rice CM. Formation and intracellular localization of hepatitis C virus envelope glycoprotein complexes expressed by recombinant vaccinia and Sindbis viruses. *J Virol.* 1994;68: 6147–6160.
97. Miyanari Y, Atsuzawa K, Usuda N, et al. The lipid droplet is an important organelle for hepatitis C virus production. *Nat Cell Biol.* 2007;9:1089–1097.
98. Jirasko V, Montserret R, Lee JY, et al. Structural and functional studies of nonstructural protein 2 of the hepatitis C virus reveal its key role as organizer of virion assembly. *PLoS Pathog.* 2010;6:e1001233.
99. Ma Y, Anantpadma M, Timpe JM, et al. Hepatitis C virus NS2 protein serves as a scaffold for virus assembly by interacting with both structural and nonstructural proteins. *J Virol.* 2011;85:86–97.
100. Popescu CI, Callens N, Trinel D, et al. NS2 protein of hepatitis C virus interacts with structural and non-structural proteins towards virus assembly. *PLoS Pathog.* 2011;7:e1001278.
101. Stapleford KA, Lindenbach BD. Hepatitis C virus NS2 coordinates virus particle assembly through physical interactions with the E1-E2 glycoprotein and NS3-NS4A enzyme complexes. *J Virol.* 2011;85:1706–1717.
102. Suzuki R, Matsuda M, Watashi K, et al. Signal peptidase complex subunit 1 participates in the assembly of hepatitis C virus through an interaction with E2 and NS2. *PLoS Pathog.* 2013;9:e1003589.
103. Gentzsch J, Brohm C, Steinmann E, et al. Hepatitis C virus p7 is critical for capsid assembly and envelopment. *PLoS Pathog.* 2013;9:e1003355.

104. Nahmias Y, Goldwasser J, Casali M, et al. Apolipoprotein B-dependent hepatitis C virus secretion is inhibited by the grapefruit flavonoid naringenin. *Hepatology*. 2008;47:1437–1445.
105. Da Costa D, Turek M, Felmlee DJ, et al. Reconstitution of the entire hepatitis C virus life cycle in nonhepatic cells. *J Virol*. 2012;86:11919–11925.
106. Hueging K, Doepke M, Vieyres G, et al. Apolipoprotein E codetermines tissue tropism of hepatitis C virus and is crucial for viral cell-to-cell transmission by contributing to a postenvelopment step of assembly. *J Virol*. 2014;88:1433–1446.
107. Coller KE, Heaton NS, Berger KL, Cooper JD, Saunders JL, Randall G. Molecular determinants and dynamics of hepatitis C virus secretion. *PLoS Pathog*. 2012;8: e1002466.
108. Wozniak AL, Griffin S, Rowlands D, et al. Intracellular proton conductance of the hepatitis C virus p7 protein and its contribution to infectious virus production. *PLoS Pathog*. 2010;6:e1001087.
109. Krey T, d'Alayer J, Kikuti CM, et al. The disulfide bonds in glycoprotein E2 of hepatitis C virus reveal the tertiary organization of the molecule. *PLoS Pathog*. 2010;6:e1000762.
110. Albecka A, Montserret R, Krey T, et al. Identification of new functional regions in hepatitis C virus envelope glycoprotein E2. *J Virol*. 2011;85:1777–1792.
111. Maurin G, Fresquet J, Granio O, Wychowski C, Cosset FL, Lavillette D. Identification of interactions in the E1E2 heterodimer of hepatitis C virus important for cell entry. *J Biol Chem*. 2011;286:23865–23876.
112. Douam F, Dao Thi VL, Maurin G, et al. Critical interaction between E1 and E2 glycoproteins determines binding and fusion properties of hepatitis C virus during cell entry. *Hepatology*. 2014;59:776–788.
113. Kong L, Giang E, Nieusma T, et al. Hepatitis C virus E2 envelope glycoprotein core structure. *Science*. 2013;342:1090–1094.
114. Khan AG, Whidby J, Miller MT, et al. Structure of the core ectodomain of the hepatitis C virus envelope glycoprotein 2. *Nature*. 2014;509:381–384.
115. Flint M, Maidens C, Loomis-Price LD, et al. Characterization of hepatitis C virus E2 glycoprotein interaction with a putative cellular receptor, CD81. *J Virol*. 1999; 73:6235–6244.
116. Patel AH, Wood J, Penin F, Dubuisson J, McKeating JA. Construction and characterization of chimeric hepatitis C virus E2 glycoproteins: analysis of regions critical for glycoprotein aggregation and CD81 binding. *J Gen Virol*. 2000;81:2873–2883.
117. Owsianka A, Clayton RF, Loomis-Price LD, McKeating JA, Patel AH. Functional analysis of hepatitis C virus E2 glycoproteins and virus-like particles reveals structural dissimilarities between different forms of E2. *J Gen Virol*. 2001;82: 1877–1883.
118. Owsianka AM, Timms JM, Tarr AW, et al. Identification of conserved residues in the E2 envelope glycoprotein of the hepatitis C virus that are critical for CD81 binding. *J Virol*. 2006;80:8695–8704.
119. Clayton RF, Owsianka A, Aitken J, Graham S, Bhella D, Patel AH. Analysis of antigenicity and topology of E2 glycoprotein present on recombinant hepatitis C virus-like particles. *J Virol*. 2002;76:7672–7682.
120. Roccasecca R, Ansuini H, Vitelli A, et al. Binding of the hepatitis C virus E2 glycoprotein to CD81 is strain specific and is modulated by a complex interplay between hypervariable regions 1 and 2. *J Virol*. 2003;77:1856–1867.
121. Drummer HE, Boo I, Maerz AL, Poumbourios P. A conserved Gly436-Trp-Leu-Ala-Gly-Leu-Phe-Tyr motif in hepatitis C virus glycoprotein E2 is a determinant of CD81 binding and viral entry. *J Virol*. 2006;80:7844–7853.

122. Rothwangl KB, Manicassamy B, Uprichard SL, Rong L. Dissecting the role of putative CD81 binding regions of E2 in mediating HCV entry: putative CD81 binding region 1 is not involved in CD81 binding. *Virol J*. 2008;5:46.
123. Witteveldt J, Evans MJ, Bitzegeio J, et al. CD81 is dispensable for hepatitis C virus cell-to-cell transmission in hepatoma cells. *J Gen Virol*. 2009;90:48–58.
124. Boo I, teWierik K, Douam F, Lavillette D, Poumbourios P, Drummer HE. Distinct roles in folding, CD81 receptor binding and viral entry for conserved histidine residues of hepatitis C virus glycoprotein E1 and E2. *Biochem J*. 2012;443:85–94.
125. Lavie M, Sarrazin S, Montserret R, et al. Identification of conserved residues in HCV envelope glycoprotein E2 modulating virus dependence on CD81 and SRB1 entry factors. *J Virol*. 2014;88:10584–10597.
126. Pileri P, Uematsu Y, Campagnoli S, et al. Binding of hepatitis C virus to CD81. *Science*. 1998;282:938–941.
127. Russell RS, Kawaguchi K, Meunier JC, et al. Mutational analysis of the hepatitis C virus E1 glycoprotein in retroviral pseudoparticles and cell-culture-derived H77/JFH1 chimeric infectious virus particles. *J Viral Hepat*. 2009;16: 621–632.
128. van Spriel AB, Figdor CG. The role of tetraspanins in the pathogenesis of infectious diseases. *Microbes Infect*. 2010;12:106–112.
129. Monk PN, Partridge LJ. Tetraspanins: gateways for infection. *Infect Disord Drug Targets*. 2012;12:4–17.
130. Drummer HE, Wilson KA, Poumbourios P. Identification of the hepatitis C virus E2 glycoprotein binding site on the large extracellular loop of CD81. *J Virol*. 2002;76:11143–11147.
131. Bartosch B, Bukh J, Meunier JC, et al. In vitro assay for neutralizing antibody to hepatitis C virus: evidence for broadly conserved neutralization epitopes. *Proc Natl Acad Sci USA*. 2003;100:14199–14204.
132. Kapadia SB, Barth H, Baumert T, McKeating JA, Chisari FV. Initiation of hepatitis C virus infection is dependent on cholesterol and cooperativity between CD81 and scavenger receptor B type I. *J Virol*. 2007;81:374–383.
133. Molina S, Castet V, Pichard-Garcia L, et al. Serum-derived hepatitis C virus infection of primary human hepatocytes is tetraspanin CD81 dependent. *J Virol*. 2008;82: 569–574.
134. Flint M, von Hahn T, Zhang J, et al. Diverse CD81 proteins support hepatitis C virus infection. *J Virol*. 2006;80:11331–11342.
135. Koutsoudakis G, Herrmann E, Kallis S, Bartenschlager R, Pietschmann T. The level of CD81 cell surface expression is a key determinant for productive entry of hepatitis C virus into host cells. *J Virol*. 2007;81:588–598.
136. Ploss A, Evans MJ, Gaysinskaya VA, et al. Human occludin is a hepatitis C virus entry factor required for infection of mouse cells. *Nature*. 2009;457:882–886.
137. Bitzegeio J, Bankwitz D, Hueging K, et al. Adaptation of hepatitis C virus to mouse CD81 permits infection of mouse cells in the absence of human entry factors. *PLoS Pathog*. 2010;6:e1000978.
138. Brazzoli M, Bianchi A, Filippini S, et al. CD81 is a central regulator of cellular events required for hepatitis C virus infection of human hepatocytes. *J Virol*. 2008;82: 8316–8329.
139. Harris HJ, Farquhar MJ, Mee CJ, et al. CD81 and claudin 1 coreceptor association: role in hepatitis C virus entry. *J Virol*. 2008;82:5007–5020.
140. Harris HJ, Davis C, Mullins JG, et al. Claudin association with CD81 defines hepatitis C virus entry. *J Biol Chem*. 2010;285:21092–21102.
141. Farquhar MJ, Hu K, Harris HJ, et al. Hepatitis C virus induces CD81 and claudin-1 endocytosis. *J Virol*. 2012;86:4305–4316.

142. Liu Z, Tian Y, Machida K, et al. Transient activation of the PI3K-AKT pathway by hepatitis C virus to enhance viral entry. *J Biol Chem*. 2012;287:41922–41930.
143. Lupberger J, Zeisel MB, Xiao F, et al. EGFR and EphA2 are host factors for hepatitis C virus entry and possible targets for antiviral therapy. *Nat Med*. 2011;17: 589–595.
144. Diao J, Pantua H, Ngu H, et al. Hepatitis C virus induces epidermal growth factor receptor activation via CD81 binding for viral internalization and entry. *J Virol*. 2012;86:10935–10949.
145. Zona L, Lupberger J, Sidahmed-Adrar N, et al. HRas signal transduction promotes hepatitis C virus cell entry by triggering assembly of the host tetraspanin receptor complex. *Cell Host Microbe*. 2013;13:302–313.
146. Rocha-Perugini V, Montpellier C, Delgrange D, et al. The CD81 partner EWI-2wint inhibits hepatitis C virus entry. *PLoS One*. 2008;3:e1866.
147. Montpellier C, Tews BA, Poitrimole J, et al. Interacting regions of CD81 and two of its partners, EWI-2 and EWI-2wint, and their effect on hepatitis C virus infection. *J Biol Chem*. 2011;286:13954–13965.
148. Potel J, Rassam P, Montpellier C, et al. EWI-2wint promotes CD81 clustering that abrogates hepatitis C virus entry. *Cell Microbiol*. 2013;15:1234–1252.
149. Evans MJ, von Hahn T, Tscherne DM, et al. Claudin-1 is a hepatitis C virus co-receptor required for a late step in entry. *Nature*. 2007;446:801–805.
150. Gunzel D, Fromm M. Claudins and other tight junction proteins. *Compr Physiol*. 2012;2:1819–1852.
151. Yang W, Qiu C, Biswas N, et al. Correlation of the tight junction-like distribution of Claudin-1 to the cellular tropism of hepatitis C virus. *J Biol Chem*. 2008;283: 8643–8653.
152. Liu S, Yang W, Shen L, Turner JR, Coyne CB, Wang T. Tight junction proteins claudin-1 and occludin control hepatitis C virus entry and are downregulated during infection to prevent superinfection. *J Virol*. 2009;83:2011–2014.
153. Davis C, Harris HJ, Hu K, et al. In silico directed mutagenesis identifies the CD81/claudin-1 hepatitis C virus receptor interface. *Cell Microbiol*. 2012;14:1892–1903.
154. Krieger SE, Zeisel MB, Davis C, et al. Inhibition of hepatitis C virus infection by anti-claudin-1 antibodies is mediated by neutralization of E2-CD81-claudin-1 associations. *Hepatology*. 2010;51:1144–1157.
155. Haid S, Windisch MP, Bartenschlager R, Pietschmann T. Mouse-specific residues of claudin-1 limit hepatitis C virus genotype 2a infection in a human hepatocyte cell line. *J Virol*. 2010;84:964–975.
156. Benedicto I, Molina-Jimenez F, Barreiro O, et al. Hepatitis C virus envelope components alter localization of hepatocyte tight junction-associated proteins and promote occludin retention in the endoplasmic reticulum. *Hepatology*. 2008;48:1044–1053.
157. Liu S, Kuo W, Yang W, et al. The second extracellular loop dictates Occludin-mediated HCV entry. *Virology*. 2010;407:160–170.
158. Sourisseau M, Michta ML, Zony C, et al. Temporal analysis of hepatitis C virus cell entry with occludin directed blocking antibodies. *PLoS Pathog*. 2013;9:e1003244.
159. Michta ML, Hopcraft SE, Narbus CM, et al. Species-specific regions of occludin required by hepatitis C virus for cell entry. *J Virol*. 2010;84:11696–11708.
160. Mee CJ, Grove J, Harris HJ, Hu K, Balfe P, McKeating JA. Effect of cell polarization on hepatitis C virus entry. *J Virol*. 2008;82:461–470.
161. Mee CJ, Harris HJ, Farquhar MJ, et al. Polarization restricts hepatitis C virus entry into HepG2 hepatoma cells. *J Virol*. 2009;83:6211–6221.
162. Coller KE, Berger KL, Heaton NS, Cooper JD, Yoon R, Randall G. RNA interference and single particle tracking analysis of hepatitis C virus endocytosis. *PLoS Pathog*. 2009;5:e1000702.

163. Blanchard AA, Watson PH, Shiu RP, et al. Differential expression of claudin 1, 3, and 4 during normal mammary gland development in the mouse. *DNA Cell Biol.* 2006;25:79–86.
164. Meertens L, Bertaux C, Dragic T. Hepatitis C virus entry requires a critical post-internalization step and delivery to early endosomes via clathrin-coated vesicles. *J Virol.* 2006;80:11571–11578.
165. Martin DN, Uprichard SL. Identification of transferrin receptor 1 as a hepatitis C virus entry factor. *Proc Natl Acad Sci USA.* 2013;110:10777–10782.
166. Timpe JM, Stamataki Z, Jennings A, et al. Hepatitis C virus cell-cell transmission in hepatoma cells in the presence of neutralizing antibodies. *Hepatology.* 2008;47:17–24.
167. Brimacombe CL, Grove J, Meredith LW, et al. Neutralizing antibody-resistant hepatitis C virus cell-to-cell transmission. *J Virol.* 2011;85:596–605.
168. Catanese MT, Loureiro J, Jones CT, Dorner M, von Hahn T, Rice CM. Different requirements for scavenger receptor class B type I in hepatitis C virus cell-free versus cell-to-cell transmission. *J Virol.* 2013;87:8282–8293.
169. Timpe JM, McKeating JA. Hepatitis C virus entry: possible targets for therapy. *Gut.* 2008;57:1728–1737.
170. Sainz Jr B, Barretto N, Martin DN, et al. Identification of the Niemann-Pick C1-like 1 cholesterol absorption receptor as a new hepatitis C virus entry factor. *Nat Med.* 2012;18:281–285.
171. Barretto N, Sainz Jr B, Hussain S, Uprichard SL. Determining the involvement and therapeutic implications of host cellular factors in hepatitis C virus cell-to-cell spread. *J Virol.* 2014;88:5050–5061.
172. Xiao F, Fofana I, Heydmann L, et al. Hepatitis C virus cell–cell transmission and resistance to direct-acting antiviral agents. *PLoS Pathog.* 2014;10:e1004128.
173. Sharma NR, Mateu G, Dreux M, Grakoui A, Cosset FL, Melikyan GB. Hepatitis C virus is primed by CD81 protein for low pH-dependent fusion. *J Biol Chem.* 2011;286:30361–30376.
174. Kobayashi M, Bennett MC, Bercot T, Singh IR. Functional analysis of hepatitis C virus envelope proteins, using a cell–cell fusion assay. *J Virol.* 2006;80:1817–1825.
175. Haid S, Pietschmann T, Pecheur EI. Low pH-dependent hepatitis C virus membrane fusion depends on E2 integrity, target lipid composition, and density of virus particles. *J Biol Chem.* 2009;284:17657–17667.
176. Tscherne DM, Jones CT, Evans MJ, Lindenbach BD, McKeating JA, Rice CM. Time- and temperature-dependent activation of hepatitis C virus for low-pH-triggered entry. *J Virol.* 2006;80:1734–1741.
177. Mothes W, Boerger AL, Narayan S, Cunningham JM, Young JA. Retroviral entry mediated by receptor priming and low pH triggering of an envelope glycoprotein. *Cell.* 2000;103:679–689.
178. Dreux M, Pietschmann T, Granier C, et al. High density lipoprotein inhibits hepatitis C virus-neutralizing antibodies by stimulating cell entry via activation of the scavenger receptor BI. *J Biol Chem.* 2006;281:18285–18295.
179. Carette JE, Raaben M, Wong AC, et al. Ebola virus entry requires the cholesterol transporter Niemann-Pick C1. *Nature.* 2011;477:340–343.
180. Stiasny K, Koessl C, Heinz FX. Involvement of lipids in different steps of the flavivirus fusion mechanism. *J Virol.* 2003;77:7856–7862.
181. Kielian M. Class II virus membrane fusion proteins. *Virology.* 2006;344:38–47.
182. Modis Y. Class II fusion proteins. *Adv Exp Med Biol.* 2013;790:150–166.
183. Modis Y, Ogata S, Clements D, Harrison SC. Structure of the dengue virus envelope protein after membrane fusion. *Nature.* 2004;427:313–319.
184. Backovic M, Jardetzky TS. Class III viral membrane fusion proteins. *Adv Exp Med Biol.* 2011;714:91–101.

185. Drummer HE, Boo I, Poumbourios P. Mutagenesis of a conserved fusion peptide-like motif and membrane-proximal heptad-repeat region of hepatitis C virus glycoprotein E1. *J Gen Virol.* 2007;88:1144–1148.
186. El Omari K, Iourin O, Harlos K, Grimes JM, Stuart DI. Structure of a pestivirus envelope glycoprotein E2 clarifies its role in cell entry. *Cell Rep.* 2013;3:30–35.
187. Li Y, Wang J, Kanai R, Modis Y. Crystal structure of glycoprotein E2 from bovine viral diarrhea virus. *Proc Natl Acad Sci USA.* 2013;110:6805–6810.
188. Hulst MM, Moormann RJ. Inhibition of pestivirus infection in cell culture by envelope proteins E(rns) and E2 of classical swine fever virus: E(rns) and E2 interact with different receptors. *J Gen Virol.* 1997;78(Pt 11):2779–2787.
189. Krey T, Thiel HJ, Rumenapf T. Acid-resistant bovine pestivirus requires activation for pH-triggered fusion during entry. *J Virol.* 2005;79:4191–4200.

CHAPTER FOUR

The Evolution of HIV-1 Interactions with Coreceptors and Mannose C-Type Lectin Receptors

Marie Borggren*, Marianne Jansson†,‡,1

*Virus Research and Development Laboratory, Department of Microbiological Diagnostics and Virology, Statens Serum Institut, Copenhagen, Denmark

†Department of Laboratory Medicine, Lund University, Lund, Sweden

‡Department of Microbiology, Tumor and Cell biology, Karolinska Institute, Stockholm, Sweden

[1]Corresponding author: e-mail address: Marianne.Jansson@med.lu.se

Contents

Abstract

The phenotype of human immunodeficiency virus type 1 (HIV-1) commonly evolves between and within infected individuals, at virus transmission, and during disease progression. This evolution includes altered interactions between the virus and its coreceptors, i.e., chemokine receptors, as well as mannose C-type lectin receptors (CLRs). Transmitted/founder viruses are predominantly restricted to CCR5, whereas the subsequent intrapatient evolution of HIV-1 coreceptor use during progressive disease can be subdivided into two distinct pathways. Accordingly, the CCR5-restricted virus population is either gradually replaced by virus variants able to use CXCR4 or evolves toward an altered, more flexible use of CCR5. Despite a strong dependency on these coreceptors for host cell entry, HIV-1 also interacts with other cell surface molecules during target cell attachment, including the CLRs. The virus interaction with the

Progress in Molecular Biology and Translational Science, Volume 129
ISSN 1877-1173
http://dx.doi.org/10.1016/bs.pmbts.2014.10.004

CLRs may result either in the efficient transfer of virus to CD4^{+} T cells or in the degradation of the virus in endosomal compartments. The determinants of the diverse outcomes depend on which CLR is engaged and also on the glycan makeup of the envelope glycoproteins, which may evolve with the strength of the immune pressure during the disease course. With the current clinical introduction of CCR5 antagonists and the development of additional entry inhibitors, knowledge on the evolution and baseline characteristics of HIV-1 interactions with coreceptor and CLR interactions may play important roles for individualized and optimized treatment strategies. This review summarizes our current understanding of the evolution of HIV-1 interactions with these receptors.

1. INTRODUCTION

The combination of extreme genetic diversity, high viral turnover, immune pressure, and flexibility of the viral envelope glycoproteins (Env) fuels the evolution of human immunodeficiency virus type 1 (HIV-1) interactions with several cell surface proteins during target cell encounters. CD4 was the first HIV-1 receptor to be identified,[1] and subsequently seven-transmembrane G-protein-coupled chemokine receptors, mainly CCR5 and CXCR4, were recognized as essential coreceptors for target cell entry.[2–7] Although HIV-1 clearly depends on these receptors for host cell entry, it also interacts with a wide range of other cell surface molecules during target cell attachment. One group of such infection enhancing HIV-1 receptors is the mannose C-type lectin receptors (CLRs), which may support the efficient transfer of virus to CD4^{+} T cells.[8,9] In this review, we describe HIV-1 interactions with coreceptors and CLRs and discuss recent progress in the understanding of the evolution, kinetics, and the proportions of HIV-1 populations able to use these receptors. Different aspects of the impact that these receptor interactions have on virus transmission and disease progression are also elaborated on.

2. CHEMOKINE RECEPTORS AS CRITICAL HIV-1 CORECEPTORS

A number of observations, including the lack of HIV-1 entry into murine cells expressing human CD4,[10] replication patterns of HIV-1 in primary cells and cell lines,[11–14] and the inhibitory capacity of certain chemokines against virus *in vitro* infections,[15] spurred the identification of chemokine receptors, CCR5 and CXCR4, as HIV-1 coreceptors.[2–7] A viral classification based on coreceptor use was established,[16] in which HIV-1 variants monotropic for CCR5 or CXCR4 today are named R5 and X4, respectively, and viruses able to use both coreceptors are called R5X4. Dualtropic virus variants that preferentially infect CCR5 or CXCR4 cells and are accordingly called Dual-R or Dual-X, whereas the isolates containing mixed virus populations are designated dual/mixed (D/M).[17,18] In addition to CCR5 and CXCR4, several other related chemokine receptors have been described as alternative HIV-1 coreceptors (reviewed in Ref. 19). CCR3 especially has been implicated in HIV-1 neurotropism[20,21] and Env cloned directly from the peripheral blood of individuals infected with HIV-1 subtype B have been reported to frequently mediate CCR3, as well as CCR5, use.[22] Still, the recognition that individuals homozygous for a 32-bp deletion in the *CCR5* gene (*CCR5Δ32*), resulting in the lack of CCR5 surface expression, are highly resistant to HIV-1 infection,[23,24] and that those rare cases of HIV-1 infection detected in such individuals carry CXCR4-using virus variants[25] strongly supported the *in vivo* relevance of CCR5 and CXCR4 as the main HIV-1 coreceptors. Similar to HIV-1, the vast majority of HIV-2 strains thus far characterized use CCR5, and certain HIV-2 variants, especially those derived from viremic patients, can also utilize CXCR4.[26] However, findings from *in vitro* infections suggest that the coreceptor use of HIV-2 is more promiscuous than that of HIV-1, with a broader range of tropism including the alternative coreceptors GPR15 (BOB), CXCR6, and CCR9 (D6),[26–28] Nevertheless, the broadened use of alternative coreceptors does not appear to be associated with more aggressive disease.

The natural function of CCR5 and CXCR4, being seven-transmembrane G-protein-coupled receptors, is to respond to chemokine gradients, and in this way trigger cells to migrate in a certain direction. Both CCR5 and CXCR4 can be found at different levels on $CD4^+$ T cells, macrophages, and dendritic cells (DCs), and expression patterns of these receptors contribute to the tropism of HIV-1.[29] Distinct subsets of $CD4^+$ T cells also express different levels of these receptors; CCR5 expression is highly elevated on $CD4^+$ memory T cells, whereas CXCR4 predominates on naïve $CD4^+$ T cells.[30] Chemokine ligands to CCR5 include CCL3, CCL4, and CCL5 (previously known as MIP-1α, MIP-1β, and RANTES, respectively), and the ligand to CXCR4 is SDF-1α. These chemokines inhibits R5 and X4/R5X4 virus replication *in vitro*.[15,31–33]

3. EVOLUTION OF CORECEPTOR USE DURING VIRUS TRANSMISSION AND ESTABLISHMENT IN THE NEW HOST

Several studies have established that the population of HIV-1 variants emerging in the newly infected host, known as the transmitted/founder (TF) population, is homogenous[34,35]; however, the mechanisms whereby TF viruses are transmitted through this bottleneck remain not fully understood. The difficulty in clarifying this lies in the fact that primary HIV-1 infection is usually not detected and the sampling of specimens is performed weeks or months after the transmission. However, it has been thoroughly established that R5 viruses preferential are detected during the primary and early phases of the HIV-1 infection independent of the transmission route, even though the donor may have harbored mixed virus populations with different coreceptor usage[36–40] (Fig. 1). Several mechanisms, working one by one or in concert, have been suggested to govern the selection against the transmission and establishment of CXCR4-using HIV-1 variants into

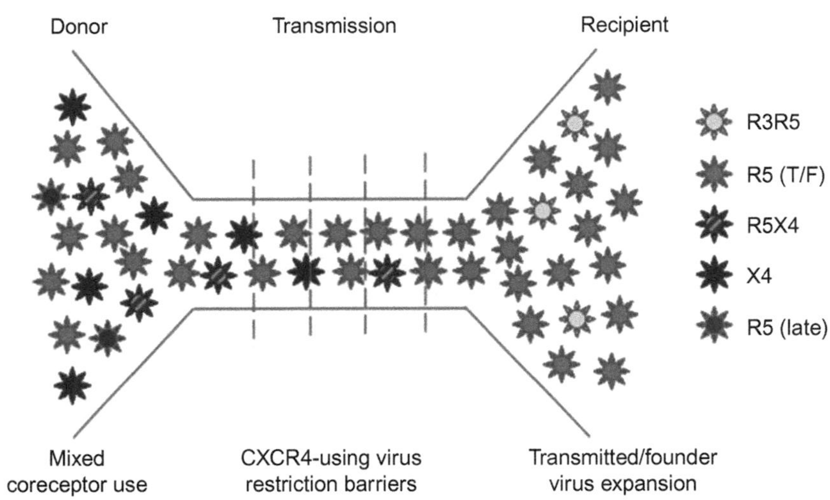

Figure 1 HIV-1 coreceptor evolution during transmission between individuals. The population of HIV-1 variants emerging in the newly infected host, known as the transmitted/founder population, is homogenous and several barriers of selection may impact the phenotype of the virus transmitted and established in the new host. Viruses of the R5 phenotype preferential are detected during the primary and early phases of the infection, though the donor may have harbored mixed virus populations with different coreceptor usage.

the new host (for review, see Ref. 41). These gatekeeping barriers have been suggested to include soluble factors of mucus, such as the CXCR4-ligand SDF-1α and polyanionic mucus components, that would preferentially inhibit the CXCR4-using viruses containing more positively charged Env. The availability of $CD4^+$ target cells that express abundant amounts of CCR5, such as $CD4^+$ T cells and DCs/Langerhans cells in the mucosa, at the site of transmission can also contribute to the establishment of R5 virus transmission and amplification. Furthermore, it has been demonstrated that intestinal epithelial cells selectively transfer R5 viruses via transcytosis[42] and that DCs may preferentially sample R5 viruses across the intestinal epithelia.[43] Resident macrophages may also play a role in the preferential outgrowth of R5 viruses because these cells express high levels of CCR5. However, T/F HIV-1 variants with strong macrophage tropism, as tested *in vitro*, are rarely detected soon after transmission.[44,45] The dominance of R5 virus variants in the primary infection phase may be explained by the fact that amplification of the virus replication primarily takes place in the gut-associated lymphoid tissue (GALT), in which the $CCR5^+CD4^+$ memory T cells are abundant.[46,47] It was also suggested that homing of the virus to GALT depends on the interaction between Env and the gut homing integrin α4β7 expressed on peripheral T cells.[48] However, studies with infectious molecular clones derived from T/F and chronic subtype C HIV-1 revealed that saturating concentrations of anti-α4β7 antibodies failed to inhibit infection and replication of either of these virus variants.[49] Moreover, findings from studies of SHIV infections in macaque models indicate that control of CXCR4-using viruses may be easier than control of R5 viruses in the immunocompetent host. Support for this notion was obtained by the predominance of R5 SHIV in immunocompetent macaques dually infected with a mixture of R5 and X4 SHIV strains, and the emergence of X4 SHIV in animals where $CD8^+$ T cells were depleted.[50] However, it should be noted that the CXCR4-using virus barriers are not absolute. The transmission of CXCR4-using virus strains into individuals carrying homozygous alleles of the Δ32 CCR5 deletion has been documented in a few cases.[25] Moreover, several strains of CXCR4-using SHIVs readily infect macaques via different routes (reviewed in Ref. 51). Although R5 viruses are predominately detected in children infected by their mothers, it is not uncommon that CXCR4-using strains are identified early after birth.[52,53] Thus, the transmission of CXCR4-using viruses appears to be more frequent in vertical transmission than in other routes of transmission (reviewed in Ref. 54). The reason for this is not known, but it should be considered that HIV-1 vertical transmission is complex and includes *in utero*, perinatal, and breastfeeding transmission, and that the pool of HIV-1 target cells in the fetus/newborn may have a different composition compared with adults.

4. INTRAPATIENT EVOLUTION OF HIV-1 CORECEPTOR USE

The evolution of the HIV-1 biological phenotype during disease progression within infected individuals was recognized early,[55–57] and the mechanisms behind this evolution, including differences in cell tropism, were greatly illuminated due to the identification HIV-1 coreceptors.[2–7] Hence, R5 viruses predominate early after transmission and also during the asymptomatic chronic phase of the infection.[35–38] During progressive disease and AIDS development, the evolution of HIV-1 coreceptor use can instead be divided into two distinct pathways, one where the R5 virus population evolves, or switches, toward CXCR4-tropism, either exclusively or in addition to CCR5,[33,58] and the other in which the R5 virus population remains CCR5-restricted throughout the whole course of the disease[59,60] (Fig. 2).

5. THE SWITCH PATHWAY

The proportion of individuals developing CXCR4-tropic viruses is associated with the HIV-1 subtype that the individual is infected with (reviewed in Ref. 61). It was previously estimated that approximately 50% of infected individuals develop CXCR4-using viruses.[62] However, it is currently clear that there are large differences in the prevalence among HIV-1 subtypes. Thus, the proportion of individuals developing CXCR4-using viruses varies between 13% and 76% among the HIV-1 subtypes A–F, G, CRF01, and CRF02 (reviewed in Ref. 61). For example, AIDS patients infected with subtypes C and G rarely yield CXCR4-tropic viruses (only 13% and 12%, respectively), whereas CXCR4-using viruses are detected in a large majority of AIDS patients infected with subtype D and CRF01 viruses (70% and 76%, respectively). Furthermore, accumulating observations, especially among subtypes C[63] and CRF02,[64] infected individuals suggest that the proportion of individuals who will develop CXCR4-tropic viruses may increase and evolve over time.[61]

The underlying selection forces and mechanisms behind the intrapatient evolution of virus coreceptor use remain poorly understood; however,

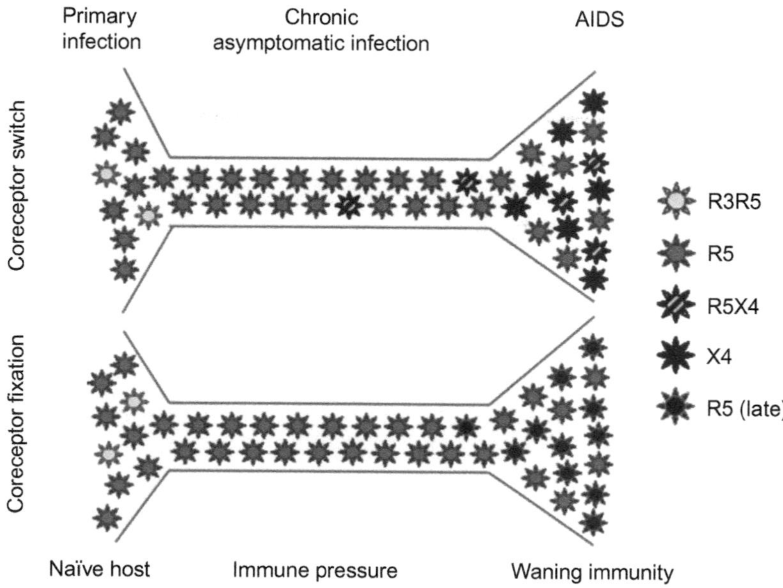

Figure 2 Two pathways of intratype HIV-1 coreceptor evolution in relation to disease progression and immune pressure. R5 viruses predominate early after transmission and during the asymptomatic chronic phase of the infection. During progressive disease, the evolution of HIV-1 coreceptor use can instead be divided into two distinct pathways, one where the R5 virus population evolves, or switches, toward CXCR4-tropism, and the other where the R5 virus population remains to be CCR5 restricted during the whole disease course. The proportion of individuals developing CXCR4-tropic viruses is, however, associated with the HIV-1 subtype.

different hypotheses have been put forward. Although it has repeatedly been observed that the emergence of CXCR4-tropic viruses coincides with accelerated disease progression,[55,56,65] it remains questionable if evolution toward CXCR4-tropic viruses is the cause or consequence of disease progression. In the macaque model, SIV infection rarely results in the emergence of CXCR4-tropic virus variants, instead, whereas in R5 SHIV-infected macaques the development of CXCR4-using variants has in several cases been associated with rapid disease progression (reviewed in Ref. 51). Leading hypotheses (reviewed in Ref. 66) on mechanisms behind the coreceptor switch are based on (i) transmission–mutation, (ii) immune control, or (iii) target cell availability. The transmission–mutation hypothesis is based on the preferential transmission and establishment of R5 viruses early in infection and the impact of gradual mutations triggering the emergence of CXCR4-using viruses. The immune control hypothesis proposes that

CXCR4-tropic viruses are easier to control during the phase of relative immunocompetence than R5 viruses. Thus, CXCR4-tropic viruses would only emerge as opportunists when the immune system is severely damaged. The target cell-based hypothesis suggests that the dynamics of CCR5 and CXCR4 on different subsets of potential target cells, i.e., naïve and memory $CD4^+$ T cells, during the course of the infection would determine the timing of the coreceptor switch. Although each of these hypotheses may contribute to coreceptor evolution, it is likely that they also act in concert. Another hypothesis put forward is in part based on a combination of the above hypothesis and suggests that chimeric viruses, being recombinants of coexisting R5 and X4 populations may be involved in the emergence and expansion of CXCR4-tropic viruses.[67] The chimeric viruses, consisting of an R5-derived gp120 region (V1–V2) that is plausibly less sensitive to neutralizing antibodies and an X4-derived gp120 region (V3) that facilitates expanded target cell tropism, would enable the more immune sensitive CXCR4-using viruses to expand without being targeted by neutralizing antibodies. Subsequently, it has also been shown in the macaque model that the gp120 V3 region of an X4 virus may be transferred into an R5 backbone by recombination in animals coinfected with R5 and X4 SHIV strains.[68] However, it is commonly accepted that CXCR4-tropic viruses may evolve from founder R5 viruses during the disease course. In the macaque model, it was shown that increased rate of SHIV evolution and reduction in the availability of target cells with optimal CD4 expression in addition to the absence of immunological restrictions coincided with the emergence of CXCR4 virus variants with a more opened Env.[69,70] Recent findings also suggest that HIV-1 coreceptor switch is associated with higher evolutionary rates on both the synonymous and nonsynonymous substitution level, but not with dN/dS ratio rates.[71] A stepwise mutational pathway in the gp120 V3 region during transition from R5 to CXCR4-tropic viruses was revealed via ultradeep sequencing of sequentially obtained samples obtained near the time point of the coreceptor switch.[72] Once CXCR4-using viruses have emerged, they usually remain in circulation of untreated individuals; however, the transient appearance of CXCR4-tropic variants has been reported, highlighting that this process is dynamic and not absolutely irreversible.[73] Moreover, results from ultradeep sequencing suggest that the CXCR4-using population may even diversify in cell reservoirs of patients on long-term suppressive ART.[74]

The observation that children vertically infected with R5 HIV-1 are predisposed to the development of CXCR4-using viruses if their mothers also carry CXCR4-tropic viruses[75] implies that coreceptor switch may be linked to specific virus properties of the transmitted virus or that similarity in genetics may aid such evolution. However, because the prevalence of coreceptor switch is strongly correlated with the HIV-1 subtype, a virus-specific trait appears to be a stronger determinant of CXCR4-tropic virus evolution than host genetics. The reason for differences in the prevalence of CXCR4-using viruses among HIV-1 subtypes remains unknown. A recent study proposed that the evolution of CXCR4-using viruses from R5 strains requires more mutations in subtype C infected individuals compared with those infected with other subtypes.[76] Thus, it is plausible that structural differences in Env of various HIV-1 strains may contribute to the likelihood of the development of CXCR4-use.

6. THE CCR5-RESTRICTED PATHWAY

This pathway constitutes adaptive changes of the R5 virus population, while maintaining CCR5 dependency[59,60] (Fig. 2). These adaptive changes include both phenotypic and molecular alterations of Env, in parallel with disease progression.[32,59,77–87] As of today, the CCR5-restricted pathway appears to be more prevalent because HIV-1 subtype C infections dominate globally (>50%), and it has been estimated that more than 85% of AIDS patients only harbor the R5 phenotype.[61,88,89,64,90]

The alteration of the R5 virus population has been studied using sequential HIV-1 samples collected during disease progression. The results show that R5 virus variants emerging after the onset of AIDS display augmented *in vitro* fitness.[77,78,83,85] Increased infectivity in cultures of cell lines and primary T lymphocytes, as well as augmented fusogenicity and dominance in competition assays have been found to be traits of late R5 variants.[77,78,83,85] Other reports have established that the late R5 viruses also exhibit increased macrophage tropism (reviewed in Ref. 91). Moreover, late R5 viruses have been found to be more cytopathic and induce apoptosis to a greater extent.[81,87]

More detailed studies on the interaction between R5 viruses and different domains of CCR5 were explored using a panel of cell lines expressing chimeric CCR5/CXCR4 receptors, in which successively larger portions of CCR5 were replaced by the corresponding regions of CXCR4.[92] Findings from these studies revealed that early R5 isolates exclusively used wild-type CCR5 ($R5^{narrow}$ phenotype), whereas late R5 viruses were able to infect via one or more of the chimeric receptors ($R5^{broad1-3}$ phenotype), indicating intrapatient evolution over time to a more flexible use of CCR5.[79,93] These studies also revealed that the late R5 viruses could to a greater extent infect target cells independent of the CCR5 N-terminus. In line with this, the late R5 viruses were also more resistant to inhibition by the natural CCR5 ligands[32,59,79] that bind to the CCR5 N-terminus.[94] Evolution in the mode of HIV-1 CCR5 use occurs during disease progression in both adults and children.[79,93,95–97] This evolution parallels decline of $CD4^+$ T cells, in that individuals with the most severe immune suppression harbor R5 viruses with the most flexible CCR5 use ($R5^{broad3}$). Most importantly, an altered mode of CCR5 interaction also translates into intrapatient evolution of sensitivity to entry inhibitors. Thus, in those who are naïve to entry inhibitor treatment, the R5 variants that emerge after AIDS exhibit increased resistance to both natural CCR5 ligands, such as CCL5, and the small-molecule CCR5 antagonist TAK-779[32,59,83] in parallel with a more flexible CCR5 interaction.[79,97] Interestingly, it was recently reported that R5 Envs from viruses obtained during the chronic phase displayed increased ability to interact with drug-bound CCR5 as compared with R5 Envs of TF viruses.[98] This suggests that a more promiscuous use of CCR5 is selected against during transmission, while selected for during the course of the infection.

R5 virus evolution has been linked to structural modifications of Env, in which late R5 virus variants emerging after AIDS onset display reduced surface envelope glycoprotein, gp120, glycosylation.[77,99,100] Env glycosylation has been implicated in shielding against humoral immune defense, including neutralizing antibodies. In line with this hypothesis, potent neutralizing activity in sera obtained early in the infection correlates with the accumulation of potential N-linked glycosylation sites.[100] In contrast, it was noted that R5 viruses with reduced Env glycosylation obtained at the end-stage infection were more sensitive to broadly neutralizing antibodies.[99] Thus, it is plausible that in the severely immune-deficient host, the selection forces that favor a dense glycosylation shield are lost and that other Env modifications may be selected for. Indeed, R5 viruses detected in the immune-deficient host display gp120 with increased net positive charge.[84,101] Here, the mutations leading to an altered amino acid charge were mapped mainly to the variable regions, excluding the V3 loop. Thus, CCR5-restricted target cell entry appears not to enable a V3 region with an increased charge, in contrast to CXCR4-mediated entry.[102] An increased positive charge of R5 Env was found to correlate with elevated infectivity and increased cell attachment.[84] Combining the effects of an elevated number of positively charged amino acids and reduced numbers of negatively charged sugar residues may thus significantly alter the Env charge of late R5 virus variants. Interestingly, the initial step of retrovirus infections is primarily thought to be adsorption provided by nonspecific electrostatic interactions.[103] Thus, an elevation in the positive charge of R5 Env could reduce the electrostatic repulsion between the viral and cell membranes leading to increased adherence to the cell surface. Accordingly, at the time of failing immunity, HIV-1 R5 viruses with increased fitness for direct target cell infection, displaying Env modifications including altered glycosylation and charge, may emerge in an opportunistic manner.

7. CLRs IN HIV-1 INFECTION

In addition to the main receptor, CD4, and coreceptors, HIV-1 can interact with several CLRs, including DC-specific ICAM-3 grabbing nonintegrin (DC-SIGN), mannose receptor, Langerin and DC-SIGN homologs, expressed by DC, macrophages and endothelial cells.[9] The primary purpose of these receptors is to bind to carbohydrate domains, which are present on pathogens or host tissue, via their carbohydrate-recognition domains. Binding to host sugar motifs promotes the trafficking of these innate immune cells and the formation of the immunological synapse between T cells and antigen-presenting cells.[104] Furthermore, binding to sugar motifs on pathogens can lead to the internalization, degradation, and subsequent loading of antigens onto MHC molecules. However, several pathogens, including HIV-1, have developed the ability to bind and utilize these receptors for enhanced infection of target cells. Below is a short description of the currently known CLRs that can interact with HIV-1 (Fig. 3).

Since its discovery, DC-SIGN has attracted significant attention as an enhancing receptor for HIV-1[8] and a potential receptor for HIV-2 and SIV.[105] Its expression at mucosal sites has suggested a role in facilitating transmission and systemic dissemination. DC-SIGN-expressing cells have been demonstrated in female endo- and ectocervix,[106,107] both sites considered to be target sites for the sexual transmission of HIV-1. Foreskin tissue has demonstrated abundant DC-SIGN-expressing cells in close proximity to potential $CD4^+$ HIV-1 target cells.[108] Rectal DC-SIGN-expressing cells have been demonstrated to bind and efficiently transport HIV.[109] The presence of DC-SIGN-expressing maternal and fetal macrophages in the placenta imposes a role in vertical transmission.[110] Additionally, intestinal DCs expressing DC-SIGN have the ability to transport HIV-1 through

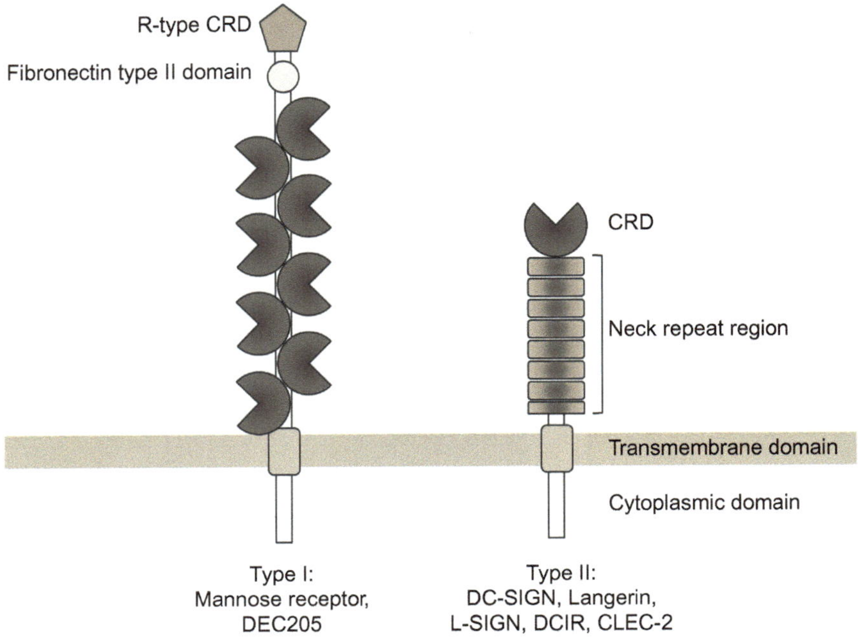

Figure 3 The membrane-bound C-type lectin receptors are classified into two groups based on their molecular structure. Type I contains several carbohydrate-recognition domains (CRDs) and an extracellular amino-terminal with R-type CRD followed by a fibronectin type II domain. Type II carries a single CRD domain at the carboxy-terminal extracellular domain. Both DC-SIGN and L-SIGN harbor a neck repeat region with several tandem repeats which can vary in number in the population. The type II receptors form tetramers on the cellular membrane.

intestinal mucosa to viral target cells.[111] The potential of HIV to bind to DC-SIGN via the highly glycosylated gp120 has been thoroughly established, but the outcome of this binding appears to be variable. The virus can simply hijack the DC-SIGN-expressing cell for transfer to T cells via an infectious synapse, a process known as *trans*-infection.[8] How the DCs transport infectious virus is not completely clear, whether mediated through internalized compartments,[8,112] binding to the surface[113], or surface-accessible compartments.[114] Another outcome of the DC-SIGN–HIV interaction is a productive infection of DCs and *de novo* synthesized viral particles.[115–117] The DC-SIGN interaction with HIV-1 glycoproteins has been implicated in enhancing *cis*-infection,[118] and signaling mediated by DC-SIGN binding has been demonstrated to be involved in the productive replication of virus in DCs.[119,120] The different outcomes of HIV interactions with DC-SIGN are illustrated in Fig. 4.

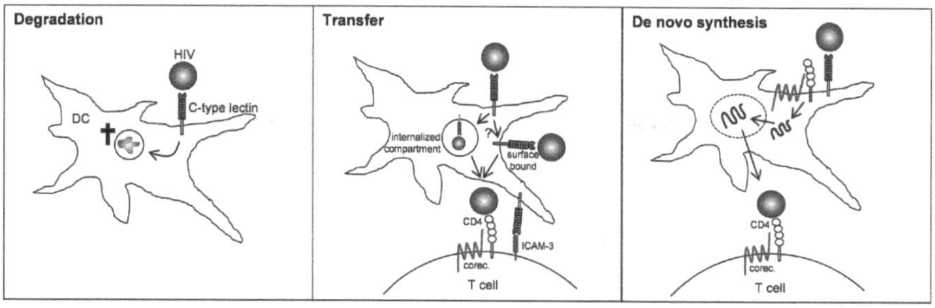

Figure 4 HIV-1 interaction with C-type lectin receptors may result in different outcomes. (A) Degradation of the virus results in antigen presentation by the DC and an antiviral immune response. (B) Virus is preserved in nonlysosomal compartments and DC migrates to the lymph nodes where HIV is transferred to T cells in the infectious synapse. (C) A productive viral infection of the DC results in *de novo* viral production and subsequent transmission to T cells.

DC-SIGNR, also known as L-SIGN, is closely related to DC-SIGN[121] and shares carbohydrate-recognition properties with DC-SIGN.[122] However, the expression pattern differs with DC-SIGNR being expressed on endothelial cells in the liver, lymph nodes, and placenta.[122–124] By binding to oligosaccharides on HIV-1 Env, DC-SIGNR can enhance infection of T cells in *trans*,[125,126] but DC-SIGNR can also function as an HIV-1 antigen-capture receptor.[127]

Another CLR, Langerin, expressed on epidermal Langerhans cells at mucosal sites, is reported to be among the first receptors to capture the virus.[128,129] HIV-1 binding to Langerin appears to occur via slightly different sugar motifs on gp120 than the residues that DC-SIGN preferentially binds to.[130] The fate of HIV-1 after interacting with Langerin may also differ from binding to DC-SIGN. Langerin that is expressed by skin Langerhans cells appears to serve as a natural barrier to HIV-1 transmission by directing endocytosed HIV-1 to Birbeck granules for degradation.[131] In contrast, vaginal Langerhans cells seem to promote *trans*-infection, although the involvement of Langerin in this process is not clarified.[132] HIV-1 has also been shown to effectively infect Langerhans cells.[133,134]

The mannose receptor is expressed by DC, macrophages, and epithelial cells, including those in the vagina, with the potential to capture HIV-1 at the site of transmission.[135] The low expression of mannose receptor in vaginal epithelial cells has been correlated with the prevention of sexual transmission.[136] The mannose receptor is also able to mediate the transmission of bound HIV to T cells in coculture conditions.[137] Alternatively, the mannose receptor-mediated uptake of HIV-1 can also lead to degradation and antigen presentation.[138]

The dendritic cell receptor (DCIR) is expressed on plasmacytoid DCs and possesses the ability to bind HIV-1, followed by enhancement of both *cis*- and *trans*-infection.[139,140] DCIR appears to play an additional role in HIV infection, because the virus itself can induce expression of the receptor on $CD4^+$ T cells, which potentially could be advantageous for infection.[141] The physiological functions of DCIR are not fully understood, but additional activities that promote virus replication can be involved because internalized DCIR inhibits the antiviral response after TLR8 and TLR9 signaling.[142,143]

Furthermore, the C-type lectin-like receptor 2 is expressed by lymphatic endothelial cells and platelets and has been demonstrated to capture and transfer infectious HIV-1 in association with DC-SIGN.[144]

HIV-1 has been detected in renal cells, which lack the expression of the main receptors CD4, CCR5, and CXCR4. Here, it appears that the C-type lectin DEC-205 can act as an HIV-1 receptor that mediates the internalization of the virus into renal tubular cells, leading to the establishment of a nonproductive infection.[145] A small fraction of incoming virus escapes degradation and can be rescued by the *trans*-infection of T cells.[146] Thus, renal tubular cells have been suggested to serve as reservoir for HIV-1.

8. CLRs AND HIV-1 INTERACTIONS DURING VIRUS TRANSMISSION

HIV-1 has evolved to interact with CLRs in various complex manners, implying that these interactions are important for the infection and potentially influence disease progression. How the virus evolves CLR use during disease progression has not been extensively studied. Virus–CLR interactions may result in different outcomes, depending on when during the infection such interactions occur.

A role for CLRs in HIV-1 transmission into a new host has been suggested, but little evidence except for the expression of CLRs at sites of transmission has been presented.[106,108–111,147–152] However, several reports have implicated genetic polymorphisms of CLRs with a risk of transmission. In particular mother to child transmission (MTCT) has gained interest concerning the genetic polymorphism of CLRs because placental cells, both on the maternal and fetal side, express the receptors.[110] Specific promoter variants of the DC-SIGN gene in infants have been associated with an increased risk of MTCT.[153] The genetic variants reduced the DC-SIGN expression on fetal macrophages but at the same time enhanced the capture and transmission of HIV-1 to $CD4^+$ T cells. Similarly, the polymorphism of the DC-SIGNR gene, which is expressed by fetal capillary endothelial cells in the placenta, has been associated with increased infectivity during pregnancy and at birth.[154] In this scenario, the DC-SIGNR variants resulted in a decrease in placental receptor expression but increased risk of transmission. A possible explanation for the relationship between fewer CLRs and increased transmission could be that the receptors have a protective role of the placental barrier by degrading the virus instead of transferring it. Less CLR would favor HIV-1 binding to CD4 and CCR5 instead and possibly facilitate migration from infected maternal cells to fetal target cells. In contrast, other promoter variants ofDC-SIGN also associated with a decrease in expression have been demonstrated to protect from MTCT.[155] Thus, even though genetic polymorphism of CLRs appear to have an effect on MTCT, the functional effects of the polymorphisms are not clear. We have previously demonstrated that virus isolated early after birth from children infected either *in utero* or *intrapartum* used DC-SIGN for *trans*-infection less efficiently than the corresponding maternal virus obtained close to delivery.[156] This suggests that the efficient use of DC-SIGN is not an advantage during MTCT. Our results are in agreement with the findings of Baan *et al.* that viruses transferred to children are not selected based on their capacity for DC-SIGN binding.[157]

The effect of a polymorphism in the DC-SIGN gene has also been associated with other routes of infection, which has been correlated both with protection and increased risk of infection. A certain genotype has been associated with an increased likelihood of infection via parenteral infection, but not mucosal infection.[158] The same genotype was associated with protection in a different population[159] or in the context of MTCT.[155]

DC-SIGNR differs from DC-SIGN receptors in the neck repeated region of the extracellular domain of the receptor, where DC-SIGNR is highly polymorphic in repetitions and DC-SIGN is more conserved. The polymorphism varies in the population and has been widely studied with regard to association with HIV-1 infection. Many studies are contradictive, and a metaanalysis summarizing the results of ten studies concludes that a polymorphism of the DC-SIGNR neck region is associated with susceptibility to HIV-1 infection.[160] However, amino acid substitutions in other regions of the DC-SIGNR gene may also have an effect on HIV-1 infection, possibly through reducing receptor expression.[161,162]

CLR interactions can indirectly affect an HIV-1 infection by binding to host glycoproteins and thereby competing with the virus at sites of infection instead of HIV-1. An example is vertical transmission via breastfeeding where the CLR can potentially bind and transfer HIV-1. Interestingly, several components of human milk have been found to inhibit HIV-1 infection by binding to CLR *in vitro*.[163–170] Similarly, host glycoproteins in seminal plasma can bind to DC-SIGN and prevent a viral interaction.[171,172]

Taken together, the effect of CLR at transmission is not clear, but the fact that *in vivo* expression of Langerin, DC-SIGN, and mannose receptors is higher in the mucosa of a high-risk population[107] suggests that the receptors play a role at the site of infection. Whether this increased expression is protective for the host or facilitates HIV-1 infection may depend on viral and host genetics and the timing of interactions during infection.

9. CLRS AND HIV-1 INTERACTIONS DURING THE CHRONIC INFECTION PHASE

In addition to its role at time of transmission, HIV-1 interaction with CLRs appears to have a potential role during disease progression. The expression level of DC-SIGN in HIV-1-infected individuals has not been correlated with disease progression,[173] but we have demonstrated that the viral use of DC-SIGN varies during the different phases of disease progression[77] (Fig. 5). CCR5-using HIV-1 variants emerging late at the end-stage of the disease course, after AIDS onset, were found to be less efficient in using DC-SIGN for *trans*-infection compared with corresponding virus isolated during the chronic phase.[77] This would suggest that HIV-1 DC-SIGN interactions are of greater importance in the earlier/chronic phases of the disease and that DC-SIGN could serve as an immune evasion mechanism that is no longer necessary late in the infection. It has been suggested that DC-SIGN serves as an escape from neutralizing antibodies, either through immune evasion from the neutralizing effects[174] or by blocking the neutralizing epitopes.[175] Indeed, myeloid DCs expressing DC-SIGN have been shown to efficiently transfer the virus, and at the same time, protect it from antibody neutralization.[176]

Changes in Env along with disease progression influence HIV-1 interactions with CLRs. The glycan composition of Env appears to determine, at least in part, the outcome of the DC-SIGN interaction via viral degradation or transmission to target cells.[177] Reduced Env glycosylation after the development of AIDS[99,100] most likely has an effect on interactions with

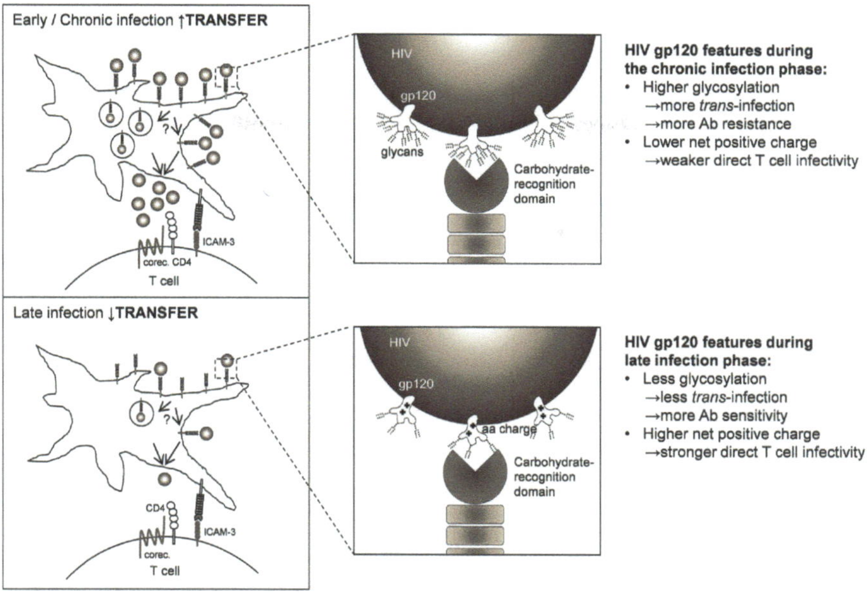

Figure 5 HIV-1 interactions with CLRs evolve during the different phases of disease and the outcome of such interaction may differ. During the early and chronic phases of the infection in the immunocompetent host, the virus envelope increases the glycan shield which not only protects from antibodies but also increases the binding to CLRs. Efficient use of CLRs does not only help the virus in spreading the infection but also seems to be an immune evasion mechanism. During late-stage disease the use of CLRs is diminished. Waning immune response allows the virus to reduce the glycan shield, which results in weaker use of CLRs and elevated antibody sensitivity. Instead the viruses emerging during late-stages are more infectious in a direct target cell infection manner.

CLRs because a decrease in glycan sites was shown to correlate with reduced ability to mediate *trans*-infection.[157] We and others reported that the presence of a specific potential N-linked glycosylation site in the V2 region of gp120 correlated it with efficient DC-SIGN use[77,178] Interestingly, emerging end-stage R5 viruses with a diminished use of DC-SIGN for *trans*-infection were also associated with a loss of this specific N-linked glycosylation site.[77] Glycans by themselves may not be sufficient for optimal CLR use; it has been reported that multiple modifications of gp120, including V1/V2 length and V3 charge, in combination with the N-linked glycosylation pattern affected DC-SIGN use.[179] The coreceptor use of HIV-1 may also have an effect on the interaction with C-type lectins because CXCR4-using viruses have been shown to use DC-SIGN mediated *trans*-infection more efficiently than R5 viruses.[180]

A few studies have associated the genetic polymorphism of DC-SIGN and its ligands with disease progression.[181,182] A certain DC-SIGN variant was associated with accelerated AIDS progression because it appeared at a higher frequency in fast progressors than in slow progressors.[181] Polymorphism within the bile salt-stimulated lipase gene, a ligand of DC-SIGN, appears to be associated with both binding to DC-SIGN and with the rate of disease progression.[182]

10. CLINICAL ASPECTS OF VIRUS EVOLUTION AT THE INTERFACE OF CORECEPTORS AND MANNOSE CLR

New treatment strategies based on the inhibition of virus entry have been prompted by the emergence of multidrug-resistant HIV-1 variants in infected individuals where the current standard antiretroviral treatment is failing.[183] Consequently, research on the development of new drugs that could block either the HIV-1 coreceptors or the mannose CLRs has been intense and led to a wide range of small-molecule CCR5 antagonists that effectively block both chemokine function and R5 HIV-1 entry.[184] Five such antagonists, including the currently only licensed CCR5 antagonist Maraviroc (MVC), were studied in more detail. Despite the fact that they are different in structure, shape, and electrostatic potential, they all fit into the same binding pocket formed by the transmembrane domains of CCR5.[184] However, each of these antagonists interacted in a different manner with the amino acids lining the pocket. These differences in the interaction may reflect unique alterations in the conformation of CCR5 after antagonist binding and predict the distinct inhibition sensitivity of different R5 virus variants. In the future, this may spur the development of individualized drug regiments for an optimal HIV-1 therapy. Accordingly, detailed studies on baseline binding characteristics of HIV-1 variants naturally evolving in the infected individual should be prioritized.

MVC is currently clinically used as a complement to other antiretroviral drugs, usually in patients with late-stage disease who experience virological failure with other regimens.[185] However, the introduction of MVC in the later stages of the disease may not be beneficial because the likelihood of the emergence of CXCR4-using virus variants as well as R5 viruses with reduced baseline sensitivity to entry inhibitors may already exist at baseline in such patients. Today either virus phenotypic (including the Trofile™, Monogram Biosciences) or genotypic (sequencing linked to bioinformatic algorithms) assays are used to determine if patients are eligible for MVC treatment (reviewed in Refs. 186,187) Next-generation sequencing platforms are also available. However, these assays only predict the virological response in 70–80% of those treated with MVC.[188] Clinical studies have revealed that the emergence of CXCR4-tropic viruses predominates in patients where MVC treatment fails.[185,189,190] Thus, although not detected at the initiation of MVC treatment it is likely that minor CXCR4-using virus variants are present and selected during treatment. Indeed, MVC has been demonstrated to efficiently block R5 viruses derived from the asymptomatic phase (reviewed in Ref. 185). This could indicate that CCR5 antagonist treatment strategies may be more effective in HIV-1 subtypes where the emergence of CXCR4-tropic viruses are rare, i.e., subtypes C and G.

Nonetheless, both *in vitro* R5 virus passage experiments and *in vivo* observations have revealed that MVC-resistant R5 variants can be selected for. In this case, R5 viruses were able to infect target cells via the MVC-bound CCR5, and resistance was due to changes in the virus CCR5-binding properties.[191,192] Studies on MVC as well as vicriviroc and AD101, which are two other CCR5 antagonists, have demonstrated that R5 viruses able to utilize drug-bound CCR5 have developed strong CCR5 N-terminus dependency.[193–196] Here, Envs from MVC-resistant R5 viruses have been shown to exhibit increased reliance on sulfated tyrosine residues in the CCR5 N-terminus.[193]

Although the role of HIV-1 interactions with CLRs has not been clarified to be pro- or antiviral, it is appealing to interfere with these interactions as a novel antiviral strategy. Both promoting and blocking the interactions could be of interest for different CLRs. Increasing Langerin involvement at cell entry and directing HIV-1 toward the degradation pathway could be a useful strategy. Blocking the interaction of HIV-1 with DC-SIGN during disease progression could possibly diminish the infection and viral immune evasion.

Carbohydrate-containing drugs (glycan decoys) or carbodhydrate-binding agents (CBAs) can directly block the viral interaction with CLRs. Glycan decoys can bind to specific CLRs and outcompete other ligands such as HIV-1. Glycan decoys containing mannose residues have been shown to block the binding of HIV-1 to DC-SIGN[197–200] and inhibit the *trans*-infection to T cells.[201–203] CBAs conversely bind to the virus and thereby block the CLR interaction. Several CBAs are effective in preventing the viral binding to DC-SIGN and the *trans*-infection.[204–209] However, the prolonged exposure of HIV-1 to CBAs selects for virus variants, which have lost glycosylation sites in the viral Env and developed resistance against

CBAs.[210–215] It appears as if the degree of resistance is increasing with increasing numbers of lost glycan sites in Env.[210–212] Interestingly, the loss of glycan sites also makes the mutated HIV-1 less infective with a decrease in viral fitness.[216] In addition, a loss in the glycan shield of HIV-1 gp120 means that neutralizing antibody can more easily access the viral glycoproteins and block the virus.[217–219] Thus, the potential role of CBAs (summarized in reviews 220 and 221) could not only be to block the CLR interactions and subsequent *trans*-infection, but also could increase immunogenicity of the virus and cause a reduction in viral fitness.

ACKNOWLEDGMENTS

We are indebted to Anders Fomsgaard and Eva Maria Fenyö for discussions and critical comments on the manuscript. Grant sources were received from the Swedish Research Council and the Crafoord Foundation.

REFERENCES

1. Dalgleish AG, Beverley PC, Clapham PR, Crawford DH, Greaves MF, Weiss RA. The CD4 (T4) antigen is an essential component of the receptor for the AIDS retrovirus. *Nature*. 1984;312(5996):763–767.
2. Feng Y, Broder CC, Kennedy PE, Berger EA. HIV-1 entry cofactor: functional cDNA cloning of a seven-transmembrane, G protein-coupled receptor. *Science*. 1996;272(5263): 872–877.
3. Dragic T, Litwin V, Allaway GP, et al. HIV-1 entry into CD4+ cells is mediated by the chemokine receptor CC-CKR-5. *Nature*. 1996;381(6584):667–673.
4. Doranz BJ, Rucker J, Yi Y, et al. A dual-tropic primary HIV-1 isolate that uses fusin and the beta-chemokine receptors CKR-5, CKR-3, and CKR-2b as fusion cofactors. *Cell*. 1996;85(7):1149–1158.
5. Choe H, Farzan M, Sun Y, et al. The beta-chemokine receptors CCR3 and CCR5 facilitate infection by primary HIV-1 isolates. *Cell*. 1996;85(7):1135–1148.
6. Alkhatib G, Locati M, Kennedy PE, Murphy PM, Berger EA. HIV-1 coreceptor activity of CCR5 and its inhibition by chemokines: independence from G protein signaling and importance of coreceptor downmodulation. *Virology*. 1997;234(2):340–348.
7. Deng H, Liu R, Ellmeier W, et al. Identification of a major co-receptor for primary isolates of HIV-1. *Nature*. 1996;381(6584):661–666.
8. Geijtenbeek TB, Kwon DS, Torensma R, et al. DC-SIGN, a dendritic cell-specific HIV-1-binding protein that enhances trans-infection of T cells. *Cell*. 2000;100(5):587–597.
9. Wu L, KewalRamani VN. Dendritic-cell interactions with HIV: infection and viral dissemination. *Nat Rev Immunol*. 2006;6(11):859–868.
10. Maddon PJ, Dalgleish AG, McDougal JS, Clapham PR, Weiss RA, Axel R. The T4 gene encodes the AIDS virus receptor and is expressed in the immune system and the brain. *Cell*. 1986;47(3):333–348.
11. Tersmette M, de Goede RE, Al BJ, et al. Differential syncytium-inducing capacity of human immunodeficiency virus isolates: frequent detection of syncytium-inducing isolates in patients with acquired immunodeficiency syndrome (AIDS) and AIDS-related complex. *J Virol*. 1988;62(6):2026–2032.

12. Cheng-Mayer C, Seto D, Tateno M, Levy JA. Biologic features of HIV-1 that correlate with virulence in the host. *Science*. 1988;240(4848):80–82.
13. Asjo B, Morfeldt-Manson L, Albert J, et al. Replicative capacity of human immunodeficiency virus from patients with varying severity of HIV infection. *Lancet*. 1986;2(8508):660–662.
14. Fenyo EM, Morfeldt-Manson L, Chiodi F, et al. Distinct replicative and cytopathic characteristics of human immunodeficiency virus isolates. *J Virol*. 1988;62(11): 4414–4419.
15. Cocchi F, DeVico AL, Garzino-Demo A, Arya SK, Gallo RC, Lusso P. Identification of RANTES, MIP-1 alpha, and MIP-1 beta as the major HIV-suppressive factors produced by CD8+ T cells. *Science*. 1995;270(5243):1811–1815.
16. Berger EA, Doms RW, Fenyo EM, et al. A new classification for HIV-1. *Nature*. 1998;391(6664):240.
17. Irlbeck DM, Amrine-Madsen H, Kitrinos KM, Labranche CC, Demarest JF. Chemokine (C-C motif) receptor 5-using envelopes predominate in dual/mixed-tropic HIV from the plasma of drug-naive individuals. *AIDS*. 2008;22(12):1425–1431.
18. Yi Y, Isaacs SN, Williams DA, et al. Role of CXCR4 in cell-cell fusion and infection of monocyte-derived macrophages by primary human immunodeficiency virus type 1 (HIV-1) strains: two distinct mechanisms of HIV-1 dual tropism. *J Virol*. 1999;73(9):7117–7125.
19. Pollakis G, Paxton WA. Use of (alternative) coreceptors for HIV entry. *Curr Opin HIV AIDS*. 2012;7(5):440–449.
20. He J, Chen Y, Farzan M, et al. CCR3 and CCR5 are co-receptors for HIV-1 infection of microglia. *Nature*. 1997;385(6617):645–649.
21. Karlsson U, Antonsson L, Ljungberg B, et al. Dual R3R5 tropism characterizes cerebrospinal fluid HIV-1 isolates from individuals with high cerebrospinal fluid viral load. *AIDS*. 2012;26(14):1739–1744.
22. Aasa-Chapman MM, Aubin K, Williams I, McKnight A. Primary CCR5 only using HIV-1 isolates does not accurately represent the in vivo replicating quasi-species. *Virology*. 2006;351(2):489–496.
23. Liu R, Paxton WA, Choe S, et al. Homozygous defect in HIV-1 coreceptor accounts for resistance of some multiply-exposed individuals to HIV-1 infection. *Cell*. 1996;86(3):367–377.
24. Samson M, Libert F, Doranz BJ, et al. Resistance to HIV-1 infection in caucasian individuals bearing mutant alleles of the CCR-5 chemokine receptor gene. *Nature*. 1996;382(6593):722–725.
25. O'Brien TR, Winkler C, Dean M, et al. HIV-1 infection in a man homozygous for CCR5 delta 32. *Lancet*. 1997;349(9060):1219.
26. Morner A, Bjorndal A, Albert J, et al. Primary human immunodeficiency virus type 2 (HIV-2) isolates, like HIV-1 isolates, frequently use CCR5 but show promiscuity in coreceptor usage. *J Virol*. 1999;73(3):2343–2349.
27. Blaak H, Boers PH, Gruters RA, Schuitemaker H, van der Ende ME, Osterhaus AD. CCR5, GPR15, and CXCR6 are major coreceptors of human immunodeficiency virus type 2 variants isolated from individuals with and without plasma viremia. *J Virol*. 2005;79(3):1686–1700.
28. Neil SJ, Aasa-Chapman MM, Clapham PR, Nibbs RJ, McKnight A, Weiss RA. The promiscuous CC chemokine receptor D6 is a functional coreceptor for primary isolates of human immunodeficiency virus type 1 (HIV-1) and HIV-2 on astrocytes. *J Virol*. 2005;79(15):9618–9624.
29. Loetscher P, Moser B, Baggiolini M. Chemokines and their receptors in lymphocyte traffic and HIV infection. *Adv Immunol*. 2000;74:127–180.

30. Bleul CC, Wu L, Hoxie JA, Springer TA, Mackay CR. The HIV coreceptors CXCR4 and CCR5 are differentially expressed and regulated on human T lymphocytes. *Proc Natl Acad Sci U S A*. 1997;94(5):1925–1930.
31. Bleul CC, Farzan M, Choe H, et al. The lymphocyte chemoattractant SDF-1 is a ligand for LESTR/fusin and blocks HIV-1 entry. *Nature*. 1996;382(6594):829–833.
32. Jansson M, Popovic M, Karlsson A, et al. Sensitivity to inhibition by beta-chemokines correlates with biological phenotypes of primary HIV-1 isolates. *Proc Natl Acad Sci U S A*. 1996;93(26):15382–15387.
33. Bjorndal A, Deng H, Jansson M, et al. Coreceptor usage of primary human immunodeficiency virus type 1 isolates varies according to biological phenotype. *J Virol*. 1997;71(10):7478–7487.
34. Scarlatti G, Leitner T, Halapi E, et al. Comparison of variable region 3 sequences of human immunodeficiency virus type 1 from infected children with the RNA and DNA sequences of the virus populations of their mothers. *Proc Natl Acad Sci U S A*. 1993;90(5):1721–1725.
35. Salazar-Gonzalez JF, Bailes E, Pham KT, et al. Deciphering human immunodeficiency virus type 1 transmission and early envelope diversification by single-genome amplification and sequencing. *J Virol*. 2008;82(8):3952–3970.
36. Zhu T, Mo H, Wang N, et al. Genotypic and phenotypic characterization of HIV-1 patients with primary infection. *Science*. 1993;261(5125):1179–1181.
37. van't Wout AB, Kootstra NA, Mulder-Kampinga GA, et al. Macrophage-tropic variants initiate human immunodeficiency virus type 1 infection after sexual, parenteral, and vertical transmission. *J Clin Invest*. 1994;94(5):2060–2067.
38. Long EM, Rainwater SM, Lavreys L, Mandaliya K, Overbaugh J. HIV type 1 variants transmitted to women in Kenya require the CCR5 coreceptor for entry, regardless of the genetic complexity of the infecting virus. *AIDS Res Hum Retrovir*. 2002;18(8): 567–576.
39. Salazar-Gonzalez JF, Salazar MG, Keele BF, et al. Genetic identity, biological phenotype, and evolutionary pathways of transmitted/founder viruses in acute and early HIV-1 infection. *J Exp Med*. 2009;206(6):1273–1289.
40. Keele BF, Giorgi EE, Salazar-Gonzalez JF, et al. Identification and characterization of transmitted and early founder virus envelopes in primary HIV-1 infection. *Proc Natl Acad Sci U S A*. 2008;105(21):7552–7557.
41. Grivel JC, Shattock RJ, Margolis LB. Selective transmission of R5 HIV-1 variants: where is the gatekeeper? *J Transl Med*. 2011;9(suppl 1):S6.
42. Meng G, Wei X, Wu X, et al. Primary intestinal epithelial cells selectively transfer R5 HIV-1 to CCR5+ cells. *Nat Med*. 2002;8(2):150–156.
43. Cavarelli M, Foglieni C, Rescigno M, Scarlatti G. R5 HIV-1 envelope attracts dendritic cells to cross the human intestinal epithelium and sample luminal virions via engagement of the CCR5. *EMBO Mol Med*. 2013;5(5):776–794.
44. Zhang Z, Schuler T, Zupancic M, et al. Sexual transmission and propagation of SIV and HIV in resting and activated CD4+ T cells. *Science*. 1999;286(5443):1353–1357.
45. Ochsenbauer C, Edmonds TG, Ding H, et al. Generation of transmitted/founder HIV-1 infectious molecular clones and characterization of their replication capacity in CD4 T lymphocytes and monocyte-derived macrophages. *J Virol*. 2012;86(5): 2715–2728.
46. Brenchley JM, Schacker TW, Ruff LE, et al. CD4+ T cell depletion during all stages of HIV disease occurs predominantly in the gastrointestinal tract. *J Exp Med*. 2004;200(6):749–759.
47. Veazey RS, DeMaria M, Chalifoux LV, et al. Gastrointestinal tract as a major site of CD4+ T cell depletion and viral replication in SIV infection. *Science*. 1998;280(5362): 427–431.

48. Arthos J, Cicala C, Martinelli E, et al. HIV-1 envelope protein binds to and signals through integrin alpha4beta7, the gut mucosal homing receptor for peripheral T cells. *Nat Immunol.* 2008;9(3):301–309.
49. Parrish NF, Wilen CB, Banks LB, et al. Transmitted/founder and chronic subtype C HIV-1 use CD4 and CCR5 receptors with equal efficiency and are not inhibited by blocking the integrin alpha4beta7. *PLoS Pathog.* 2012;8(5):e1002686.
50. Harouse JM, Buckner C, Gettie A, et al. CD8+ T cell-mediated CXC chemokine receptor 4-simian/human immunodeficiency virus suppression in dually infected rhesus macaques. *Proc Natl Acad Sci U S A.* 2003;100(19):10977–10982.
51. Sina ST, Ren W, Cheng-Mayer C. Coreceptor use in nonhuman primate models of HIV infection. *J Transl Med.* 2011;9(suppl 1):S7.
52. Scarlatti G, Hodara V, Rossi P, et al. Transmission of human immunodeficiency virus type 1 (HIV-1) from mother to child correlates with viral phenotype. *Virology.* 1993;197(2):624–629.
53. Huang W, Eshleman SH, Toma J, et al. Vertical transmission of X4-tropic and dual-tropic HIV-1 in five Ugandan mother-infant pairs. *AIDS.* 2009;23(14):1903–1908.
54. Cavarelli M, Scarlatti G. HIV-1 co-receptor usage: influence on mother-to-child transmission and pediatric infection. *J Transl Med.* 2011;9(suppl 1):S10.
55. Karlsson A, Parsmyr K, Sandstrom E, Fenyo EM, Albert J. MT-2 cell tropism as prognostic marker for disease progression in human immunodeficiency virus type 1 infection. *J Clin Microbiol.* 1994;32(2):364–370.
56. Koot M, Keet IP, Vos AH, et al. Prognostic value of HIV-1 syncytium-inducing phenotype for rate of CD4+ cell depletion and progression to AIDS. *Ann Intern Med.* 1993;118(9):681–688.
57. Richman DD, Bozzette SA. The impact of the syncytium-inducing phenotype of human immunodeficiency virus on disease progression. *J Infect Dis.* 1994;169(5):968–974.
58. Connor RI, Sheridan KE, Ceradini D, Choe S, Landau NR. Change in coreceptor use correlates with disease progression in HIV-1–infected individuals. *J Exp Med.* 1997;185(4):621–628.
59. Jansson M, Backstrom E, Bjorndal A, et al. Coreceptor usage and RANTES sensitivity of non-syncytium-inducing HIV-1 isolates obtained from patients with AIDS. *J Hum Virol.* 1999;2(6):325–338.
60. de Roda Husman AM, van Rij RP, Blaak H, Broersen S, Schuitemaker H. Adaptation to promiscuous usage of chemokine receptors is not a prerequisite for human immunodeficiency virus type 1 disease progression. *J Infect Dis.* 1999;180(4):1106–1115.
61. Fenyo EM, Esbjornsson J, Medstrand P, Jansson M. Human immunodeficiency virus type 1 biological variation and coreceptor use: from concept to clinical significance. *J Intern Med.* 2011;270(6):520–531.
62. Spijkerman IJ, Koot M, Prins M, et al. Lower prevalence and incidence of HIV-1 syncytium-inducing phenotype among injecting drug users compared with homosexual men. *AIDS (London, England).* 1995;9(9):1085–1092.
63. Connell BJ, Michler K, Capovilla A, Venter WD, Stevens WS, Papathanasopoulos MA. Emergence of X4 usage among HIV-1 subtype C: evidence for an evolving epidemic in South Africa. *AIDS.* 2008;22(7):896–899.
64. Esbjornsson J, Mansson F, Martinez-Arias W, et al. Frequent CXCR4 tropism of HIV-1 subtype A and CRF02_AG during late-stage disease—indication of an evolving epidemic in West Africa. *Retrovirology.* 2010;7(1):23.
65. Shepherd JC, Jacobson LP, Qiao W, et al. Emergence and persistence of CXCR4-tropic HIV-1 in a population of men from the multicenter AIDS cohort study. *J Infect Dis.* 2008;198(8):1104–1112.

66. Regoes RR, Bonhoeffer S. The HIV coreceptor switch: a population dynamical perspective. *Trends Microbiol.* 2005;13(6):269–277.
67. Mild M, Esbjornsson J, Fenyo EM, Medstrand P. Frequent intrapatient recombination between human immunodeficiency virus type 1 R5 and X4 envelopes: implications for coreceptor switch. *J Virol.* 2007;81(7):3369–3376.
68. Nishimura Y, Shingai M, Lee WR, et al. Recombination-mediated changes in coreceptor usage confer an augmented pathogenic phenotype in a nonhuman primate model of HIV-1-induced AIDS. *J Virol.* 2011;85(20):10617–10626.
69. Zhuang K, Finzi A, Tasca S, et al. Adoption of an "open" envelope conformation facilitating CD4 binding and structural remodeling precedes coreceptor switch in R5 SHIV-infected macaques. *PLoS One.* 2011;6(7):e21350.
70. Zhuang K, Finzi A, Toma J, et al. Identification of interdependent variables that influence coreceptor switch in R5 SHIV(SF162P3N)-infected macaques. *Retrovirology.* 2012;9:106.
71. Mild M, Gray RR, Kvist A, et al. High intrapatient HIV-1 evolutionary rate is associated with CCR5-to-CXCR4 coreceptor switch. *Infect Genet Evol.* 2013;19:369–377.
72. Bunnik EM, Swenson LC, Edo-Matas D, et al. Detection of inferred CCR5- and CXCR4-using HIV-1 variants and evolutionary intermediates using ultra-deep pyrosequencing. *PLoS Pathog.* 2011;7(6):e1002106.
73. Ida S, Gatanaga H, Shioda T, et al. HIV type 1 V3 variation dynamics in vivo: long-term persistence of non-syncytium-inducing genotypes and transient presence of syncytium-inducing genotypes during the course of progressive AIDS. *AIDS Res Hum Retrovir.* 1997;13(18):1597–1609.
74. Raymond S, Saliou A, Delobel P, et al. Evolution of HIV-1 quasispecies and coreceptor use in cell reservoirs of patients on suppressive antiretroviral therapy. *J Antimicrob Chemother.* 2014;69(9):2527–2530.
75. Casper CH, Clevestig P, Carlenor E, et al. Link between the X4 phenotype in human immunodeficiency virus type 1-infected mothers and their children, despite the early presence of R5 in the child. *J Infect Dis.* 2002;186(7):914–921.
76. Coetzer M, Nedellec R, Cilliers T, Meyers T, Morris L, Mosier DE. Extreme genetic divergence is required for coreceptor switching in HIV-1 subtype C. *J Acquir Immune Defic Syndr.* 2011;56(1):9–15.
77. Borggren M, Repits J, Kuylenstierna C, et al. Evolution of DC-SIGN use revealed by fitness studies of R5 HIV-1 variants emerging during AIDS progression. *Retrovirology.* 2008;5:28.
78. Gray L, Sterjovski J, Churchill M, et al. Uncoupling coreceptor usage of human immunodeficiency virus type 1 (HIV-1) from macrophage tropism reveals biological properties of CCR5-restricted HIV-1 isolates from patients with acquired immunodeficiency syndrome. *Virology.* 2005;337(2):384–398.
79. Karlsson I, Antonsson L, Shi Y, et al. Coevolution of RANTES sensitivity and mode of CCR5 receptor use by human immunodeficiency virus type 1 of the R5 phenotype. *J Virol.* 2004;78(21):11807–11815.
80. Koning FA, Kwa D, Boeser-Nunnink B, et al. Decreasing sensitivity to RANTES (regulated on activation, normally T cell-expressed and -secreted) neutralization of CC chemokine receptor 5-using, non-syncytium-inducing virus variants in the course of human immunodeficiency virus type 1 infection. *J Infect Dis.* 2003;188(6):864–872.
81. Kwa D, Vingerhoed J, Boeser B, Schuitemaker H. Increased in vitro cytopathicity of CC chemokine receptor 5-restricted human immunodeficiency virus type 1 primary isolates correlates with a progressive clinical course of infection. *J Infect Dis.* 2003;187(9):1397–1403.
82. Mild M, Kvist A, Esbjornsson J, Karlsson I, Fenyo EM, Medstrand P. Differences in molecular evolution between switch (R5 to R5X4/X4-tropic) and non-switch

(R5-tropic only) HIV-1 populations during infection. *Infect Genet Evol.* 2010;10(3): 356–364.
83. Repits J, Oberg M, Esbjornsson J, et al. Selection of human immunodeficiency virus type 1 R5 variants with augmented replicative capacity and reduced sensitivity to entry inhibitors during severe immunodeficiency. *J Gen Virol.* 2005;86(pt 10):2859–2869.
84. Repits J, Sterjovski J, Badia-Martinez D, et al. Primary HIV-1 R5 isolates from end-stage disease display enhanced viral fitness in parallel with increased gp120 net charge. *Virology.* 2008;379(1):125–134.
85. Sterjovski J, Churchill MJ, Ellett A, et al. Asn 362 in gp120 contributes to enhanced fusogenicity by CCR5-restricted HIV-1 envelope glycoprotein variants from patients with AIDS. *Retrovirology.* 2007;4:89.
86. Sterjovski J, Roche M, Churchill MJ, et al. An altered and more efficient mechanism of CCR5 engagement contributes to macrophage tropism of CCR5-using HIV-1 envelopes. *Virology.* 2010;404(2):269–278.
87. Wade J, Sterjovski J, Gray L, et al. Enhanced CD4+ cellular apoptosis by CCR5-restricted HIV-1 envelope glycoprotein variants from patients with progressive HIV-1 infection. *Virology.* 2010;396(2):246–255.
88. Abebe A, Demissie D, Goudsmit J, et al. HIV-1 subtype C syncytium- and non-syncytium-inducing phenotypes and coreceptor usage among Ethiopian patients with AIDS. *AIDS.* 1999;13(11):1305–1311.
89. Bjorndal A, Sonnerborg A, Tscherning C, Albert J, Fenyo EM. Phenotypic characteristics of human immunodeficiency virus type 1 subtype C isolates of Ethiopian AIDS patients. *AIDS Res Hum Retrovir.* 1999;15(7):647–653.
90. Tscherning C, Alaeus A, Fredriksson R, et al. Differences in chemokine coreceptor usage between genetic subtypes of HIV-1. *Virology.* 1998;241(2):181–188.
91. Duenas-Decamp MJ, Peters PJ, Repik A, et al. Variation in the biological properties of HIV-1 R5 envelopes: implications of envelope structure, transmission and pathogenesis. *Futur Virol.* 2010;5(4):435–451.
92. Antonsson L, Boketoft A, Garzino-Demo A, Olde B, Owman C. Molecular mapping of epitopes for interaction of HIV-1 as well as natural ligands with the chemokine receptors, CCR5 and CXCR4. *AIDS.* 2003;17(18):2571–2579.
93. Karlsson I, Antonsson L, Shi Y, et al. HIV biological variability unveiled: frequent isolations and chimeric receptors reveal unprecedented variation of coreceptor use. *AIDS.* 2003;17(18):2561–2569.
94. Zhou N, Luo Z, Hall JW, Luo J, Han X, Huang Z. Molecular modeling and site-directed mutagenesis of CCR5 reveal residues critical for chemokine binding and signal transduction. *Eur J Immunol.* 2000;30(1):164–173.
95. Cavarelli M, Karlsson I, Ripamonti C, Plebani A, Fenyo EM, Scarlatti G. Flexible use of CCR5 in the absence of CXCR4 use explains the immune deficiency in HIV-1 infected children. *AIDS.* 2010;24(16):2527–2533.
96. Cavarelli M, Karlsson I, Zanchetta M, et al. HIV-1 with multiple CCR5/CXCR4 chimeric receptor use is predictive of immunological failure in infected children. *PLoS One.* 2008;3(9):e3292.
97. Karlsson U, Antonsson L, Repits J, et al. Mode of coreceptor use by R5 HIV type 1 correlates with disease stage: a study of paired plasma and cerebrospinal fluid isolates. *AIDS Res Hum Retrovir.* 2009;25(12):1297–1305.
98. Parker ZF, Iyer SS, Wilen CB, et al. Transmitted/founder and chronic HIV-1 envelope proteins are distinguished by differential utilization of CCR5. *J Virol.* 2013;87(5):2401–2411.
99. Borggren M, Repits J, Sterjovski J, et al. Increased sensitivity to broadly neutralizing antibodies of end-stage disease R5 HIV-1 correlates with evolution in Env glycosylation and charge. *PLoS One.* 2011;6(6):e20135.

100. Bunnik EM, Pisas L, van Nuenen AC, Schuitemaker H. Autologous neutralizing humoral immunity and evolution of the viral envelope in the course of subtype B human immunodeficiency virus type 1 infection. *J Virol.* 2008;82(16):7932–7941.
101. Seclen E, Soriano V, Del Mar Gonzalez M, Gonzalez-Lahoz J, Poveda E. Severe immune suppression in patients infected with R5-tropic HIV-1 strains is associated with increased gp120 net charge at variable regions. *AIDS Res Hum Retrovir.* 2011;27(9):965–967.
102. Fouchier RA, Groenink M, Kootstra NA, et al. Phenotype-associated sequence variation in the third variable domain of the human immunodeficiency virus type 1 gp120 molecule. *J Virol.* 1992;66(5):3183–3187.
103. Davis HE, Rosinski M, Morgan JR, Yarmush ML. Charged polymers modulate retrovirus transduction via membrane charge neutralization and virus aggregation. *Biophys J.* 2004;86(2):1234–1242.
104. van Kooyk Y, Geijtenbeek TB. DC-SIGN: escape mechanism for pathogens. *Nat Rev Immunol.* 2003;3(9):697–709.
105. Pohlmann S, Baribaud F, Lee B, et al. DC-SIGN interactions with human immunodeficiency virus type 1 and 2 and simian immunodeficiency virus. *J Virol.* 2001;75(10):4664–4672.
106. Hirbod T, Kaldensjo T, Broliden K. In situ distribution of HIV-binding CCR5 and C-type lectin receptors in the human endocervical mucosa. *PLoS One.* 2011;6(9): e25551.
107. Hirbod T, Kaldensjo T, Lopalco L, et al. Abundant and superficial expression of C-type lectin receptors in ectocervix of women at risk of HIV infection. *J Acquir Immune Defic Syndr.* 2009;51(3):239–247.
108. Hirbod T, Bailey RC, Agot K, et al. Abundant expression of HIV target cells and C-type lectin receptors in the foreskin tissue of young Kenyan men. *Am J Pathol.* 2010;176(6):2798–2805.
109. Gurney KB, Elliott J, Nassanian H, et al. Binding and transfer of human immunodeficiency virus by DC-SIGN+ cells in human rectal mucosa. *J Virol.* 2005;79(9): 5762–5773.
110. Soilleux EJ, Morris LS, Lee B, et al. Placental expression of DC-SIGN may mediate intrauterine vertical transmission of HIV. *J Pathol.* 2001;195(5):586–592.
111. Shen R, Smythies LE, Clements RH, Novak L, Smith PD. Dendritic cells transmit HIV-1 through human small intestinal mucosa. *J Leukoc Biol.* 2010;87(4):663–670.
112. Kwon DS, Gregorio G, Bitton N, Hendrickson WA, Littman DR. DC-SIGN-mediated internalization of HIV is required for trans-enhancement of T cell infection. *Immunity.* 2002;16(1):135–144.
113. Cavrois M, Neidleman J, Kreisberg JF, Greene WC. In vitro derived dendritic cells trans-infect CD4 T cells primarily with surface-bound HIV-1 virions. *PLoS Pathog.* 2007;3(1):e4.
114. Yu HJ, Reuter MA, McDonald D. HIV traffics through a specialized, surface-accessible intracellular compartment during trans-infection of T cells by mature dendritic cells. *PLoS Pathog.* 2008;4(8):e1000134.
115. Burleigh L, Lozach PY, Schiffer C, et al. Infection of dendritic cells (DCs), not DC-SIGN-mediated internalization of human immunodeficiency virus, is required for long-term transfer of virus to T cells. *J Virol.* 2006;80(6):2949–2957.
116. Nobile C, Petit C, Moris A, et al. Covert human immunodeficiency virus replication in dendritic cells and in DC-SIGN-expressing cells promotes long-term transmission to lymphocytes. *J Virol.* 2005;79(9):5386–5399.
117. Turville SG, Santos JJ, Frank I, et al. Immunodeficiency virus uptake, turnover, and 2-phase transfer in human dendritic cells. *Blood.* 2004;103(6):2170–2179.
118. Hijazi K, Wang Y, Scala C, et al. DC-SIGN increases the affinity of HIV-1 envelope glycoprotein interaction with CD4. *PLoS One.* 2011;6(12):e28307.

119. Gringhuis SI, van der Vlist M, van den Berg LM, den Dunnen J, Litjens M, Geijtenbeek TB. HIV-1 exploits innate signaling by TLR8 and DC-SIGN for productive infection of dendritic cells. *Nat Immunol.* 2010;11(5):419–426.
120. van der Vlist M, van der Aar AM, Gringhuis SI, Geijtenbeek TB. Innate signaling in HIV-1 infection of dendritic cells. *Curr Opin HIV AIDS.* 2011;6(5):348–352.
121. Soilleux EJ, Barten R, Trowsdale J. DC-SIGN; a related gene, DC-SIGNR; and CD23 form a cluster on 19p13. *J Immunol.* 2000;165(6):2937–2942.
122. Pohlmann S, Soilleux EJ, Baribaud F, et al. DC-SIGNR, a DC-SIGN homologue expressed in endothelial cells, binds to human and simian immunodeficiency viruses and activates infection in trans. *Proc Natl Acad Sci U S A.* 2001;98(5):2670–2675.
123. Engering A, van Vliet SJ, Hebeda K, et al. Dynamic populations of dendritic cell-specific ICAM-3 grabbing nonintegrin-positive immature dendritic cells and liver/lymph node-specific ICAM-3 grabbing nonintegrin-positive endothelial cells in the outer zones of the paracortex of human lymph nodes. *Am J Pathol.* 2004;164(5):1587–1595.
124. Jeffers SA, Tusell SM, Gillim-Ross L, et al. CD209L (L-SIGN) is a receptor for severe acute respiratory syndrome coronavirus. *Proc Natl Acad Sci U S A.* 2004;101(44):15748–15753.
125. Feinberg H, Mitchell DA, Drickamer K, Weis WI. Structural basis for selective recognition of oligosaccharides by DC-SIGN and DC-SIGNR. *Science.* 2001;294(5549):2163–2166.
126. Bashirova AA, Geijtenbeek TB, van Duijnhoven GC, et al. A dendritic cell-specific intercellular adhesion molecule 3-grabbing nonintegrin (DC-SIGN)-related protein is highly expressed on human liver sinusoidal endothelial cells and promotes HIV-1 infection. *J Exp Med.* 2001;193(6):671–678.
127. Snyder GA, Ford J, Torabi-Parizi P, et al. Characterization of DC-SIGN/R interaction with human immunodeficiency virus type 1 gp120 and ICAM molecules favors the receptor's role as an antigen-capturing rather than an adhesion receptor. *J Virol.* 2005;79(8):4589–4598.
128. Hladik F, Sakchalathorn P, Ballweber L, et al. Initial events in establishing vaginal entry and infection by human immunodeficiency virus type-1. *Immunity.* 2007;26(2):257–270.
129. Turville SG, Cameron PU, Handley A, et al. Diversity of receptors binding HIV on dendritic cell subsets. *Nat Immunol.* 2002;3(10):975–983.
130. Holla A, Skerra A. Comparative analysis reveals selective recognition of glycans by the dendritic cell receptors DC-SIGN and Langerin. *Protein Eng Des Sel.* 2011;24(9):659–669.
131. de Witte L, Nabatov A, Pion M, et al. Langerin is a natural barrier to HIV-1 transmission by Langerhans cells. *Nat Med.* 2007;13(3):367–371.
132. Ballweber L, Robinson B, Kreger A, et al. Vaginal langerhans cells nonproductively transporting HIV-1 mediate infection of T cells. *J Virol.* 2011;85(24):13443–13447.
133. Kawamura T, Gulden FO, Sugaya M, et al. R5 HIV productively infects Langerhans cells, and infection levels are regulated by compound CCR5 polymorphisms. *Proc Natl Acad Sci U S A.* 2003;100(14):8401–8406.
134. Peressin M, Holl V, Schmidt S, et al. HIV-1 replication in Langerhans and interstitial dendritic cells is inhibited by neutralizing and Fc-mediated inhibitory antibodies. *J Virol.* 2011;85(2):1077–1085.
135. Fanibunda SE, Modi DN, Gokral JS, Bandivdekar AH. HIV gp120 binds to mannose receptor on vaginal epithelial cells and induces production of matrix metalloproteinases. *PLoS One.* 2011;6(11):e28014.
136. Jadhav SK, Velhal SM, Deshpande A, Bandivdekar AH. Association of human mannose receptor in sexual transmission of human immunodeficiency virus in serodiscordant couples. *AIDS Res Hum Retrovir.* 2013;29(1):156–163.

137. Nguyen DG, Hildreth JE. Involvement of macrophage mannose receptor in the binding and transmission of HIV by macrophages. *Eur J Immunol.* 2003;33(2):483–493.
138. Trujillo JR, Rogers R, Molina RM, et al. Noninfectious entry of HIV-1 into peripheral and brain macrophages mediated by the mannose receptor. *Proc Natl Acad Sci U S A.* 2007;104(12):5097–5102.
139. Lambert AA, Gilbert C, Richard M, Beaulieu AD, Tremblay MJ. The C-type lectin surface receptor DCIR acts as a new attachment factor for HIV-1 in dendritic cells and contributes to trans- and cis-infection pathways. *Blood.* 2008;112(4):1299–1307.
140. Lambert AA, Barabe F, Gilbert C, Tremblay MJ. DCIR-mediated enhancement of HIV-1 infection requires the ITIM-associated signal transduction pathway. *Blood.* 2011;117(24):6589–6599.
141. Lambert AA, Imbeault M, Gilbert C, Tremblay MJ. HIV-1 induces DCIR expression in CD4+ T cells. *PLoS Pathog.* 2010;6(11):e1001188.
142. Meyer-Wentrup F, Benitez-Ribas D, Tacken PJ, et al. Targeting DCIR on human plasmacytoid dendritic cells results in antigen presentation and inhibits IFN-alpha production. *Blood.* 2008;111(8):4245–4253.
143. Meyer-Wentrup F, Cambi A, Joosten B, et al. DCIR is endocytosed into human dendritic cells and inhibits TLR8-mediated cytokine production. *J Leukoc Biol.* 2009;85(3):518–525.
144. Chaipan C, Soilleux EJ, Simpson P, et al. DC-SIGN and CLEC-2 mediate human immunodeficiency virus type 1 capture by platelets. *J Virol.* 2006;80(18):8951–8960.
145. Hatsukari I, Singh P, Hitosugi N, et al. DEC-205-mediated internalization of HIV-1 results in the establishment of silent infection in renal tubular cells. *J Am Soc Nephrol.* 2007;18(3):780–787.
146. Mikulak J, Teichberg S, Faust T, Schmidtmayerova H, Singhal PC. HIV-1 harboring renal tubular epithelial cell interaction with T cells results in T cell trans-infection. *Virology.* 2009;385(1):105–114.
147. de Witte L, de Vries RD, van der Vlist M, et al. DC-SIGN and CD150 have distinct roles in transmission of measles virus from dendritic cells to T-lymphocytes. *PLoS Pathog.* 2008;4(4):e1000049.
148. Engering A, Van Vliet SJ, Geijtenbeek TB, Van Kooyk Y. Subset of DC-SIGN(+) dendritic cells in human blood transmits HIV-1 to T lymphocytes. *Blood.* 2002;100(5):1780–1786.
149. Jameson B, Baribaud F, Pohlmann S, et al. Expression of DC-SIGN by dendritic cells of intestinal and genital mucosae in humans and rhesus macaques. *J Virol.* 2002;76(4): 1866–1875.
150. Kaldensjo T, Petersson P, Tolf A, Morgan G, Broliden K, Hirbod T. Detection of intraepithelial and stromal Langerin and CCR5 positive cells in the human endometrium: potential targets for HIV infection. *PLoS One.* 2011;6(6):e21344.
151. Lore K, Sonnerborg A, Brostrom C, et al. Accumulation of DC-SIGN+CD40+ dendritic cells with reduced CD80 and CD86 expression in lymphoid tissue during acute HIV-1 infection. *AIDS.* 2002;16(5):683–692.
152. te Velde AA, van Kooyk Y, Braat H, et al. Increased expression of DC-SIGN+IL-12 +IL-18+ and CD83+IL-12-IL-18- dendritic cell populations in the colonic mucosa of patients with Crohn's disease. *Eur J Immunol.* 2003;33(1):143–151.
153. Boily-Larouche G, Milev MP, Zijenah LS, et al. Naturally-occurring genetic variants in human DC-SIGN increase HIV-1 capture, cell-transfer and risk of mother-to-child transmission. *PLoS One.* 2012;7(7):e40706.
154. Boily-Larouche G, Iscache AL, Zijenah LS, et al. Functional genetic variants in DC-SIGNR are associated with mother-to-child transmission of HIV-1. *PLoS One.* 2009;4(10):e7211.

155. da Silva RC, Segat L, Zanin V, Arraes LC, Crovella S. Polymorphisms in DC-SIGN and L-SIGN genes are associated with HIV-1 vertical transmission in a Northeastern Brazilian population. *Hum Immunol.* 2012;73(11):1159–1165.
156. Borggren M, Naver L, Casper C, Ehrnst A, Jansson M. R5 human immunodeficiency virus type 1 with efficient DC-SIGN use is not selected for early after birth in vertically infected children. *J Gen Virol.* 2013;94(pt 4):767–773.
157. Baan E, de Ronde A, Stax M, et al. HIV-1 autologous antibody neutralization associates with mother to child transmission. *PLoS One.* 2013;8(7):e69274.
158. Martin MP, Lederman MM, Hutcheson HB, et al. Association of DC-SIGN promoter polymorphism with increased risk for parenteral, but not mucosal, acquisition of human immunodeficiency virus type 1 infection. *J Virol.* 2004;78(24):14053–14056.
159. Selvaraj P, Alagarasu K, Swaminathan S, Harishankar M, Narendran G. CD209 gene polymorphisms in South Indian HIV and HIV-TB patients. *Infect Genet Evol.* 2009;9(2):256–262.
160. Li H, Yu XM, Wang JX, Hong ZH, Tang NL. The VNTR polymorphism of the DC-SIGNR gene and susceptibility to HIV-1 infection: a meta-analysis. *PLoS One.* 2012;7(9):e42972.
161. Wichukchinda N, Kitamura Y, Rojanawiwat A, et al. The polymorphisms in DC-SIGNR affect susceptibility to HIV type 1 infection. *AIDS Res Hum Retrovir.* 2007;23(5):686–692.
162. Zhu D, Kawana-Tachikawa A, Iwamoto A, Kitamura Y. Influence of polymorphism in dendritic cell-specific intercellular adhesion molecule-3-grabbing nonintegrin-related (DC-SIGNR) gene on HIV-1 trans-infection. *Biochem Biophys Res Commun.* 2010;393(4):598–602.
163. Groot F, Geijtenbeek TB, Sanders RW, et al. Lactoferrin prevents dendritic cell-mediated human immunodeficiency virus type 1 transmission by blocking the DC-SIGN–gp120 interaction. *J Virol.* 2005;79(5):3009–3015.
164. Hong P, Ninonuevo MR, Lee B, Lebrilla C, Bode L. Human milk oligosaccharides reduce HIV-1-gp120 binding to dendritic cell-specific ICAM3-grabbing non-integrin (DC-SIGN). *Br J Nutr.* 2009;101(4):482–486.
165. Naarding MA, Dirac AM, Ludwig IS, et al. Bile salt-stimulated lipase from human milk binds DC-SIGN and inhibits human immunodeficiency virus type 1 transfer to CD4+ T cells. *Antimicrob Agents Chemother.* 2006;50(10):3367–3374.
166. Naarding MA, Ludwig IS, Groot F, et al. Lewis X component in human milk binds DC-SIGN and inhibits HIV-1 transfer to CD4+ T lymphocytes. *J Clin Invest.* 2005;115(11):3256–3264.
167. Requena M, Bouhlal H, Nasreddine N, et al. Inhibition of HIV-1 transmission in trans from dendritic cells to CD4+ T lymphocytes by natural antibodies to the CRD domain of DC-SIGN purified from breast milk and intravenous immunoglobulins. *Immunology.* 2008;123(4):508–518.
168. Saeland E, de Jong MA, Nabatov AA, Kalay H, Geijtenbeek TB, van Kooyk Y. MUC1 in human milk blocks transmission of human immunodeficiency virus from dendritic cells to T cells. *Mol Immunol.* 2009;46(11–12):2309–2316.
169. Stax MJ, Naarding MA, Tanck MW, et al. Binding of human milk to pathogen receptor DC-SIGN varies with bile salt-stimulated lipase (BSSL) gene polymorphism. *PLoS One.* 2011;6(2):e17316.
170. Yagi Y, Watanabe E, Watari E, et al. Inhibition of DC-SIGN-mediated transmission of human immunodeficiency virus type 1 by Toll-like receptor 3 signalling in breast milk macrophages. *Immunology.* 2010;130(4):597–607.
171. Sabatte J, Faigle W, Ceballos A, et al. Semen clusterin is a novel DC-SIGN ligand. *J Immunol.* 2011;187(10):5299–5309.

172. Stax MJ, van Montfort T, Sprenger RR, et al. Mucin 6 in seminal plasma binds DC-SIGN and potently blocks dendritic cell mediated transfer of HIV-1 to CD4(+) T-lymphocytes. *Virology*. 2009;391(2):203–211.
173. Vazquez-Guillen JM, Garcia-Jacobo PJ, Zapata-Benavides P, et al. Expression of DC-SIGN in peripheral blood dendritic cells of patients with typical, slow, and rapid progression to AIDS. *Arch Med Res*. 2009;40(2):132–135.
174. van Montfort T, Nabatov AA, Geijtenbeek TB, Pollakis G, Paxton WA. Efficient capture of antibody neutralized HIV-1 by cells expressing DC-SIGN and transfer to CD4 + T lymphocytes. *J Immunol*. 2007;178(5):3177–3185.
175. Marzi A, Mitchell DA, Chaipan C, et al. Modulation of HIV and SIV neutralization sensitivity by DC-SIGN and mannose-binding lectin. *Virology*. 2007;368(2):322–330.
176. Ganesh L, Leung K, Lore K, et al. Infection of specific dendritic cells by CCR5-tropic human immunodeficiency virus type 1 promotes cell-mediated transmission of virus resistant to broadly neutralizing antibodies. *J Virol*. 2004;78(21):11980–11987.
177. van Montfort T, Eggink D, Boot M, et al. HIV-1 N-glycan composition governs a balance between dendritic cell-mediated viral transmission and antigen presentation. *J Immunol*. 2011;187(9):4676–4685.
178. Lue J, Hsu M, Yang D, Marx P, Chen Z, Cheng-Mayer C. Addition of a single gp120 glycan confers increased binding to dendritic cell-specific ICAM-3-grabbing nonintegrin and neutralization escape to human immunodeficiency virus type 1. *J Virol*. 2002;76(20):10299–10306.
179. Nabatov AA, van Montfort T, Geijtenbeek TB, Pollakis G, Paxton WA. Interaction of HIV-1 with dendritic cell-specific intercellular adhesion molecule-3-grabbing nonintegrin-expressing cells is influenced by gp120 envelope modifications associated with disease progression. *FEBS J*. 2006;273(21):4944–4958.
180. van Montfort T, Thomas AA, Pollakis G, Paxton WA. Dendritic cells preferentially transfer CXCR4-using human immunodeficiency virus type 1 variants to CD4+ T lymphocytes in trans. *J Virol*. 2008;82(16):7886–7896.
181. Koizumi Y, Kageyama S, Fujiyama Y, et al. RANTES -28G delays and DC-SIGN -139C enhances AIDS progression in HIV type 1-infected Japanese hemophiliacs. *AIDS Res Hum Retrovir*. 2007;23(5):713–719.
182. Stax MJ, Kootstra NA. van 't Wout AB, et al. HIV-1 disease progression is associated with bile-salt stimulated lipase (BSSL) gene polymorphism. *PLoS One*. 2012;7(3): e32534.
183. MacArthur RD, Novak RM. Reviews of anti-infective agents: maraviroc: the first of a new class of antiretroviral agents. *Clin Infect Dis*. 2008;47(2):236–241.
184. Kondru R, Zhang J, Ji C, et al. Molecular interactions of CCR5 with major classes of small-molecule anti-HIV CCR5 antagonists. *Mol Pharmacol*. 2008;73(3):789–800.
185. Perry CM. Maraviroc: a review of its use in the management of CCR5-tropic HIV-1 infection. *Drugs*. 2010;70(9):1189–1213.
186. Raymond S, Delobel P, Izopet J. Phenotyping methods for determining HIV tropism and applications in clinical settings. *Curr Opin HIV AIDS*. 2012;7(5):463–469.
187. Obermeier M, Symons J, Wensing AM. HIV population genotypic tropism testing and its clinical significance. *Curr Opin HIV AIDS*. 2012;7(5):470–477.
188. Archer J, Weber J, Henry K, et al. Use of four next-generation sequencing platforms to determine HIV-1 coreceptor tropism. *PLoS One*. 2012;7(11):e49602.
189. Fatkenheuer G, Nelson M, Lazzarin A, et al. Subgroup analyses of maraviroc in previously treated R5 HIV-1 infection. *N Engl J Med*. 2008;359(14):1442–1455.
190. Symons J, van Lelyveld SF, Hoepelman AI, et al. Maraviroc is able to inhibit dual-R5 viruses in a dual/mixed HIV-1-infected patient. *J Antimicrob Chemother*. 2010;66(4): 890–895.

191. Moore JP, Kuritzkes DR. A piece de resistance: how HIV-1 escapes small molecule CCR5 inhibitors. *Curr Opin HIV AIDS*. 2009;4(2):118–124.
192. Westby M. Resistance to CCR5 antagonists. *Curr Opin HIV AIDS*. 2007;2(2): 137–144.
193. Roche M, Salimi H, Duncan R, et al. A common mechanism of clinical HIV-1 resistance to the CCR5 antagonist maraviroc despite divergent resistance levels and lack of common gp120 resistance mutations. *Retrovirology*. 2013;10:43.
194. Berro R, Sanders RW, Lu M, Klasse PJ, Moore JP. Two HIV-1 variants resistant to small molecule CCR5 inhibitors differ in how they use CCR5 for entry. *PLoS Pathog*. 2009;5(8):e1000548.
195. Ogert RA, Ba L, Hou Y, et al. Structure-function analysis of human immunodeficiency virus type 1 gp120 amino acid mutations associated with resistance to the CCR5 coreceptor antagonist vicriviroc. *J Virol*. 2009;83(23):12151–12163.
196. Tilton JC, Wilen CB, Didigu CA, et al. A maraviroc-resistant HIV-1 with narrow cross-resistance to other CCR5 antagonists depends on both N-terminal and extracellular loop domains of drug-bound CCR5. *J Virol*. 2010;84(20):10863–10876.
197. Wang SK, Liang PH, Astronomo RD, et al. Targeting the carbohydrates on HIV-1: Interaction of oligomannose dendrons with human monoclonal antibody 2G12 and DC-SIGN. *Proc Natl Acad Sci U S A*. 2008;105(10):3690–3695.
198. Luallen RJ, Fu H, Agrawal-Gamse C, et al. A yeast glycoprotein shows high-affinity binding to the broadly neutralizing human immunodeficiency virus antibody 2G12 and inhibits gp120 interactions with 2G12 and DC-SIGN. *J Virol*. 2009;83(10): 4861–4870.
199. Martinez-Avila O, Hijazi K, Marradi M, et al. Gold manno-glyconanoparticles: multivalent systems to block HIV-1 gp120 binding to the lectin DC-SIGN. *Chemistry*. 2009;15(38):9874–9888.
200. Becer CR, Gibson MI, Geng J, et al. High-affinity glycopolymer binding to human DC-SIGN and disruption of DC-SIGN interactions with HIV envelope glycoprotein. *J Am Chem Soc*. 2010;132(43):15130–15132.
201. Martinez-Avila O, Bedoya LM, Marradi M, Clavel C, Alcami J, Penades S. Multivalent manno-glyconanoparticles inhibit DC-SIGN-mediated HIV-1 trans-infection of human T cells. *Chembiochem*. 2009;10(11):1806–1809.
202. Sattin S, Daghetti A, Thepaut M, et al. Inhibition of DC-SIGN-mediated HIV infection by a linear trimannoside mimic in a tetravalent presentation. *ACS Chem Biol*. 2010;5(3):301–312.
203. Berzi A, Reina JJ, Ottria R, et al. A glycomimetic compound inhibits DC-SIGN-mediated HIV infection in cellular and cervical explant models. *AIDS*. 2012;26(2):127–137.
204. Balzarini J, Francois KO, Van Laethem K, et al. Pradimicin S, a highly soluble nonpeptidic small-size carbohydrate-binding antibiotic, is an anti-HIV drug lead for both microbicidal and systemic use. *Antimicrob Agents Chemother*. 2010;54(4):1425–1435.
205. Balzarini J, Van Herrewege Y, Vermeire K, Vanham G, Schols D. Carbohydrate-binding agents efficiently prevent dendritic cell-specific intercellular adhesion molecule-3-grabbing nonintegrin (DC-SIGN)-directed HIV-1 transmission to T lymphocytes. *Mol Pharmacol*. 2007;71(1):3–11.
206. Bertaux C, Daelemans D, Meertens L, et al. Entry of hepatitis C virus and human immunodeficiency virus is selectively inhibited by carbohydrate-binding agents but not by polyanions. *Virology*. 2007;366(1):40–50.
207. Hoorelbeke B, Van Damme EJ, Rouge P, et al. Differences in the mannose oligomer specificities of the closely related lectins from Galanthus nivalis and Zea mays strongly determine their eventual anti-HIV activity. *Retrovirology*. 2011;8(1):10.

208. Alexandre KB, Gray ES, Mufhandu H, et al. The lectins griffithsin, cyanovirin-N and scytovirin inhibit HIV-1 binding to the DC-SIGN receptor and transfer to CD4(+) cells. *Virology*. 2012;423(2):175–186.
209. Turville SG, Vermeire K, Balzarini J, Schols D. Sugar-binding proteins potently inhibit dendritic cell human immunodeficiency virus type 1 (HIV-1) infection and dendritic-cell-directed HIV-1 transfer. *J Virol*. 2005;79(21):13519–13527.
210. Balzarini J, Van Laethem K, Hatse S, et al. Profile of resistance of human immunodeficiency virus to mannose-specific plant lectins. *J Virol*. 2004;78(19):10617–10627.
211. Balzarini J, Van Laethem K, Hatse S, et al. Carbohydrate-binding agents cause deletions of highly conserved glycosylation sites in HIV GP120: a new therapeutic concept to hit the achilles heel of HIV. *J Biol Chem*. 2005;280(49):41005–41014.
212. Balzarini J, Van Laethem K, Hatse S, et al. Marked depletion of glycosylation sites in HIV-1 gp120 under selection pressure by the mannose-specific plant lectins of Hippeastrum hybrid and Galanthus nivalis. *Mol Pharmacol*. 2005;67(5):1556–1565.
213. Balzarini J, Van Laethem K, Peumans WJ, et al. Mutational pathways, resistance profile, and side effects of cyanovirin relative to human immunodeficiency virus type 1 strains with N-glycan deletions in their gp120 envelopes. *J Virol*. 2006;80(17):8411–8421.
214. Balzarini J. The alpha(1,2)-mannosidase I inhibitor 1-deoxymannojirimycin potentiates the antiviral activity of carbohydrate-binding agents against wild-type and mutant HIV-1 strains containing glycan deletions in gp120. *FEBS Lett*. 2007;581(10): 2060–2064.
215. Witvrouw M, Fikkert V, Hantson A, et al. Resistance of human immunodeficiency virus type 1 to the high-mannose binding agents cyanovirin N and concanavalin A. *J Virol*. 2005;79(12):7777–7784.
216. Balzarini J, Van Laethem K, Daelemans D, et al. Pradimicin A, a carbohydrate-binding nonpeptidic lead compound for treatment of infections with viruses with highly glycosylated envelopes, such as human immunodeficiency virus. *J Virol*. 2007;81(1): 362–373.
217. Reitter JN, Means RE, Desrosiers RC. A role for carbohydrates in immune evasion in AIDS. *Nat Med*. 1998;4(6):679–684.
218. Kang SM, Quan FS, Huang C, et al. Modified HIV envelope proteins with enhanced binding to neutralizing monoclonal antibodies. *Virology*. 2005;331(1):20–32.
219. Bolmstedt A, Hinkula J, Rowcliffe E, Biller M, Wahren B, Olofsson S. Enhanced immunogenicity of a human immunodeficiency virus type 1 env DNA vaccine by manipulating N-glycosylation signals. Effects of elimination of the V3 N306 glycan. *Vaccine*. 2001;20(3–4):397–405.
220. Balzarini J. Targeting the glycans of glycoproteins: a novel paradigm for antiviral therapy. *Nat Rev Microbiol*. 2007;5(8):583–597.
221. Francois KO, Balzarini J. Potential of carbohydrate-binding agents as therapeutics against enveloped viruses. *Med Res Rev*. 2012;32(2):349–387.

CHAPTER FIVE

A Game of Numbers: The Stoichiometry of Antibody-Mediated Neutralization of Flavivirus Infection

Theodore C. Pierson*,[1], Michael S. Diamond†,[1]

*Viral Pathogenesis Section, National Institute of Allergy and Infectious Diseases, National Institutes of Health, Bethesda, Maryland, USA

†Departments of Medicine, Molecular Microbiology, Pathology & Immunology, Center for Human Immunology and Immunotherapy Programs, Washington University School of Medicine, St. Louis, Missouri, USA

[1]Corresponding authors: e-mail address: piersontc@niaid.nih.gov; diamond@borcim.wustl.edu

Contents

Abstract

The humoral response contributes to the protection against viral pathogens. Although antibodies have the potential to inhibit viral infections via several mechanisms, an ability to neutralize viruses directly may be particularly important. Neutralizing antibody titers

Progress in Molecular Biology and Translational Science, Volume 129
ISSN 1877-1173
http://dx.doi.org/10.1016/bs.pmbts.2014.10.005

are commonly used as predictors of protection from infection, especially in the context of vaccine responses and immunity. Despite the simplicity of the concept, how antibody binding results in virus inactivation is incompletely understood despite decades of research. Flaviviruses have been an attractive system in which to seek a structural and quantitative understanding of how antibody interactions with virions modulate infection because of the contribution of antibodies to both protection and pathogenesis. This review will present a stoichiometric model of antibody-mediated neutralization of flaviviruses and discuss how these concepts can inform the development of vaccines and antibody-based therapeutics.

1. INTRODUCTION

Flaviviruses are positive-stranded RNA viruses that have the potential to cause significant morbidity and mortality in humans. Many viruses of this genus have a global impact on human health including the mosquito-borne dengue virus (DENV), yellow fever virus (YFV), Japanese encephalitis virus (JEV) and West Nile virus (WNV), and the tick-borne encephalitis viruses (TBEVs). Flaviviruses are endemic in many regions of the globe. For example, it has been estimated that 390 million human DENV infections occur each year, with 3.6 billion people at risk of infection in more than 100 countries.[1] DENV is now the leading arthropod-borne viral disease in the world. Sporadic intense local transmission of flaviviruses also may be a serious threat to public health as illustrated by WNV activity in the United States over the past decade.[2] Flaviviruses cause a variety of disease manifestations including encephalitis and paralysis, massive hepatic injury, and hemorrhagic and plasma leakage syndromes associated with visceral organ injury. At present, there is no specific therapy to treat flavivirus infections; only vaccines have proven effective at blunting the impact of these viruses on public health.

Multiple strategies have been employed for generating effective flavivirus vaccines (reviewed in Ref. 3). The live-attenuated YFV-17D vaccine, generated by Max Theiler and colleagues in 1938, was created by extensive passage of the virulent Asibi strain *ex vivo* (reviewed in Refs. 4,5). While more than 600 million doses of this highly effective vaccine have been administered, several hundred thousand human infections (and ~30,000 deaths) still occur annually, primarily in Africa and South America.[6,7] Numerous vaccines have been created for JEV (reviewed in Ref. 8). First-generation vaccine strategies used inactivated mouse brain preparations of antigen. The live-attenuated SA14-14-2 was developed in China in the late 1980s and has been deployed extensively in several JEV-endemic countries. The development of second- and third-generation JEV vaccines remains an active area of study as reflected by the introductions of a formalin-inactivated Vero cell-derived vaccine (IXIARO)[9] and the licensing of a live-attenuated JEV-YF-17D chimeric vaccine (IMOJEV).[10] The clinical impact of TBEV has been reduced substantially in regions where the vaccine is used; a formalin-inactivated viral vaccine has been administered widely in Austria and is estimated to be 96–99% effective.[11,12] Combined with the success of veterinary vaccines against WNV,[13] these experiences suggest flavivirus-induced disease can be prevented. However, vaccines are not yet available for all flaviviruses that impact human health. Despite numerous promising candidates, no vaccine for WNV is available yet for use in humans (reviewed in Ref. 13). Critically, vaccines are not yet approved to protect against the four groups (serotypes) of dengue viruses circulating among much of the human population. Indeed, the most advanced tetravalent live-attenuated DENV vaccine candidate showed a poor efficacy rate in DENV-naïve individuals in a recently published phase 3 clinical trial.[14]

Neutralizing antibodies have been shown to correlate with protection against several flavivirus infections following vaccination[4,15] and are a critical component of immunity to natural infections (reviewed in Ref. 16). However, the role of antibodies in DENV pathogenesis is more complex. Pioneering studies by Sabin demonstrated that DENV infection could be prevented by vaccination with a single serotype.[17] Although short-lived (~6 months) protection was observed against all four DENV serotypes, long-term protection was generated only against the DENV of the same serotype. Because several epidemiological studies identify secondary heterologous DENV infection as a significant risk factor for severe, potentially fatal, clinical manifestations of disease (reviewed in Ref. 18), monovalent vaccines are not tenable and a tetravalent approach is considered necessary. The current generation of live-attenuated, inactivated, and subunit-based DENV vaccine candidates target all four serotypes for antigenic responses (reviewed in Ref. 19).

Although an understanding of the underlying mechanisms by which DENV infection by a given serotype sensitizes an individual to more severe disease after infection with a heterologous DENV serotype is incomplete, several lines of evidence support a pathogenic role of pre-existing or rapidly induced antibodies (reviewed in Ref. 20). Antibody-dependent enhancement (ADE) of infection describes a marked increase in the efficiency of infection of cells expressing Fcγ-receptors in the presence of subneutralizing amounts of antibody.[21,22] Cross-reactive mouse, monkey, and human antibodies against the structural proteins have been shown to increase virus burden in mouse and primate models of DENV.[23–27] Because a vaccine-induced humoral response has at least the theoretical potential to contribute to disease, understanding how antibodies interact with flavivirus virions is a critical area of study. In this review, we discuss how advances in our quantitative understanding of antibody neutralization of flavivirus infection provide insight into the types of antibodies that protect against or contribute to pathogenesis.

2. FLAVIVIRUS STRUCTURE

The positive-sense genomic RNA of flaviviruses encodes a single open reading frame that is processed co- and posttranslationally by cellular and viral proteases into 10 different proteins. Flaviviruses are spherical virions (~50 nm diameter) composed of the structural proteins capsid (C), envelope (E), and premembrane (prM), a lipid membrane derived from the endoplasmic reticulum, and a ~11 kb genomic RNA (reviewed in Ref. 28). High-resolution atomic structures have been solved of the three structural proteins that comprise the virus particle (reviewed in Ref. 29). The flavivirus E protein is an elongated molecule with three domains composed principally of β-strands that are connected to the viral membrane by a helical stem and two transmembrane domains (Fig. 1). The E protein has a central role in virus attachment to cells, entry, and membrane fusion. Domain III (DIII) is an immunoglobulin-like domain hypothesized to interact with cellular attachment factors that enhance the efficiency of virus entry, such as heparin sulfate.[30,31] Domain II (DII) is composed of two elongated fingers that contribute many of the contacts required for the dimerization of E proteins on the surface of mature virus particles.[32] Importantly, the distal end of DII contains a hydrophobic fusion loop (DII-FL) that is highly conserved among flaviviruses.[33,34] A central domain I (DI) is connected to both DII and DIII via flexible linkers that facilitates rotation among the three domains required for E protein function. The E protein may be glycosylated at one or two positions (DI or DI and DII, respectively) in some flavivirus strains.

During the virion assembly process, spikes composed of three prM-E heterodimers are incorporated into immature virions in the lumen of the endoplasmic reticulum (Fig. 1).[35,36] Fully immature virus particles are non-infectious.[37] prM functions to restrict structural transitions and control the

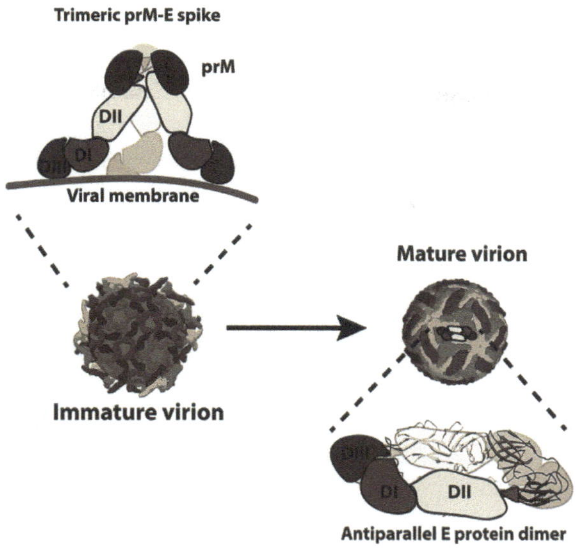

Figure 1 The structure of the flavivirus envelope proteins and their organization on the virus particle. Flaviviruses are small spherical virions that incorporate a dense array of prM and E proteins that function to promote virus assembly, budding, and entry. The E protein is an elongated three domain molecule that is the principal target of neutralizing antibodies. Domain III (DIII, shown in blue) is thought to interact with receptors on target cells. The conserved 13-amino acid fusion loop (shown in green) is located at the distal end of domain II (DII, shown in yellow). DIII and DII are connected by the central domain I (DI, shown in red). In this cartoon, the prM protein is depicted as a purple spherical oval. Flaviviruses assemble as immature virus particles on which the E protein exists as 60 trimers of prM-E dimers that project as spikes away from the virion surface. Virus maturation is mediated by cleavage of prM by a cellular furin protease. On mature virions, the E protein exists as antiparallel homodimers that lie flat against the surface on the virion. The arrangement of E proteins on the mature DENV virion is depicted. Each virus particle is composed of 30 rafts of three antiparallel dimers in a herringbone pattern. (See the color plate.)

oligomeric state of the E protein and thereby prevent adventitious fusion and inactivation of the virus particle during egress from infected cells.[38,39] Flaviviruses undergo a pH-dependent maturation process during transit through the *trans*-Golgi network defined by the cleavage of prM by a host furin-like serine protease.[40,41] The product of the maturation process is a mature virus particle that retains only the M peptide; the larger "pr" portion of prM remains with the particle throughout the secretory pathway and then dissociates from the virus particle upon release into the neutral pH of the extracellular space.[39,40]

The structure of mature flavivirus virions has been studied extensively by cryoelectron microscopy.[28] Kuhn and colleagues demonstrated that DENV virions are relatively smooth virus particles, with E protein dimers arranged in an unusual herringbone fashion.[42] In this configuration, the 180 E proteins incorporated into the virion exist in three different chemical environments defined by their proximity to the three icosahedral symmetry axes of the virus particle. This dense arrangement of E proteins on the surface of the virion represents a complex surface for antibody recognition. Not all epitopes recognized by neutralizing antibodies are equally accessible on the E proteins in the three different symmetry environments of the mature virion. For example, the WNV DIII-reactive monoclonal antibody (MAb) E16 cannot bind its epitope on E proteins at the fivefold symmetry axis due to steric constraints.[43,44] Thus, only 120 of 180 E proteins on the mature WNV virion can bind Fab fragments of the MAb E16 at saturation. Several potently neutralizing MAbs against WNV and DENV recently have been shown to recognize complex epitopes formed from residues on more than one E protein of the mature virus particle[45–48]; these antibodies typically do not have the capacity to bind monomeric E proteins. Studies of murine and human MAbs have revealed that prM and all three domains of E can be recognized by antibodies although the frequency with which specific epitopes are targeted by the humoral response varies among species.[49–52] Moreover, the functional properties of anti-E antibodies that bind the virus particle vary substantially (for example, see Refs. 49,53–55).

3. A MULTIPLE-HIT MODEL FOR THE NEUTRALIZATION OF FLAVIVIRUSES

Interest in the mechanism and stoichiometry of antibody-mediated neutralization dates back almost 100 years, long before the concept and identity of antibody molecules was understood (reviewed in Ref. 56). Early debate focused on the stoichiometric requirements for neutralization: how many antibodies are required to neutralize virus infectivity? One concept was that viruses could be neutralized following engagement by a single antibody molecule.[57,58] This "one-hit" hypothesis rested principally upon negative data obtained from kinetic neutralization experiments; the absence of a measureable lag phase of virus neutralization following the addition of antibody was interpreted as a requirement for binding by only a single antibody. From this perspective, virions were thought to contain sites of vulnerability that result in noninfectious virions upon binding by antibody. Several limitations of this model have been discussed, including that antibodies bind antigens rapidly relative to the rate at which biological outcomes are measured.[59] As an example, high-affinity WNV-reactive antibodies bind virus in solution in seconds, whereas the half-maximal binding of WNV to cells expressing a highly efficient virus attachment factor is 45 min![60,61] The alternative "multiple-hit" model assumes that virions become decorated with antibody and are neutralized only at a critical occupancy.[21,59] The number of antibodies required to neutralize viruses with different structures has been hypothesized to vary markedly (reviewed by Refs. 56,62).

The early literature describing antibody-mediated neutralization of flaviviruses arose from efforts to distinguish among a growing number of antigenically related viruses.[63,64] Many early observations supported the concept that the docking of multiple antibody molecules was required for the neutralization of flaviviruses. This evidence was presented in an outstanding review by Della-Porta and Westaway.[59] Since that time, the large number of studies on flavivirus interaction with antibodies has improved our understanding of the basis of neutralization and refined models of the relationship between antibody occupancy and virus inactivation.

3.1. A neutralization-resistant population of flaviviruses

One prediction of a multiple-hit model of neutralization is that infectious virions can be decorated by antibody with a stoichiometry insufficient for neutralization even under conditions of antibody excess or saturation. Neutralization-resistant viruses may express epitopes in small numbers (small number of viral protein targets/virion) or display them in an inaccessible manner. Several factors may limit epitope accessibility on the virion, such as steric constraints among densely arranged viral proteins,[43,65] proximity to the viral membrane,[66] or the presence of carbohydrates that shield antibody-binding determinants.[67] Neutralization profiles of some flavivirus-reactive immune sera or MAbs reveal a plateau effect in which a subset of virions remains resistant to neutralization despite saturating antibody concentrations (Fig. 2A). In this context, a fraction of virus particles are not neutralized despite the presence of high concentrations of flavivirus-reactive immune sera or MAbs, whereas the remaining fraction of virions is neutralized. The existence of a neutralization-resistant fraction of virions could be explained by structural heterogeneity that translates into differences in the maximal number of antibodies bound to an individual virion, or the existence of a subset of viruses in the stock encoding mutation(s) at epitopes recognized by neutralizing antibodies.

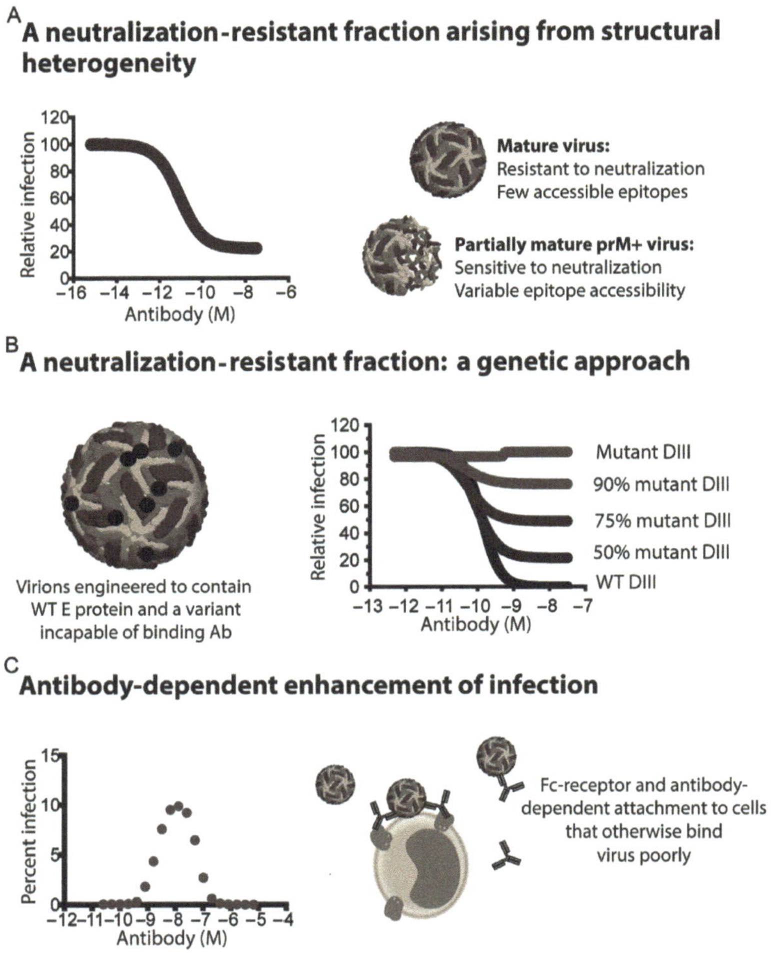

Figure 2 Evidence in support of a multiple-hit requirement for neutralization. Several types of experimental studies support a multiple-hit model of antibody neutralization of flaviviruses. (A) In some instances, neutralization assays reveal a plateau effect at high concentrations of antibody. For flaviviruses, the neutralization-resistant fraction is a consequence of the significant structural heterogeneity of viruses released from infected cells. Mature virions in the population that do not display enough epitopes for a given antibody with a stoichiometry sufficient for neutralization are resistant to inhibition. Thus, infectious virions can be decorated by antibody with a stoichiometry insufficient for neutralization even under conditions of antibody excess or saturation. (B) A fraction of WNV virions resistant to neutralization can be created experimentally by reducing the number of epitopes on the average virion using genetic approaches in which E proteins with mutations (in this instance on DIII) that do not bind a given antibody are increasingly expressed on virions. (C) ADE describes a phenomenon in which antibody binds at an occupancy that is insufficient for neutralization yet capable of enhancing the efficiency of virus attachment and entry into cells via interactions with Fcγ-receptors expressed on cells. Because incubation with subneutralizing concentrations of antibody can enhance the infectivity via ADE, docking of a single antibody to the virion is insufficient for neutralization. (See the color plate.)

Insight into the mechanisms responsible for the neutralization-resistant fraction of flaviviruses was obtained through a series of studies with reporter virus particles (RVPs). RVPs are pseudo-infectious virions produced by genetic complementation of a subgenomic replicon with the structural proteins of the virus *in trans*.[68,69] Importantly, the production of RVPs by complementation with cDNA expression constructs eliminates the possibility that neutralization escape mutations are present in virus stocks. A neutralization-resistant fraction of WNV and DENV RVPs has been documented in experiments with many E- and prM-reactive MAbs. For example, studies with the WNV E protein-specific MAb E53 indicated that a significant fraction of WNV RVPs produced in HEK-293T cells were resistant to neutralization at concentrations of antibody sufficient to result in antibody occupancy of all accessible epitopes on the virus particle.[70] The size of the E53-resistant fraction varied among different RVP preparations[70] and when assayed using different target cells.[71] The resistant fraction could be manipulated by changing the completeness of virion maturation, which modulates the number of prM and DII-fusion loop epitopes (among others) accessible for antibody recognition. The appearance of a resistant fraction reflects the subset of virions in the population that do not display epitopes with a stoichiometry sufficient to support neutralization (Fig. 2A). In agreement with this interpretation, a resistant fraction of WNV can be created experimentally by reducing the number of epitopes on the average virion using genetic approaches[21] (Fig. 2B). Collectively, these findings suggest that neutralization of infectious virions requires engagement of flaviviruses by multiple antibody molecules.

3.2. ADE of flavivirus infection

Early studies by Halstead and colleagues demonstrated that under some circumstances flavivirus infection is enhanced by virus-reactive antibodies.[72,73] Initially, DENV immune sera were shown to enhance the infection of primary blood mononuclear cells in a concentration-dependent fashion. ADE was dependent upon the Fc-portion of the antibody molecule and could be blocked by antibodies against Fcγ-receptors.[74] The cellular mechanism of ADE relates to an increase in the efficiency of virus attachment and entry into cells via interactions between antibodies bound to virions and Fcγ-receptors expressed on cells (Fig. 2C)[75–77]; antibody-dependent signaling through Fcγ-receptors also has been proposed to affect the permissiveness of the target cell by modulating antiviral and proinflammatory responses through a process termed "intrinsic ADE" (reviewed in Ref. 78). That incubation with non-neutralizing concentrations of antibody may enhance infectivity via antibody-mediated attachment to target cells demonstrates that docking of a single antibody to the virion is not sufficient for neutralization and is in agreement with a "multiple-hit" concept of antibody-mediated neutralization. The stoichiometric requirements for ADE will be discussed in more detail below.

4. THE STOICHIOMETRY OF NEUTRALIZATION AND ENHANCEMENT OF FLAVIVIRUSES

Although early studies provided compelling evidence in support of a "multiple-hit" model for neutralization of flaviviruses,[59] the stoichiometric requirements for neutralization of these viruses remained unknown. Quantitative insights into the requirements for neutralization are of value for defining characteristics of antibodies with significant neutralization potential, as well as those that might readily enhance infection of cells expressing Fcγ-receptors. To estimate the number of antibody molecules required for neutralization of WNV, two complementary experimental approaches were employed using a small panel of E DIII-reactive antibodies.[21]

4.1. The relationship between antibody occupancy and neutralization

The fraction of epitopes on a virus particle bound by antibody at any given concentration by antibody can be estimated simply if the strength of binding to the virus particle can be measured. The fractional occupancy can be calculated for any antibody concentration using the following equation, provided assays are performed under conditions where free antibody remains in large excess over informative points of the antibody dose–response curve

$$\text{Ab bound}/\text{Ab bound}_{\max} = [\text{Ab}]/([\text{Ab}] + K_d).$$

The epitope occupancy requirement for neutralization of WNV was examined by integrating neutralization dose–response profiles of a panel of MAbs with estimates of their functional affinity for virions.[21] These results revealed that the fraction of epitopes bound by antibody when WNV was neutralized varied considerably among MAbs that recognized distinct epitopes on the virus particle. Potent neutralization by some MAbs was observed even when a small fraction of epitopes was bound; in these instances, complete neutralization of the population of viruses studied was achieved. Antibodies with these characteristics have a "low" occupancy requirement for neutralization. For example, the WNV-specific MAb E16 completely neutralized WNV infection when less than half of the DIII-lateral ridge (DIII-LR) epitopes were bound.[21] Alternatively, neutralization by other antibodies was observed only when a large fraction of accessible epitopes on the virion was engaged. In many instances, neutralization was not possible even when all of the accessible epitopes on an individual virion were bound. Antibodies with this type of "high" occupancy requirement may be incapable of neutralizing all the virions added to a neutralization experiment even when added at saturating concentrations. DENV-reactive antibodies with both "low" and "high" occupancy requirements also have been characterized in a series of biochemical and genetic studies by Barrett and colleagues.[79,80]

Epitopes recognized by a large number of mouse, monkey, and human anti-flavivirus MAbs have been mapped using biochemical, genetic, and neutralization escape approaches (Refs. 49,50,81,82 and references within). Many of these epitopes are not predicted to be accessible for antibody recognition using static models of the mature virion structure.[83–86] Occupancy requirements for neutralization by an antibody and the predicted accessibility of its corresponding epitope on the mature virus particle revealed an inverse correlation. Antibodies that bind epitopes predicted to be readily accessible on the surface of virions (such as the DIII-LR) neutralize infection by binding a small fraction of the constituent E proteins. In contrast, antibodies with "high" occupancy requirements for neutralization typically bind cryptic epitopes defined by the contribution of amino acids not predicted to be accessible for interaction with the antibody paratope. For many antibodies, the mode of epitope recognition cannot be explained using existing models of virion structure. As detailed below, understanding the structural basis for antibody recognition has expanded our understanding of the ensemble of structures of infectious flaviviruses.

4.2. Estimating the stoichiometry of WNV neutralization using mixed virion particles

Cryoelectron microscopic (cryo-EM) reconstruction studies of Fab fragments of WNV E16 docked on WNV revealed that this antibody can bind 120 of 180 E proteins on the virus particle,[44] confirming earlier predictions.[43] As mentioned above, the DIII-LR epitope recognized by E16 is positioned too close together at the fivefold symmetry axis of the mature virion, so that it cannot be bound due to steric constraints. To estimate the fraction of DIII-LR epitopes bound by E16 when WNV is neutralized, populations of RVPs were produced composed of varying proportions of wild-type (WT) E proteins and a variant encoding a single amino acid substitution (T332K), which prevents antibody binding and epitope recognition[60]; indeed, structural studies demonstrated that the T332 is a principal contact residue for E16 recognition.[43] Accordingly, antibody dose–response curves from studies with populations of virions composed of both WT and T332K E proteins resulted in the appearance of a fraction of viruses resistant to neutralization.[21] The size of the resistant fraction changed as a function of the number of T332K E proteins assembled into the virion (Fig. 2B), which were incapable of binding E16 antibody. Inspection of these curves revealed that roughly 50% of virions were sensitive to neutralization when one-quarter of the E proteins had an intact DIII-LR epitope, corresponding to a requirement for 30 antibodies for neutralization of WNV (25% of 120 accessible sites). Similar results were obtained with a second DIII-LR antibody, E24.[75]

4.3. Is 30 antibodies a reasonable number?

The number of antibodies required to neutralize animal viruses has been studied using multiple approaches (reviewed in Ref. 56). In many instances, the number of antibodies required for neutralization correlates with the size of the virus particle. A requirement for 30 MAbs to neutralize WNV agrees well with the number of antibodies predicted by the antibody "coating theory" of neutralization.[62] Unfortunately, a threshold number of 30 antibody molecules by itself does not provide insight into structural or mechanistic requirements for neutralizing virus infectivity, as discussed below.

4.4. Experimental and conceptual limitations

In our view, several independent lines of evidence support a multiple-hit requirement for antibody neutralization of flaviviruses. Our estimates of the precise stoichiometry required for neutralization arose from two complementary experimental approaches to study the interaction of antibodies recognizing a single epitope.[21] However, some caveats and unanswered questions exist:

(i) *Measurement of antibody affinity*. Analysis of the occupancy requirements for neutralization requires an accurate estimate of the monovalent antibody affinity for the virion. While highly quantitative measures of antibody binding to soluble recombinant viral envelope proteins have been obtained and been informative in many contexts, these reductionist approaches do not capture the full complexity of the virion structure. With few exceptions, biosensors have not been used to measure the kinetics of antibody binding to flaviviruses due to the challenges of capturing intact virions or the requirement for having precision and costly instruments in a high-level containment facility.[87,88] Instead, a common strategy to study the strength of antibody binding to viral antigens employs an ELISA. Limitations of this approach include the potential to distort the structure of the virion due to the method used to capture virus to plastic, the potential for artifactual bivalent antibody engagement arising from plating density, and quantitative limitations in the assay format. Some of the issues can be mitigated through the use of Fab fragments that are directly labeled (e.g., ^{125}I) to ensure the linearity of the assay.

(ii) *How many accessible epitopes are present on the virion?* Structural studies of the DIII-LR epitope suggest that residues critical for recognition by E16-like antibodies are accessible for binding at 120 locations on the mature virus particle. Sixty DIII-LR epitopes are not bound by Fab fragments of E16 due to steric clashes arising from close proximity of DIII in the fivefold symmetry environment.[43,44] An estimate of the number of DIII-LR antibodies capable of simultaneously binding the virion was a critical piece of information required to estimate the number of antibody-binding events that define the stoichiometric threshold.[21] However, the number of intact antibodies capable of binding the virion may differ from the maximal occupancy of Fab fragments due to their increased bulk. Indeed, biochemical studies suggest that the number of IgG molecules capable of simultaneously binding virions may be reduced as compared to Fabs.[89] A role for the heavy chain of the antibody molecule in defining the stoichiometry of neutralization was hypothesized to contribute to the maturation state-sensitive pattern of neutralization by a DIII-LR antibody.[90] To date, the stoichiometry of neutralization by a Fab fragment has not been solved.

(iii) *Beyond the DIII-LR epitope?* Estimates for a stoichiometric threshold of 30 antibodies arise from experiments with antibodies that bind a single epitope.[21] To date, experiments with antibodies of differing specificity (e.g., DII-FL MAbs) have been complicated by other factors including whether it is experimentally possible to change the maximal number of epitopes accessible for antibody recognition and retain infectivity.

(iv) *Relating stoichiometry and function.* Antibodies have the potential to inhibit virus infection via numerous mechanisms. For flaviviruses, antibodies that block virus attachment have been characterized[43] and may be an important component of the overall neutralizing activity of human sera.[91] Antibodies also may inhibit conformational changes in the E protein required for fusion and thus have the ability to neutralize viruses already attached to cells. Several antibodies, including the WNV-reactive MAb E16, are capable of inhibiting the fusion of antibody-decorated virus particles with synthetic liposomes.[88,92,93] During infection, virions decorated by E16 are unable to escape from endosomal compartments, consistent with an inability to fuse with membranes.[93] Cryo-EM studies of E16 Fab-decorated virions incubated at the acidic conditions that are required to trigger the fusion process suggested that at high concentrations, Fab fragments trap the virus particle at an intermediate step in the fusion process.[94] This exciting result highlights the potential for antibodies as tools to understand the mechanisms of a dynamic and quite rapid fusion process.

While considerable insight into the mechanisms of neutralization exists, why ~30 antibodies are required for neutralization is unclear. The surface area contacted directly when 30 antibodies engage the virion is modest (<10%).[21,43] While Fab fragments have neutralizing activity, the potency of neutralizing antibodies may be influenced significantly by contributions from the heavy chain. Furthermore, antibodies may neutralize via distinct mechanisms when decorated at different occupancies. While E16 has been shown to modestly block attachment at high concentrations,[43,61] at the EC50 inhibition is mediated largely at a postattachment step.[43] Paradoxically, fusion loop antibodies that bind the virion only in the presence of prM efficiently block attachment.[43] The relationships between epitope location, mechanism of neutralization, and antibody occupancy require further study.

5. FACTORS THAT MODULATE THE STOICHIOMETRY OF NEUTRALIZATION

Considering the stoichiometric requirements for neutralization provides a useful platform to explore factors that modulate the potency of neutralizing antibodies. In many instances, epitope mapping studies alone of neutralizing antibodies do not explain how antibodies engage the virus particle. In fact, several well-characterized anti-flavivirus antibodies bind epitopes not predicted to be accessible for recognition using static models of the mature virus particle.[42,85,86] Epitope accessibility is a critical aspect for defining conditions that allow for occupancy of the virion with a stoichiometry sufficient for neutralization, as detailed above.[95] Two properties to the virion have been characterized with the potential to regulate the accessibility of epitopes on the virion.

5.1. Virion maturation

Biochemical studies of the composition of flaviviruses commonly reveal the presence of uncleaved prM in virus preparations (reviewed in Ref. 96). The extent of uncleaved prM may vary as a function of the cell type in which virus was propagated and the time of harvest (Ref. 97; S. Mukherjee and T. Pierson, unpublished data). The significance of prM in virus stocks cannot be inferred solely from biochemical studies, as they are incapable of distinguishing between two possibilities: uncleaved prM is found solely on noninfectious immature virions or uncleaved prM may be present on virus particles capable of productive entry into cells. Several observations identify the existence of infectious partially mature virions: (i) prM can promote virus attachment to cells during entry,[98] (ii) changes in the efficiency of prM cleavage modulate the pH at which infectious virus is rendered inactive,[99] (iii) prM-reactive antibodies may enhance infection via ADE,[49,82] and (iv) the presence of prM alters the sensitivity of virions to neutralization by anti-E protein antibodies.[70]

We refer to partially mature viruses as virions on which prM is cleaved incompletely. Despite considerable progress, the structures of this heterogeneous population of virions are not fully understood. Cryo-EM tomographic reconstructions suggest partially mature virions are a mosaic of regions with structural similarity to immature virions (heterotrimeric spikes of prM and E) or mature virus particles (antiparallel E proteins arranged in a herringbone pattern).[100] Immunoprecipitation studies of DENV suggest that most of the E protein in virus-containing supernatants from insect cells can be bound by prM-reactive antibodies, indicating that under these culture conditions, a majority of secreted virions retain prM.[101] The extent of prM cleavage of viruses produced *in vivo* remains unknown.

The prM content of infectious virions alters the sensitivity of the virus particle to neutralization by many classes of E protein-reactive antibodies (reviewed in Ref. 96). In all instances to date, viruses that retain uncleaved prM are more sensitive to neutralization than fully mature virions; thus, the process of virion maturation also could serve as an independent immune evasion function. Because the epitopes recognized by maturation state-dependent antibodies are predicted to be poorly accessible on mature virions, it has been hypothesized that this pattern of maturation state-sensitive neutralization reflects changes in epitope accessibly.[70] For example, the antibodies that bind the highly conserved DII-FL only neutralize WNV and DENV particles retaining a significant amount of uncleaved prM. Structural studies confirm that the DII-FL MAb E53 only binds prM^+ virions.[102] In contrast, the DIII-LR reactive MAb E16 epitope is not sensitive to the maturation state of the virion.[70] Of significance, polyclonal mixtures of antibodies also may display maturation state-sensitive patterns of neutralization. Half the recipients of two different candidate WNV vaccines mounted an antibody response less capable of neutralizing mature virions as compared to those that retain uncleaved prM.[70] Cross-reactive, but not type-specific, DENV neutralizing antibody responses have been shown to be maturation sensitive.[71,103]

Unexpectedly, a recent study reported that the neutralization potency of a DIII-LR MAb (E33) also is modulated by uncleaved prM on the virion (unlike E16).[90] Like all maturation state-sensitive MAbs characterized to date, E33 neutralizes mature virions at higher concentrations than needed for partially mature virions retaining significant prM. However, Fab and Fab′2 fragments of E33 neutralize WNV in a maturation state-independent manner. While E16 and E33 bind the virion with similar affinity, extensive epitope mapping experiments suggest they engage the DIII-LR in subtly different manner. These data indicate that the sensitivity of E33 to the prM content of WNV is explained by steric constraints imposed by the Fc domain of the antibody, which in turn governs the availability of an otherwise accessible DIII-LR epitope. The structural basis for this phenomenon requires further study, but highlights how factors beyond epitope accessibility may contribute to maturation state-sensitive patterns of neutralization.

5.2. The structural dynamics of virions

Most of the well-characterized epitopes recognized by antibody are not expected to be accessible for binding using current models of the mature flavivirus structure. As detailed above, change in virus structure associated with virion maturation is one mechanism for modulating epitope accessibility. Changes in epitope accessibility also may arise from changes in virus structure occurring at equilibrium. Proteins are not static structures; their dynamic nature may be a key facet for understanding how proteins (including antibodies) bind their targets.[104] Several elegant approaches have been used to demonstrate that picornaviruses sample multiple structures at equilibrium, a process referred to herein as viral "breathing."[105–107] For example, reversible changes in capsid structure have been shown to regulate neutralization by antibodies that target the VP4 protein of poliovirus.[108] A role for flavivirus "breathing" initially was suggested by the observation that the complex-specific DENV MAb 1A1D2 binds virions in a temperature-dependent manner and appears to trap E proteins on the virus particle in a unique conformation.[65]

A more general role for viral "breathing" in antibody recognition of flaviviruses has been established. Using modified neutralization assay conditions with prolonged incubations in the presence of antibody, the contribution of "breathing" to the overall neutralization sensitivity can be explored.[61] Briefly, viruses and antibody are incubated for roughly 1 h to allow for steady-state binding, at which time the fractional occupancy of epitopes on the virion is defined by antibody concentration and affinity. Further increases in antibody binding will not occur unless the number or presentation of epitopes on the virion changes with time. Experiments with both WNV and DENV demonstrate that further increases in either the time or temperature of incubation result in a marked increase in neutralization potency; the magnitude of this change correlates well with the predicted accessibility of the epitope on the mature virus particle. This pattern of neutralization was not explained by the kinetics of antibody binding, virion aggregation, or the action of complement.[61]

To explain these data, it is hypothesized that the accessibility of epitopes changes as viruses sample different ensembles of conformations during viral breathing. Given enough time, even antibodies with limited to almost no neutralizing activity in standard neutralization assays have the potential to dock on the virion with a stoichiometry sufficient for neutralization. In addition to exposing more epitopes, time-dependent changes in the antigenic surface of the virus particle also may allow antibody engagement of the virion with increased avidity via bivalent interactions among E proteins in conformations not present in the average state, as well as cooperative effects during antibody binding.[61] A requirement for the presence of antibody during the prolonged incubation suggests that conformational changes that result in increases in epitope exposure on mature virions are reversible,[109] as has been reported for picornaviruses.[108] Of interest, strain-dependent patterns of neutralization may reflect differences in the structural ensembles sampled by related flaviviruses.[53,84]

The impact of the dynamic exposure of viral epitopes *in vivo* remains uncertain. Virtually nothing is known about the relevant concentrations and volumes that govern antibody–virion interactions in tissues and blood where the key events in the pathogenesis of these viruses occur. Because the kinetics of neutralization are increased by an elevated temperature, it is tempting to speculate that certain classes of antibodies, such as those recognizing the DII-FL epitope commonly observed in infected individuals, may function better than anticipated in the context of an acute febrile response.

5.3. Complement

The complement system is a family of serum and cell-surface proteins that recognize pathogens and activate inflammatory responses.[110] Activation of complement contributes to protection against viral infection through the direct lysis of virions or infected cells, release of inflammatory peptides, opsonization of viruses, and by facilitating antigen presentation and immune cell priming. Opsonization of viruses with classical pathway complement components (C1q, C4b, and C3b) in an antibody-dependent manner can inhibit virus infection by blocking receptor attachment, entry, and fusion or by promoting terminal complement pathway activation and deposition of the C5b-C9 membrane attack components that results in virolysis. Indeed, the absence of classical pathway complement components leads to more severe primary WNV infection.[111] Complement has been shown to increase the neutralization potency of antibodies in many systems, including flaviviruses.[59,75,112–114] Not all IgG subclasses bind C1q avidly and thus complement-augmented neutralization is restricted to certain antibody isotypes (mouse IgG2a and 2b; human IgG1 and 3). Two lines of evidence suggest that C1q has the potential to alter the stoichiometric requirements for neutralization. First, C1q has the potential to markedly reduce the size of (or eliminate altogether) the neutralization-resistant fraction of flaviviruses observed in studies of antibodies with a maturation state-dependent phenotype.[75] Second, the resistant fraction observed with populations of virions engineered to display varying numbers of intact antibody epitopes also may be reduced.[75] Experiments of this type led to the conclusion that the stoichiometric threshold for neutralization of WNV was reduced roughly 50% in the presence of C1q.[75] How C1q functions in this regard is not completely understood, but likely involves its ability to cross-link antibodies on the surface of the virion. This mechanism is consistent with the analogous outcome of enhanced neutralization obtained by addition of cross-linking goat anti-mouse IgG secondary antibodies. In addition to its effects on neutralization, complement impacts conditions that support ADE by virtue of its ability to shift the stoichiometric threshold of neutralization below the minimal number of antibodies required to support ADE (discussed below). C1q inhibits ADE by WNV and DENV *in vitro* on cells that express the Fcγ-receptor CD32 and *in vivo*.[123]

6. THE STOICHIOMETRY OF ADE

Insight into the stoichiometric requirements for neutralization provides a biochemical rationale to understand ADE. Analysis of large panels of flavivirus-reactive MAbs suggests that virtually all antibodies that bind the virion and neutralize infection have the potential to promote ADE *in vitro*, irrespective of their epitope specificity.[21,86] For human cells expressing the Fcγ-receptor CD32, concentrations of antibody that promote maximal enhancement of infection *in vitro* are similar to those that neutralize half of the virions in parallel studies.[21] Thus, neutralization and ADE are phenomena related by the number of antibodies bound to the virion. When half of the virions are engaged by antibody with a stoichiometry sufficient to inactivate virus infection, the other half is not, and remains infectious despite being decorated by antibody. Antibodies that bind the flavivirus virion yet never reach a stoichiometric threshold sufficient for neutralization broadly enhance infection over a wide range of antibody concentrations. Thus, at the upper limit, the number of antibodies that promote ADE is defined by the stoichiometric threshold of neutralization. Because the dose–response curves of virus–antibody binding to cells and ADE are very similar, the minimal number of antibodies required for ADE is likely defined by the requirements for avid attachment to Fcγ-receptor. However, significant gaps in our knowledge exist. Recent studies suggest that neutralization potency is impacted by expression of Fcγ-receptors of target cells, which in turn translates into changes in the antibody concentrations that support ADE.[115] Although not yet identified, it remains possible that antibodies exist which bind the surface E or prM protein in a specific geometry (e.g., parallel to the plane of the virus surface) that allows for neutralization but are not accessible to Fcy-receptors and thus cannot support ADE.

7. INSIGHTS INTO VACCINES AND THERAPEUTICS

During the past decade, structural and quantitative functional studies with MAbs have contributed to a detailed understanding of how antibodies engage virions and block infection of flaviviruses (reviewed in Ref. 95). This information is now being exploited to identify antibodies with the most useful functional properties[24,116,117] and to deconstruct the complexity of polyclonal antibody responses to infection and vaccination.[49,54,103,118,119] The limited accessibility of many epitopes recognized by antibodies on flavivirus virions is a critical determinant of the neutralization potency of antibodies. Antibodies that bind accessible epitopes on the virion neutralize infection at relatively low fractional occupancy and thus low antibody concentrations.[21] In contrast, epitopes that are poorly accessible require engagement of a larger fraction of these binding determinants in order to neutralize infection. Finally, not all epitopes support neutralization even at full occupancy (e.g., prM epitopes on a subset of partially mature virions and DII-FL epitopes on mature virions); these classes of antibodies have the potential to support ADE despite high concentrations of antibody.[21,70,82]

Identifying readily accessible epitopes on the virus particle is an important practical goal as these represent promising targets for antibody-based therapeutics and vaccine responses. For example, the MAb E16 neutralizes infection at a very low fractional occupancy of its accessible DIII-LR epitope,[21] is highly protective in passive transfer experiments,[60] and has shown promise in animal models as a postexposure therapeutic agent.[117] E16 is capable of binding WNV irrespective of the maturation state of the virus particle.[70] Unfortunately, humans do not appear to produce a functionally significant titer of antibodies that recognize this epitope in the context of natural infection.[70,120] Whether the structural and stoichiometric information about flaviviruses can be harnessed using recombinant antigens with novel platforms of epitope display remains to be determined.[121,122] If such programs were successful, safer and more effective vaccines likely could be generated against several species of flaviviruses.

ACKNOWLEDGMENTS

This work was supported by the intramural program of the National Institute of Allergy and Infectious Diseases, the Burroughs Wellcome Fund, and NIH grants R01AI073755 and R01A1089591. We apologize to those colleagues whose work we were unable to cite due to the limited scope of this review.

REFERENCES

1. Bhatt S, Gething PW, Brady OJ, et al. The global distribution and burden of dengue. *Nature*. 2013;496:504–507.
2. Beasley DW, Barrett AD, Tesh RB. Resurgence of West Nile neurologic disease in the United States in 2012: what happened? What needs to be done? *Antivir Res*. 2013;99:1–5.
3. Heinz FX, Stiasny K. Flaviviruses and flavivirus vaccines. *Vaccine*. 2012;30:4301–4306.
4. Monath TP, Cetron MS, Teuwen DE. Yellow fever vaccine. In: Plotkin SA, Orenstein WA, Offit PA, eds. *Vaccines*. 5th ed. Philadelphia: Saunders Elsevier; 2008.
5. Monath TP. Yellow fever vaccine. *Expert Rev Vaccines*. 2005;4:553–574.
6. Thomas RE, Lorenzetti DL, Spragins W, Jackson D, Williamson T. The safety of yellow fever vaccine 17D or 17DD in children, pregnant women, HIV+ individuals, and older persons: systematic review. *Am J Trop Med Hyg*. 2012;86:359–372.
7. Carrington CV, Auguste AJ. Evolutionary and ecological factors underlying the tempo and distribution of yellow fever virus activity. *Infect Genet Evol*. 2013;13:198–210.
8. Halstead SB, Thomas SJ. New Japanese encephalitis vaccines: alternatives to production in mouse brain. *Expert Rev Vaccines*. 2011;10:355–364.
9. Jelinek T. IXIARO updated: overview of clinical trials and developments with the inactivated vaccine against Japanese encephalitis. *Expert Rev Vaccines*. 2013;12:859–869.
10. Appaiahgari MB, Vrati S. IMOJEV((R)): a Yellow fever virus-based novel Japanese encephalitis vaccine. *Expert Rev Vaccines*. 2010;9:1371–1384.
11. Heinz FX, Stiasny K, Holzmann H, et al. Vaccination and tick-borne encephalitis, central Europe. *Emerg Infect Dis*. 2013;19:69–76.
12. Heinz FX, Holzmann H, Essl A, Kundi M. Field effectiveness of vaccination against tick-borne encephalitis. *Vaccine*. 2007;25:7559–7567.
13. Beasley DW. Vaccines and immunotherapeutics for the prevention and treatment of infections with West Nile virus. *Immunotherapy*. 2011;3:269–285.
14. Capeding MR, Tran NH, Hadinegoro SR, et al. Clinical efficacy and safety of a novel tetravalent dengue vaccine in healthy children in Asia: a phase 3, randomised, observer-masked, placebo-controlled trial. *Lancet*. 2014.
15. Hombach J, Solomon T, Kurane I, Jacobson J, Wood D. Report on a WHO consultation on immunological endpoints for evaluation of new Japanese encephalitis vaccines, WHO, Geneva, 2–3 September, 2004. *Vaccine*. 2005;23:5205–5211.
16. Suthar MS, Diamond MS, Gale Jr M. West Nile virus infection and immunity. *Nat Rev Microbiol*. 2013;11:115–128.
17. Sabin AB. Research on dengue during World War II. *Am J Trop Med Hyg*. 1952;1:30–50.
18. Guzman MG, Alvarez M, Halstead SB. Secondary infection as a risk factor for dengue hemorrhagic fever/dengue shock syndrome: an historical perspective and role of antibody-dependent enhancement of infection. *Arch Virol*. 2013;158:1445–1459.
19. Yauch LE, Shresta S. Dengue virus vaccine development. *Adv Virus Res*. 2014;88:315–372.
20. Halstead SB. Neutralization and antibody-dependent enhancement of dengue viruses. *Adv Virus Res*. 2003;60:421–467.
21. Pierson TC, Xu Q, Nelson S, et al. The stoichiometry of antibody-mediated neutralization and enhancement of West Nile virus infection. *Cell Host Microbe*. 2007;1:135–145.
22. Morens DM, Halstead SB, Marchette NJ. Profiles of antibody-dependent enhancement of dengue virus type 2 infection. *Microb Pathog*. 1987;3:231–237.
23. Halstead SB. In vivo enhancement of dengue virus infection in rhesus monkeys by passively transferred antibody. *J Infect Dis*. 1979;140:527–533.

24. Balsitis SJ, Williams KL, Lachica R, et al. Lethal antibody enhancement of dengue disease in mice is prevented by Fc modification. *PLoS Pathog.* 2010;6:e1000790.
25. Zellweger RM, Prestwood TR, Shresta S. Enhanced infection of liver sinusoidal endothelial cells in a mouse model of antibody-induced severe dengue disease. *Cell Host Microbe.* 2010;7:128–139.
26. Goncalvez AP, Engle RE, St Claire M, Purcell RH, Lai CJ. Monoclonal antibody-mediated enhancement of dengue virus infection in vitro and in vivo and strategies for prevention. *Proc Natl Acad Sci U S A.* 2007;104:9422–9427.
27. Ng JK, Zhang SL, Tan HC, et al. First experimental in vivo model of enhanced dengue disease severity through maternally acquired heterotypic dengue antibodies. *PLoS Pathog.* 2014;10:e1004031.
28. Kaufmann B, Rossmann MG. Molecular mechanisms involved in the early steps of flavivirus cell entry. *Microbes Infect.* 2011;13:1–9.
29. Perera R, Kuhn RJ. Structural proteomics of dengue virus. *Curr Opin Microbiol.* 2008;11:369–377.
30. Rey FA, Heinz FX, Mandl C, Kunz C, Harrison SC. The envelope glycoprotein from tick-borne encephalitis virus at 2 A resolution. *Nature.* 1995;375:291–298.
31. Chen Y, Maguire T, Hileman RE, et al. Dengue virus infectivity depends on envelope protein binding to target cell heparan sulfate. *Nat Med.* 1997;3:866–871.
32. Luca VC, AbiMansour J, Nelson CA, Fremont DH. Crystal structure of the Japanese encephalitis virus envelope protein. *J Virol.* 2012;86:2337–2346.
33. Li L, Lok SM, Yu IM, et al. The flavivirus precursor membrane-envelope protein complex: structure and maturation. *Science.* 2008;319:1830–1834.
34. Allison SL, Schalich J, Stiasny K, Mandl CW, Heinz FX. Mutational evidence for an internal fusion peptide in flavivirus envelope protein E. *J Virol.* 2001;75:4268–4275.
35. Zhang Y, Corver J, Chipman PR, et al. Structures of immature flavivirus particles. *EMBO J.* 2003;22:2604–2613.
36. Zhang Y, Kaufmann B, Chipman PR, Kuhn RJ, Rossmann MG. Structure of immature West Nile virus. *J Virol.* 2007;81:6141–6145.
37. Elshuber S, Allison SL, Heinz FX, Mandl CW. Cleavage of protein prM is necessary for infection of BHK-21 cells by tick-borne encephalitis virus. *J Gen Virol.* 2003;84:183–191.
38. Moesker B, Rodenhuis-Zybert IA, Meijerhof T, Wilschut J, Smit JM. Characterization of the functional requirements of West Nile virus membrane fusion. *J Gen Virol.* 2010;91:389–393.
39. Yu IM, Holdaway HA, Chipman PR, Kuhn RJ, Rossmann MG, Chen J. Association of the pr peptides with dengue virus at acidic pH blocks membrane fusion. *J Virol.* 2009;83:12101–12107.
40. Yu IM, Zhang W, Holdaway HA, et al. Structure of the immature dengue virus at low pH primes proteolytic maturation. *Science.* 2008;319:1834–1837.
41. Stadler K, Allison SL, Schalich J, Heinz FX. Proteolytic activation of tick-borne encephalitis virus by furin. *J Virol.* 1997;71:8475–8481.
42. Kuhn RJ, Zhang W, Rossmann MG, et al. Structure of dengue virus: implications for flavivirus organization, maturation, and fusion. *Cell.* 2002;108:717–725.
43. Nybakken GE, Oliphant T, Johnson S, Burke S, Diamond MS, Fremont DH. Structural basis of West Nile virus neutralization by a therapeutic antibody. *Nature.* 2005;437:764–769.
44. Kaufmann B, Nybakken GE, Chipman PR, et al. West Nile virus in complex with the Fab fragment of a neutralizing monoclonal antibody. *Proc Natl Acad Sci U S A.* 2006;103:12400–12404.
45. Teoh EP, Kukkaro P, Teo EW, et al. The structural basis for serotype-specific neutralization of dengue virus by a human antibody. *Sci Transl Med.* 2012;4:139ra183.

46. Fibriansah G, Tan JL, Smith SA, et al. A potent anti-dengue human antibody preferentially recognizes the conformation of E protein monomers assembled on the virus surface. *EMBO Mol Med.* 2014;6:358–371.
47. de Alwis R, Smith SA, Olivarez NP, et al. Identification of human neutralizing antibodies that bind to complex epitopes on dengue virions. *Proc Natl Acad Sci U S A.* 2012;109:7439–7444.
48. Kaufmann B, Vogt MR, Goudsmit J, et al. Neutralization of West Nile virus by cross-linking of its surface proteins with Fab fragments of the human monoclonal antibody CR4354. *Proc Natl Acad Sci U S A.* 2010;107:18950–18955.
49. Beltramello M, Williams KL, Simmons CP, et al. The human immune response to dengue virus is dominated by highly cross-reactive antibodies endowed with neutralizing and enhancing activity. *Cell Host Microbe.* 2010;8:271–283.
50. Smith SA, de Alwis AR, Kose N, et al. The potent and broadly neutralizing human dengue virus-specific monoclonal antibody 1C19 reveals a unique cross-reactive epitope on the bc loop of domain II of the envelope protein. *mBio.* 2013;4: e00873–e00913.
51. Lin HE, Tsai WY, Liu IJ, et al. Analysis of epitopes on dengue virus envelope protein recognized by monoclonal antibodies and polyclonal human sera by a high throughput assay. *PLoS Negl Trop Dis.* 2012;6:e1447.
52. Crill WD, Hughes HR, Delorey MJ, Chang GJ. Humoral immune responses of dengue fever patients using epitope-specific serotype-2 virus-like particle antigens. *PLoS One.* 2009;4:e4991.
53. Sukupolvi-Petty S, Brien JD, Austin SK, et al. Functional analysis of antibodies against dengue virus type 4 reveals strain-dependent epitope exposure that impacts neutralization and protection. *J Virol.* 2013;87:8826–8842.
54. de Alwis R, Beltramello M, Messer WB, et al. In-depth analysis of the antibody response of individuals exposed to primary dengue virus infection. *PLoS Negl Trop Dis.* 2011;5:e1188.
55. Brien JD, Austin SK, Sukupolvi-Petty S, et al. Genotype-specific neutralization and protection by antibodies against dengue virus type 3. *J Virol.* 2010;84:10630–10643.
56. Burton DR, Saphire EO, Parren PW. A model for neutralization of viruses based on antibody coating of the virion surface. *Curr Top Microbiol Immunol.* 2001;260:109–143.
57. Dulbecco R, Vogt M, Strickland AG. A study of the basic aspects of neutralization of two animal viruses, western equine encephalitis virus and poliomyelitis virus. *Virology.* 1956;2:162–205.
58. Taylor HP, Armstrong SJ, Dimmock NJ. Quantitative relationships between an influenza virus and neutralizing antibody. *Virology.* 1987;159:288–298.
59. Della-Porta AJ, Westaway EG. A multi-hit model for the neutralization of animal viruses. *J Gen Virol.* 1978;38:1–19.
60. Oliphant T, Engle M, Nybakken GE, et al. Development of a humanized monoclonal antibody with therapeutic potential against West Nile virus. *Nat Med.* 2005;11:522–530.
61. Dowd KA, Jost CA, Durbin AP, Whitehead SS, Pierson TC. A dynamic landscape for antibody binding modulates antibody-mediated neutralization of West Nile virus. *PLoS Pathog.* 2011;7:e1002111.
62. Klasse PJ, Burton DR. Antibodies to West Nile virus: a double-edged sword. *Cell Host Microbe.* 2007;1:87–89.
63. Calisher CH, Karabatsos N, Dalrymple JM, et al. Antigenic relationships between flaviviruses as determined by cross-neutralization tests with polyclonal antisera. *J Gen Virol.* 1989;70(pt 1):37–43.
64. Calisher CH. Antigenic classification and taxonomy of flaviviruses (family Flaviviridae) emphasizing a universal system for the taxonomy of viruses causing tick-borne encephalitis. *Acta Virol.* 1988;32:469–478.

65. Lok SM, Kostyuchenko V, Nybakken GE, et al. Binding of a neutralizing antibody to dengue virus alters the arrangement of surface glycoproteins. *Nat Struct Mol Biol.* 2008;15:312–317.
66. Gach JS, Leaman DP, Zwick MB. Targeting HIV-1 gp41 in close proximity to the membrane using antibody and other molecules. *Curr Top Med Chem.* 2011;11:2997–3021.
67. Pantophlet R. Antibody epitope exposure and neutralization of HIV-1. *Curr Pharm Des.* 2010;16:3729–3743.
68. Pierson TC, Sanchez MD, Puffer BA, et al. A rapid and quantitative assay for measuring antibody-mediated neutralization of West Nile virus infection. *Virology.* 2006; 346:53–65.
69. Khromykh AA, Varnavski AN, Westaway EG. Encapsidation of the flavivirus Kunjin replicon RNA by using a complementation system providing Kunjin virus structural proteins in trans. *J Virol.* 1998;72:5967–5977.
70. Nelson S, Jost CA, Xu Q, et al. Maturation of West Nile virus modulates sensitivity to antibody-mediated neutralization. *PLoS Pathog.* 2008;4:e1000060.
71. Mukherjee S, Dowd KA, Manhart CJ, et al. Mechanism and significance of cell type-dependent neutralization of flaviviruses. *J Virol.* 2014;88:7210–7220.
72. Halstead SB, O'Rourke EJ. Dengue viruses and mononuclear phagocytes. I. Infection enhancement by non-neutralizing antibody. *J Exp Med.* 1977;146:201–217.
73. Peiris JS, Porterfield JS. Antibody-mediated enhancement of Flavivirus replication in macrophage-like cell lines. *Nature.* 1979;282:509–511.
74. Peiris JS, Gordon S, Unkeless JC, Porterfield JS. Monoclonal anti-Fc receptor IgG blocks antibody enhancement of viral replication in macrophages. *Nature.* 1981;289:189–191.
75. Mehlhop E, Nelson S, Jost CA, et al. Complement protein C1q reduces the stoichiometric threshold for antibody-mediated neutralization of West Nile virus. *Cell Host Microbe.* 2009;6:381–391.
76. Gollins SW, Porterfield JS. Flavivirus infection enhancement in macrophages: radioactive and biological studies on the effect of antibody on viral fate. *J Gen Virol.* 1984;65(pt 8):1261–1272.
77. Sun P, Bauza K, Pal S, et al. Infection and activation of human peripheral blood monocytes by dengue viruses through the mechanism of antibody-dependent enhancement. *Virology.* 2011;421:245–252.
78. Halstead SB, Mahalingam S, Marovich MA, Ubol S, Mosser DM. Intrinsic antibody-dependent enhancement of microbial infection in macrophages: disease regulation by immune complexes. *Lancet Infect Dis.* 2010;10:712–722.
79. Gromowski GD, Barrett ND, Barrett AD. Characterization of dengue virus complex-specific neutralizing epitopes on envelope protein domain III of dengue 2 virus. *J Virol.* 2008;82:8828–8837.
80. Gromowski GD, Barrett AD. Characterization of an antigenic site that contains a dominant, type-specific neutralization determinant on the envelope protein domain III (ED3) of dengue 2 virus. *Virology.* 2007;366:349–360.
81. Diamond MS, Pierson TC, Roehrig JT. Antibody therapeutics against flaviviruses. In: Shi PY, ed. *Molecular Virology and Control of Flaviviruses.* Poole, United Kingdom: Caister Academic Press; 2012.
82. Dejnirattisai W, Jumnainsong A, Onsirisakul N, et al. Cross-reacting antibodies enhance dengue virus infection in humans. *Science.* 2010;328:745–748.
83. Cockburn JJ, Navarro Sanchez ME, Fretes N, et al. Mechanism of dengue virus broad cross-neutralization by a monoclonal antibody. *Structure.* 2012;20:303–314.
84. Austin SK, Dowd KA, Shrestha B, et al. Structural basis of differential neutralization of DENV-1 genotypes by an antibody that recognizes a cryptic epitope. *PLoS Pathog.* 2012;8:e1002930.

85. Stiasny K, Kiermayr S, Holzmann H, Heinz FX. Cryptic properties of a cluster of dominant flavivirus cross-reactive antigenic sites. *J Virol.* 2006;80:9557–9568.
86. Oliphant T, Nybakken GE, Engle M, et al. Antibody recognition and neutralization determinants on domains I and II of West Nile Virus envelope protein. *J Virol.* 2006;80:12149–12159.
87. Edeling MA, Austin SK, Shrestha B, et al. Potent dengue virus neutralization by a therapeutic antibody with low monovalent affinity requires bivalent engagement. *PLoS Pathog.* 2014;10:e1004072.
88. Costin JM, Zaitseva E, Kahle KM, et al. Mechanistic study of broadly neutralizing human monoclonal antibodies against dengue virus that target the fusion loop. *J Virol.* 2013;87:52–66.
89. Kiermayr S, Stiasny K, Heinz FX. Impact of quaternary organization on the antigenic structure of the tick-borne encephalitis virus envelope glycoprotein E. *J Virol.* 2009;83:8482–8491.
90. Lee PD, Mukherjee S, Edeling MA, et al. The Fc region of an antibody impacts the neutralization of West Nile viruses in different maturation states. *J Virol.* 2013;87:13729–13740.
91. He RT, Innis BL, Nisalak A, et al. Antibodies that block virus attachment to Vero cells are a major component of the human neutralizing antibody response against dengue virus type 2. *J Med Virol.* 1995;45:451–461.
92. Stiasny K, Brandler S, Kossl C, Heinz FX. Probing the flavivirus membrane fusion mechanism by using monoclonal antibodies. *J Virol.* 2007;81:11526–11531.
93. Thompson BS, Moesker B, Smit JM, Wilschut J, Diamond MS, Fremont DH. A therapeutic antibody against West Nile virus neutralizes infection by blocking fusion within endosomes. *PLoS Pathog.* 2009;5:e1000453.
94. Kaufmann B, Chipman PR, Holdaway HA, et al. Capturing a flavivirus pre-fusion intermediate. *PLoS Pathog.* 2009;5:e1000672.
95. Dowd KA, Pierson TC. Antibody-mediated neutralization of flaviviruses: a reductionist view. *Virology.* 2011;411:306–315.
96. Pierson TC, Diamond MS. Degrees of maturity: the complex structure and biology of flaviviruses. *Curr Opin Virol.* 2012;2:168–175.
97. Murray JM, Aaskov JG, Wright PJ. Processing of the dengue virus type 2 proteins prM and C-prM. *J Gen Virol.* 1993;74(pt 2):175–182.
98. Davis CW, Nguyen HY, Hanna SL, Sanchez MD, Doms RW, Pierson TC. West Nile virus discriminates between DC-SIGN and DC-SIGNR for cellular attachment and infection. *J Virol.* 2006;80:1290–1301.
99. Guirakhoo F, Bolin RA, Roehrig JT. The Murray Valley encephalitis virus prM protein confers acid resistance to virus particles and alters the expression of epitopes within the R2 domain of E glycoprotein. *Virology.* 1992;191:921–931.
100. Plevka P, Battisti AJ, Junjhon J, et al. Maturation of flaviviruses starts from one or more icosahedrally independent nucleation centres. *EMBO Rep.* 2011;12:602–606.
101. Junjhon J, Edwards TJ, Utaipat U, et al. Influence of pr-M cleavage on the heterogeneity of extracellular dengue virus particles. *J Virol.* 2010;84:8353–8358.
102. Cherrier MV, Kaufmann B, Nybakken GE, et al. Structural basis for the preferential recognition of immature flaviviruses by a fusion-loop antibody. *EMBO J.* 2009;28:3269–3276.
103. VanBlargan LA, Mukherjee S, Dowd KA, Durbin AP, Whitehead SS, Pierson TC. The type-specific neutralizing antibody response elicited by a dengue vaccine candidate is focused on two amino acids of the envelope protein. *PLoS Pathog.* 2013; 9:e1003761.
104. Boehr DD, Wright PE. Biochemistry. How do proteins interact? *Science.* 2008;320:1429–1430.

105. Phelps DK, Rossky PJ, Post CB. Influence of an antiviral compound on the temperature dependence of viral protein flexibility and packing: a molecular dynamics study. *J Mol Biol*. 1998;276:331–337.
106. Lewis JK, Bothner B, Smith TJ, Siuzdak G. Antiviral agent blocks breathing of the common cold virus. *Proc Natl Acad Sci U S A*. 1998;95:6774–6778.
107. Bothner B, Dong XF, Bibbs L, Johnson JE, Siuzdak G. Evidence of viral capsid dynamics using limited proteolysis and mass spectrometry. *J Biol Chem*. 1998;273:673–676.
108. Li Q, Yafal AG, Lee YM, Hogle J, Chow M. Poliovirus neutralization by antibodies to internal epitopes of VP4 and VP1 results from reversible exposure of these sequences at physiological temperature. *J Virol*. 1994;68:3965–3970.
109. Dowd KA, Mukherjee S, Kuhn RJ, Pierson TC. Combined effects of the structural heterogeneity and dynamics of flaviviruses on antibody recognition. *J Virol*. 2014;.
110. Carroll MC. The complement system in regulation of adaptive immunity. *Nat Immunol*. 2004;5:981–986.
111. Mehlhop E, Fuchs A, Engle M, Diamond MS. Complement modulates pathogenesis and antibody-dependent neutralization of West Nile virus infection through a C5-independent mechanism. *Virology*. 2009;393:11–15.
112. Benhnia MR, McCausland MM, Moyron J, et al. Vaccinia virus extracellular enveloped virion neutralization in vitro and protection in vivo depend on complement. *J Virol*. 2009;83:1201–1215.
113. Mozdzanowska K, Feng J, Eid M, Zharikova D, Gerhard W. Enhancement of neutralizing activity of influenza virus-specific antibodies by serum components. *Virology*. 2006;352:418–426.
114. Mayer AE, Johnson JB, Parks GD. The neutralizing capacity of antibodies elicited by parainfluenza virus infection of African Green Monkeys is dependent on complement. *Virology*. 2014;460–461:23–33.
115. Rodrigo WW, Block OK, Lane C, et al. Dengue virus neutralization is modulated by IgG antibody subclass and Fcgamma receptor subtype. *Virology*. 2009;394:175–182.
116. Williams KL, Sukupolvi-Petty S, Beltramello M, et al. Therapeutic efficacy of antibodies lacking Fcgamma receptor binding against lethal dengue virus infection is due to neutralizing potency and blocking of enhancing antibodies [corrected]. *PLoS Pathog*. 2013;9:e1003157.
117. Morrey JD, Siddharthan V, Olsen AL, et al. Defining limits of treatment with humanized neutralizing monoclonal antibody for West Nile virus neurological infection in a hamster model. *Antimicrob Agents Chemother*. 2007;51:2396–2402.
118. Messer WB, de Alwis R, Yount BL, et al. Dengue virus envelope protein domain I/II hinge determines long-lived serotype-specific dengue immunity. *Proc Natl Acad Sci U S A*. 2014;111:1939–1944.
119. Williams KL, Wahala WM, Orozco S, de Silva AM, Harris E. Antibodies targeting dengue virus envelope domain III are not required for serotype-specific protection or prevention of enhancement in vivo. *Virology*. 2012;429:12–20.
120. Oliphant T, Nybakken GE, Austin SK, et al. Induction of epitope-specific neutralizing antibodies against West Nile virus. *J Virol*. 2007;81:11828–11839.
121. Zhou T, Zhu J, Yang Y, et al. Transplanting supersites of HIV-1 vulnerability. *PLoS One*. 2014;9:e99881.
122. Correia BE, Bates JT, Loomis RJ, et al. Proof of principle for epitope-focused vaccine design. *Nature*. 2014;507:201–206.
123. Mehlhop E, Ansarah-Sobrinho C, Johnson S, et al. Complement protein C1q inhibits antibody-dependent enhancement of flavivirus infection in an IgG subclass-specific manner. *Cell Host Microbe*. 2007;2(6):417–426.

CHAPTER SIX

TRIM21-Dependent Intracellular Antibody Neutralization of Virus Infection

William A. McEwan*, Leo C. James*,[1]

*Protein and Nucleic Acid Chemistry Division, Medical Research Council Laboratory of Molecular Biology, Cambridge, United Kingdom

[1]Corresponding author: e-mail address: lcj@mrc-lmb.cam.ac.uk

Contents

Abstract

The ability of antibodies to prevent viral infection has long been recognized. *In vitro* neutralization assays, which take place in the absence of professional immune effector mechanisms, have demonstrated that the process of neutralization can occur by a variety of molecular mechanisms. Most known mechanisms involve the blocking of an event essential for infection, for instance, the steric inhibition of attachment to entry receptors. As such, neutralization is often thought of as a passive process that can occur without the need for host effector machinery. In contrast to this view, it has recently been demonstrated that neutralization can depend on the widely expressed cytosolic Fc binding protein TRIM21. This unique and novel Ig receptor directs the ubiquitin and proteasome-dependent degradation of intracellular antibody-bound viral particles and prevents infection. It has been further demonstrated that detection of cytosolic antibody by TRIM21 activates inflammatory signaling pathways and promotes the production of cytokines and chemokines. Studies in a TRIM21-null mouse demonstrate the

Progress in Molecular Biology and Translational Science, Volume 129
ISSN 1877-1173
http://dx.doi.org/10.1016/bs.pmbts.2014.10.006

importance of these activities: homozygous knockouts suffer fatal viral infection where wild-type mice survive. Though there is much to be learned about the role of TRIM21 in immunity, it is clear that there is a hitherto unappreciated role for antibodies in the intracellular environment.

1. INTRODUCTION

The adaptive immune system can mount specific responses to a near limitless range of antigens. The relative role of antibodies versus cell-mediated and other forms of immunity in overcoming viral infection has been a source of investigation over the past several decades. However, determining the true correlates of immunity for any particular pathogen is complicated by the redundant nature of the immune system and the potential for misleading co-correlation.[1] The natural experiment of transcytosis of IgG from mother to infant provides evidence that antibody-mediated immunity can be protective against a range of viruses including poliovirus, influenza virus, hepatitis A, and hepatitis B viruses. Furthermore, animal studies strongly support a role for antibodies in antiviral protection. For instance, mice unable to produce antibody are acutely susceptible to adenovirus infection while passive transfer of immune serum to naïve animals is sufficient to restore protection.[2]

Since the very first experiments studying the effect of antibodies on animal viruses, it has been clear that antibodies are capable of reducing viral titre during *in vitro* infections of permissive cells. This process, termed neutralization, applies specifically to such *in vitro* activity, thus excluding many alternative mechanisms by which antibodies may exert antiviral activity *in vivo*. It is due to the belief that the production of neutralizing antibodies is a correlate of (or causative of) antiviral protection that it is routinely measured as marker for pathogen-specific immunity. Nonetheless, the mechanisms by which antibodies exert their neutralization effect are often not known and have been a source of controversy over previous decades.

In the past few years, our laboratory has shown that IgG can be carried into the cell during infection by human adenovirus. Once within the cytosol, antibody is bound by the atypical Fc receptor TRIM21, which mediates the proteasome-dependent destruction of virus particles. As such we have demonstrated that neutralization can be an active process that depends on host cell machinery. This review summarizes these findings and discusses them in the context of both antibody neutralization and other intracellular pathogen restriction factors.

2. THE TRIPARTITE MOTIF FAMILY

TRIM21, also known as Ro52 and SS-A, belongs to the large family of tripartite motif proteins of which there are approximately 100 members in the human genome.[3,4] Members of the TRIM family share a RING, one or two B Box domains, and a coiled-coil domain, known collectively as the RBCC. Approximately half of TRIM proteins, including TRIM21 and the antiretroviral factor TRIM5α, possess an additional C-terminal PRYSPRY domain that mediates interactions with target proteins (Fig. 1A). The RING and B Box are Zn^{2+}-coordinating alpha-helical domains with the RING domain capable of E3 ubiquitin ligase activity. The coiled-coil mediates homo-dimerization of TRIM molecules in an antiparallel orientation.[5] The PRYSPRY is a fusion of two more ancient

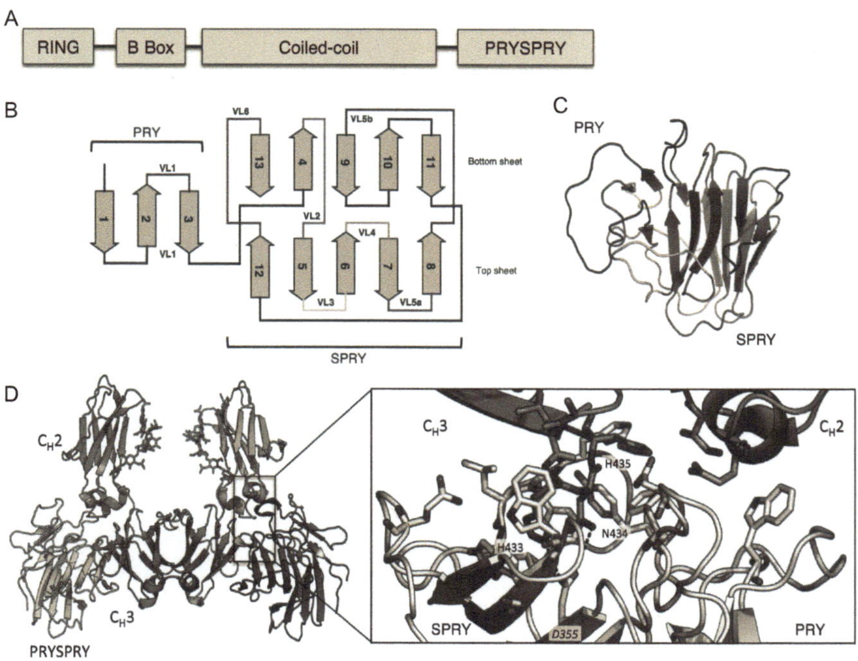

Figure 1 TRIM21 structure and antibody binding mechanism. (A) TRIM21 domain structure. (B) Schematic diagram of TRIM21 PRYSPRY secondary structure in which the β-sheets and flexible binding site loops (VLs) are numbered. (C) Structure of TRIM21 PRYSPRY domain with VLs depicted in the same color scheme as in (B). (D) Structure of TRIM21 PRYSPRY:IgG Fc based on PDB 2IWG. The Fc is indicated in green and cyan and the PRYSPRY in yellow and gray. The boxed region is a close-up view of the interface. The "HNH" motif from Fc is marked. (See the color plate.)

subdomains, the PRY and the SPRY comprising approximately 60 and 150 amino acids, respectively. Crystal structures of the PRYSPRY reveal a twisted beta-sheet with variable loop (VL) regions that, between TRIM members, vary in both primary sequence and length, and are the major determinants of ligand specificity (Fig. 1B and C).[6] The TRIM family is evolutionarily ancient with homologues of TRIM37 found in diverse taxa including fungi, plants, and various protozoans. However, TRIM proteins with PRYSPRY domains are uniquely present in vertebrates, with a large expansion of such genes in mammals resulting from gene duplication events.[7] As a member of a recently expanded cluster of *TRIM* genes on human chromosome 11, *TRIM21* is therefore likely to have arisen in the same epoch as IgG, which also only occurs in mammals.[8]

3. TRIM21 IS A HIGH-AFFINITY CYTOSOLIC Fc RECEPTOR

A crystal structure of the TRIM21 PRYSPRY:Fc complex has been solved, which shows that the PRYSPRY domain forms a deep binding pocket surrounded by flexible loops (VL2–6) and supported at the base by a concave beta-sheet (Fig. 1D). This binding pocket accommodates the Fc $H_{433}N_{434}H_{435}$ motif that protrudes from the C_H3 domain of IgG. This HNH motif has been found to be important for TRIM21-mediated neutralization, since its mutation to alanine reduces the ability of a monoclonal antibody to prevent adenovirus infection.[9] Hydrogen bonds are also made between Fc H_{433} and N_{434} to PRYSPRY D_{355}, which is buried deep within the PRYSPRY binding pocket, an interaction also found to be important for neutralization (Fig. 1D). Additional interactions are mediated by residues in the PRYSPRY VL1 and VL6, which form contacts with the IgG C_H2 domain. The TRIM21 binding site on Fc overlaps with that of the superantigen *Staphylococcus aureus* protein A, and has been shown to compete for binding,[6] but is distinct from the FcγRI binding site. Isothermal titration calorimetry (ITC) studies reveal that TRIM21 PRYSPRY binds with a 2:1 stoichiometric ratio to Fc, meaning that a dimer of TRIM21 is able to interact with a single antibody molecule. TRIM21 therefore does not cross-link antibody, with the possible exception of the case where an antibody is directed against TRIM21 itself. This is of potential medical significance given that TRIM21 is a major target for antibodies in the autoimmune diseases Sjogren's syndrome and systemic lupus erythematosus.[10,11] These diseases are characterized by circulating and deposited immune complexes that contribute to their pathophysiology. An involvement of TRIM21 in bipolar bridging of antibody suggests a potential mechanism by which immune complexes may form and be stabilized.

The dissociation constant of monomeric PRYSPRY and Fc has been determined by ITC to be 37 nM[6] and the avidity effects of TRIM21 dimerization increase affinity further to approximately 600 pM.[12] Alongside IgεR, TRIM21 is therefore the highest affinity antibody receptor in the human body and the highest affinity receptor for IgG. The TRIM21:Fc interaction is widely conserved among mammals: mouse antibodies bind to mouse TRIM21 PRYSPRY with a similar affinity as the cognate human TRIM21:human Fc interaction.[13] The interaction is also species cross-reactive: human and mouse TRIM21 PRYSPRY domains can bind IgG from diverse mammalian host species. These findings imply that evolution acts to maintain the high-affinity binding of IgG to TRIM21. Finally, in addition to binding IgG, TRIM21 has been shown to bind directly to IgM, albeit with a reduced affinity.

4. TRIM21 MEDIATES ANTIBODY-DEPENDENT INTRACELLULAR NEUTRALIZATION

TRIM21 has been shown to mediate the neutralization of a replication-deficient human adenovirus type 5 vector (Fig. 2). When TRIM21 was depleted in HeLa cells, a cell line used frequently for such studies over the past half century, neutralization by pooled human serum IgG was diminished by approximately 10-fold.[12] In agreement with molecular interaction data, human TRIM21 was shown to provide the majority of neutralization for anti-adenovirus antibodies raised in goats and mice.[12,14] Moreover, TRIM21 was found to be crucial for efficient neutralization in mouse cells as fibroblasts from TRIM21$^{-/-}$ mice permitted only low levels of neutralization when compared to wild-type fibroblasts.[9] This activity of TRIM21 is referred to as "antibody-dependent intracellular neutralization" or ADIN.[15]

TRIM21 mediates neutralization by targeting antibody-coated viruses for degradation in the proteasome. The intracellular adenovirus:antibody complex is degraded within 2 h of infection and degradation can be prevented by depletion of TRIM21.[12] Furthermore, chemical inhibitors of the proteasome prevent degradation and restore infectivity of antibody-coated virus. TRIM21 is an interferon-stimulated gene and its expression is substantially upregulated by type I interferons. Under conditions of IFN stimulation, antibodies are capable of mediating more potent

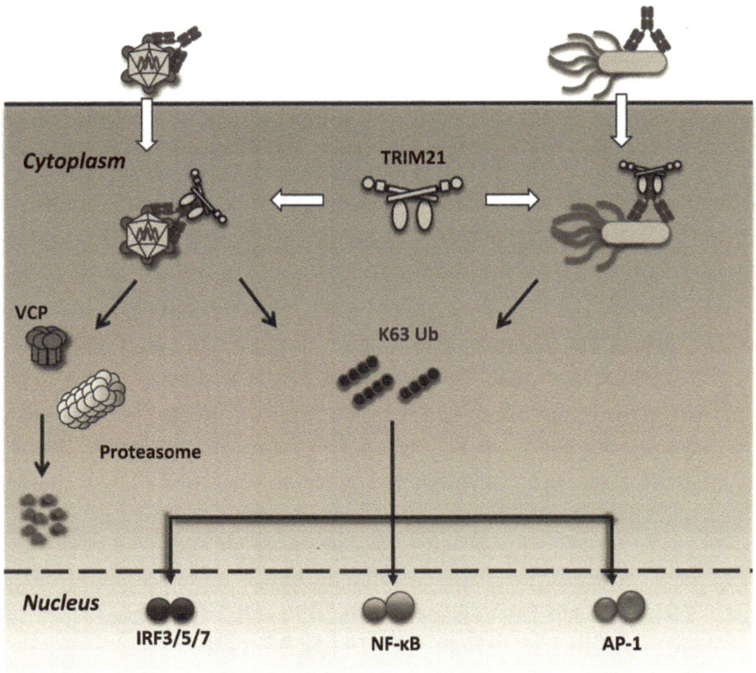

Figure 2 TRIM21 detects antibody-coated pathogens in the cytosol and mediates signaling and neutralization pathways.

neutralization and TRIM21 depletion reverses this effect. Such observations are not unprecedented: potentiation of neutralization by interferon has been previously demonstrated for adenovirus type 3 as well as other nonenveloped viruses.[16] The involvement of the proteasome in neutralization raises the question of how a challenging substrate such as an adenovirus particle can be processed: the proteasomal pore is no greater than 2 nm in diameter[17] yet the adenovirus capsid is approximately 90 nm.[18] Indeed it has been demonstrated that TRIM21 neutralization requires the AAA ATPase valosin-containing protein (VCP, also known as p97 or Cdc48 in yeast), a homo-hexameric barrel-shaped enzyme involved in the disassembly of protein complexes and their proteasomal degradation.[19,20,21] Inhibition of VCP by the chemical inhibitor DBeQ[22] or its depletion by RNAi severely diminishes neutralization.[23] Interestingly, degradation of IgG heavy chain expressed in the cytoplasm of cells is prevented by proteasome inhibition but not VCP inhibition. This suggests that it is the type of substrate that stimulates the recruitment of VCP rather than VCP being a constitutive auxiliary in TRIM21-mediated degradation. This hypothesis is supported by findings that VCP is recruited to stalled proteasomes.[20]

5. TRIM21 IS A SENSOR FOR CYTOPLASMIC ANTIBODY

Infection can be sensed via the detection of pathogen-associated molecular patterns by germline-encoded molecules of the innate immune system.[24] Alternatively, infection can be sensed by the mislocalization of host molecules or danger-associated molecular patterns (DAMPs).[25] The transfer of antibodies from the extracellular milieu to the cytoplasm is detected by TRIM21, which initiates a signaling cascade activating NF-κB, AP-1, and IRF transcription factors (Fig. 2).[14] These findings demonstrate that antibodies in the cytosolic environment act as a DAMP, signaling a breach of membrane integrity. Stimulation of TRIM21 signaling with adenovirus:antibody complexes resulted in the production of proinflammatory chemokines and cytokines including IL-6, tumor necrosis factor and IFN-β. When supernatant from cells stimulated in this manner was transferred to fresh cells, an antiviral state was induced. This state included the upregulation of TRIM21 itself, a mechanism ensuring that bystander cells are primed to exhibit high levels of virus neutralization.

The ability to signal in response to intracellular antibody is not limited to professional immune cells but also occurs in cells that are not of myeloid or lymphoid lineage such as fibroblasts, airway epithelium, and several commonly used cell lines.[14] Indeed TRIM21 exhibits widespread tissue expression, and may therefore represent a prevalent sensor for intracellular antibody. Experiments using the replication competent mouse adenovirus-1 (MAV-1) demonstrated that neutralization and signaling occur simultaneously to reduce viral titers.[26] This ability to both neutralize and sense infection makes TRIM21 a dual effector-sensor receptor, a class of molecules to which the antiretroviral factors TRIM5α and tetherin also belong.[27,28] It is likely that other pathogen-specific "restriction factors" will also be found to display such dual capabilities. In a similar fashion, multiple serum proteins are known to attach to pathogens in circulation and may act alongside antibody as DAMPs during infection.

6. TRIM21 FUNCTIONS ARE UBIQUITIN DEPENDENT

In addition to binding antibodies, TRIM21 is an active E3 ubiquitin ligase and autoubiquitination can be observed *in vitro*. TRIM21-mediated ubiquitination is required for both neutralization and signaling as inhibition of the proteasome or deletion of the RING and B Box domains abolishes these activities. TRIM21 has been shown to catalyze the formation of ubiquitin polymers with different chain linkages and function with different E2 enzymes *in vitro*. In the presence of the promiscuous E2 enzyme UbcH5, TRIM21 undergoes autoubiquitination with K48-ubiquitin chains, while using the E2 enzyme pair Ubc13/UEV1A TRIM21 synthesizes the formation of unanchored K63 chains. These different chain types appear to be important for TRIM21 activity. K63-linked ubiquitin chains can promote NF-κB signaling via the TAK1 complex[28,29] and IKKα/β/NEMO complex.[30,31] Inhibition of these pathways by (5Z)-7-oxozeaenol, IKK VII or panepoxydone—inhibitors of TAK1, IKKα, and IκB, respectively—is sufficient to prevent TRIM21-mediated signaling in response to antibody-coated virus. From these data, it is clear that ubiquitination is crucial to TRIM21 function; however, the specifics are still unclear, including the complete range of ubiquitin chains that are synthesized and the order in which they are made and the substrates that undergo modification. Moreover, there is a clear need for these ubiquitination activities to be strictly regulated in order to act as a trigger for effector and sensor pathways. How this is accomplished is not currently understood but intuitively must be related to the binding of TRIM21 to antibody-coated virus.

7. *IN VIVO* RELEVANCE

Neutralizing antibodies, which inhibit infection *in vitro*, are a principal correlate of protection for infectious disease (for review, see Refs. 15,32). Their elicitation is used to measure vaccine efficacy and to determine prior pathogen exposure. Antibodies that are not neutralizing *in vitro* can also be protective *in vivo*, either because they require effector mechanisms (such as ADCC, complement, or phagocytosis) or because they require epitopes or specific physiological conditions that are only present *in vivo* (for discussion, see Ref. 33). The importance of antibody protection can be studied *in vivo* by genetic ablation of genes required for antibody production and maturation or that encode Fc receptors and by passively transferring antibodies to naïve animals. The importance of TRIM21 in antiviral immunity has been studied using knockout animals challenged with the model nonenveloped virus (MAV-1). In these studies, it was shown that naïve mice lacking TRIM21 had a higher viral load and increased mortality, with 25% of knockout animals succumbing to virally induced encephalomyelitis within the first week of infection. As these experiments were carried out in naïve animals, this protection was presumably a result of the use of IgM by TRIM21. Passive transfer experiments in TRIM21 knockout animals have also been carried out, in which the ability of antisera against MAV-1 to protect against a high viral dose was determined. Antiserum was shown to fully protect wild-type animals given a lethal virus dose; however, the same sera failed to protect most TRIM21 knockout mice. As with the naïve experiments, the observed protection by TRIM21 correlated with a reduction in viremia. Importantly, no difference in mortality was observed between genotypes when neutralizing F(ab′)2 fragments were administered. These results illustrate the importance of both antibodies and TRIM21 in antiviral protection. Furthermore they show that, in the absence of TRIM21, antibodies are still protective but with reduced efficacy while in the absence of antibody, TRIM21 gives no antiviral protection. The latter finding is consistent with the role of TRIM21 being entirely dependent upon antibody.

8. VIRAL DETERMINANTS OF TRIM21-MEDIATED NEUTRALIZATION

The recent identification of TRIM21 neutralization means that the spectrum of pathogens that are susceptible to ADIN is still being determined. Susceptibility is likely to be dependent upon the particulars of the pathogen replication cycle and whether the pathogen has evolved a mechanism of antagonism. However, the basic requirement for ADIN sensitivity can be defined as an infection event in which the pathogen enters the cytosol with an attached antibody. Fulfilling this criterion is sufficient to provoke an immune response, as demonstrated in experiments in which antibody-coated latex beads transfected into the cell are rapidly bound by TRIM21 and activate immune signaling.[12,14] Thus, any cytoplasmic antibody-bound particle may, in principle, be subject to the activities of TRIM21. Immunofluorescent microscopy has demonstrated that adenovirus particles are capable of entering the cell with antibody attached and are bound by TRIM21 within 30 min of infection.[12] Similar findings were demonstrated for the facultatively intracellular bacterium *Salmonella enterica typhimurium*, which may reside either within vesicles or free in the cytoplasm.[34,35] However, the capability of different pathogens to traffic antibodies into the cytoplasm will depend on their mechanism of entry and replication.

Nonenveloped viruses are presumed to be the major target of ADIN because antibodies against them bind directly to their capsid. This means that unless the virus divests itself of its capsid before entering the cytosol it will be visible to TRIM21. For adenovirus, entry occurs after binding of viral fiber protein to the cell surface receptors coxsackie and adenovirus receptor or CD46.[36,37] Subsequent infectious entry is via clathrin-mediated endocytosis though macropinocytosis is also stimulated.[38,39,40] Once within the endosome, adenovirus escapes into the cytosol through breakage of the endosomal membrane. This process relies on the partial uncoating of the virus particle and the release of the interior capsid protein VI, which induces membrane damage.[40] The subviral particle is then free in the cytosol and uses microtubule-associated motor proteins to access the nuclear pore complex where the viral DNA genome and its associated proteins are released from the capsid and enter the nucleus.[41] This method of entry to the cell allows antibodies directed to the major capsid protein hexon to remain associated with the subviral particle, permitting direct contact with the cytosol. Adenovirus may therefore represent a virus that is particularly susceptible to TRIM21.

Feline calicivirus was shown to elicit TRIM21 signaling when incubated with feline IgG.[14] Comparatively little is known about the entry of this virus into the cytosol, though it has been demonstrated that infection by FCV results in the permeabilization of the cell membrane to the 17 kDa toxin α-sarcin.[42] Given this finding, it is likely that some exposure of endosomal contents to the cytosol occurs, which may explain the ability of TRIM21 to detect antibody. Other positive-sense RNA viruses such as the picornaviruses are thought to translocate their genomes through pores in the viral capsid while remaining in endosomes. For instance, human rhinovirus 2 is triggered at low pH to insert amphipathic peptides contained within proteins VP1 and VP4 into the endosomal membrane.[43] A pore approximately 10 Å in diameter is then formed at the pseudo threefold axis of the capsid which permits translocation of viral RNA into the cytosol.[44] Capsid proteins remain within endosomes and are subsequently degraded within lysosomes[45] and only low molecular weight dyes are able to penetrate to the cytosol.[46] Direct contact between capsid and cytosol is therefore prevented. HRV-2 and other viruses such as poliovirus that use a similar strategy may therefore evade detection by TRIM21. However, rhinoviruses belonging to the major group such as HRV-14 have been shown to enter the cell by endosomal lysis with subviral particles detectable in cytoplasmic fractions.[47] Future experiments are required to determine exactly how different entry pathways affect the ability of TRIM21 to detect viral infection.

Enveloped viruses are not believed to be susceptible to ADIN because the antibodies that target them bind not to their capsid but to antigens embedded in the lipid membrane that surrounds them. When enveloped viruses infect cells they do so through a fusion event between their envelope and either the plasma membrane or, if they have entered through endocytosis, an endosomal membrane. This has the result of leaving any attached antibodies on the outside of the cell or inside an endosome and therefore not accessible to TRIM21 in the cytosol. As an example of this, the plasma-membrane-fusing respiratory syncytial virus has been shown to evade TRIM21 signaling.[14] However, there are reports that cellular membranes become permeable to macromolecules even during enveloped virus entry[48] and this could potentially permit the entry of antibody and facilitate TRIM21 sensing. Future work will determine whether this is the case.

9. TRIM21 EXERTS HIGHLY EFFICIENT INCREMENTAL NEUTRALIZATION

The number of antibodies required for virus neutralization has been a source of considerable investigation and controversy since some of the earliest experiments on virus neutralization.[49] Controversy has centered on the single-hit and multi-hit hypotheses, under which viruses are neutralized by a single antibody or a threshold number (greater than 1) of antibodies, respectively. Such models assume "all-or-nothing" neutralization where the remaining level of infectivity is governed solely by the proportion of viruses with subthreshold numbers of antibodies attached. During single-hit neutralization only viruses without any antibody molecules bound will be infectious. Remaining infectivity of virus will decrease linearly when plotted on a logarithmic scale against the number of antibodies per virus. The curve will pass through e^{-1}, or 37% remaining infectivity, when a single antibody is bound and e^{-2}, 14% remaining infectivity, when two antibodies are bound (Fig. 3A). For multi-hit neutralization, an initial "lag" is predicted that becomes more pronounced as the threshold number of antibodies increases.

The observation that poliovirus is neutralized with first-order kinetics was interpreted under the assumptions of all-or-nothing neutralization and has been proposed as evidence in support of single-hit neutralization.[49] Such assertions appeared to be at least partially substantiated by experimental data when it was shown that both poliovirus and adenovirus can be neutralized by relatively few (about 5 and 1.4, respectively) antibodies per

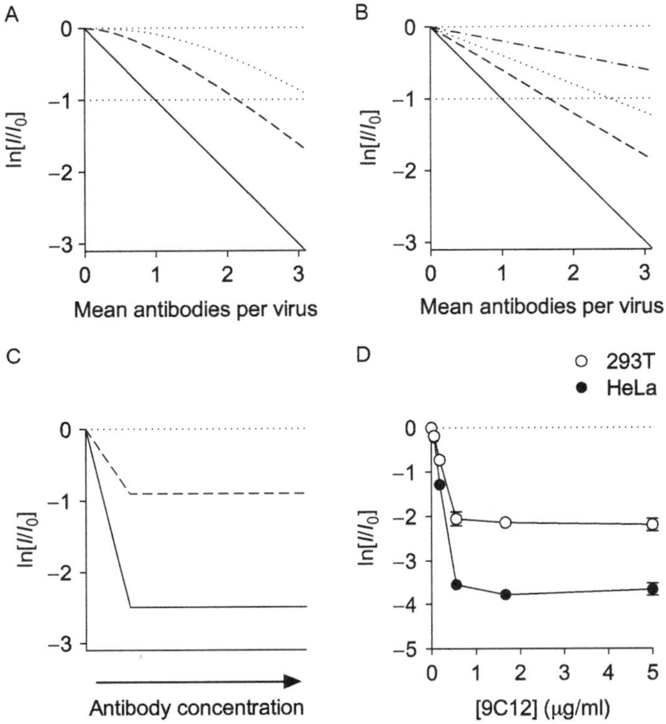

Figure 3 Features of neutralization curves that are influenced by TRIM21. (A) All-or-nothing neutralization curves. If it is assumed that antibodies are distributed among viruses according to a Poisson distribution, then the proportion of viruses that remain infectious will decline as the number of antibody molecules per virus (*r*) increases above a threshold number (*n*) of antibodies required for neutralization. Remaining infectivity (I/I_0) is thus equal to $I/I_0 = \sum_{r=0}^{n-1}(m^r e^{-m})/r!$ where *m* is the average number of antibodies per virus. The remaining infectivity of virus is plotted against a natural logarithmic scale (solid line). Under a single-hit model ($n=1$), a straight-line curve is observed, whereas under a multi-hit model ($n=2$, dashed and $n=3$, dotted), a lag is observed. (B) Incremental neutralization. Neutralization by TRIM21 results in straight-line curves of differing slopes. This suggests that antibodies contribute incrementally to neutralization, rather than providing all-or-nothing neutralization. As TRIM21 levels increase, so the efficiency of neutralization increases toward the theoretical maximal efficiency of neutralization where a single antibody is sufficient for neutralization, depicted by the solid line. (C) Experimental observations over the past several decades have often demonstrated the existence of a proportion of viruses that is resistant to neutralization, the persistent fraction (PF), marked by a leveling off in neutralization curves. Under conditions where TRIM21 levels are high, the PF is low (solid line). However, depletion of TRIM21 or inhibition of VCP or the proteasome increases the PF (dashed line). (D) It has previously been demonstrated that the PF varies between cell lines in a similar manner to changing TRIM21 levels (see text). This observation is also true for adenovirus, which has a markedly different PF on two commonly used human cell lines when treated with mouse monoclonal antibody 9C12.

virus particle.[50,51] However, these findings have proven controversial and potentially flawed.[52] More recently it has been argued that apparent first-order kinetics cannot be interpreted as evidence for single-antibody neutralization.[53]

The mouse monoclonal antibody 9C12, which binds the major capsid protein, hexon,[54] does not block entry of adenovirus 5 and acts in concert with TRIM21 to neutralize infection.[15] 9C12 was found by immunogold electron microscopy to bind adenovirus particles approximating a Poisson distribution.[9] No evidence for a lag was observed and neutralization curves were found to be approximately linear. The gradient of the slope was determined by TRIM21 levels, with a steep gradient observed when TRIM21 levels were high. Parallel infection and quantification experiments, where virus was simultaneously examined for infectivity and stoichiometry of binding, demonstrated that few antibodies were sufficient for neutralization. Under conditions of high TRIM21 expression (after IFN stimulation) in mouse cells, neutralization by 9C12 approached its theoretical maximal efficiency, requiring 1.6 antibody molecules per virus. Under conditions of low TRIM21 expression (obtained by shRNA), approximately fivefold greater levels of antibody were required to obtain the same level of neutralization. These findings argue against an all-or-nothing model of neutralization, given that multiple straight-line curves were observed with differing slopes. Rather the findings suggest that TRIM21 exerts incremental neutralization whereby each antibody that binds contributes to neutralization and that the magnitude of this contribution is in turn determined by the levels of TRIM21 (Fig. 3B). When experiments were performed using polyclonal anti-adenovirus sera or pooled human IgG, where antibodies that neutralize via multiple mechanisms are likely to be present, neutralization was found to shift from straight-line curves to multi-hit curves as TRIM21 was depleted.[9] The shape of a neutralization curve, in particular the presence or absence of a lag, may therefore provide useful information, not perhaps as originally envisaged as a way of illustrating that few antibodies can neutralize and that neutralization is all-or-nothing, but concerning the mechanism by which neutralization takes place (e.g., entry blocking or TRIM21 mediated).

The finding that neutralization by TRIM21 can be exceedingly efficient suggests that antibodies may exert a protective role even when present at very low concentrations or low affinities. Such low stoichiometric complexes are likely to fail to elicit effector responses that rely on cross-linking Fc receptors such as phagocytosis, but may nonetheless contribute to clearance of viral infection via TRIM21. This may be particularly important in situations where pathogen-specific antibodies may be at low concentrations, for example, in the initial stages of a primary humoral response or during secondary challenge before activation of memory B cells.

10. THE PERSISTENT FRACTION

The addition of high concentrations of antibody, or prolonged incubation with excess concentrations of antibody, frequently results in a fraction of viruses that remain resistant to neutralization. These "non-neutralizable" viruses are referred to as the persistent fraction (PF; Fig. 3C). Mechanisms proposed to underpin the PF include the cross-linking of virus particles, dissociation of virus antibody complex upon dilution and the slow attainment of steady-state conditions.[52,55,56] Addition of 9C12 to adenovirus resulted in PF occurring at stoichiometries of approximately 50 antibodies per virion, a level well below saturation binding, which occurred at stoichiometries of approximately 200.[9] This implies that factors other than antibody become limiting, resulting in the PF. Depletion of TRIM21 was found to increase the PF from 5% of remaining infectivity to 30% demonstrating that limiting levels of TRIM21 can be a cause of the PF. Similarly, depletion or chemical inhibition of TRIM21 co-factor VCP also increased the PF.[23] These findings indicate that the PF can be influenced by the intracellular environment. It is interesting to note that the PF of enterovirus 71 was found to vary with cell type.[57] In the human rhabdomyosarcoma cell line RD, 1% of viruses were resistant to neutralization but on green monkey kidney cells 10% of viruses treated with the same neutralizing serum were resistant. It is tempting to speculate that saturation of intracellular neutralization machinery was the underlying cause of the PF in these experiments and that altered levels of PF between the cell lines are a result of cell type-specific variations in such machinery. Though the role of TRIM21 in the neutralization of picornaviruses has yet to be investigated, the variable PF observed by Kjellen *et al.* is observed during adenovirus infection of commonly used human cell lines (Fig. 3D).

11. COMPARISON OF TRIM21 WITH TRIM5α

While many of the cofactors and basic principles necessary for TRIM21-mediated signaling and neutralization have been identified, the detailed molecular mechanisms of these processes remain to be elucidated. In attempting to understand these mechanisms it is useful to compare TRIM21 to other related antiviral proteins. In particular, TRIM21 shares many properties with the antiretroviral restriction factor TRIM5α. Both proteins have dual-antiviral activities, mediating both effector and sensor pathways. They are also similar in their mechanism of function. TRIM5α and TRIM21 have the same domain architecture and both are E3 ubiquitin ligases. In each case, their ability to activate innate immunity is thought to be dependent upon their catalysis of free K63-ubiquitin chains.

TRIM5α was identified as a host factor that prevents the infection of rhesus macaque cells by HIV-1.[58] While the human homologue is inactive against HIV, it retains activity against other retroviruses. In contrast to TRIM21, the block to infection provided by TRIM5α is not reversed by proteasome inhibitors (though such compounds do relieve the block to reverse transcription).[59] While deletion of the TRIM5α RING domain does not prevent HIV restriction,[60] deletion, or mutation of the B Box abolishes TRIM5α inhibition of HIV infection.[61] Together this suggests that TRIM5α restriction of HIV is independent of ubiquitination but dependent upon the B Box. The B Box domain of TRIM5α has been shown to mediate higher order oligomerisation and this activity correlates with its ability to bind to *in vitro* assembled capsid protein.[62] The current model by which TRIM5α is believed to inhibit HIV-1 replication is by forming a lattice that encases the viral capsid, thereby preventing proper disassembly. Formation of this TRIM5α lattice is hypothesized to require initial binding of the PRYSPRY domain to the viral capsid followed by intermolecular B Box interactions. In support of this, in cells where the proteasome has been inhibited, TRIM5α forms large stable cytoplasmic bodies localized to viral particles.[63] Meanwhile, cryo-EM experiments have demonstrated that TRIM5α forms a lattice on two-dimensional arrays of hexagonal capsid protein.[64]

Whether TRIM21 utilizes a mechanism like TRIM5α to bind and/or capture antibody-coated virus is unknown, however, current evidence suggests that this is not the case. In domain swap experiments it was found that the RING, B Box and coiled-coil domains of TRIM5α could be substituted, and restriction maintained, with those from TRIM6 and TRIM34 but not TRIM21.[65] However, the RING, B Box and coiled-coil domains of TRIM21 could be substituted for those from TRIMCyp, a variant of TRIM5α that also restricts HIV-1.[66] The transplanted RING and B Box domains in TRIM21Cyp were shown to be active as inhibition of the proteasome rescued the block to reverse transcription. However, deletion of the RING and B Box domains from TRIMCyp does not rescue infection, unlike with TRIM5α.[67] Thus, the ability of the B Box to mediate higher order assembly is not required for TRIMCyp restriction. The fact that the TRIM21 B Box can substitute for TRIMCyp but not TRIM5α suggests that it does not perform this function and only has a role in recruiting the proteasome.

The above analysis suggests that despite the obvious similarities between TRIM5α and TRIM21 they utilize different mechanisms of restriction. Experiments into the relative stoichiometric requirements of each protein also support this conclusion. As discussed earlier, TRIM21 can restrict viruses that are bound by <2 antibody molecules. The necessary stoichiometry for TRIM5α and TRIMCyp activity has also been estimated, by testing chimeric viruses that have different ratios of restriction-sensitive capsid protein. In these studies, most of the block to infection was lost once the ratio of sensitive protein reached $\leq 25\%$,[68] approximately 375 capsid monomers per virion. This contrast in the necessary stoichiometries between TRIM5α and TRIM21 further supports the notion that restriction by the former is a direct consequence of binding, whereas the latter requires a catalytic step and the involvement of the proteasome.

If it is assumed that TRIM5α forms a more-or-less complete lattice around incoming viral capsids, whereas TRIM21 is more sparsely recruited then there are also implications for their respective mechanisms of regulation. As both proteins have a role in innate immune signaling via the production of K63-ubiquitin chains, there must be a switch that triggers or upregulates this activity upon viral infection. In the case of TRIM5α, the template-dependent assembly of higher order oligomers on the capsid surface provides an attractive model of regulation. Indeed, it has been shown that the addition of virus greatly increases K63-chain formation by TRIM5α.[28] However, the cross-linking of multiple molecules as a mechanism of regulation is harder to imagine in the case of TRIM21 given that only a few copies per virus are sufficient for activity. Cross-linking or higher order assembly of TRIM21 is also not observed either when unliganded or in the presence of IgG Fc, with which it forms a stable 1:1 complex. Finally, the different propensity of TRIM5α and TRIM21 to form higher order multimers is suggested by the fact that while TRIM5α overexpression is sufficient to activate innate immune pathways, TRIM21 overexpression either does not,[69] or appears to be inhibited at higher levels of expression.[70]

12. CONCLUSIONS

In addition to being the highest affinity Fc receptor yet identified, TRIM21 represents a very different type of antibody receptor to those previously described. It is structurally unrelated to other mammalian receptors, which all belong to the immunoglobulin superfamily, widely expressed rather than confined to professional immune cells, cytosolically located rather than membrane-associated, neither subtype nor isotype specific, being able to bind both IgM and IgG, and in possession of its own intrinsic enzymatic activity. TRIM21 does not fit into conventional immunological classifications. It is an immune-stimulated gene and therefore part of innate immunity but it uses antibodies to target pathogens, which are part of adaptive immunity. TRIM21 possess dual signaling and effector functions and these events appear to occur simultaneously, as opposed to a signaling event proceeded by an effector response as is the typical paradigm in immunology. The discovery of TRIM21 also challenges widespread assumptions specifically concerning humoral immunity. It illustrates that neutralization is not necessarily a receptor-independent event and that antibodies mediate important immune functions inside cells, within the cytoplasm, as well as in the extracellular space. It remains to be seen whether TRIM21 is the only protein that possesses the many unexpected properties summarized here or whether other conceptually related receptors exist.

ACKNOWLEDGMENTS

We thank P. J. Klasse and C. Rosie Williams for insightful comments on the chapter. This work was supported by Medical Research Council Grant U105181010 and European Research Council Grant 281627-IAI.

REFERENCES

1. Plotkin SA. Vaccines: correlates of vaccine-induced immunity. *Clin Infect Dis.* 2008;47(3):401–409.
2. Moore ML, McKissic EL, Brown CC, Wilkinson JE, Spindler KR. Fatal disseminated mouse adenovirus type 1 infection in mice lacking B cells or Bruton's tyrosine kinase. *J Virol.* 2004;78(11):5584–5590.
3. Han K, Lou DI, Sawyer SL. Identification of a genomic reservoir for new TRIM genes in primate genomes. *PLoS Genet.* 2011;7(12):e1002388.
4. Reymond A, Meroni G, Fantozzi A, Merla G, Cairo S, Luzi L, et al. The tripartite motif family identifies cell compartments. *EMBO Jl.* 2001;20(9):2140–2151.
5. Sanchez JG, Okreglicka K, Chandrasekaran V, Welker JM, Sundquist WI, Pornillos O. The tripartite motif coiled-coil is an elongated antiparallel hairpin dimer. *Proc Natl Acad Sci USA.* 2014;111(7):2494–2499.

6. James LC, Keeble AH, Khan Z, Rhodes DA, Trowsdale J. Structural basis for PRYSPRY-mediated tripartite motif (TRIM) protein function. *Proc Natl Acad Sci U S A*. 2007;104(15):6200–6205.
7. Marin I. Origin and diversification of TRIM ubiquitin ligases. *PLoS One*. 2012;7(11): e50030.
8. Flajnik MF. Comparative analyses of immunoglobulin genes: surprises and portents. *Nat Rev Immunol*. 2002;2(9):688–698.
9. McEwan WA, Hauler F, Williams CR, Bidgood SR, Mallery DL, Crowther RA, et al. Regulation of virus neutralization and the persistent fraction by TRIM21. *J Virol*. 2012;86(16):8482–8491.
10. Chan EK, Hamel JC, Buyon JP, Tan EM. Molecular definition and sequence motifs of the 52-kD component of human SS-A/Ro autoantigen. *J Clin Invest*. 1991;87(1):68–76.
11. Wahren-Herlenius M, Muller S, Isenberg D. Analysis of B-cell epitopes of the Ro/SS-A autoantigen. *Immunol Today*. 1999;20(5):234–240.
12. Mallery DL, McEwan WA, Bidgood SR, Towers GJ, Johnson CM, James LC. Antibodies mediate intracellular immunity through tripartite motif-containing 21 (TRIM21). *Proc Natl Acad Sci USA*. 2010;107(46):19985–19990.
13. Keeble AH, Khan Z, Forster A, James LC. TRIM21 is an IgG receptor that is structurally, thermodynamically, and kinetically conserved. *Proc Natl Acad Sci USA*. 2008;105(16):6045–6050.
14. McEwan WA, Tam JC, Watkinson RE, Bidgood SR, Mallery DL, James LC. Intracellular antibody-bound pathogens stimulate immune signaling via the Fc receptor TRIM21. *Nat Immunol*. 2013;14(4):327–336.
15. McEwan WA, Mallery DL, Rhodes DA, Trowsdale J, James LC. Intracellular antibody-mediated immunity and the role of TRIM21. *Bioessays*. 2011;33(11):803–809.
16. Langford MP, Villarreal AL, Stanton GJ. Antibody and interferon act synergistically to inhibit enterovirus, adenovirus, and herpes simplex virus infection. *Infect Immun*. 1983;41(1):214–218.
17. Gallastegui N, Groll M. The 26S proteasome: assembly and function of a destructive machine. *Trends Biochem Sci*. 2010;35(11):634–642.
18. Reddy VS, Natchiar SK, Stewart PL, Nemerow GR. Crystal structure of human adenovirus at 3.5 A resolution. *Science*. 2010;329(5995):1071–1075.
19. Zhang X, Shaw A, Bates PA, Newman RH, Gowen B, Orlova E, et al. Structure of the AAA ATPase p97. *Mol Cell*. 2000;6(6):1473–1484.
20. Isakov E, Stanhill A. Stalled proteasomes are directly relieved by P97 recruitment. *J Biol Chem*. 2011;286(35):30274–30283.
21. Meyer H, Bug M, Bremer S. Emerging functions of the VCP/p97 AAA-ATPase in the ubiquitin system. *Nat Cell Biol*. 2012;14(2):117–123.
22. Chou TF, Brown SJ, Minond D, Nordin BE, Li K, Jones AC, et al. Reversible inhibitor of p97, DBeQ, impairs both ubiquitin-dependent and autophagic protein clearance pathways. *Proc Natl Acad Sci USA*. 2011;108(12):4834–4839.
23. Hauler F, Mallery DL, McEwan WA, Bidgood SR, James LC. AAA ATPase p97/VCP is essential for TRIM21-mediated virus neutralization. *Proc Natl Acad Sci USA*. 2012;109(48):19733–19738.
24. Medzhitov R, Janeway Jr CA. Decoding the patterns of self and nonself by the innate immune system. *Science*. 2002;296(5566):298–300.
25. Matzinger P. Tolerance, danger, and the extended family. *Annu Rev Immunol*. 1994;12:991–1045.
26. Watkinson RE, Tam JC, Vaysburd MJ, James LC. Simultaneous neutralization and innate immune detection of a replicating virus by TRIM21. *J Virol*. 2013;87(13):7309–7313.

27. Galao RP, Le Tortorec A, Pickering S, Kueck T, Neil SJ. Innate sensing of HIV-1 assembly by Tetherin induces NFkappaB-dependent proinflammatory responses. *Cell Host Microbe*. 2012;12(5):633–644.
28. Pertel T, Hausmann S, Morger D, Zuger S, Guerra J, Lascano J, et al. TRIM5 is an innate immune sensor for the retrovirus capsid lattice. *Nature*. 2011;472(7343):361–365.
29. Xia ZP, Sun L, Chen X, Pineda G, Jiang X, Adhikari A, et al. Direct activation of protein kinases by unanchored polyubiquitin chains. *Nature*. 2009;461(7260):114–119.
30. Ea CK, Deng L, Xia ZP, Pineda G, Chen ZJ. Activation of IKK by TNFalpha requires site-specific ubiquitination of RIP1 and polyubiquitin binding by NEMO. *Mol Cell*. 2006;22(2):245–257.
31. Wu CJ, Conze DB, Li T, Srinivasula SM, Ashwell JD. Sensing of Lys 63-linked polyubiquitination by NEMO is a key event in NF-kappaB activation [corrected]. *Nat Cell Biol*. 2006;8(4):398–406.
32. Burton DR. Antibodies, viruses and vaccines. *Nat Rev Immunol*. 2002;2(9):706–713.
33. Watkinson RE, McEwan WA, James LC. Intracellular antibody immunity. *J Clin Immunol*. 2014;34(Suppl. 1):S30–S34.
34. Beuzon CR, Salcedo SP, Holden DW. Growth and killing of a Salmonella enterica serovar Typhimurium sifA mutant strain in the cytosol of different host cell lines. *Microbiology*. 2002;148(Pt. 9):2705–2715.
35. Birmingham CL, Brumell JH. Autophagy recognizes intracellular Salmonella enterica serovar Typhimurium in damaged vacuoles. *Autophagy*. 2006;2(3):156–158.
36. Bergelson JM, Cunningham JA, Droguett G, Kurt-Jones EA, Krithivas A, Hong JS, et al. Isolation of a common receptor for Coxsackie B viruses and adenoviruses 2 and 5. *Science*. 1997;275(5304):1320–1323.
37. Gaggar A, Shayakhmetov DM, Lieber A. CD46 is a cellular receptor for group B adenoviruses. *Nat Med*. 2003;9(11):1408–1412.
38. Roelvink PW, Lizonova A, Lee JG, Li Y, Bergelson JM, Finberg RW, et al. The coxsackievirus-adenovirus receptor protein can function as a cellular attachment protein for adenovirus serotypes from subgroups A, C, D, E, and F. *J Virol*. 1998;72(10):7909–7915.
39. Meier O, Boucke K, Hammer SV, Keller S, Stidwill RP, Hemmi S, et al. Adenovirus triggers macropinocytosis and endosomal leakage together with its clathrin-mediated uptake. *J Cell Biol*. 2002;158(6):1119–1131.
40. Wiethoff CM, Wodrich H, Gerace L, Nemerow GR. Adenovirus protein VI mediates membrane disruption following capsid disassembly. *J Virol*. 2005;79(4):1992–2000.
41. Greber UF, Suomalainen M, Stidwill RP, Boucke K, Ebersold MW, Helenius A. The role of the nuclear pore complex in adenovirus DNA entry. *EMBO J*. 1997;16(19):5998–6007.
42. Stuart AD, Brown TD. Entry of feline calicivirus is dependent on clathrin-mediated endocytosis and acidification in endosomes. *J Virol*. 2006;80(15):7500–7509.
43. Weiss VU, Subirats X, Pickl-Herk A, Bilek G, Winkler W, Kumar M, et al. Characterization of rhinovirus subviral A particles via capillary electrophoresis, electron microscopy and gas-phase electrophoretic mobility molecular analysis: part I. *Electrophoresis*. 2012;33(12):1833–1841.
44. Hewat EA, Neumann E, Blaas D. The concerted conformational changes during human rhinovirus 2 uncoating. *Mol Cell*. 2002;10(2):317–326.
45. Brabec-Zaruba M, Pfanzagl B, Blaas D, Fuchs R. Site of human rhinovirus RNA uncoating revealed by fluorescent in situ hybridization. *J Virol*. 2009;83(8):3770–3777.
46. Brabec M, Schober D, Wagner E, Bayer N, Murphy RF, Blaas D, et al. Opening of size-selective pores in endosomes during human rhinovirus serotype 2 in vivo uncoating monitored by single-organelle flow analysis. *J Virol*. 2005;79(2):1008–1016.

47. Schober D, Kronenberger P, Prchla E, Blaas D, Fuchs R. Major and minor receptor group human rhinoviruses penetrate from endosomes by different mechanisms. *J Virol.* 1998;72(2):1354–1364.
48. Madan V, Sanz MA, Carrasco L. Requirement of the vesicular system for membrane permeabilization by Sindbis virus. *Virology.* 2005;332(1):307–315.
49. Dulbecco R, Vogt M, Strickland AG. A study of the basic aspects of neutralization of two animal viruses, western equine encephalitis virus and poliomyelitis virus. *Virology.* 1956;2(2):162–205.
50. Icenogle J, Shiwen H, Duke G, Gilbert S, Rueckert R, Anderegg J. Neutralization of poliovirus by a monoclonal antibody: kinetics and stoichiometry. *Virology.* 1983;127(2):412–425.
51. Wohlfart C. Neutralization of adenoviruses: kinetics, stoichiometry, and mechanisms. *J Virol.* 1988;62(7):2321–2328.
52. Klasse PJ, Sattentau QJ. Occupancy and mechanism in antibody-mediated neutralization of animal viruses. *J Gen Virol.* 2002;83(Pt. 9):2091–2108.
53. Klasse PJ, Sattentau QJ. Mechanisms of virus neutralization by antibody. *Curr Top Microbiol Immunol.* 2001;260:87–108.
54. Varghese R, Mikyas Y, Stewart PL, Ralston R. Postentry neutralization of adenovirus type 5 by an antihexon antibody. *J Virol.* 2004;78(22):12320–12332.
55. Parren PW, Burton DR. The antiviral activity of antibodies in vitro and in vivo. *Adv Immunol.* 2001;77:195–262.
56. Magnus C, Regoes RR. Restricted occupancy models for neutralization of HIV virions and populations. *J Theor Biol.* 2011;283(1):192–202.
57. Kjellen L. A hypothesis accounting for the effect of the host cell on neutralization-resistant virus. *J Gen Virol.* 1985;66(Pt. 10):2279–2283.
58. Stremlau M, Owens CM, Perron MJ, Kiessling M, Autissier P, Sodroski J. The cytoplasmic body component TRIM5alpha restricts HIV-1 infection in Old World monkeys. *Nature.* 2004;427(6977):848–853.
59. Wu X, Anderson JL, Campbell EM, Joseph AM, Hope TJ. Proteasome inhibitors uncouple rhesus TRIM5alpha restriction of HIV-1 reverse transcription and infection. *Proc Natl Acad Sci USA.* 2006;103(19):7465–7470.
60. Diaz-Griffero F, Li X, Javanbakht H, Song B, Welikala S, Stremlau M, et al. Rapid turnover and polyubiquitylation of the retroviral restriction factor TRIM5. *Virology.* 2006;349(2):300–315.
61. Javanbakht H, Diaz-Griffero F, Stremlau M, Si Z, Sodroski J. The contribution of RING and B-box 2 domains to retroviral restriction mediated by monkey TRIM5alpha. *J Biol Chem.* 2005;280(29):26933–26940.
62. Diaz-Griffero F, Qin XR, Hayashi F, Kigawa T, Finzi A, Sarnak Z, et al. A B-box 2 surface patch important for TRIM5alpha self-association, capsid binding avidity, and retrovirus restriction. *J Virol.* 2009;83(20):10737–10751.
63. Campbell EM, Perez O, Anderson JL, Hope TJ. Visualization of a proteasome-independent intermediate during restriction of HIV-1 by rhesus TRIM5alpha. *J Cell Biol* 2008;180(3):549–561.
64. Ganser-Pornillos BK, Chandrasekaran V, Pornillos O, Sodroski JG, Sundquist WI, Yeager M. Hexagonal assembly of a restricting TRIM5alpha protein. *Proc Natl Acad Sci USA.* 2011;108(2):534–539.
65. Li X, Li Y, Stremlau M, Yuan W, Song B, Perron M, et al. Functional replacement of the RING, B-box 2, and coiled-coil domains of tripartite motif 5alpha (TRIM5alpha) by heterologous TRIM domains. *J Virol.* 2006;80(13):6198–6206.
66. Chan E, Schaller T, Eddaoudi A, Zhan H, Tan CP, Jacobsen M, et al. Lentiviral gene therapy against human immunodeficiency virus type 1, using a novel human TRIM21-cyclophilin A restriction factor. *Hum Gene Ther.* 2012;23(11):1176–1185.

67. Diaz-Griffero F, Kar A, Lee M, Stremlau M, Poeschla E, Sodroski J. Comparative requirements for the restriction of retrovirus infection by TRIM5alpha and TRIMCyp. *Virology*. 2007;369(2):400–410.
68. Shi J, Friedman DB, Aiken C. Retrovirus restriction by TRIM5 proteins requires recognition of only a small fraction of viral capsid subunits. *J Virol*. 2013;87(16):9271–9278.
69. Uchil PD, Hinz A, Siegel S, Coenen-Stass A, Pertel T, Luban J, et al. TRIM protein-mediated regulation of inflammatory and innate immune signaling and its association with antiretroviral activity. *J Virol*. 2013;87(1):257–272.
70. Versteeg GA, Rajsbaum R, Sanchez-Aparicio MT, Maestre AM, Valdiviezo J, Shi M, et al. The E3-ligase TRIM family of proteins regulates signaling pathways triggered by innate immune pattern-recognition receptors. *Immunity*. 2013;38(2):384–398.

CHAPTER SEVEN

Picornavirus–Host Interactions to Construct Viral Secretory Membranes

Alexander L. Greninger[1]
School of Medicine, University of California, San Francisco, California, USA
[1]Corresponding author: e-mail address: gerbix@gmail.com

Contents

Abstract

Picornaviruses are positive-stranded RNA viruses of significant disease burden and ubiquitous global reach. Microscopic examination of picornavirus-infected cells has long revealed a drastic reordering of intracellular membranes. Through a confluence of candidate-based approaches and genomic and proteomic screens, the past decade has seen great leaps in understanding how picornaviruses usurp intracellular membranes for their own replication. The growing cast of assembled characters allows for a rich plot in the upcoming years. With their widespread genomic divergence, the number of potential mechanisms for RNA virus vesicogenesis for driving membrane formation and lipid synthesis required for viral replication is broad, but the overall story arch remains surprisingly recognizable. This chapter reviews the major discoveries associating picornavirus pathogenic interactions with the secretory system and highlights important questions and opportunities for future study.

Progress in Molecular Biology and Translational Science, Volume 129
ISSN 1877-1173
http://dx.doi.org/10.1016/bs.pmbts.2014.10.007

1. BACK ON THE RADAR

Picornaviruses are ready again for their close-up. After decades of focus on poliovirus as the model species of choice for RNA viruses, the declining worldwide incidence of poliovirus combined with the recognition of the significant burden of hepatitis C virus led to a relative diminution of the picornavirus family as a research focus for drug discovery and basic biological understanding. Arguably, for the past 20 years, the flaviviruses have driven both the basic and the applied research agenda in positive-stranded RNA virus biology. With the success of recent hepatitis C protease inhibitors in potentially curing the hepacivirus scurge, the time is nigh for picornaviruses.[1,2] In a time of billion dollar valuations for mobile phone apps, we still lack the proverbial cure for the common cold, the majority of which are caused by rhinoviruses.[3] Interest in epidemiological associations with enteroviruses and autoimmune diseases such as type 1 diabetes has become resurgent.[4–6] And the genomics of the past decade has generously attended to the picornaviruses, leading to a doubling of the known diversity of the Picornaviridae family from 19 to 46 species and 8 to 26 genera as of 2014[7–11] ("Picornavirus Home", http://www.picornaviridae.com visited November 15, 2014 (Fig. 1)).

Indeed, the burden and diversity of picornaviruses combined with their tight genome size in this early stage of synthetic biology offers an excellent opportunity and rationale for studying the principles of RNA virus biology. Reorganization of the cellular membranes has long been associated with RNA virus replication. It can be seen as a hallmark of viral–host interactions required for RNA virus replication, along with entry, uncoating, translational control, innate immunoevasion, packaging, and exit. A number of excellent reviews have been written about these steps of RNA virus replication,[12–16] including secretory system reorganization and involvement of PI4KB.[17–19] With the advent of a variety of mammalian genome-wide screens and the continuation of brilliant, fastidious candidate biology, recently much progress has been made in the area of picornavirus interactions with the host proteins and lipids that allow them to remodel cellular membranes. This review highlights some of those advances and offer thoughts on potentially worthwhile, outstanding questions in the field.

2. GETTING TO 3A

Poliovirus was long noted to dramatically reorganize intracellular membranes and associate its replication with these membranous structures

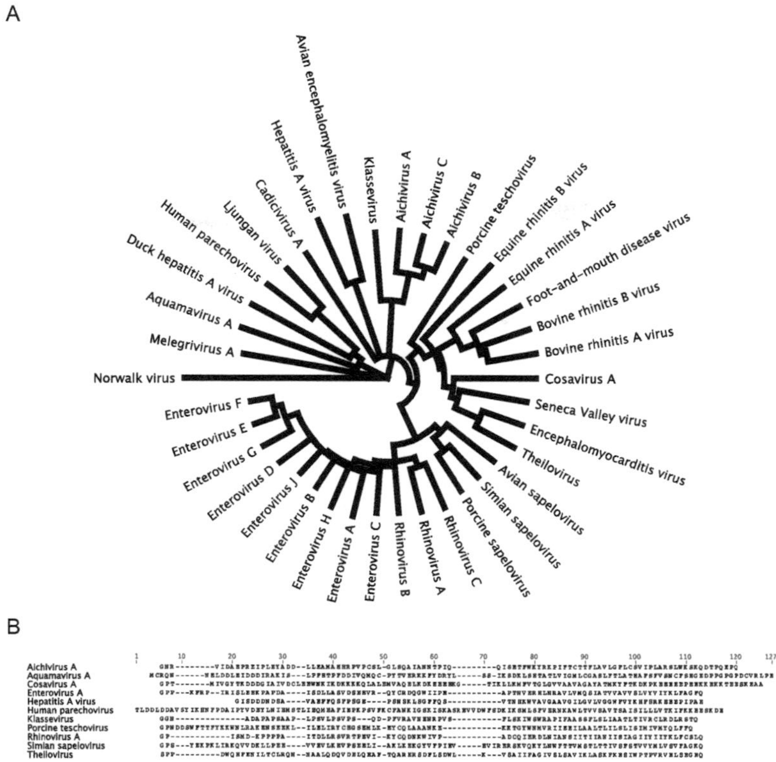

Figure 1 A growing picornavirus family. (A) Deep sequencing has vastly expanded the numbers of known genera *Picornaviridae* family in the last decade as shown in this 2013 phylogeny. (B) Alignment of the 3A proteins from select picornaviruses illustrates their short though varying length and incredibly sequence diversity. Each 3A protein has approximately 30–40% amino acid identity between genera.

in a process that required *de novo* lipid synthesis to reorganize membranes.[20–22] Inhibition of *de novo* lipid synthesis with oleic acid or cerulenin was associated with decreased replication of the poliovirus genome.[23,24] The focus on internal membranes also lead to the discovery that Brefeldin A, a fungal metabolite that also disrupted cellular secretion, inhibited poliovirus but not encephalomyocarditis virus replication in cells.[25,26] Interestingly, cells with a BFA-resistant secretory pathway still supported poliovirus replication, indicating that a specific host factor was targeted by the virus for replication.[27] Subsequently, poliovirus infection was shown to inhibit cellular transport, and this property was localized to the 2B and 3A proteins, with the 3A protein inhibiting ER–Golgi transport.[28] Both genes encode small viral proteins with transmembrane domains that are critical for replication. The secretory phenotype of the 3A protein was isolated to its far N-terminus, which happened to associate at the same location as the 3A-2 temperature-sensitive viral replication mutant that had been isolated many years prior.[29,30] To that point, variations in the 3A gene had also been associated with resistance to the enviroxime family of compounds and Brefeldin A, although it was found that 3A did not directly bind enviroxime.[31–33] Viral replication was associated with a beneficial "just so story" of secretory pathway interruption leading to adaptive immunoevasion.[34–36]

3. GBF1

A flurry of papers from the van Kuppeveld and Ehrenfeld laboratories in the aughts identified the first host protein involved in picornavirus membrane reorganization. Leveraging the correlation of the morphological effects of 3A with the inhibition of secretion mediated by Brefeldin A and dominant-negative Arf1, detailed biochemical assays demonstrated 3A inhibition of Arf1 activation through direct binding of GBF1, a known guanine exchange factor (GEF) for Arf1.[30] Through mutation analysis, binding of GBF1 could be assigned to the same N-terminal residues of 3A as the secretory inhibition phenotype.[37,38] Using independent methods, the cell type-specific inhibition of poliovirus replication by Brefeldin A was shown to be due to cell type-dependent mutations in GBF1.[39] Most importantly, depletion of GBF1 completely abrogated enterovirus replication in the same way as Brefeldin A.

Interestingly, although GBF1 is best known for its function as a GEF for Arf1, rescue of secretory inhibition and poliovirus replication by GBF1 mutants indicated that this GEF activity of GBF1 is not required.[40,41] Furthermore, different GBF1-inhibiting drugs such as Golgicide A and AG1847 show differential activity against enteroviruses.[42] This suggests dependence by multiple enteroviruses on GBF1 function(s), but also to a differing degree, perhaps utilizing alternative mechanisms. GBF1 is a large multidomain protein, with its GEF activity catalyzed by a C-terminal Sec7 domain. Neither overexpression of Rab1B nor Arf1 rescues enterovirus replication in the presence of Brefeldin A, suggesting viral recruitment of GBF1 is not necessarily related to disruption of known GBF1 functions.[43] It appears that there is likely an additional biochemical function of the N-terminal half of GBF1 that is yet to be recognized. Indeed recent work has ascribed lipid-binding functions to other conserved portions of GBF1.[44,45] It remains to be seen whether other lipid-binding domains exist

N-terminal to the Sec7 domain. Nonetheless, ample room exists to better isolate the molecular determinants of both the 3A–GBF1 and Rab1B–GBF1 interactions at the single amino acid level on the GBF1 side.

An intriguing possibility is that the enteroviral 3A acts as a Rab1B-GTP analog or mimetic—a so-called viral *vesicogene* in the manner of DNA virus and retroviral oncogenes—based on the known functions of Rab1B in binding GBF1. Alignment of enterovirus 3A with Rab1B shows conservation of the known critical residues for 3A–GBF1 binding, and the binding sites for each correspond to the far N-terminus of GBF1.[46] GTP binding to Rab proteins involves a subtle change in protein morphology that exposes a single helix that corresponds in size to the portion of enterovirus 3A responsible for binding GBF1. The 3A protein may mimic or interfere with Rab1B-GTP binding to the Arf1–GBF1 complex. This idea would also be consistent with the fact that knockdown of Rab1B does not alter poliovirus replication. Given the role of the Rab1B–GBF1–Arf1 for recruiting COP-I-like capsid coats for vesicle budding, a corollary to this hypothesis might include that positive-stranded RNA viruses evolved from rogue, infectious vesicles.[47] The many potential mechanisms for viral vesicogenesis are reflected in the diversity of 3A genes across the picornaviruses (Fig. 1B).

4. PI4KB

Diligent candidate approaches informed by ER and Golgi biology—for example, immunostaining for localization of GBF1-related genes in poliovirus replication complexes—allowed for the critical discovery that the lipid kinase, PI4KB, was recruited to picornaviral replication organelles by the viral 3A protein.[48] Genetic and chemical inhibition demonstrated that PI4KB activity was required for viral replication.[48] Further work demonstrated that levels of the product of PI4KB, phosphatidylinositol-4-phosphate (PI4P), increase during enterovirus replication, and furthermore, the viral RNA polymerase, 3D, contains a PI4P-binding site.[48] The polymerase PI4P-binding site combined with 3A's role in recruiting and/or activating PI4KB suggested a potential closed circuit model of P3 protein activity as the polymerase had already been shown to physically interact with the 3AB protein to initiate viral genome replication.[49–55]

This work led to interrogation of the enviroxime family of antienteroviral compounds, which were shown to be direct inhibitors of PI4KB.[56,57] Resistance to a variety of PI4KB inhibitors in enteroviruses including the enviroxime family mapped to the C-terminal half of the 3A protein further unifying the theory of 3A recruitment and/or activation of PI4KB.[32,56–58] Detailed work on Coxsackieviruses harboring a 3A mutation making them resistant to enviroxime-like PI4KB inhibitors beautifully demonstrated that these inhibitors only block the known enzymatic function of PI4KB and not an alternative function of the kinase or the localization of the enzyme to the Golgi apparatus.[57] The involvement of PI4KB in enteroviral replication dovetailed nicely with multiple recent siRNA screens that implicated PI4KA signaling in replication of the distantly related flavivirus, Hepatitis C virus.[59–61] Thirty years of research in candidate genes had seemingly come to a surprisingly unified mechanism of positive-stranded RNA virus replication.

The discovery of PI4KB involvement in enterovirus replication opened up a number of important questions. It was still hard to determine whether inhibition of PI4KB indirectly affected viral growth by globally disrupting Golgi function or directly inhibited viral growth. Was the *in vitro* binding of PI4P by the 3D polymerase important for localization or did it affect activation of RNA polymerase activity? Furthermore, the mechanism by which 3A recruited and/or activated PI4KB was not clear, as Arf1-related functions of GBF1 had previously been found to be dispensible in viral replication.

Further work on enteroviruses and PI4KB inhibitors has complicated the picture and hope of the original model. PI4KB inhibitor-resistant Coxsackieviruses with mutations in the 3A protein were no longer sensitive to knockdown of PI4KB, and immunofluorescence studies of virally infected cells were unable to detect cellular PI4P.[57] These results are also consistent with a model in which only local increases in PI4P are required for viral replication, as it has also been noted that knockdown of PI4KB needs to be near complete to see a strong viral inhibition.[56,62] However, the major conclusion of these results is that the therapeutic window of antivirals that relied on inhibition of PI4KB is likely quite small. Although desirable from a *prima facie* evolutionary standpoint, nearly complete inhibition of the PI4KB host protein function may require toxic levels of inhibitor and under such selective pressure could lead to resistance.

Within a year, two studies on new PI4KB inhibitors *in vivo* had offered mixed results.[58,63] Both screens of antivirals against enteroviruses recovered specific PI4KB inhibitors whose resistance again mapped to the C-terminal half of 3A. Both new inhibitors demonstrated strong activity within the same log EC50 across a variety of enteroviruses from across the genus (but not cardioviruses or aphthoviruses in one study[58]), suggesting the possibility of a pan-enteroviral inhibitor. However, in one study, attempts to try

PI4KB inhibitors *in vivo* were stymied by toxicity, as control mice treated with low-dose inhibitors demonstrated weakness and ruffled fur, and few mice could even survive high doses of the inhibitors. Similarly, Novartis stopped the development of highly specific inhibitors of PI4KB that showed efficacy against hepatitis C virus after they demonstrated antiproliferative effects against lymphocytes *in vitro*.[64] To date, only one PI4KB inhibitor has shown no adverse effects in mice and dose-dependent efficacy against a CVB4 mouse model of pancreatitis.[58]

5. ACBD3

Additional functions were yet to be assigned to the tiny 3A protein. Unbiased interaction mapping for the picornaviral 3A proteins has recently been accomplished using two hybrid and affinity purification-mass spectrometry (AP-MS) approaches, producing a systematic ranking of interacting proteins and yielding a variety of discoveries.[62,65] For example, although most picornavirus proteins begin with glycine due to the 3C picornain protease cleavage specificity, 3A was the first nonstructural picornavirus protein to be shown to carry a myristoylation modification at its N-terminus, as revealed in the course of AP-MS analyses. Without any prior knowledge of decades of candidate biology outlined in the preceding text, both GBF1 and PI4KB were the top-ranked specific hits for enterovirus and kobuvirus 3A proteins, respectively. PI4KA but not PI4KB was specifically found in the interaction data for a variety of cardiovirus 3A proteins. It is also worth noting that AP-MS interaction data for picornaviral proteins other than 3A were surprisingly quiet of specific host interactors.

Most notably, however, both enterovirus and kobuvirus 3A proteins specifically bound the Golgi-associated protein, acyl-CoA-binding domain-containing protein 3 (ACBD3), despite the fact that these 3A proteins share less than 30% amino acid identity.[62,65,66] Subsequent AP-MS, AP-Western, and mammalian two-hybrid studies of ACBD3 revealed that PI4KB was a highly ranked specific, native interacting protein. Although picornaviruses required the activity of PI4KB for replication and drastically increased cellular PI4P levels, no picornavirus 3A protein bound directly to PI4KB but instead bound a PI4KB-interacting protein. ACBD3 could also be found enriched on viral replication organelles. Two-hybrid studies revealed that ACBD3 interacted with a broad array of picornaviral 3A proteins, including those of the kobuviruses, enteroviruses, parechoviruses, Hepatitis

A virus, and klassevirus/salivirus, but not the apthovirus or cardioviruses, which had specifically been affinity purified with PI4KA and whose replication was not inhibited by specific PI4KB inhibitors.[58,67]

Kobuviruses demonstrated the same dependence on PI4KB inhibition as enteroviruses, but kobuviral replication was not affected by GBF1 inhibition. This suggests that kobuviruses had evolved a functionally separate mechanism of interacting with ACBD3 that correlates with their sequence divergence. Detailed mechanistic analysis of the 3A–ACBD3 complex revealed that kobuvirus 3A promoted a stable 3A–ACBD3–PI4KB complex by competing off the Rab33 GTPase-activating protein, TBC1D22A/B, a direct competitor to PI4KB–ACBD3 binding (Fig. 2).[67] Almost the entirety of the kobuvirus 3A protein was involved in the interaction with ACBD3, as alanine scanning along the length of the Aichi virus 3A protein revealed only two residues that abrogated binding to ACBD3. Most importantly, biochemical reconstitution of the wild-type Aichi virus 3A–ACBD3–PI4KB complex stimulated PI4KB activity while 3A mutants that affected Aichi virus replication were unable to stimulate PI4P generation, which is consistent with increased sensitization of mutant 3As to PI4KB inhibition.[62,68]

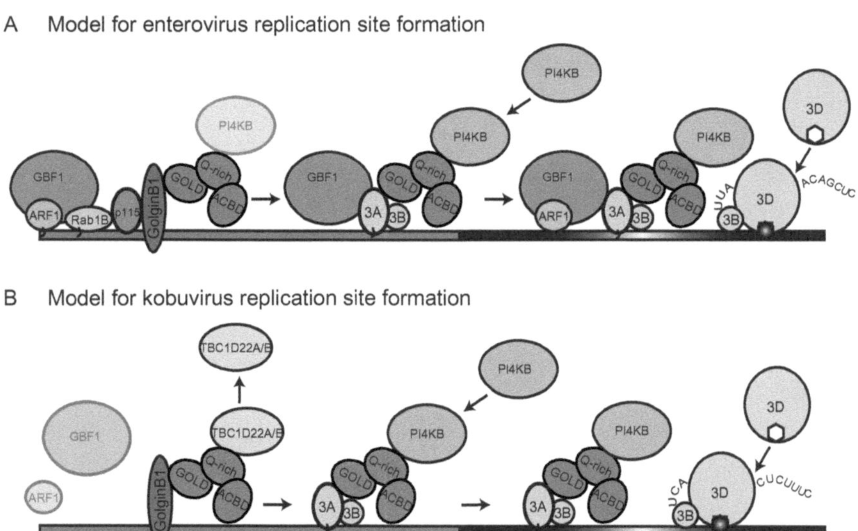

Figure 2 A potential model by which 3A proteins promote replication in enteroviruses and kobuviruses. (A) Enterovirus 3A strongly binds GBF1 and ACBD3 to activate PI4KB which in turn activates the 3D polymerase. (B) Kobuvirus 3A binds only to ACBD3, promoting a stable ACBD3–PI4KB complex by allosterically inhibiting TBC1D22A/B from binding the same site on ACBD3 as PI4KB, resulting in activation of PI4KB and deposition of PI4P.

Biochemical studies of the enterovirus 3A showed an ACBD3-interacting domain that was separate and more C-terminal from the GBF1-interacting domain, suggesting the potential for interacting with both proteins at once. Notably, mutations identified in enterovirus 3A associated with enviroxime resistance mapped to the exact location of ACBD3 binding by 3A.[57,67] However, when studied *in vitro*, binding of enterovirus 3A does not affect TBC1D22A/B binding or promote stable association of the PI4KB–ACBD3 complex. These studies suggested a potential model of convergent evolution of 3A–ACBD3–PI4KB complexes to drive PI4P generation, in which enteroviruses "push the gas" by recruiting GBF1 to the 3A–ACBD3–PI4KB complex and kobuviruses "cut the brakes" by creating a stable 3A–ACBD3–PI4KB complex in which TBC1D22A/B cannot compete off PI4KB (Fig. 2).

Other recent studies of ACBD3 have complicated the role of the protein in enterovirus replication. Most notably, while knockdown of ACBD3 resulted in an inhibition of Aichi virus replication, two reports now demonstrate that enterovirus replication is slightly enhanced by genetic inhibition of ACBD3.[69,70] Also, immunofluorescence studies of PI4KB recruitment to enterovirus replication organelles were independent of ACBD3, GBF1, or Arf1.[69] In these models, ACBD3 is either an inhibiting factor for enterovirus replication or does not make significant contributions. How to resolve the differing roles of ACBD3 and PI4KB in enterovirus versus kobuvirus replication remains a central question to be answered in the coming years.

Our understanding of the function of ACBD3 in picornavirus replication is hampered by our lack of understanding of ACBD3's complete cellular function.[71] Indeed, no biochemical function for the GOLD domain is known, other than being associated with Golgi localization through binding to the Golgi anchor protein giantin in the case of ACBD3. The acyl-CoA-binding (ACB) domain in ACBD3 is notably longer than that of other ACB proteins in all metazoans that contain an ACBD3, although there are no data on its lipid-binding preference. Intriguingly, the main yeast gene involved in ACB and transport, a similarly named ACBP, was shown in genome-wide screen of brome mosaic virus (BMV) replication to be required for the assembly and activity of BMV replication complexes.[72] Knockdown of ACBD3 results in dispersion of PI4KB in the cytoplasm, suggesting ACBD3 is responsible for Golgi localization of PI4KB.[65] However, it is unclear what effect ACBD3 has on PI4P generation. Could ACBD3 serve as a lipid cofactor that colocalizes PI4KB and its phosphatidylinositol substrate in close proximity to catalyze PI4P generation? Or is the interaction of ACBD3 and PI4KB doing the opposite and sequestering PI4KB from the membrane? Data from PI4P generation from biochemical reconstitution of the Aichi virus 3A–ACBD3–PI4KB complex are consistent with the former, while the enterovirus replication data suggest the latter.

6. CHOLESTEROL

Work in the last 2 years has demonstrated that, in addition to PI4P, enteroviruses accumulate cholesterol on their replication organelles. Again, flaviviruses led the way with the identification that HCV and dengue virus replication required cholesterol metabolism.[73,74] Fluorescent microscopy using lipid-sensing domain proteins has correlated increasing cholesterol levels with enterovirus replication, while chemical inhibition with statins is associated with decreased viral replication.[75] Unesterified cholesterol was rerouted from plasma membrane stores and rapidly accumulated at the sites of viral replication. Rab11 recycling endosomes were recruited to viral replication membranes, possibly since Rab11 is a known physical interactor of PI4KB. Whether these effects on cholesterol distribution and synthesis are directly related to viral replication or indirectly as a byproduct of GBF1 inhibition or PI4KB redistribution remain to be seen.

The role of cholesterol sorting in enterovirus replication was also implicated in a screen for a mechanism of viral replication inhibition by minor enviroxime-like compounds. Whereas major enviroxime-like compounds had been shown to exert their antiviral effect by direct inhibition of PI4KB, the target for minor enviroxime-like compounds was as yet unknown. Notably, minor enviroxime compounds showed more activity against hepatitis C virus than those from the major group, although resistance to both sets of compounds is associated with the same alanine-to-threonine mutation near the C-terminus of the enterovirus 3A protein within its transmembrane domain. Using the same strategy for target identification for the major enviroxime-like compounds, knockdown of oxysterol-binding protein (OSBP) family members OSBP1 and OSBP2 demonstrated sensitization to the minor enviroxime-like compounds.[76] A high-affinity ligand of OSBP, 25-hydroxycholesterol, also inhibited poliovirus replication in the same manner as the minor enviroxime-like compounds. In contrast to the major group, minor enviroxime compounds decreased PI4P deposition as well as PI4KB localization to the Golgi.[76]

OSBP may be responsible for keeping separate cholesterol from PI4P and ensuring the partition of lipid species in the membrane. Alternatively, OSBP recruitment by PI4P to replication organelles could simply be a byproduct of the enhanced local PI4P concentration driven by viral replication. Given that ACBD3 appears to be a major regulator of PI4KB localization to the Golgi, an indirect interaction *vis-à-vis* ACBD3 and OSBP seems possible, although a direct physical interaction between OSBP and any of these proteins has not been recovered to date. It should be noted that VAPA, a known physical interactor of OSBP, was a highly specific prey in AP-MS studies of porcine kobuvirus 3A along with ACBD3, also suggesting a link between these proteins for cholesterol and lipid sorting regulation.[62] A cholesterol-recognition motif at the very C-terminus was also critically dependent for ACBD3 homomerization and its interactions with 3A proteins.[67]

7. 2B–2C PORE FORMING WITH ER–GOLGI MEMBRANES

While much of the work on picornavirus interaction with Golgi membranes focuses on 3A, the P2 proteins have also shown an important role in permeabilizing and altering ER–Golgi membranes for viral replication. In enteroviruses, the 2B protein is a small pore-forming, integral membrane protein with two transmembrane spans. One of these transmembrane domains contains a cationic amphipathic helix usually found in lytic polypeptides, while in other picornaviruses, this amphipathic helix is not present.[77] 2B is found both as its own mature protein and attached to 2C, a larger viral protein that has helicase-like domains such as Walker AB motifs. 2B has been shown to bind to itself, creating homomultimers that allow for formation of a viroporin.[78,79] Mutation of the two hydrophobic domains results in decreased viral production as well as inability for 2B to localize to the Golgi.[80–82] Formation of the pore promotes efflux of ions including Ca^{2+} from the ER and Golgi, resulting in inhibition of protein trafficking through the Golgi.[28,83] The Ca^{2+} flux is thought to be beneficial for viral replication both through the inhibition of vesicular protein transport and antiapoptotic effects on host cells.[84] Alternatively, the extraintracellular Ca^{2+} could be required for correct folding and assembly of viral capsids as structures of enteroviral capsids include calcium ions.[85–87] Ca^{2+} flux may also assist in activation of PI4KB, as the neuronal calcium sensor-1 protein has been shown to form an activation network in yeast and human cells with PI4KB and Arf1.[88–91]

Of note, other picornaviral 2B proteins do not uniformly localize to the ER–Golgi, inhibit secretion, or alter ion flux.[77] HAV 2B localizes to the ER but does not affect Ca^{2+} flux or inhibit the secretory pathway. FMDV 2BC but not 3A protein was shown to inhibit cellular secretion.[92] Of course, picornaviral 2B's interfering with protein transport could simply be collateral damage of a required viral function. Data from the 1a protein of BMV that corresponds to the picornaviral 2BC protein suggests a model whereby binding of the membrane via a similar amphipathic helix in 1a allows the helicase-like activity to bend the membrane to create spherules where negative-stranded RNA genomes are simultaneously recruited for genome replication and protect dsRNA from detection by the cellular innate immune system.[93–96] These 1a-induced arrangements of ER membranes also require interaction with host reticulon proteins, an interaction that has also been reported with enterovirus 71 2C.[97–99] The role of host proteins in how 2BC promotes replication of both the negative- and positive-stranded RNA is deserving of more attention in the coming years.

8. NEXT STEPS

In less than a decade, a number of important cellular players have been identified that help us to understand why picornaviruses are obligate intracellular parasites. The work highlighted in the preceding text is impressive in its both granularity and scope, given the awesome diversity and disease burden that picornaviruses comprise. The mechanisms initially worked out for RNA virus type species have yielded novel insights into viral replication that now seem surprisingly broad in their relevance, especially the reliance on lipid synthesis and PI4 kinases. However, no discussion of our understanding of picornavirus interactions with the secretory system would be complete without some mention of future opportunities and questions still left unanswered.

8.1. Leveraging genomics

The first opportunity in future studies of picornaviruses lies in the breadth of viruses studied. Although enteroviruses constitute much of the human picornaviral disease burden and have long been the focus of picornaviral study, genomics has greatly increased the number of picornaviruses and RNA viruses available for study. To date, all 3A proteins from picornaviruses that rely on PI4KB for replication interact with ACBD3 while those that do not use PI4KB (and likely use PI4KA) do not show a significant interaction.

Given the wide-sequence divergence of the 3A proteins, this suggests many mechanisms of interaction. With the increasing number of picornaviruses, new genera and species will need to be tested to examine whether this trend is maintained. In addition to sequencing genomes, establishing models of replication for the new picornaviruses will increasingly be important.[100]

At the very least, laboratories that study enteroviruses should consider adding kobuviruses to the list of viruses studied. The type species Aichi virus has a number of advantages that the enteroviruses share, including a reverse genetics system, freely available clones, rapid replication in a wide variety of cell lines, and reliance on many of the basic replication mechanisms that enteroviruses share. It also has a number of differences from enteroviruses, including the presence of an L protein, a nonproteolytic 2A protein and a seemingly restricted evolution given the paucity of serotypes compared to enteroviruses. Although not reaching the level of enteroviruses, the global reach and burden of kobuviruses are also increasingly being recognized. Twenty-five years after the discovery of Aichi virus, one laboratory in the world has done the yeoman's work establishing the basis of Aichi virus as a rich system for understanding picornavirus biology.[65,101–103] It is worth continuing their work to include Aichi virus in screens of antiviral compounds and understanding the mechanistic basis of resistance formation. For instance, despite the wide array of mutants available, no resistant clones to PI4KB inhibitors or other antivirals have been isolated, leading to a lack of understanding of basic mechanisms of viral–host interaction.

8.2. Proteomic screening

With the growing number of viruses, studying their cellular interactions will require high-throughput unbiased methods. Here, the development of AP-MS has been especially helpful at identifying viral protein interactors. Because many picornaviruses have short replication cycles and exert translational control mechanisms whereby the vast majority of newly synthesized protein is of viral origin, studies of their cellular interactions are especially amenable to a transient transfection approach, as that is closest to their "native" expression. For the purposes of rapid and high-throughput screening, transient transfection of an epitope-tagged viral protein is a much more efficient approach that is amenable to genetic manipulation through mutagenesis or alternate tagging strategies.[104] Other laboratories have taken the strategy of tagging proteins within the whole viral genome, for example, human RSV,[105] poliovirus,[106,107] or measles virus.[108] This approach has the advantage of recapitulating host–viral interactions in the context of the entire viral protein assembly. However, finding a suitable location for the tag and the generation of stable viruses are both laborious, as many tagged viruses are either nonviable or at a significant selective disadvantage to untagged virus.

Of course, AP-MS is fundamentally a hypothesis-generating experiment. These hypotheses must be tested for relation to the viral biology of interest through knockdown genetic approaches and probing of structure–function relationships. With picornaviral proteins being so small and gene synthesis decreasing in cost, alanine scanning of entire viral proteins is possible. Where proteins are too large for unbiased scanning, 3D modeling of structures can inform which interfaces are involved in the structure. With the many genomes now available, phylogeny of both the host and the viruses can inform on critical residues required for conserved interaction. New developments in mass spectrometry are also able to perform native complex analysis on intact protein complexes. This requires modifications to commercial instrumentation and has only recently become available in the form of a high-mass Orbitrap, which incorporates high-pressure desolvation for ionization of large proteins.[109] Chemical crosslinking techniques can enhance sensitive detection of transient or weak interactions.[110,111] Since viral proteins are not native to a host cell, mapping of the viral–host protein interaction on the host protein can be especially informative to understand whether a native interactor is displaced by the viral protein. Finally, knock-ins and rescues of deleted host factors with mutant host proteins via new genetic manipulation tools such as TALNs and CRISPR will allow for true tests of the various functions of host factors that are required for viral replication and membrane reorganization. This is especially true when the effects of knocking out or overexpressing the proteins are lethal or toxic.

8.3. Complex biochemistry

The work reviewed in the preceding text also highlights the opportunity to move toward a more molecular, structural understanding of how picornaviral replication organelles are formed. Structural studies of picornaviral proteins have to date focused on capsids, polymerases, and proteases. Understandably, it has been more difficult to obtain structures for the membrane-associated viral proteins. For instance, a solution NMR structure of poliovirus 3A has been determined, although the entire structure is driven by the dimerization of 3A *in vitro*.[112] The transmembrane region of 3A had to be removed for the structure determination and even then the C-terminal half was completely unstructured. Whether dimerization of 3A is part of viral replication is not clear, as the residues that are required for dimerization overlap with those required for binding GBF1. While an important first step, the identification of host cellular proteins that interact with 3A such as GBF1 and ACBD3 might yield for a more rigid conformation of 3A of that will allow for determination of its molecular structure. That said, currently, we also do not have structures available for GBF1 or ACBD3—only modeling of small conserved domains from yeast proteins.

The identification of the players involved in host–viral interactions and the importance of lipid biosynthesis also creates an opportunity for biochemistry. This was most recently highlighted in the study of reconstituted Aichi virus 3A–ACBD3–PI4KB complex that demonstrated increased PI4P deposition when all three proteins were present.[68] This study also made important use of critical 3A mutants that had been shown to alter ACBD3 binding to show those 3A mutants did not affect PI4P deposition, indicating that 3A–ACBD3 interactions may directly influence PI4KB activity. Due to the small size of the viral proteins, opportunity for reverse genetics, and rapid evolutionary landscape sampled by RNA viruses, studies of the molecular determinants of the various cellular and viral phenotypes have long taken advantage of isolating point mutants that specifically inhibit singular functions of these viral proteins, beginning with the discovery of the 3A-2 poliovirus genotype.[29,38,40,62,113] Indeed, a bevy of mutations have been isolated in the 3A gene associated with PI4KB inhibitor resistance. These mutants create important tools for studying the molecular determinants of viral replication and host protein interaction. To date, it is not known how these mutants affect PI4KB recruitment and activation, PI4P deposition, and ACBD3/GBF1 binding.

The ability to define cellular proteins involved in picornaviral replication creates an excellent basis for funding of basic mechanisms into membrane oxysterol and phospholipid regulation in mammals and other model organisms. Given the extraordinarily conserved nature of these proteins for structure of the metazoan secretory apparatus, it is unlikely that we will find many human mutations that hasten understanding of their organismal phenotype as in the cholesterol pathways that aided the development of the statins. However, it is an opportunity for studying these proteins in other model organisms. Understanding the interaction of poliovirus 3A with GBF1 was enhanced by basic research into secretory mechanisms that had elucidated Sec7 domain-containing proteins, and yet there are still unannotated phenotypes for GBF1 that remain to be discovered. Similarly, our understanding of ACBD3's role in the secretory system and potentially viral replication has been limited by the lack of a yeast ortholog and no previously apparent role in human disease. The development of its interactome along with that of PI4KB will hopefully help better flesh out their biochemical role in PI4P deposition and organizing the early and late Golgi.[67,114] The role of completely unannotated proteins such as C10orf76 presents an exciting venue for further research.

The same rationale for better identifying and understanding the mechanisms of PI4P deposition also apply to understanding PI4KA recruitment and activity at various membranes. These mechanisms will likely be important for understanding non-PI4KB-dependent viral replication mechanisms including those of aphthovirus and cardiovirus. Our knowledge of PI4KA was recently advanced by correct annotation of its translational start site, which allowed for purification of two interactors whose orthologs also regulate its plasma membrane association and regulation in yeast.[115] The development of specific inhibitors against PI4KA has yielded important insights about the toxicity of targeting that enzyme for hepatitis C virus, much in the same way that a flurry of specific PI4KB inhibitors have elucidated its problematic role as a target for picornaviruses.[116]

Complex biochemistry can also be deconvoluted at the cellular level through higher throughput methods such lipidomics. Lipids are used in control of membrane dynamics, essential to viral biology in forming replication sites or budding, thus increasing interest in lipidomic approaches for identifying lipid interactions with viral proteins.[117] Lipid separations and detection methods are rapidly growing, coupled with high-mass accuracy mass spectrometry for unequivocal identification of novel species.[118,119] These tools will be especially important for the unbiased monitoring and discovery of changes in potentially novel lipid species that are altered during viral replication.

9. CONCLUSION

So much has been accomplished in the past decade in our understanding of the list of players involved in picornavirus interaction with secretory membranes. The stage is set for another decade of discoveries to elucidate how these proteins affect lipid metabolism and alter cellular membrane organization. While targeting of picornavirus-host lipid regulators is unlikely to offer an adequate therapeutic window, there is the possibility of developing direct 3A inhibitors that block host protein binding and activation in the same way as hepacivirus NS5A inhibitors. The mechanisms by which these viruses usurp cellular vesicle formation by recruiting positive regulators and inhibiting negative regulators will simultaneously enrich our understanding of native cellular membrane organization, much in the same way that viruses informed our understanding of oncogenesis through the discovery of native oncogenes and tumor suppressor genes. Given the vast number of Rab- and Sec-related proteins involved in localized vesicogenesis throughout the cell, the broad potential for vesicogene evolution will keep picornavirologists and cell biologists busy for a long time.

ACKNOWLEDGMENTS

Thanks to Giselle Knudsen for reviewing this chapter and Joseph DeRisi for helpful discussions.

REFERENCES

1. Jacobson IM, Gordon SC, Kowdley KV, et al. Sofosbuvir for hepatitis C genotype 2 or 3 in patients without treatment options. *N Engl J Med.* 2013;368(20):1867–1877. http://dx.doi.org/10.1056/NEJMoa1214854.
2. Lawitz E, Mangia A, Wyles D, et al. Sofosbuvir for previously untreated chronic hepatitis C infection. *N Engl J Med.* 2013;368(20):1878–1887. http://dx.doi.org/10.1056/NEJMoa1214853.
3. Mäkelä MJ, Puhakka T, Ruuskanen O, et al. Viruses and bacteria in the etiology of the common cold. *J Clin Microbiol.* 1998;36(2):539–542.
4. Hammerstad SS, Tauriainen S, Hyöty H, Paulsen T, Norheim I, Dahl-Jørgensen K. Detection of enterovirus in the thyroid tissue of patients with graves' disease. *J Med Virol.* 2013;85(3):512–518. http://dx.doi.org/10.1002/jmv.23476.
5. Hober D, Alidjinou EK. Enteroviral pathogenesis of type 1 diabetes: queries and answers. *Curr Opin Infect Dis.* 2013;26(3):263–269. http://dx.doi.org/10.1097/QCO.0b013e3283608300.
6. Yeung W-CG, Rawlinson WD, Craig ME. Enterovirus infection and type 1 diabetes mellitus: systematic review and meta-analysis of observational molecular studies. *BMJ.* 2011;342:d35.
7. Chiu CY, Greninger AL, Kanada K, et al. Identification of cardioviruses related to Theiler's murine encephalomyelitis virus in human infections. *Proc Natl Acad Sci USA.* 2008;105(37):14124–14129. http://dx.doi.org/10.1073/pnas.0805968105.
8. Greninger AL, Holtz L, Kang G, Ganem D, Wang D, DeRisi JL. Serological evidence of human klassevirus infection. *Clin Vaccine Immunol.* 2010;17(10):1584–1588. http://dx.doi.org/10.1128/CVI.00152-10.
9. Greninger AL, Runckel C, Chiu CY, et al. The complete genome of klassevirus—a novel picornavirus in pediatric stool. *Virol J.* 2009;6:82. http://dx.doi.org/10.1186/1743-422X-6-82.
10. Holtz LR, Finkbeiner SR, Zhao G, et al. Klassevirus 1, a previously undescribed member of the family Picornaviridae, is globally widespread. *Virol J.* 2009;6:86. http://dx.doi.org/10.1186/1743-422X-6-86.
11. Li L, Victoria J, Kapoor A, et al. A novel picornavirus associated with gastroenteritis. *J Virol.* 2009;83(22):12002–12006. http://dx.doi.org/10.1128/JVI.01241-09.

12. Chase AJ, Semler BL. Viral subversion of host functions for picornavirus translation and RNA replication. *Futur Virol.* 2012;7(2):179–191.
13. Richards AL, Jackson WT. How positive-strand RNA viruses benefit from autophagosome maturation. *J Virol.* 2013;87(18):9966–9972. http://dx.doi.org/10.1128/JVI.00460-13.
14. Lin J-Y, Chen T-C, Weng K-F, Chang S-C, Chen L-L, Shih S-R. Viral and host proteins involved in picornavirus life cycle. *J Biomed Sci.* 2009;16:103. http://dx.doi.org/10.1186/1423-0127-16-103.
15. Koonin EV, Wolf YI, Nagasaki K, Dolja VV. The Big Bang of picorna-like virus evolution antedates the radiation of eukaryotic supergroups. *Nat Rev Microbiol.* 2008;6(12):925–939. http://dx.doi.org/10.1038/nrmicro2030.
16. Nagy PD, Pogany J. The dependence of viral RNA replication on co-opted host factors. *Nat Rev Microbiol.* 2012;10(2):137–149. http://dx.doi.org/10.1038/nrmicro2692.
17. Delang L, Paeshuyse J, Neyts J. The role of phosphatidylinositol 4-kinases and phosphatidylinositol 4-phosphate during viral replication. *Biochem Pharmacol.* 2012;84(11):1400–1408. http://dx.doi.org/10.1016/j.bcp.2012.07.034.
18. Altan-Bonnet N, Balla T. Phosphatidylinositol 4-kinases: hostages harnessed to build panviral replication platforms. *Trends Biochem Sci.* 2012;37(7):293–302. http://dx.doi.org/10.1016/j.tibs.2012.03.004.
19. Belov GA, van Kuppeveld FJM. (+)RNA viruses rewire cellular pathways to build replication organelles. *Curr Opin Virol.* 2012;2(6):740–747. http://dx.doi.org/10.1016/j.coviro.2012.09.006.
20. Caliguiri LA, Tamm I. The role of cytoplasmic membranes in poliovirus biosynthesis. *Virology.* 1970;42(1):100–111. http://dx.doi.org/10.1016/0042-6822(70)90242-4.
21. Caliguiri LA, Tamm I. Characterization of poliovirus-specific structures associated with cytoplasmic membranes. *Virology.* 1970;42(1):112–122. http://dx.doi.org/10.1016/0042-6822(70)90243-6.
22. Mosser AG, Caliguiri LA, Scheid AS, Tamm I. Chemical and enzymatic characteristics of cytoplasmic membranes of poliovirus-infected HeLa cells. *Virology.* 1972;47(1):30–38.
23. Guinea R, Carrasco L. Effects of fatty acids on lipid synthesis and viral RNA replication in poliovirus-infected cells. *Virology.* 1991;185(1):473–476. http://dx.doi.org/10.1016/0042-6822(91)90802-I.
24. Molla A, Paul AV, Wimmer E. Effects of temperature and lipophilic agents on poliovirus formation and RNA synthesis in a cell-free system. *J Virol.* 1993;67(10):5932–5938.
25. Irurzun A, Perez L, Carrasco L. Involvement of membrane traffic in the replication of poliovirus genomes: effects of brefeldin A. *Virology.* 1992;191(1):166–175.
26. Maynell LA, Kirkegaard K, Klymkowsky MW. Inhibition of poliovirus RNA synthesis by brefeldin A. *J Virol.* 1992;66(4):1985–1994.
27. Doedens J, Maynell LA, Klymkowsky MW, Kirkegaard K. Secretory pathway function, but not cytoskeletal integrity, is required in poliovirus infection. *Arch Virol Suppl.* 1994;9:159–172.
28. Doedens JR, Kirkegaard K. Inhibition of cellular protein secretion by poliovirus proteins 2B and 3A. *EMBO J.* 1995;14(5):894–907.
29. Doedens JR, Giddings TH, Kirkegaard K. Inhibition of endoplasmic reticulum-to-Golgi traffic by poliovirus protein 3A: genetic and ultrastructural analysis. *J Virol.* 1997;71(12):9054–9064.
30. Wessels E, Duijsings D, Niu T-K, et al. A viral protein that blocks Arf1-mediated COP-I assembly by inhibiting the guanine nucleotide exchange factor GBF1. *Dev Cell.* 2006;11(2):191–201. http://dx.doi.org/10.1016/j.devcel.2006.06.005.

31. Brown-Augsburger P, Vance LM, Malcolm SK, Hsiung H, Smith DP, Heinz BA. Evidence that enviroxime targets multiple components of the rhinovirus 14 replication complex. *Arch Virol.* 1999;144(8):1569–1585. http://dx.doi.org/10.1007/s007050050611.
32. Heinz BA, Vance LM. The antiviral compound enviroxime targets the 3A coding region of rhinovirus and poliovirus. *J Virol.* 1995;69(7):4189–4197.
33. Crotty S, Saleh M-C, Gitlin L, Beske O, Andino R. The poliovirus replication machinery can escape inhibition by an antiviral drug that targets a host cell protein. *J Virol.* 2004;78(7):3378–3386.
34. Deitz SB, Dodd DA, Cooper S, Parham P, Kirkegaard K. MHC I-dependent antigen presentation is inhibited by poliovirus protein 3A. *Proc Natl Acad Sci USA.* 2000;97(25):13790–13795.
35. Dodd DA, Giddings TH, Kirkegaard K. Poliovirus 3A protein limits interleukin-6 (IL-6), IL-8, and beta interferon secretion during viral infection. *J Virol.* 2001;75(17):8158–8165. http://dx.doi.org/10.1128/JVI.75.17.8158-8165.2001.
36. Neznanov N, Kondratova A, Chumakov KM, et al. Poliovirus protein 3A inhibits tumor necrosis factor (TNF)-induced apoptosis by eliminating the TNF receptor from the cell surface. *J Virol.* 2001;75(21):10409–10420. http://dx.doi.org/10.1128/JVI.75.21.10409-10420.2001.
37. Wessels E, Duijsings D, Lanke KHW, et al. Effects of picornavirus 3A proteins on protein transport and GBF1-dependent COP-I recruitment. *J Virol.* 2006;80(23):11852–11860. http://dx.doi.org/10.1128/JVI.01225-06.
38. Wessels E, Duijsings D, Notebaart RA, Melchers WJG, van Kuppeveld FJM. A proline-rich region in the coxsackievirus 3A protein is required for the protein to inhibit endoplasmic reticulum-to-golgi transport. *J Virol.* 2005;79(8):5163–5173. http://dx.doi.org/10.1128/JVI.79.8.5163-5173.2005.
39. Belov GA, Feng Q, Nikovics K, Jackson CL, Ehrenfeld E. A critical role of a cellular membrane traffic protein in poliovirus RNA replication. *PLoS Pathog.* 2008;4(11):e1000216. http://dx.doi.org/10.1371/journal.ppat.1000216.
40. Wessels E, Duijsings D, Lanke KHW, Melchers WJG, Jackson CL, van Kuppeveld FJM. Molecular determinants of the interaction between coxsackievirus protein 3A and guanine nucleotide exchange factor GBF1. *J Virol.* 2007;81(10):5238–5245. http://dx.doi.org/10.1128/JVI.02680-06.
41. Belov GA, Kovtunovych G, Jackson CL, Ehrenfeld E. Poliovirus replication requires the N-terminus but not the catalytic Sec7 domain of ArfGEF GBF1. *Cell Microbiol.* 2010;12(10):1463–1479. http://dx.doi.org/10.1111/j.1462-5822.2010.01482.x.
42. Van der Linden L, van der Schaar HM, Lanke KHW, Neyts J, van Kuppeveld FJM. Differential effects of the putative GBF1 inhibitors Golgicide A and AG1478 on enterovirus replication. *J Virol.* 2010;84(15):7535–7542. http://dx.doi.org/10.1128/JVI.02684-09.
43. Lanke KHW, van der Schaar HM, Belov GA, et al. GBF1, a guanine nucleotide exchange factor for Arf, is crucial for coxsackievirus B3 RNA replication. *J Virol.* 2009;83(22):11940–11949. http://dx.doi.org/10.1128/JVI.01244-09.
44. Bouvet S, Golinelli-Cohen M-P, Contremoulins V, Jackson CL. Targeting of the Arf-GEF GBF1 to lipid droplets and Golgi membranes. *J Cell Sci.* 2013;126(20):4794–4805. http://dx.doi.org/10.1242/jcs.134254.
45. Mazaki Y, Nishimura Y, Sabe H. GBF1 bears a novel phosphatidylinositol-phosphate binding module, BP3K, to link PI3Kγ activity with Arf1 activation involved in GPCR-mediated neutrophil chemotaxis and superoxide production. *Mol Biol Cell.* 2012;23(13):2457–2467. http://dx.doi.org/10.1091/mbc.E12-01-0062.

46. Monetta P, Slavin I, Romero N, Alvarez C. Rab1b interacts with GBF1 and modulates both ARF1 dynamics and COPI association. *Mol Biol Cell.* 2007;18(7):2400–2410. http://dx.doi.org/10.1091/mbc.E06-11-1005.
47. Alvarez C, Garcia-Mata R, Brandon E, Sztul E. COPI recruitment is modulated by a Rab1b-dependent mechanism. *Mol Biol Cell.* 2003;14(5):2116–2127. http://dx.doi.org/10.1091/mbc.E02-09-0625.
48. Hsu N-Y, Ilnytska O, Belov G, et al. Viral reorganization of the secretory pathway generates distinct organelles for RNA replication. *Cell.* 2010;141(5):799–811. http://dx.doi.org/10.1016/j.cell.2010.03.050.
49. Xiang W, Cuconati A, Hope D, Kirkegaard K, Wimmer E. Complete protein linkage map of poliovirus P3 proteins: interaction of polymerase 3Dpol with VPg and with genetic variants of 3AB. *J Virol.* 1998;72(8):6732–6741.
50. Lama J, Sanz MA, Rodríguez PL. A role for 3AB protein in poliovirus genome replication. *J Biol Chem.* 1995;270(24):14430–14438.
51. Plotch SJ, Palant O. Poliovirus protein 3AB forms a complex with and stimulates the activity of the viral RNA polymerase, 3Dpol. *J Virol.* 1995;69(11):7169–7179.
52. Hope DA, Diamond SE, Kirkegaard K. Genetic dissection of interaction between poliovirus 3D polymerase and viral protein 3AB. *J Virol.* 1997;71(12):9490–9498.
53. Chen C, Wang Y, Shan C, et al. Crystal structure of enterovirus 71 RNA-dependent RNA polymerase complexed with its protein primer VPg: implication for a trans mechanism of VPg uridylylation. *J Virol.* 2013;87(10):5755–5768. http://dx.doi.org/10.1128/JVI.02733-12.
54. Sun Y, Wang Y, Shan C, et al. Enterovirus 71 VPg uridylation uses a two-molecular mechanism of 3D polymerase. *J Virol.* 2012;86(24):13662–13671. http://dx.doi.org/10.1128/JVI.01712-12.
55. Gruez A, Selisko B, Roberts M, et al. The crystal structure of coxsackievirus B3 RNA-dependent RNA polymerase in complex with its protein primer VPg confirms the existence of a second VPg binding site on Picornaviridae polymerases. *J Virol.* 2008;82(19):9577–9590. http://dx.doi.org/10.1128/JVI.00631-08.
56. Arita M, Kojima H, Nagano T, Okabe T, Wakita T, Shimizu H. Phosphatidylinositol 4-kinase III beta is a target of enviroxime-like compounds for antipoliovirus activity. *J Virol.* 2011;85(5):2364–2372. http://dx.doi.org/10.1128/JVI.02249-10.
57. Van der Schaar HM, van der Linden L, Lanke KHW, et al. Coxsackievirus mutants that can bypass host factor PI4KIIIβ and the need for high levels of PI4P lipids for replication. *Cell Res.* 2012;22(11):1576–1592. http://dx.doi.org/10.1038/cr.2012.129.
58. Van der Schaar HM, Leyssen P, Thibaut HJ, et al. A novel, broad-spectrum inhibitor of enterovirus replication that targets host cell factor phosphatidylinositol 4-kinase IIIβ. *Antimicrob Agents Chemother.* 2013;57(10):4971–4981. http://dx.doi.org/10.1128/AAC.01175-13.
59. Berger KL, Cooper JD, Heaton NS, et al. Roles for endocytic trafficking and phosphatidylinositol 4-kinase III alpha in hepatitis C virus replication. *Proc Natl Acad Sci USA.* 2009;106(18):7577–7582. http://dx.doi.org/10.1073/pnas.0902693106.
60. Borawski J, Troke P, Puyang X, et al. Class III phosphatidylinositol 4-kinase alpha and beta are novel host factor regulators of hepatitis C virus replication. *J Virol.* 2009;83(19):10058–10074. http://dx.doi.org/10.1128/JVI.02418-08.
61. Tai AW, Benita Y, Peng LF, et al. A functional genomic screen identifies cellular cofactors of hepatitis C virus replication. *Cell Host Microbe.* 2009;5(3):298–307. http://dx.doi.org/10.1016/j.chom.2009.02.001.
62. Greninger AL, Knudsen GM, Betegon M, Burlingame AL, Derisi JL. The 3A protein from multiple picornaviruses utilizes the golgi adaptor protein ACBD3 to recruit PI4KIIIβ. *J Virol.* 2012;86(7):3605–3616. http://dx.doi.org/10.1128/JVI.06778-11.

63. Spickler C, Lippens J, Laberge M-K, et al. Phosphatidylinositol 4-kinase III beta is essential for replication of human rhinovirus and its inhibition causes a lethal phenotype in vivo. *Antimicrob Agents Chemother.* 2013;57(7):3358–3368. http://dx.doi.org/10.1128/AAC.00303-13.
64. LaMarche MJ, Borawski J, Bose A, et al. Anti-hepatitis C virus activity and toxicity of type III phosphatidylinositol-4-kinase beta inhibitors. *Antimicrob Agents Chemother.* 2012;56(10):5149–5156. http://dx.doi.org/10.1128/AAC.00946-12.
65. Sasaki J, Ishikawa K, Arita M, Taniguchi K. ACBD3-mediated recruitment of PI4KB to picornavirus RNA replication sites. *EMBO J.* 2012;31(3):754–766. http://dx.doi.org/10.1038/emboj.2011.429.
66. *Department of Microbiology & Immunology.* http://www.microbiology.columbia.edu/faculty/racaniello.html; Accessed May 31, 2014.
67. Greninger AL, Knudsen GM, Betegon M, Burlingame AL, DeRisi JL. ACBD3 interaction with TBC1 domain 22 protein is differentially affected by enteroviral and kobuviral 3A protein binding. *mBio.* 2013;4(2). http://dx.doi.org/10.1128/mBio.00098-13, e00098-00013.
68. Ishikawa-Sasaki K, Sasaki J, Taniguchi K. A complex comprising phosphatidylinositol 4-kinase IIIβ, ACBD3, and Aichi virus proteins enhances phosphatidylinositol 4-phosphate synthesis and is critical for formation of the viral replication complex. *J Virol.* 2014;88(12):6586–6598. http://dx.doi.org/10.1128/JVI.00208-14.
69. Dorobantu CM, van der Schaar HM, Ford LA, et al. Recruitment of PI4KIIIβ to coxsackievirus B3 replication organelles is independent of ACBD3, GBF1, and Arf1. *J Virol.* 2014;88(5):2725–2736. http://dx.doi.org/10.1128/JVI.03650-13.
70. Téoulé F, Brisac C, Pelletier I, et al. The Golgi protein ACBD3, an interactor for poliovirus protein 3A, modulates poliovirus replication. *J Virol.* 2013;87(20):11031–11046. http://dx.doi.org/10.1128/JVI.00304-13.
71. Fan J, Liu J, Culty M, Papadopoulos V. Acyl-coenzyme A binding domain containing 3 (ACBD3; PAP7; GCP60): an emerging signaling molecule. *Prog Lipid Res.* 2010;49(3):218–234. http://dx.doi.org/10.1016/j.plipres.2009.12.003.
72. Zhang J, Diaz A, Mao L, Ahlquist P, Wang X. Host acyl coenzyme A binding protein regulates replication complex assembly and activity of a positive-strand RNA virus. *J Virol.* 2012;86(9):5110–5121. http://dx.doi.org/10.1128/JVI.06701-11.
73. Ye J. Reliance of host cholesterol metabolic pathways for the life cycle of hepatitis C virus. *PLoS Pathog.* 2007;3(8):e108. http://dx.doi.org/10.1371/journal.ppat.0030108.
74. Rothwell C, LeBreton A, Young Ng C, et al. Cholesterol biosynthesis modulation regulates dengue viral replication. *Virology.* 2009;389(1–2):8–19. http://dx.doi.org/10.1016/j.virol.2009.03.025.
75. Ilnytska O, Santiana M, Hsu N-Y, et al. Enteroviruses harness the cellular endocytic machinery to remodel the host cell cholesterol landscape for effective viral replication. *Cell Host Microbe.* 2013;14(3):281–293. http://dx.doi.org/10.1016/j.chom.2013.08.002.
76. Arita M, Kojima H, Nagano T, Okabe T, Wakita T, Shimizu H. Oxysterol-binding protein family I is the target of minor enviroxime-like compounds. *J Virol.* 2013;87(8):4252–4260. http://dx.doi.org/10.1128/JVI.03546-12.
77. De Jong AS, de Mattia F, Van Dommelen MM, et al. Functional analysis of picornavirus 2B proteins: effects on calcium homeostasis and intracellular protein trafficking. *J Virol.* 2008;82(7):3782–3790. http://dx.doi.org/10.1128/JVI.02076-07.
78. Cuconati A, Xiang W, Lahser F, Pfister T, Wimmer E. A protein linkage map of the P2 nonstructural proteins of poliovirus. *J Virol.* 1998;72(2):1297–1307.
79. Agirre A, Barco A, Carrasco L, Nieva JL. Viroporin-mediated membrane permeabilization. Pore formation by nonstructural poliovirus 2B protein. *J Biol Chem.* 2002;277(43):40434–40441.

80. De Jong AS, Wessels E, Dijkman HBPM, et al. Determinants for membrane association and permeabilization of the coxsackievirus 2B protein and the identification of the Golgi complex as the target organelle. *J Biol Chem.* 2003;278(2):1012–1021. http://dx.doi.org/10.1074/jbc.M207745200.
81. Van Kuppeveld FJ, Galama JM, Zoll J, Melchers WJ. Genetic analysis of a hydrophobic domain of coxsackie B3 virus protein 2B: a moderate degree of hydrophobicity is required for a cis-acting function in viral RNA synthesis. *J Virol.* 1995;69(12): 7782–7790.
82. Van Kuppeveld FJ, Galama JM, Zoll J, van den Hurk PJ, Melchers WJ. Coxsackie B3 virus protein 2B contains cationic amphipathic helix that is required for viral RNA replication. *J Virol.* 1996;70(6):3876–3886.
83. de Jong AS, Visch H-J, de Mattia F, et al. The coxsackievirus 2B protein increases efflux of ions from the endoplasmic reticulum and Golgi, thereby inhibiting protein trafficking through the Golgi. *J Biol Chem.* 2006;281(20):14144–14150. http://dx.doi.org/10.1074/jbc.M511766200.
84. Zhou Y, Frey TK, Yang JJ. Viral calciomics: interplays between Ca2+ and virus. *Cell Calcium.* 2009;46(1):1–17. http://dx.doi.org/10.1016/j.ceca.2009.05.005.
85. Kim SS, Smith TJ, Chapman MS, et al. Crystal structure of human rhinovirus serotype 1A (HRV1A). *J Mol Biol.* 1989;210(1):91–111.
86. Muckelbauer JK, Kremer M, Minor I, et al. The structure of coxsackievirus B3 at 3.5 A resolution. *Structure.* 1995;3(7):653–667.
87. Zhao R, Pevear DC, Kremer MJ, et al. Human rhinovirus 3 at 3.0 A resolution. *Structure.* 1996;4(10):1205–1220.
88. Taverna E, Francolini M, Jeromin A, Hilfiker S, Roder J, Rosa P. Neuronal calcium sensor 1 and phosphatidylinositol 4-OH kinase β interact in neuronal cells and are translocated to membranes during nucleotide-evoked exocytosis. *J Cell Sci.* 2002;115(20):3909–3922. http://dx.doi.org/10.1242/jcs.00072.
89. Petko JA, Kabbani N, Frey C, et al. Proteomic and functional analysis of NCS-1 binding proteins reveals novel signaling pathways required for inner ear development in zebrafish. *BMC Neurosci.* 2009;10(1):27. http://dx.doi.org/10.1186/1471-2202-10-27.
90. Haynes LP, Sherwood MW, Dolman NJ, Burgoyne RD. Specificity, promiscuity and localization of ARF protein interactions with NCS-1 and phosphatidylinositol-4 kinase-IIIβ. *Traffic.* 2007;8(8):1080–1092. http://dx.doi.org/10.1111/j.1600-0854.2007.00594.x.
91. Hendricks KB, Qing Wang B, Schnieders EA, Thorner J. Yeast homologue of neuronal frequenin is a regulator of phosphatidylinositol-4-OH kinase. *Nat Cell Biol.* 1999;1(4):234–241. http://dx.doi.org/10.1038/12058.
92. Moffat K, Knox C, Howell G, et al. Inhibition of the secretory pathway by foot-and-mouth disease virus 2BC protein is reproduced by coexpression of 2B with 2C, and the site of inhibition is determined by the subcellular location of 2C. *J Virol.* 2007;81(3):1129–1139. http://dx.doi.org/10.1128/JVI.00393-06.
93. Den Boon JA, Ahlquist P. Organelle-like membrane compartmentalization of positive-strand RNA virus replication factories. *Annu Rev Microbiol.* 2010;64:241–256. http://dx.doi.org/10.1146/annurev.micro.112408.134012.
94. Liu L, Westler WM, den Boon JA, et al. An amphipathic α-helix controls multiple roles of brome mosaic virus protein 1a in RNA replication complex assembly and function. *PLoS Pathog.* 2009;5(3):e1000351. http://dx.doi.org/10.1371/journal.ppat.1000351.
95. Schwartz M, Chen J, Janda M, Sullivan M, den Boon J, Ahlquist P. A positive-strand RNA virus replication complex parallels form and function of retrovirus capsids. *Mol Cell.* 2002;9(3):505–514.

96. Kopek BG, Perkins G, Miller DJ, Ellisman MH, Ahlquist P. Three-dimensional analysis of a viral RNA replication complex reveals a virus-induced mini-organelle. *PLoS Biol.* 2007;5(9):e220. http://dx.doi.org/10.1371/journal.pbio.0050220.
97. Diaz A, Ahlquist P. Role of host reticulon proteins in rearranging membranes for positive-strand RNA virus replication. *Curr Opin Microbiol.* 2012;15(4):519–524. http://dx.doi.org/10.1016/j.mib.2012.04.007.
98. Diaz A, Wang X, Ahlquist P. Membrane-shaping host reticulon proteins play crucial roles in viral RNA replication compartment formation and function. *Proc Natl Acad Sci U S A.* 2010;107(37):16291–16296. http://dx.doi.org/10.1073/pnas.1011105107.
99. Tang W-F, Yang S-Y, Wu B-W, et al. Reticulon 3 binds the 2C protein of enterovirus 71 and is required for viral replication. *J Biol Chem.* 2007;282(8):5888–5898. http://dx.doi.org/10.1074/jbc.M611145200.
100. Chiu CY, Greninger AL, Chen EC, et al. Cultivation and serological characterization of a human Theiler's-like cardiovirus associated with diarrheal disease. *J Virol.* 2010;84(9):4407–4414. http://dx.doi.org/10.1128/JVI.02536-09.
101. Sasaki J, Kusuhara Y, Maeno Y, et al. Construction of an infectious cDNA clone of Aichi virus (a new member of the family Picornaviridae) and mutational analysis of a stem-loop structure at the 5′ end of the genome. *J Virol.* 2001;75(17):8021–8030. http://dx.doi.org/10.1128/JVI.75.17.8021-8030.2001.
102. Sasaki J, Nagashima S, Taniguchi K. Aichi virus leader protein is involved in viral RNA replication and encapsidation. *J Virol.* 2003;77(20):10799–10807. http://dx.doi.org/10.1128/JVI.77.20.10799-10807.2003.
103. Sasaki J, Taniguchi K. Aichi virus 2A protein is involved in viral RNA replication. *J Virol.* 2008;82(19):9765–9769. http://dx.doi.org/10.1128/JVI.01051-08.
104. Jäger S, Gulbahce N, Cimermancic P, et al. Purification and characterization of HIV-human protein complexes. *Methods (San Diego, Calif).* 2011;53(1):13–19. http://dx.doi.org/10.1016/j.ymeth.2010.08.007.
105. Hastie ML, Headlam MJ, Patel NB, et al. The human respiratory syncytial virus nonstructural protein 1 regulates type I and type II interferon pathways. *Mol Cell Proteomics.* 2012;11(5):108–127. http://dx.doi.org/10.1074/mcp.M111.015909.
106. Teterina NL, Lauber C, Jensen KS, Levenson EA, Gorbalenya AE, Ehrenfeld E. Identification of tolerated insertion sites in poliovirus non-structural proteins. *Virology.* 2011;409(1):1–11. http://dx.doi.org/10.1016/j.virol.2010.09.028.
107. Teterina NL, Pinto Y, Weaver JD, Jensen KS, Ehrenfeld E. Analysis of poliovirus protein 3A interactions with viral and cellular proteins in infected cells. *J Virol.* 2011;85(9):4284–4296. http://dx.doi.org/10.1128/JVI.02398-10.
108. Komarova AV, Combredet C, Meyniel-Schicklin L, et al. Proteomic analysis of virus-host interactions in an infectious context using recombinant viruses. *Mol Cell Proteomics.* 2011;10(12). http://dx.doi.org/10.1074/mcp.M110.007443, M110.007443.
109. Rose RJ, Damoc E, Denisov E, Makarov A, Heck AJR. High-sensitivity Orbitrap mass analysis of intact macromolecular assemblies. *Nat Methods.* 2012;9(11):1084–1086. http://dx.doi.org/10.1038/nmeth.2208.
110. Herzog F, Kahraman A, Boehringer D, et al. Structural probing of a protein phosphatase 2A network by chemical cross-linking and mass spectrometry. *Science.* 2012;337(6100):1348–1352. http://dx.doi.org/10.1126/science.1221483.
111. Stengel F, Aebersold R, Robinson CV. Joining forces: integrating proteomics and cross-linking with the mass spectrometry of intact complexes. *Mol Cell Proteomics.* 2012;11(3). http://dx.doi.org/10.1074/mcp.R111.014027, R111.014027.
112. Strauss DM, Glustrom LW, Wuttke DS. Towards an understanding of the poliovirus replication complex: the solution structure of the soluble domain of the poliovirus 3A protein. *J Mol Biol.* 2003;330(2):225–234.

113. Berstein HD, Baltimore D. Poliovirus mutant that contains a cold-sensitive defect in viral RNA synthesis. *J Virol.* 1988;62(8):2922–2928.
114. Jovic M, Kean MJ, Szentpetery Z, et al. Two phosphatidylinositol 4-kinases control lysosomal delivery of the Gaucher disease enzyme, β-glucocerebrosidase. *Mol Biol Cell.* 2012;23(8):1533–1545. http://dx.doi.org/10.1091/mbc.E11-06-0553.
115. Nakatsu F, Baskin JM, Chung J, et al. PtdIns4P synthesis by PI4KIIIα at the plasma membrane and its impact on plasma membrane identity. *J Cell Biol.* 2012;199(6):1003–1016. http://dx.doi.org/10.1083/jcb.201206095.
116. Bojjireddy N, Botyanszki J, Hammond G, et al. Pharmacological and genetic targeting of the PI4KA enzyme reveals its important role in maintaining plasma membrane phosphatidylinositol 4-phosphate and phosphatidylinositol 4,5-bisphosphate levels. *J Biol Chem.* 2014;289(9):6120–6132. http://dx.doi.org/10.1074/jbc.M113.531426.
117. Perera R, Riley C, Isaac G, et al. Dengue virus infection perturbs lipid homeostasis in infected mosquito cells. *PLoS Pathog.* 2012;8(3):e1002584. http://dx.doi.org/10.1371/journal.ppat.1002584.
118. Ogiso H, Taguchi R. Reversed-phase LC/MS method for polyphosphoinositide analyses: changes in molecular species levels during epidermal growth factor activation in A431 cells. *Anal Chem.* 2008;80(23):9226–9232. http://dx.doi.org/10.1021/ac801451p.
119. Wang Y, Sousa KM, Bodin K, et al. Targeted lipidomic analysis of oxysterols in the embryonic central nervous system. *Mol Biosyst.* 2009;5(5):529–541. http://dx.doi.org/10.1039/b819502a.

CHAPTER EIGHT

Retroviral Factors Promoting Infectivity

Emilia Cristiana Cuccurullo, Chiara Valentini, Massimo Pizzato[1]
Centre for Integrative Biology, University of Trento, Trento, Italy
[1]Corresponding author: e-mail address: massimo.pizzato@unitn.it

Contents

Abstract

The ability of a virus particle to establish an infectious event is a fundamental property required for viral propagation and survival. Retrovirus invasion of target cells is a multistep process that begins with entry into the cytoplasm and culminates with the integration of the proviral genome into the host DNA. Along this journey, many obstacles await the retrovirus particle and undermine its infectivity. Host–cell barriers to retrovirus infection can either be basic structural components of the eukaryotic cell or specific antiretroviral activities developed by the cell to prevent the retroviral invasion. Resulting from a long host–parasite coevolution, retroviruses have developed auxiliary factors that promote infectivity by conferring the virion the ability to overcome several cellular obstacles, which interfere with the infection process. Here, we provide an overview of different retroviral auxiliary factors that promote virion infectivity, comparing their mechanism of action and highlighting common mechanistic strategies. Special attention is given to infectivity factors that remain enigmatic in the biology of retroviruses.

Progress in Molecular Biology and Translational Science, Volume 129
ISSN 1877-1173
http://dx.doi.org/10.1016/bs.pmbts.2014.10.008

1. INTRODUCTION

In their coevolutionary journey with the host, viruses are continuously shaped by the selective pressure required to promote their replication and ultimately their survival. Viruses require both the ability to replicate inside one individual and the capacity to spread within a host population by propagating horizontally from one individual to another. With the exception of endogenous viruses, which have invaded the genome of the host's germ line and are therefore transmitted vertically, a virus survival depends on its ability to infect a new individual, replicate, and promote horizontal transmission. What is crucial is the ability of the virus to infect, i.e., invade, new cells and the infectious power of viruses is therefore key to their evolutionary survival. At the same time, the host response is to erect barriers to interfere with the infection process and the replication of a pathogenic virus. Constant selective pressure therefore drives this equilibrium in which the pathogenic virus evolves factors which promote infectivity in response to the countermeasures taken by the host to fight the invasion.

1.1. Retrovirus infection and retrovirus infectivity

1.1.1 Different modalities of retrovirus infection

Following a classical view, viruses disseminate after being released free in the extracellular environment by productively infected cells. Virions can then spread and enter into new cells located at a short or long distance from the producer cells. Alternatively, like other enveloped viruses, retroviruses can also infect new cells by direct transfer of cell-associated virus particles to target cells (for a review see Ref. 1). The so-called "cell-to-cell transfer" occurs on the site of cell-to-cell contact, when a virological synapse is formed between donor and target cells, mediating direct passage of budding virions from one cell to another. For some retroviruses, such as *Human T-lymphotropic virus* (HTLV), cell-to-cell transmission seems to be the prevalent mechanism of spreading.[2] This way of transmission was later demonstrated, at least *in vitro*, also with *Human immunodeficiency virus* (HIV)[3,4] and *Murine leukemia virus* (MLV).[5] Since cell-associated virus transfer requires close cell-to-cell contact, this mechanism could certainly be relevant for HIV transmission within the lymphoid tissue such as the lymphnodes, where CD4-positive cells are found in close contact. However, high level of circulating cell-free virions are observed in most retroviral infections *in vivo* indicating that cell-free virus infection remains a fundamental mechanism of transmission. In this chapter, we will provide an overview of retroviral factors that have been mainly investigated in the context of cell-free virus transmission. However, at least for some of them, the activity promoting infectivity was observed also in the context cell-to-cell retrovirus transfer.[6,7]

1.1.2 The infection process

We define infectivity as the intrinsic capability of the retroviral particle to establish an infectious event. Hallmarks of retrovirus infection are the ability of the viral complex to make a cDNA copy of the retrovirus RNA genome and to integrate it into the host chromosomes. The infectious process therefore culminates with the irreversible invasion of the retrovirus genome into the host DNA. To understand how infection can be altered by retroviral factors, here, we will briefly recapitulate the key events that characterize the journey of the virus toward the host genome.

Infection begins with the virus particle adhering to the cell surface and binding to a specific receptor. These two events are not necessarily coincident, as it was shown that a preliminary binding may be driven by interactions that do not depend on the retroviral envelope glycoprotein.[8] This allows the particle to adhere to the target cell and to encounter the cognate retroviral receptor. In the case of cell-to-cell transmission, it is the donor cell that adjoins the target cell, via cell adhesion determinants, and offers the budding virus particle the chance to establish a retroviral synapse through which the virus can transfer.[9] Receptor(s) recognition by the envelope glycoprotein is in any case the necessary event which triggers conformational changes throughout the molecule leading to fusion between retrovirus and cell membranes and to the formation of a fusion pore which then enlarges to allow the retroviral capsid to enter the cell. After having overcome the first physical barrier, i.e., the cell membrane, the actin cortex underlying the cell membrane may represent a second obstacle on the retrovirus journey, which the retroviral capsid must be capable of overcoming.[10] This is not a problem for those retroviruses that access the cell via preventive internalization into an endosomal vesicle, which provides an alternative delivery pathway into the cell. In this case (see for example *Avian leukosis virus*, ALV[11]), the retroviral envelope glycoprotein has evolved to exert its fusogenic activity only upon acidification of the endosomal content, thus ensuring that the retroviral core is delivered directly within the cell. The infectious process proceeds in the cytoplasm with the retroviral particle undergoing uncoating and the retroviral RNA genome being reverse transcribed. How synchronous and reciprocally dependent these two activities are remains to be fully established, although evidence suggests that timely uncoating is crucial for efficient reverse transcription and ultimately infection.[12,13] The ability of the so-called "reverse transcription complex" (RTC) to interact with and travel along microtubules is likely to be important for the movement of the RTC toward the nucleus.[14] However, the translocation of the retroviral complex across the cytoplasm and into the nucleus remains perhaps the least understood steps of the retrovirus life cycle. The retroviral complex must access the nuclear region to integrate into the host DNA. Some retroviruses, such as gammaretroviruses, lack the ability to cross the intact nuclear membrane and in this case the spontaneous dissolution of the nuclear membrane in cells undergoing mitosis is required for infection. For other retroviruses, capable of infecting nonreplicating cells, the retroviral complex penetrates the intact nuclear membrane, possibly by engaging with components of the nuclear pore complex to promote translocation across the nuclear pore.[15]

Once in close proximity to the host genome, retroviral integrase (IN) mediates the insertion of the reverse transcribed retroviral genome into the host DNA. The invasion of the retroviral genome within the host genome is the event that finalizes the irreversible infection of the cell.

The successful outcome of virus infection depends on the ability of the virus to perform all these steps of the life cycle despite the presence of several obstacles, which often are specific antiviral barriers posed by the host in response to an infection.

It is clear that most components of the virus particle have the function of sustaining virus transmission and replication and, as such, are factors that promote infectivity. However, in this chapter, here, we will only provide an overview and will focus only on factors that are not the canonical gene products of *gag*, *pol*, and *env* common to all replication-competent retroviruses. Many retroviral genomes, in fact, have developed additional genes encoding auxiliary infectivity promoting factors, which play crucial and additive roles in promoting virus replication and pathogenesis.

1.2. Retrovirus infectivity: What we mean and how we measure it

When it comes to cell-free virions, infectivity is the intrinsic ability of a retrovirus particle to establish infection of a target cell. In other words, infectivity describes the infectious power of the virion particle. However, the concept of infectivity goes "hand in hand" and is complementary with that of permissivity, which refers to the cell and indicates its predisposition to be infected by a particular virus. It sometimes becomes impossible to functionally distinguish between the two. In fact, the more a cell is permissive to infection, the more a virus is infectious for that specific cell and vice versa. It is therefore important to keep in mind that infectivity is not an absolute property of a virus particle, but is related to the target cell. Hence, a retrovirus highly infectious for one cell type can be noninfectious for another, non permissive, cell type. As we will see, retrovirus factors have often evolved to promote infectivity toward specific tissue or cell types containing antiretroviral barriers, to increase the tropism of that particular retrovirus.

It is well documented that, when infecting cells in culture with a retrovirus inoculum, not every virion particle is capable of establishing an infectious event. Only a fraction of the virion particles present in a virus inoculum will in fact succeed in producing infection, for the reasons well described and discussed in another chapter of this volume ("Molecular determinants of the ratio of inert to infectious virus particles," by Klasse). The number, or the ratio, of infectious events per physical amount of virus particles ultimately defines retrovirus infectivity. Infectivity is therefore established by normalizing the observed infectious events with a physical measure of virus particle number, often determined by quantifying the amount of capsid molecules or reverse transcriptase (RT) activity using an ELISA or an enzymatic assay (RT-assay), respectively.

Furthermore, when investigating retrovirus infectivity, it is important to limit the measurement to detect a single round of viral replication and exclude contributions from subsequent progeny virus. This is normally done using trans-complemented molecular clones (common for HIV and MLV) or by the addition of RT-inhibitors, protease inhibitors, or entry inhibitors at various time points following infection with replication-competent virus.

1.3. How do retroviral auxiliary factors promote infectivity?

As we already pointed out, during the process of infection, along the route going from the extracellular environment to the nucleus, the virion particle has to overcome many barriers. While some of these obstacles are the basic structural components of the cell (cell membrane, actin cortex, nuclear envelope) and others are specific antiretroviral activities developed by the cell to counteract infection. Retrovirus factors may therefore act by facilitating the transition of the virus complex through a default barrier (see, for example, the proposed action of viral protein R (Vpr) in facilitating the translocation of HIV-1 across the nuclear envelope) or by counteracting an antiviral cellular activity (such as the HIV-1 virus infectivity factor (Vif) protein, which counteracts the cytoplasmic deaminase Apobec3).

Despite the variability of their mechanism or their cellular target, retroviral factors that promote infectivity act by modifying the composition of the virus particle, sometimes just by simply being incorporated into it (see, for example, Vpr and Vpx). The increased infectivity stems therefore from an imprint acquired from the producer cell by the retroviral particle during its biogenesis.

Of note, while for some infectivity factors the mechanism of action has been revealed and molecularly well characterized, for others (notably Nef of primate lentiviruses) this remains unknown. In addition, and well exemplified by Nef itself, retroviral auxiliary factors have often evolved to perform numerous activities, making their precise molecular mechanism of action particularly difficult to elucidate.

Here, we will survey the main retroviral auxiliary factors that promote virion infectivity. We will briefly describe first those infectivity factors for which the mechanism of action has been better elucidated. We will then focus on a class of retroviral infectivity factors that have remained enigmatic for more than 20 years (Fig. 1).

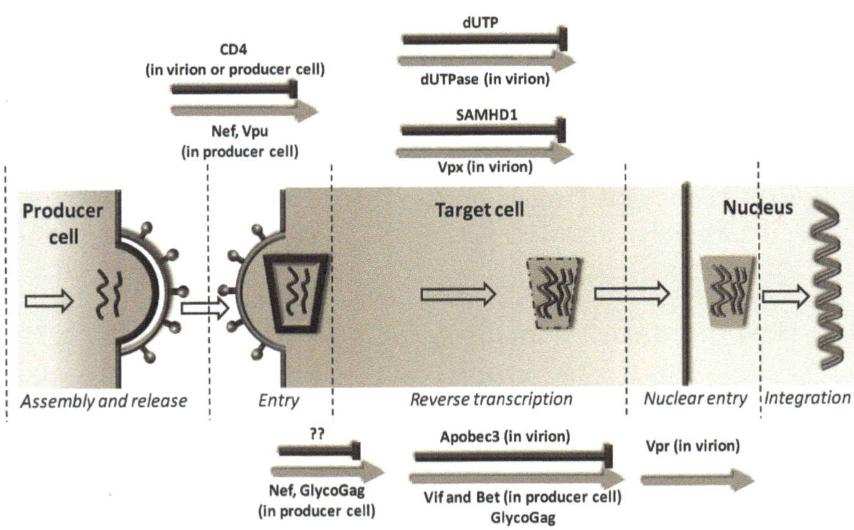

Figure 1 Steps of the infectious process and host factors targeted by retrovirus factors promoting infectivity. Retrovirus infectivity factors act by promoting the transition of the retrovirus particle through the different steps of the infectious process between virus entry and integration. The site of action of each retrovirus factor is shown together with the host factor counteracted.

2. RETROVIRAL AUXILIARY FACTORS THAT PROMOTE INFECTIVITY

2.1. Promoting infectivity by facilitating nuclear entry

2.1.1 Vpr

Vpr is a small basic and multifunctional protein conserved in all primate lentiviruses,[16] required for full replication capacity of the virus *in vivo*[17,18] and actively and abundantly incorporated into virus particles via a specific interaction with Gag (Group-specific antigen) P6.[19–21] It was specifically found to promote infection of HIV-1 toward terminally differentiated macrophages,[22–24] which suggests that Vpr functions as a virion infectivity factor, with a molecular mechanism which remains in part still obscure. Indeed, its active and specific incorporation into the viral particle and its association with the reverse transcription and preintegration complexes (RTC and PIC, respectively)[25–28] suggest a role of Vpr upon virus entry into the target cell, during the early steps of viral replication. Accordingly, several reports have suggested that Vpr may affect reverse transcription and promote HIV-1 nuclear entry (see below). Of note, the best-characterized molecular activity of Vpr is its fundamental ability to arrest infected cells in G2/M, recently found to involve activation of the structure-specific endonuclease regulator SLX4.[29]

Intuitively, the limited size of the nuclear pore requires an active mechanism for the translocation of the retroviral PIC, which exceeds the nuclear pore size.[30] However, the mechanism by which HIV and other retroviruses that infect nondividing cells enter the interphasic nucleus remains today unknown, and perhaps involves more than one retroviral component. Early studies indicated that Vpr plays a crucial role in mediating translocation of the preintegration complex through the nuclear membrane.[31] A role of Vpr as a nuclear entry factor was proposed after observing that Vpr is an integral part of the RTC/PIC and is highly karyophilic.[32–34] Accordingly, although Vpr lacks known classical nuclear localization signals (NLSs), it contains two nonconventional nuclear targeting signals[33] and nuclear export signals.[33,35–38] Further studies showed that Vpr contributes to the docking of the HIV-1 PIC to the nuclear membrane,[39,40] that Vpr interacts with the importin alpha subunit of the nuclear import receptor,[24] and that this binding is relevant (though not absolutely required) for the ability of HIV-1 to infect macrophages.[24,41] The field remains controversial, because evidence has been produced to show that in the absence of Vpr, HIV-1 remains indeed capable of infecting nondividing cells.[42] However, this could be explained by functional redundancy, as other components of the PIC, and in particular integrase, contain NLSs and may be capable of mediating nuclear translocation. In conclusion, Vpr may enhance infectivity toward macrophages by promoting nuclear import of HIV-1 by engaging directly with the nuclear transport machinery.

2.2. Promoting infectivity by protecting the stability of the retroviral genome during reverse transcription

2.2.1 Retroviral dUTPases

Because the retroviral RT cannot discriminate between dUTP and dTTP, uracil can be incorporated into the retroviral genome. Uracil can also be generated in retroviral DNA as a consequence of cytosine deamination. Collective evidence has been reported that uracil incorporation increases DNA instability caused by mutagenic U:G mispair and degradation of uracilated DNA by cellular uracil DNA glycosylases.[43]

Indeed, uracil could be considered a retroviral restriction factor and could represent a major barrier for efficient infection of resting or terminally differentiated cells.[44] Accordingly, several retroviruses have developed independently an enzymatic activity capable of protecting their genome from high levels of dUTP.[45–47] This is the case of nonprimate lentiviruses such as *Equine infectious anemia virus* (EIAV), *Caprine arthritis-encephalitis virus (CAEV), Feline immunodeficiency virus* (FIV), *Bovine immunodeficiency virus*, and betaretroviruses such as *Mason-Pfizer monkey virus* (MPMV) and *Mouse mammary tumor virus*, which encode a dUTPase from the *pol* and *pro* genes specifically, which converts dUTP to dUMP and inorganic pyrophosphate and maintains a low level of intracellular dUTP. Because they are synthesized as part of the gag-pol polyprotein, these retroviral dUTPases end up being actively and abundantly encapsidated into virions and their activity has been reported to be crucial for maintaining efficient virus infectivity.[46] The evidence that dUTPases have evolved independently in divergent lineages of retroviruses[47] highlights a fundamental infectivity promoting role in the biology of those retroviruses.

2.2.2 Does Vpr promote infectivity by affecting reverse transcription and incorporation of dUTP?

Early evidence has been reported that Vpr affects reverse transcription, as it modulates the *in vivo* mutation rate of HIV-1.[48] Though controversial, it was later suggested that this activity is related to the ability of Vpr to recruit and mediate virion incorporation of UNG2, a host uracil glycosylase that removes uracil from DNA.[49,50] Interestingly, Vpr has been reported to be important for infection of nondividing cells, such as macrophages, characterized by a high dUTP/dTTP ratio, which would favor uracil incorporation into viral cDNA.[51] This raised the question whether the Vpr–UNG2 interaction is related to the Vpr effect on macrophage infection.

The incorporation of UNG2 into virus particles by Vpr has been confirmed by several groups, but the meaning of this interaction on virus infectivity remains debated. For example, it remains unclear whether the interaction with Vpr results in degradation or promotion of UNG2 activity.[52–56] So, how can association of Vpr with UNG2 promote infectivity? According to one view, if uracil deglycosylation activity has a detrimental effect on HIV-1 infection, because it promotes degradation of the retroviral genome,[54,57] the positive effect of Vpr on virion infectivity toward macrophages derives from the ability of Vpr to protect the retroviral genome by targeting UNG2 to proteasomal degradation.[55,58,59] According to another line of research, UNG2 recruited by Vpr would rather repair the uracilated retroviral DNA, preventing its downstream degradation.[52,60] However, the studies supporting the latter hypothesis indicate that retroviral IN, rather than Vpr, is the main determinant which directs UNG2 incorporation into retroviral particles. In conclusion, the appealing model, for which Vpr plays a crucial role in maintaining the retroviral genome stability in response to uracilation of DNA, remains to be fully established. Interestingly, given that APOBEC3 promotes deamination, and therefore potential uracilation, of the retroviral genome (see below), Vpr could also play an additional role, together with Vif, in protecting the genome integrity from the activity of APOBEC3. However, this possibility also remains controversial.[53,54,61]

2.2.3 Vpx and the counteraction of SAMHD1

It has been known for a long time that retroviruses from the HIV-2/SIVsm (*Simian immunodeficiency virus* from sooty mangabeys) and from the SIVrcm (from red-capped mangebey)/SIVmnd-2 (from mandrills) lineages express an auxiliary gene (*vpx*), which is homologous to the *vpr* gene found in all primate lentiviruses,[62–66] and from which it is derived either by duplication[67] or recombination.[68] The gene product Vpx is a small protein, localized primarily in the nucleus when overexpressed.[69] Importantly, similarly to Vpr, Vpx is actively incorporated into virus particles by virtue of a specific interaction with Gag within the P6 region.[20,70,71] Vpx promotes retrovirus infectivity in a highly cell-type specific manner. While this small auxiliary protein is largely dispensable for replication in most immortalized lymphoid cell lines, intact Vpx is greatly required for efficient replication of HIV-2 and SIV in human myeloid primary cells, such as monocyte-derived macrophages and monocyte-derived dendritic cells (MDDC).[72–75] Vpx encapsidation is essential for conferring to virions an infectivity advantage. Most interesting was the observation that Vpx supplied *in trans* via transduction of target cells with SIV particles devoid of the viral genome (virus-like particles, VLPs), is capable of making MDDC fully susceptible to retroviruses (such as HIV-1), which normally infect them with low efficiency.[76] This suggested that Vpx could act on retrovirus infection by counteracting a host–cell inhibitory factor. This theory found crucial support from heterokaryon experiments, where restrictive myeloid cells were fused with nonrestrictive nonmyeloid cells demonstrating that the restrictive phenotype is dominant and caused by an inhibitory host factor.[77] The identity of such host factor was independently revealed by two research groups who used proteomics to analyze the protein complex bound by Vpx.[78,79] They identified the sterile alpha motif (SAM) and HD-domain-containing protein1 (SAMHD1) as the host factor required to inhibit retrovirus infection of myeloid cells. Furthermore, SAMHD1 was also found to be the main restriction factor that impairs HIV-1 infection of resting CD4+ T-lymphocytes.[80] SAMHD1 is a deoxynucleoside-triphosphate (dNTP) phosphohydrolase,[81] an enzyme which can deplete the pool of dNTPs necessary for the reverse transcription of the retroviral genome. Indeed, replenishing the dNTP pool in restrictive cells by supplying exogenous deoxyribonucleosides, rescues infection of lentiviruses that are restricted by SAMHD1.[82] While SAMHD1 seems to be expressed in many cell types, its inhibitory effect is observed especially in situations where dNTP levels are naturally low, such as postmitotic myeloid and quiescent CD4 T-lymphocytes.[80] The antiretroviral activity of SAMHD1 is regulated at the posttranslational level by cell-cycle dependent phosphorylation mediated by cyclin-dependent kinase 1.[83,84] Phosphorylated SAMHD1 is inactive in proliferating cells, partly explaining the cell-type specificity of its restrictive activity. Lately, it emerged that SAMHD1 can also inhibit retrovirus infection via a dNTPase-independent restriction mechanism, which possibly relies on the ability of the restriction factor to bind and degrade ssRNA and ssDNA.[85]

Long before SAMHD1 was identified as the host factor inhibiting HIV-1 in myeloid cells, both Vpx and Vpr were found capable of binding a large Cullin–RING ubiquitin ligase complex (the damage-specific DNA binding protein1 (DDB1)–Cullin 4A (CUL4A)–RBX1/ROC1 recruited through the interaction with the adaptor intermediate DCAF1 (DDB1–CUL4A-associated factor1)).[77,86,87] The ability of Vpx to recruit ubiquitin ligase activity pointed toward its role in promoting the degradation of a myeloid cell specific restriction factor. Indeed, it has now been well established that Vpx requires the interaction with DCAF1–DDB1–CUL4A to promote ubiquitination and proteasomal degradation of SAMHD1.[88,89]

Interestingly, unlike the anti-APOBEC3 activity that was evolved independently by different groups of retroviruses (see below), there is so far evidence that only HIV-2 and SIV have acquired an infectivity factor capable of counteracting SAMHD1, despite the host factor ability to inhibit a wide array of divergent endogenous and exogenous retroviruses, including *alpha-, beta-, and gamma-retroviruses.*[90] The requirement for a more efficient infection of the myeloid lineage is therefore a distinctive feature of HIV-2 and SIV.

In addition to promoting infectivity by counteracting SAMHD1, Vpx may target a yet unknown additional restrictive cellular activity expressed in the context of the antiviral state.[91,92] So, a Vpx mutant no longer capable of promoting SAMHD1 degradation retains the ability of rescuing HIV-1 infection in LPS (lipopolysaccharide)-treated MDDCs by relieving a preintegration inhibition.[93] Lastly, Vpx seems to have another, yet distinct, function on virion infectivity, as it positively affects infection of some cell types, such as activated peripheral blood mononuclear cell (PBMC), where SAMHD1 had been shown to be inactive against retrovirus infection. To explain this effect on infectivity, it has been suggested that Vpx may play a role in facilitating nuclear translocation of the preintegration complex, a feature that would depend on its karyophilic properties and which is reminiscent of the activity of Vpr.[94,95]

2.3. Protecting the retroviral genome from deamination

2.3.1 Vif

The genomes of all lentiviruses, with the exception of EIAV,[96,97] encode a small and basic auxiliary protein, which in HIV-1 was originally named Vif due to its potent effect on HIV-1 infectivity on natural target cells.[98] Initial studies unveiled how Vif is required for the replication of HIV-1 in CD4+ T-lymphocytes and macrophages and other lymphoid cell lines such as CEM, H9, and HUT78 (nonpermissive). On the other hand, permissive cells lines were also identified, which supported efficient replication of

Vif-deficient HIV-1.[99,100] Similar observations were reproduced using HIV-2, SIV, CAEV, and *Maedi visna virus*, highlighting an activity on infectivity of different lentiviruses. Importantly, it was clear that permissivity was dictated by producer cells, as nonpermissive cells can be infected by Vif-deficient HIV-1 but subsequent progeny virus is not infectious irrespectively of which target cell is used.[99] Eventually, the study of heterokaryons obtained from the fusion of permissive with nonpermissive cells revealed that the nonpermissive phenotype is dominant and suggested that an antiviral activity is present in nonpermissive producer cells.[101] This was a crucial observation, which later allowed the discovery of the host factor responsible for such a phenotype.

Using a cDNA hybridization subtraction method, Sheely and colleagues identified apolipoprotein B mRNA-editing enzyme catalytic polypeptide-like 3G (APOBEC3G) as a cell factor inhibiting HIV-1 infectivity and being counteracted by Vif.[102] Indeed, APOBEC3G is specifically present only in nonpermissive cells and is targeted by HIV-1 Vif and other functionally related retroviral molecules (see below).[102] APOBEC3G belongs to the seven-membered APOBEC3 family of editing enzymes, characterized by the presence of one or two cytidine deaminase motifs.[103] Within the family, APOBEC3G exhibits the most potent antiretroviral activity although other members such as APOBEC3D, -F, -H contribute to HIV-1 restriction in a Vif-sensitive manner.[104] In the absence of Vif and after infection of target cells, concomitantly with the process of reverse transcription, APOBEC3G is capable of catalyzing the deamination of cytosine to uracil in the minus-single-stranded DNA viral genome resulting in guanosine to adenosine transitions in the positive strand, leading to loss of genetic integrity. APOBEC3 proteins have a distinctive preference for cytosines in a specific sequence context, with, for example, APOBEC3G targeting predominantly the second C of a CC dinucleotide.[103] Deamination events can involve up to 10% of total cytidines, either resulting in severe hypermutation capable of inactivating the retroviral genome, or favoring the degradation of the uracylated cDNA intermediate possibly initiated by UNG2.[54] More recently, it has been shown that catalytically inactive APOBEC3G maintains significant antiviral activity, revealing that deamination-independent mechanisms contribute also to inhibition of virus replication.[105] Accordingly, APOBEC3G was shown to inhibit both reverse transcription and integration, with mechanisms so far not well understood.[106–109] Members of the APOBEC3 family were shown to be active against several exogenous and endogenous retroviruses,[110] including lentiviruses, alpharetroviruses,[111] betaretroviruses,[112] deltaretroviruses,[113] gammaretroviruses,[114] and foamy viruses,[115] making this host factor a key regulator of retrovirus infectivity.

A key event in the restriction of infectivity is the specific packaging of APOBEC3 into virions. APOBEC3G dimers bind to the nucleocapsid (NC) domain of the Gag polyprotein during virion assembly,[116] an interaction which requires the presence of RNA.[117] In the absence of Vif, APOBEC3 proteins are successfully packaged into virions and transferred to the target cells, where they remain associated with the RT complex to enable deamination to occur.[118] HIV-1 Vif overcomes the antiviral action of APOBEC3G by preventing its incorporation into virions. The best-known suppressive activity of Vif results in degradation of APOBEC3G in producer cells,[119] mediated by the recruitment of an E3 ubiquitin ligase complex.[120] A C-terminal SOCS-box of Vif contains a BC-box and a zinc-finger domain through which Vif interacts, respectively, with Elongin B/Elongin C and with Cullin5 that is associated to Rbx.[121] The E3 ubiquitin ligase complex causes the polyubiquitination of APOBEC3G thus directing it to 26S proteasome for degradation; furthermore, the coexpression of APOBEC3G and Vif enhances the ubiquitination of Vif, promoting the coordinated degradation of both proteins as a complex.[122] For an efficient degradation to occur, Vif hijacks the transcription cofactor CBF-β to the ubiquitin complex.[123] The Vif–CBF-β interaction promotes Vif stability allowing the correct assembly of the E3 ubiquitin ligase complex, an efficient degradation of APOBEC3 and an enhanced viral infectivity. In addition, further evidence suggests that Vif may also counteract APOBEC3G activity at the posttranscriptional level, by binding to *APOBEC3G* mRNA and downregulating its translation.[124,125]

2.3.2 Bet

As noted above, APOBEC3 restriction factors represent a widespread mechanism of defense against retroviral infections conserved among several mammalian species.[126] As a consequence, retroviruses have evolved different strategies to elude the problem, Vif being the most common. Indeed, a Vif Open Reading Frame has been found in all lentiviruses, except in EIAV, and its protein sequence is highly conserved among highly related lentiviruses.[96] On the other hand, foamy viruses produce Bet, a different accessory protein able to protect the virus against APOBEC3G. Bet is a 53-kDa cytoplasmic protein that shows no evident sequence homology to HIV-1 Vif and does not contain an obvious SOCS-box.[127,128] Like Vif, Bet prevents the encapsidation of APOBEC3G, but unlike the lentivirus counterpart, Bet does not induce APOBEC3G degradation, underlining a different counteracting mechanism. Although the exact mechanism is not known, the counteracting ability of Bet is attributed to its direct interaction with APOBEC3G, which occurs in a RNA-independent manner. The interaction impairs APOBEC3G dimerization and APOBEC3G cytosolic stability[129] preventing encapsidation into virus particles.

2.3.3 GlycoGag

Recently, studies using knock-out mice revealed a yet additional retroviral infectivity factor capable of counteracting APOBEC3. Intact glycoGag is in fact strictly required for gammaretrovirus MLV to replicate in the presence, but not in the absence, of APOBEC3.[130,131] GlycoGag is a membrane-targeted glycosylated form of the Gag protein originating from an alternative start codon upstream of the Gag polyprotein, which can affect virion infectivity in an additional way, as it will be described in more detail later in this chapter. How glycoGag protects virions from the antiviral activity of APOBEC3 remains largely unknown. This gammaretrovirus factor uses yet a distinct mechanism that differs from those used by Vif and Bet. Expression of glycoGag in fact makes MLV resistant to murine APOBEC3 without preventing its encapsidation into nascent virus particles.[131,132] It was suggested that glycoGag plays an important role for viral core stability. Increased stability and integrity of the capsid structure by glycoGag was indeed observed and seems to be essential for rendering MLV resistant to murine APOBEC3 and additionally for conferring resistance against host cytosolic sensors that recognize the retroviral genome.[131]

Remarkably, Vif, Bet, and glycoGag share no sequence homology and derive from unrelated retroviral genomic regions. Yet they share a similar role, delineating a functionally related class of retroviral factors, which must have evolved as a result of convergent evolution under a common selective pressure.

Knowing that APOBEC3 can target the replication of a wide panel of retroviruses, it remains likely that similar factors promoting infectivity have evolved also in other, less explored retroviral genomes. As seen with Vif, Bet, and glycoGag, their ability to counteract APOBEC3 is often species-specific, making the discovery of additional factors from other retroviruses more difficult as this might require unexplored cell culture systems and less annotated retroviral genomic information.

2.4. Promoting infectivity by preserving Env function on virion particles

2.4.1 Vpu, Nef, ORF-A

It is well established that continuous expression of the retroviral receptor in retrovirus-producing cells has detrimental effects on viral replication. This has been well documented in cells producing HIV-1, where CD4 may prematurely interact with the envelope glycoprotein (Env) forming complexes in intracellular compartments and at the plasma membrane, preventing processing and correct intracellular transport of the Env subunits gp120 and gp41.[133–135] This prevents the correct incorporation of Env into nascent virions and decreases virion infectivity.[136] High expression of CD4 could also result in its increased incorporation into virions, leading to inhibition of virus particle attachment to the CD4-positive cell surface.[137] Retroviruses have therefore developed molecular tools, which decrease expression of their receptors in producer cells and indirectly promote virion infectivity. The importance of the activity of receptor downregulation for the biology of some retroviruses is demonstrated by the fact that HIV has devoted two auxiliary proteins, which trigger two different and complementary mechanisms, to suppress expression of the primary receptor CD4.[138]

Nef, found in all primate lentiviruses, and Vpu, found only in HIV-1, are multifunctional auxiliary proteins both capable of abolishing CD4 expression. Nef functions as an adaptor protein, interacting with CD4 and the clathrin adaptor AP2, to remove the CD4 molecules that are already been transported to the cell surface of virus producing cells and promote their efficient internalization via clathrin-mediated endocytosis.[139–141] CD4 is then driven into lysosomes for degradation.[142] Vpu, instead, acts on newly synthesized CD4 by promoting its degradation while still in the endoplasmic reticulum and by preventing its translocation to the cell surface. To achieve this, Vpu molecules form a ternary complex together with CD4 and with β-TrCP,[143] which is required for ubiquitination and proteasomal degradation of CD4 itself. Interestingly, while Nef is expressed early during infection, Vpu is expressed at a later stage, ensuring efficient control of CD4 surface expression level at all time.[144]

Another indication that receptor downregulation is important for retroviruses comes also from the evidence that ORF-A, a small auxiliary protein of FIV, with similarities to HIV-1 Tat, exerts a similar function on CD134, which is the primary receptor of FIV. It was observed that reduction of cell surface expression of CD134 is associated with the ability of

ORF-A to promote its retention and accumulation in the Golgi apparatus of feline primary lymphocytes.[145]

3. RETROVIRUS FACTORS THAT PROMOTE VIRION INFECTIVITY WITH A YET UNKNOWN MECHANISM: THE NEF AND glycoGag ENIGMA

3.1. Nef

Nef is an auxiliary protein exclusively encoded by primate lentiviruses from multiply spliced RNA. The ORF partially overlaps with the 3′LTR to generate a 27–32 kDa protein, the bigger being encoded by SIV. Nef is expressed early and is antigenic during the course of natural infection.[146,147] Initial studies suggested that Nef was a *ne*gative *f*actor (hence the name *nef*, still in use today).[148–150] Later reports contradicted this evidence[151,152] and eventually established that an intact *nef* allele is required to maintain high viral load and to promote immunodeficiency in Rhesus macaques infected with $SIVmac_{239}$.[153] Compelling observations in patients confirmed that infection with Nef-deleted viruses results in low level of virus replication and long-term nonprogression.[154,155] Accordingly, the positive effect of Nef on HIV-1 replication *in vitro* was observed also using primary cell cultures and transformed cell lines.[156–160]

3.1.1 The protein and its multifunctional activity

Nef does not contain enzymatic activity, but exerts several cellular functions resulting from its ability to interact with numerous host factors. Nef is composed of a globular core domain flanked by a flexible N-terminal arm and a C-terminal disordered loop.[161–164] The protein is also myristoylated, a modification which, together with a stretch of basic amino acids in proximity to the N-terminus,[165] contributes to its association with membranes. Indeed, Nef can associate with the plasma and perinuclear membranes.[166–168] The protein is also encapsidated within virion particles[169–172] although, as discussed further below, the functional meaning of Nef packaging into virions remains unclear.

The most remarkable feature of Nef, considering its small molecular size, is its multifunctionality. Nef performs several activities and their functional relationship (or independence) has in some cases remained difficult to understand. The most characterized activities of Nef stem from the ability of the protein to manipulate the cellular vesicular trafficking machinery and to alter cell signaling. Briefly, Nef interacts with key players of intracellular vesicular trafficking to modulate surface expression of several cell membrane proteins,[173] among which is the primary receptor CD4 and the MHC-I complex. In addition, SIV Nef downregulates the restriction factor BST-2 (bone marrow stromal antigen 2 also known as "thetherin").[174] Crucial for these roles of Nef is the ability to form ternary complexes with cargo molecules and with adaptor or coatamer complexes. An acidic di-leucine motif (ExxxLL),[139] a di-acidic sequence (EE) located in its C-terminal loop[175] and an acidic cluster EEEE[176] or a YxxΦ motif on its N-terminal arm[177] are all determinants within the Nef sequence, which were found to be crucial for connecting with the vesicular trafficking machinery. Another prominent activity of Nef is its ability to alter the activation threshold of lymphocytes,[178–181] which depends on the capability of Nef to interact with several tyrosine kinases, allowing Nef to establish a transcriptional program which resembles that resulting from TCR stimulation, and might create an intracellular environment favoring virus replication. In addition, signaling perturbation by Nef results in the inactivation of Cofilin, which inhibits cytoskeleton rearrangement and cell motility.[182]

The feature of Nef that remains perhaps the least understood is also the most relevant for this chapter, i.e., the ability to promote the infectivity of the retrovirus particle. This activity was first reported by Guatelli and coworkers in 1994, who observed that HIV-1 lacking the ability to express Nef has lower infectivity compared to the Nef-positive counterpart.[183] Later reports confirmed this initial observation using different experimental systems. This activity of Nef is variable in magnitude, according to producer cell type, target cell type, retroviral molecular clone, and infection modality (cell-free vs. cell-to-cell).[184–190] It ranges from threefold to more than 40-fold and is most pronounced when infection is performed with cell-free HIV-1 produced from lymphoid cell lines.[191]

Analysis of Nef mutants has well established that the effect of Nef on infectivity is genetically separable from its ability to downregulate MHC-I[192] and to a major extent from the activity on T-cell activation.[193] However, there is no clear mutation in Nef that allows to genetically separate it from the ability to downregulate CD4. As we noted earlier, expression of high levels of CD4 in virus producing cells may impair the infectivity of progeny viruses by interfering with Env incorporation and function in virus particles,[136,194] either as consequence of premature Env-CD4 interaction, or CD4 incorporation into virions.[137] Accordingly, CD4 downregulation by Nef can favor HIV infectivity and replication,[136] and Nef from late stage virus isolates have both increased capacity to downregulate

CD4 and to increase virus infectivity.[195] However, Nef remains capable of increasing virus infectivity when virions are produced from CD4-negative cells and when virions are pseudotyped with envelope glycoproteins that do not interact with CD4.[184,186,196,197] Therefore, the effect on infectivity is not the sole consequence of the action of Nef on CD4, but is a separate activity and has relevance *in vivo*. In fact, a comprehensive analysis of *nef* alleles from a large panel of HIV and SIV isolates has shown that the activity on infectivity is phylogenetically highly conserved,[198] indicating that this is a generally important and required feature of Nef. Additionally, a functional analysis of *nef* alleles obtained during different stages of HIV infection, revealed that the Nef effect on infectivity is maintained by a strong selective pressure during disease progression,[199] suggesting that this activity is indispensable throughout the whole course of natural infection.

3.1.2 The mechanistic details of the Nef effect on infectivity

The mechanistic details of this function still remain to be understood. Here, we will try to guide the reader through 20 years of literature and experimental observations which often offer contradictory models to explain how Nef modulates retrovirus infectivity.

Nef is often reported to have only a few-fold effect on infectivity. This prompted the definition of such an activity as "enhancement of infectivity." However, in those situations where the requirement of Nef for the infectivity of HIV-1 is the highest, such as in some lymphoid cells,[191,200] Nef-defective virus has an extremely low residual infectivity, indicating that Nef is required to maintain, rather than to enhance, virus infectivity. What we currently know is that the effect of Nef requires its expression in producer cells rather than target cells[186] and is manifest at an early step of the infection process.[186,201] This implies that a modification is picked up by the virus from Nef-expressing cells and that such modification confers to virions the ability to overcome an early block in target cells. The questions therefore are: (1) at which step of the virus life cycle is the infectious process affected by Nef? (2) What is the nature of the Nef-dependent modification acquired by the virion?

3.1.3 Why is the infectivity of Nef-negative particles defective?

Multiple reports agree that infection with Nef-defective virions result in defective provirus synthesis, indicating that the Nef-negative virus is impaired anywhere between the initial interaction with the target cell and before viral DNA translocation to the nucleus.[186,191,201] An early report suggested that Nef enhances the affinity of RT for RNA *in vitro*.[202] However, Nef does not affect virion-associated reverse transcription *in vivo*,[186,188] making the possibility that Nef alters RT directly quite unlikely. The defect of infectivity must therefore map at an earlier step of the life cycle.

It is well established that pseudotyping HIV-1 with heterologous virus glycoproteins, which promote virus entry via endosomal uptake,[196,203,204] results in abrogation of the effect of Nef. In addition, Nef responsiveness can also be abolished by Env alleles derived from certain neutralization resistant HIV-1 isolates.[200] The difference between responsive and non-responsive HIV-1 Env was recently mapped to an epitope within the V2 region of gp120.[205] These observations indicate that one important determinant of the Nef activity is Env and therefore lies just at the point of virus entry. Therefore, the hypothesis tested independently by several groups was that Nef affects fusion between the virus and the cell membrane. Accordingly, an early study indicated that Nef enhances cytoplasmic delivery of the virus capsid.[206] However, a quantitative β-lactamase (βlam)-Vpr fusion assay has consistently shown that Nef-positive and Nef-negative viruses fuse with similar efficiencies, ruling out an effect of Nef on fusion.[184,207–209] To cover all possibilities, it has been pointed out that Nef-negative virions could be defective in enlarging, rather than just forming, the fusion pore, a condition that would not necessarily be detected by the βlam-Vpr assay, given that, unlike the retroviral capsid, the soluble βlam substrate could readily diffuse through a small fusion pore and generate positive signal.[208] However, time-resolved imaging of single viruses suggests that, though preceded by endocytic uptake of virions, entry of HIV-1 cores into the cytoplasm of target cells is not affected by Nef.[210] Therefore, while waiting for more powerful tools to study fusion between retroviruses and cells, the current view is that Nef does not enhance cytoplasmic delivery of HIV-1. The block to infection must then occur at a post-entry step.

The actin cortex, below the plasma membrane, was then suggested to act as a barrier blocking specifically Nef-defective viruses. One study provided evidence that the infectivity of Nef-negative HIV-1 can be rescued by treating target cells with agents disrupting the actin cytoskeleton.[207] The model was supported by evidence that treatment was shown not to affect HIV-1 particles pseudotyped with VSV-G (the envelope protein of the *Vesicular stomatitis virus*), which are intrinsically able to bypass the actin cortex barrier as they are internalized via the endocytic pathway. While this is an attractive possible model, it is at odds with other observations, which challenge the idea that Nef's responsiveness is a consequence of viral fusion at the plasma membrane. For example, Nef-responsiveness can be observed even when fusion takes place in endosomes, both with wild-type HIV-1[210] and in an experimental setting where cells express the pH-dependent ALV-A Env and virions are pseudotyped with the cognate Tva receptor.[197] While indeed cortical actin could represent a barrier for the Nef-defective virus, it seems likely that Nef-responsiveness is dictated by additional blocks in the target cell. Accordingly, evidence that infectivity of Nef-defective virions can be rescued by target cell treatment with proteasome inhibitors[211] suggests that a proteasome-dependent mechanism targeting the incoming virus[212] may be counteracted by Nef. However, in this case, it remains to be explained why pseudotyping virions with VSV-G makes HIV-1 particles more resistant to degradation.

3.1.4 What is the nature of the modification of the virus particle promoted by Nef which accounts for the effect on infectivity?

There are two possibilities: Nef itself, which is incorporated into particles, plays a direct role as virion component in promoting infection of target cells. Alternatively, Nef expression in producer cells promotes other modifications of the virions required to maintain full infectious potential.

3.1.5 Is Nef functioning as a virion protein?

Nef is found incorporated into virus particles and undergoes cleavage by the viral protease during virion maturation.[169,170,172,213] While this could indicate a role as virion component, intravirion packaging of Nef remains of yet unknown significance. Unlike the incorporation of Vpr and Vpx, Nef packaging seems to be rather the result of passive encapsidation resulting from its association with cellular membranes. Indeed, Nef is also incorporated into non-HIV particles (MLV), further indicating the lack of a nonspecific mechanism.[170] There have been attempts to solve this issue by searching for Nef mutants that failed to be incorporated. While most mutations affecting incorporation of Nef impairs also other functions, such as CD4 or MHC-I downregulation,[165,170,171,213,214] the 4E4Q Nef mutant was reported to enhance HIV-1 infectivity despite a partially defective association with HIV-1 particles,[193] suggesting that Nef incorporation and effect on infectivity are not directly linked. This indication was confirmed by one additional study, which showed that, despite active incorporation into virions via fusion with Vpr, Nef was not capable of affecting infectivity,[215] therefore concluding that Nef does not function within the virion but plays a role during the biogenesis of viral particles.

3.1.6 *Does Nef affect incorporation or functionality of other retroviral proteins?*

Potential quantitative or qualitative changes of retroviral proteins in the virion, promoted by Nef, have been investigated on several occasions. As mentioned earlier, an early report suggested that Nef enhances the affinity of RT for RNA *in vitro*.[202] However, despite evidence that infection of Nef-defective HIV-1 results in defective provirus synthesis in infected cells, virion-associated reverse transcription has not been found to depend on Nef,[186,188] making it an unlikely possibility.

More recently, it was suggested that Nef modulates HIV-1 proteolytic processing,[216] therefore acting as a regulator of virion maturation. However, if incomplete virion maturation is the reason for defective infectivity of Nef-negative virions, then it seems hard to explain why Nef-responsiveness depends on Env and, for example, can be abrogated by a modification in HIV-1 Env gp120 V2 loop.[205]

Finally, another possibility that has been investigated is the potential effect of Nef on the amount of Env being incorporated into virions.[217] As noted earlier, while this could indeed be the case for virions produced from cells expressing high level of CD4, no effect on Env incorporation was observed in a number of studies using CD4-postive and -negative cells, making also this option quite controversial and unlikely to explain the effect on infectivity.

3.1.7 *Does Nef affect virion incorporation of cell-derived components?*

If Nef itself is not acting as virion component, if Nef does not influence other retroviral proteins (or nucleic acid) in the virus particles, then Nef must alter the presence of components of cellular origin, which are packaged into virions. In this respect, studies have investigated the possibility that Nef promotes differential incorporation of proteins and lipids into virus particles. A comparative proteomic analysis of viral particles generated in the presence or absence of Nef was recently reported.[218] This study revealed that Ezrin and EHD4 are more abundant in Nef-defective particles, but these proteins could not account for the diminished HIV-1 infectivity in the absence of Nef. In this case, the sensitivity threshold of the proteomics technique being used is an important limit of the strategy, since it remains difficult to rule out that some components are incorporated at amounts below the mass spectrometry sensitivity. A negative result, therefore, does not exclude the cellular protein hypothesis.

Evidence has been shown that Nef facilitates virus budding from detergent resistant membranes[219] and therefore could promote incorporation of raft components into particles. Accordingly, an early report indicates that the cholesterol level is enriched in Nef-positive virions and this could explain the higher infectivity phenotype.[220] However, while a mass spectrometry analysis of the virion lipid composition has confirmed the raft character of the viral membrane in the presence of Nef,[221] sphingomyelin and not cholesterol was found to specifically associate with Nef-positive virus. But such enrichment in sphingomyelin was found not to be functionally linked to the effect on infectivity, because mutant *nef* alleles defective for the activity on infectivity remain capable of increasing the sphingomyelin level in the virion membrane. The identification of possible Nef-dependent modifications of the virion that might affect virus infectivity, therefore, awaits further insights.

3.2. Nef is not alone: glycoGag

As we mentioned earlier, a *nef* gene sequence is a prerogative of the genome of primate lentiviruses. However, we have recently discovered that a Nef-like activity on virus infectivity is also exerted by glycosylated-Gag (or glycoGag),[191] a molecule which is encoded by gammaretroviruses and shares no sequence similarity with Nef. While gammaretroviruses have the simplest genomic organization (hence they are classified as "simple" retroviruses), it has been known since early 1970s that many gammaretroviruses can produce an alternative form of Gag.[222–225] This is translated from unspliced RNA starting from an unusual and alternative CUG initiation codon, upstream of and in frame with the conventional *gag* ORF. When translation begins from this codon in MLV, 88 aa are added at the N-terminus of Gag, including a signal peptide which confers the protein a type II transmembrane topology. The protein therefore displays the C-terminus (the Gag moiety), which is glycosylated, at the exterior of the cell. The C-terminal half of the molecule is cleaved by a cellular protease and secreted, while the rest of the protein (including the MA and P12 sequences) remains associated with the cell membrane and is also incorporated into virus particles.

While glycoGag is not strictly required for MLV replication *in vitro*, it was shown to be an important virulence determinant *in vivo*, contributing to virus replication in mice infected with MLV and required for neuroinvasiveness of a murine neuropathic retrovirus.[226–231] The molecular function of glycoGag has remained for a long time unknown. We have shown that glycoGag from MoMLV (Moloney murine leukemia virus) expressed in human lymphoid producer cells can efficiently and specifically repair the infectivity defect of Nef-negative HIV-1, revealing that glycoGag acts as a Nef-like infectivity factor.[191] Accordingly, glycoGag expression is required for the infectivity of MLV produced from lymphoid cell lines, where the virus acquires a 20-fold reduction in infectivity in the absence of glycoGag. A 96aa N-terminal fragment was found sufficient to promote retroviral infectivity, indicating that the activity lies within the cytoplasmic domain of the molecule.[232] Interestingly, Nef and glycoGag share no sequence homology but yet are linked by an impressive functional similarity.[191] Both increase the yield of viral cDNA in target cells, suggesting that they affect a similarly early step in the virus life cycle. Both have identical dependence on the particular Env-pseudotype and producer cell type, being maximally required when using lymphoid cells as virus producers. The two proteins perfectly colocalize within the cell in perinuclear and membrane compartments. Importantly, simultaneous coexpression of Nef and glycoGag has no synergistic effect and glycoGag has no effect on VSV-G pseudotyped viruses, suggesting that these proteins are involved in a similar mechanism required to promote virus infectivity. Altogether, the impressive functional similarity between the abilities of Nef and glycoGag to promote infectivity provides an example of convergent evolution. Unlike Nef, glycoGag does not seem to downregulate MHC-I nor the MLV receptor, and does not contain sequences predicted to interact with cellular kinases, indicating that the functional similarity between the two proteins is restricted only to the effect on retroviral infectivity.

Of note, as we described earlier, some recent reports have unveiled another important function of glycoGag, which was shown to promote MLV infectivity and replication *in vivo* by protecting the virus from the antiviral activity of murine APOBEC3.[130,131] The mechanism by which APOBEC3 might inhibit virus replication and the way glycoGag protects the virus remain to be established, since APOBEC3 was not found to deaminate the MLV genome and intravirion packaging of APOBEC3 is not affected by glycoGag. However, while the glycoGag activity on HIV-1 and MLV infectivity can be readily observed in human cells, wild type (i.e., glycoGag-positive) MLV was reported to be highly sensitive to human APOBEC3G,[132] suggesting that glycoGag has no activity against human APOBEC3G. In addition, the effect of glycoGag on infectivity is abrogated by certain envelope glycoproteins, such as VSV-G, which do not seem to alter virion sensitivity to APOBEC3G, based on the numerous reports which used VSV-G pseudotypes to study Vif function. Finally, glycoGag affects infectivity of Nef-defective Vif-positive HIV-1, therefore in conditions in which HIV-1 is already protected from human APOBEC3G.[191] It therefore seems that the effect of glycoGag on MLV and HIV infectivity is unrelated to APOBEC3 and delineates a separate activity of glycoGag.

3.2.1 How do Nef-like factors promote retrovirus infection?

How Nef and glycoGag act to promote retrovirus infectivity remains at this stage difficult to understand. What is clear is that Nef and glycoGag modulate infectivity by acting in producer cells and that consequently virions undergo a modification, which makes them more infectious. Two hypotheses are therefore viable. By analogy with other auxiliary factors such as Vif, Vpx, and Vpu, Nef and glycoGag could counteract a common restriction factor, which inhibits infectivity of progeny virions and is prevalently expressed in lymphoid cells. Alternatively, Nef could promote, or provide, an activity that is needed for the biogenesis of fully infectious virus and is suboptimally expressed in lymphoid cells. Current experimental evidence does not allow discerning between these two possibilities. However, having discovered that two unrelated retroviral proteins, Nef and glycoGag, share a similar function, gives us the chance to compare their molecular activities in order to identify common traits, which could help identifying their mechanism of action.

With the current knowledge, the best indication is that the effect of Nef and glycoGag both require the endocytic machinery. To promote infectivity, both proteins require motifs predicted to recruit the clathrin adaptor complex AP2 (EXXLL in Nef and YxxL in glycogag).[183,205] Nef was also found to interact with dynamin2,[233] a crucial regulator of clathrin-mediated endocytosis, and require dynamin2 and clathrin in order to promote infectivity.[233] Similarly, the activity of glycoGag requires functional AP2,[205] and therefore functional clathrin-mediated vesiculation. Overall, given also the well-documented ability of Nef to efficiently modulate cell surface expression of several cellular proteins, the most tempting hypothesis is that Nef and glycoGag function by preventing the viral incorporation of a cellular inhibitor of retroviral infectivity, promoting its endosomal uptake and down-regulation in producer cells, an inhibitor which acts during early steps of the infectious process.

3.3. Are there other Nef-like factors promoting retrovirus infectivity?

HIV and MLV are both lymphotropic viruses and accordingly the requirement for Nef and glycoGag is manifest mostly in cell lines of lymphoid and myeloid origin (Ref. 191 and M. Pizzato, unpublished data), suggesting that the two proteins arose to solve a common "problem" which targets retroviruses that replicate in hematopoietic lineages. It is therefore plausible that other retroviruses, with similar tropism, have evolved additional factors functionally related to Nef and glycoGag. Accordingly, a report has shown that HTLV-I P12 can also promote a Nef-like activity on HIV-I infectivity,[234] although this was suggested to be restricted to a CXCR4-tropic HIV-1 isolate. Given that FIV causes a pathology which closely resembles human AIDS, it would be interesting to investigate whether a Nef-like infectivity factor has been developed also within its genome. Intriguingly, ORF-A, a regulatory protein encoded by Feline Immunodeficiency Virus (FIV) which has similarities to Tat, was reported to act as an infectivity factor which promotes FIV infectivity for feline PBMC.[235] To this end, it would be interesting to investigate whether ORF-A, in addition to promoting FIV infectivity by downregulating the FIV receptor (as discussed earlier), functions also as a Nef-like infectivity factor.

4. FINAL REMARKS

In conclusion, while more investigation is needed to understand how the Nef-like factors promote infectivity, we can appreciate at least four common mechanisms by which retroviral factors promote infectivity, based on whether or not the retroviral factor acts in producer or target cells, and whether it functions by facilitating directly the infection process or by contrasting an antiviral host–cell activity which interferes with infection (Fig. 2).

All auxiliary infectivity factors are required to promote or maintain retroviral replication, and therefore spread, *in vivo*. As in most cases these are important pathogenic determinants, the idea to use them as antiviral targets is appealing and has often been advocated, as current antiretroviral drugs control virus replication without eradicating the infection. In the absence of a protective or therapeutic vaccine against HIV, molecules targeting infectivity promoting factors could crucially synergize with the current therapies. However, with the exception of the dUTPases, retroviral auxiliary factors have no known enzymatic activity and drugs targeting them are

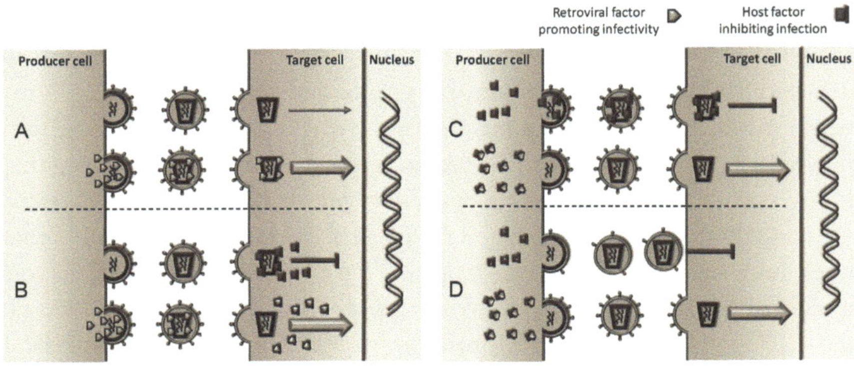

Figure 2 Mechanism of action of retroviral factors promoting infectivity. (A and B) Retroviral infectivity factors actively packaged into virion particles. (A) A packaged retroviral factor acts by facilitating directly early steps of the infection process in the target cell (like the possible effect of Vpr on nuclear import); (B) a packaged retroviral factor acts by counteracting an inhibitor present in target cells (like the counteracting activity of Vpx on SAMHD1 and retroviral dUTPases on dUTP). (C and D) Retroviral infectivity factors that are not actively packaged into virion particles and act in producer cells. (C) A retroviral factor prevents virion incorporation of a host retroviral inhibitor, which acts in target cells (like Vif and Bet which counteract Apobec3); (D) a retroviral factor counteracts a host factor that interferes with the correct biogenesis of the retroviral particles (like Nef and Vpu which counteract CD4 to prevent its potential effect on Env incorporation into virions).

designed to abolish the interaction with host cofactors. This implies a highly detailed knowledge of the factor's interactome (not always available), and a bigger challenge associated with the requirement of inhibiting a protein–protein interaction rather than an enzymatic activity.[236] However, in the last few years promising progress has been reported, with compounds that have been developed against, for example, the infectivity promoting activities of Nef[237,238] and Vif,[239] providing an important proof of principle.

REFERENCES

1. Zhong P, Agosto LM, Munro JB, Mothes W. Cell-to-cell transmission of viruses. *Curr Opin Virol.* 2013;3(1):44–50.
2. Igakura T, Stinchcombe JC, Goon PK, et al. Spread of HTLV-I between lymphocytes by virus-induced polarization of the cytoskeleton. *Science.* 2003;299(5613):1713–1716.
3. Jolly C, Kashefi K, Hollinshead M, Sattentau QJ. HIV-1 cell to cell transfer across an Env-induced, actin-dependent synapse. *J Exp Med.* 2004;199(2):283–293.
4. McDonald D, Wu L, Bohks SM, KewalRamani VN, Unutmaz D, Hope TJ. Recruitment of HIV and its receptors to dendritic cell-T cell junctions. *Science.* 2003;300(5623):1295–1297.

5. Sherer NM, Lehmann MJ, Jimenez-Soto LF, Horensavitz C, Pypaert M, Mothes W. Retroviruses can establish filopodial bridges for efficient cell-to-cell transmission. *Nat Cell Biol.* 2007;9(3):310–315.
6. Malbec M, Sourisseau M, Guivel-Benhassine F, et al. HIV-1 Nef promotes the localization of Gag to the cell membrane and facilitates viral cell-to-cell transfer. *Retrovirology.* 2013;10:80.
7. Puigdomenech I, Casartelli N, Porrot F, Schwartz O. SAMHD1 restricts HIV-1 cell-to-cell transmission and limits immune detection in monocyte-derived dendritic cells. *J Virol.* 2013;87(5):2846–2856.
8. Pizzato M, Marlow SA, Blair ED, Takeuchi Y. Initial binding of murine leukemia virus particles to cells does not require specific Env-receptor interaction. *J Virol.* 1999;73(10):8599–8611.
9. Sattentau Q. Avoiding the void: cell-to-cell spread of human viruses. *Nat Rev Microbiol.* 2008;6(11):815–826.
10. Spear M, Guo J, Wu Y. The trinity of the cortical actin in the initiation of HIV-1 infection. *Retrovirology.* 2012;9:45.
11. Mothes W, Boerger AL, Narayan S, Cunningham JM, Young JA. Retroviral entry mediated by receptor priming and low pH triggering of an envelope glycoprotein. *Cell.* 2000;103(4):679–689.
12. Hulme AE, Perez O, Hope TJ. Complementary assays reveal a relationship between HIV-1 uncoating and reverse transcription. *Proc Natl Acad Sci U S A.* 2011;108(24):9975–9980.
13. Yang Y, Fricke T, Diaz-Griffero F. Inhibition of reverse transcriptase activity increases stability of the HIV-1 core. *J Virol.* 2013;87(1):683–687.
14. McDonald D, Vodicka MA, Lucero G, et al. Visualization of the intracellular behavior of HIV in living cells. *J Cell Biol.* 2002;159(3):441–452.
15. Matreyek KA, Engelman A. Viral and cellular requirements for the nuclear entry of retroviral preintegration nucleoprotein complexes. *Viruses.* 2013;5(10):2483–2511.
16. Tristem M, Marshall C, Karpas A, Hill F. Evolution of the primate lentiviruses: evidence from vpx and vpr. *EMBO J.* 1992;11(9):3405–3412.
17. Lang SM, Weeger M, Stahl-Hennig C, et al. Importance of vpr for infection of rhesus monkeys with simian immunodeficiency virus. *J Virol.* 1993;67(2):902–912.
18. Hoch J, Lang SM, Weeger M, et al. vpr deletion mutant of simian immunodeficiency virus induces AIDS in rhesus monkeys. *J Virol.* 1995;69(8):4807–4813.
19. Accola MA, Bukovsky AA, Jones MS, Gottlinger HG. A conserved dileucine-containing motif in p6(gag) governs the particle association of Vpx and Vpr of simian immunodeficiency viruses SIV(mac) and SIV(agm). *J Virol.* 1999;73(12):9992–9999.
20. Selig L, Pages JC, Tanchou V, et al. Interaction with the p6 domain of the gag precursor mediates incorporation into virions of Vpr and Vpx proteins from primate lentiviruses. *J Virol.* 1999;73(1):592–600.
21. Kondo E, Gottlinger HG. A conserved LXXLF sequence is the major determinant in p6gag required for the incorporation of human immunodeficiency virus type 1 Vpr. *J Virol.* 1996;70(1):159–164.
22. Connor RI, Chen BK, Choe S, Landau NR. Vpr is required for efficient replication of human immunodeficiency virus type-1 in mononuclear phagocytes. *Virology.* 1995;206(2):935–944.
23. Subbramanian RA, Kessous-Elbaz A, Lodge R, et al. Human immunodeficiency virus type 1 Vpr is a positive regulator of viral transcription and infectivity in primary human macrophages. *J Exp Med.* 1998;187(7):1103–1111.
24. Vodicka MA, Koepp DM, Silver PA, Emerman M. HIV-1 Vpr interacts with the nuclear transport pathway to promote macrophage infection. *Genes Dev.* 1998;12(2):175–185.

25. Fassati A, Goff SP. Characterization of intracellular reverse transcription complexes of human immunodeficiency virus type 1. *J Virol*. 2001;75(8):3626–3635.
26. Nermut MV, Fassati A. Structural analyses of purified human immunodeficiency virus type 1 intracellular reverse transcription complexes. *J Virol*. 2003;77(15):8196–8206.
27. Farnet CM, Haseltine WA. Determination of viral proteins present in the human immunodeficiency virus type 1 preintegration complex. *J Virol*. 1991;65(4):1910–1915.
28. Bukrinsky MI, Sharova N, McDonald TL, Pushkarskaya T, Tarpley WG, Stevenson M. Association of integrase, matrix, and reverse transcriptase antigens of human immunodeficiency virus type 1 with viral nucleic acids following acute infection. *Proc Natl Acad Sci U S A*. 1993;90(13):6125–6129.
29. Laguette N, Bregnard C, Hue P, et al. Premature activation of the SLX4 complex by Vpr promotes G2/M arrest and escape from innate immune sensing. *Cell*. 2014;156(1–2):134–145.
30. Miller MD, Farnet CM, Bushman FD. Human immunodeficiency virus type 1 preintegration complexes: studies of organization and composition. *J Virol*. 1997;71(7): 5382–5390.
31. Bukrinsky MI, Haggerty S, Dempsey MP, et al. A nuclear localization signal within HIV-1 matrix protein that governs infection of non-dividing cells. *Nature*. 1993;365(6447):666–669.
32. Heinzinger NK, Bukinsky MI, Haggerty SA, et al. The Vpr protein of human immunodeficiency virus type 1 influences nuclear localization of viral nucleic acids in non-dividing host cells. *Proc Natl Acad Sci U S A*. 1994;91(15):7311–7315.
33. Jenkins Y, McEntee M, Weis K, Greene WC. Characterization of HIV-1 vpr nuclear import: analysis of signals and pathways. *J Cell Biol*. 1998;143(4):875–885.
34. Le Rouzic E, Mousnier A, Rustum C, et al. Docking of HIV-1 Vpr to the nuclear envelope is mediated by the interaction with the nucleoporin hCG1. *J Biol Chem*. 2002;277(47):45091–45098.
35. Sherman MP, de Noronha CM, Heusch MI, Greene S, Greene WC. Nucleocytoplasmic shuttling by human immunodeficiency virus type 1 Vpr. *J Virol*. 2001;75(3):1522–1532.
36. Di Marzio P, Choe S, Ebright M, Knoblauch R, Landau NR. Mutational analysis of cell cycle arrest, nuclear localization and virion packaging of human immunodeficiency virus type 1 Vpr. *J Virol*. 1995;69(12):7909–7916.
37. Yao XJ, Subbramanian RA, Rougeau N, Boisvert F, Bergeron D, Cohen EA. Mutagenic analysis of human immunodeficiency virus type 1 Vpr: role of a predicted N-terminal alpha-helical structure in Vpr nuclear localization and virion incorporation. *J Virol*. 1995;69(11):7032–7044.
38. Gallay P, Stitt V, Mundy C, Oettinger M, Trono D. Role of the karyopherin pathway in human immunodeficiency virus type 1 nuclear import. *J Virol*. 1996;70(2):1027–1032.
39. Fouchier RA, Meyer BE, Simon JH, et al. Interaction of the human immunodeficiency virus type 1 Vpr protein with the nuclear pore complex. *J Virol*. 1998;72(7):6004–6013.
40. Popov S, Rexach M, Ratner L, Blobel G, Bukrinsky M. Viral protein R regulates docking of the HIV-1 preintegration complex to the nuclear pore complex. *J Biol Chem*. 1998;273(21):13347–13352.
41. Nitahara-Kasahara Y, Kamata M, Yamamoto T, et al. Novel nuclear import of Vpr promoted by importin alpha is crucial for human immunodeficiency virus type 1 replication in macrophages. *J Virol*. 2007;81(10):5284–5293.
42. Yamashita M, Emerman M. The cell cycle independence of HIV infections is not determined by known karyophilic viral elements. *PLoS Pathog*. 2005;1(3):e18.
43. Sire J, Querat G, Esnault C, Priet S. Uracil within DNA: an actor of antiviral immunity. *Retrovirology*. 2008;5:45.

44. Mashiba M, Collins KL. Molecular mechanisms of HIV immune evasion of the innate immune response in myeloid cells. *Viruses*. 2013;5(1):1–14.
45. Elder JH, Lerner DL, Hasselkus-Light CS, et al. Distinct subsets of retroviruses encode dUTPase. *J Virol*. 1992;66(3):1791–1794.
46. Payne SL, Elder JH. The role of retroviral dUTPases in replication and virulence. *Curr Protein Pept Sci*. 2001;2(4):381–388.
47. Jern P, Sperber GO, Blomberg J. Use of endogenous retroviral sequences (ERVs) and structural markers for retroviral phylogenetic inference and taxonomy. *Retrovirology*. 2005;2:50.
48. Mansky LM. The mutation rate of human immunodeficiency virus type 1 is influenced by the vpr gene. *Virology*. 1996;222(2):391–400.
49. Chen R, Le Rouzic E, Kearney JA, Mansky LM, Benichou S. Vpr-mediated incorporation of UNG2 into HIV-1 particles is required to modulate the virus mutation rate and for replication in macrophages. *J Biol Chem*. 2004;279(27):28419–28425.
50. Mansky LM, Preveral S, Selig L, Benarous R, Benichou S. The interaction of vpr with uracil DNA glycosylase modulates the human immunodeficiency virus type 1 In vivo mutation rate. *J Virol*. 2000;74(15):7039–7047.
51. Kennedy EM, Daddacha W, Slater R, et al. Abundant non-canonical dUTP found in primary human macrophages drives its frequent incorporation by HIV-1 reverse transcriptase. *J Biol Chem*. 2011;286(28):25047–25055.
52. Priet S, Gros N, Navarro JM, et al. HIV-1-associated uracil DNA glycosylase activity controls dUTP misincorporation in viral DNA and is essential to the HIV-1 life cycle. *Mol Cell*. 2005;17(4):479–490.
53. Kaiser SM, Emerman M. Uracil DNA glycosylase is dispensable for human immunodeficiency virus type 1 replication and does not contribute to the antiviral effects of the cytidine deaminase Apobec3G. *J Virol*. 2006;80(2):875–882.
54. Yang B, Chen K, Zhang C, Huang S, Zhang H. Virion-associated uracil DNA glycosylase-2 and apurinic/apyrimidinic endonuclease are involved in the degradation of APOBEC3G-edited nascent HIV-1 DNA. *J Biol Chem*. 2007;282(16):11667–11675.
55. Eldin P, Chazal N, Fenard D, Bernard E, Guichou JF, Briant L. Vpr expression abolishes the capacity of HIV-1 infected cells to repair uracilated DNA. *Nucl Acids Res*. 2014;42(3):1698–1710.
56. Ahn J, Vu T, Novince Z, Guerrero-Santoro J, Rapic-Otrin V, Gronenborn AM. HIV-1 Vpr loads uracil DNA glycosylase-2 onto DCAF1, a substrate recognition subunit of a cullin 4A-ring E3 ubiquitin ligase for proteasome-dependent degradation. *J Biol Chem*. 2010;285(48):37333–37341.
57. Gu Y, Sundquist WI. Good to CU. *Nature*. 2003;424(6944):21–22.
58. Schrofelbauer B, Yu Q, Zeitlin SG, Landau NR. Human immunodeficiency virus type 1 Vpr induces the degradation of the UNG and SMUG uracil-DNA glycosylases. *J Virol*. 2005;79(17):10978–10987.
59. Elson HF, Dimitrov DS, Blumenthal R. A trans-dominant mutation in human immunodeficiency virus type 1 (HIV- 1) envelope glycoprotein gp41 inhibits membrane fusion when expressed in target cells. *Mol Membr Biol*. 1994;11(3):165–169.
60. Priet S, Navarro JM, Gros N, Querat G, Sire J. Functional role of HIV-1 virion-associated uracil DNA glycosylase 2 in the correction of G:U mispairs to G:C pairs. *J Biol Chem*. 2003;278(7):4566–4571.
61. Langlois MA, Neuberger MS. Human APOBEC3G can restrict retroviral infection in avian cells and acts independently of both UNG and SMUG1. *J Virol*. 2008;82(9):4660–4664.
62. Franchini G, Rusche JR, O'Keeffe TJ, Wong-Staal F. The human immunodeficiency virus type 2 (HIV-2) contains a novel gene encoding a 16 kD protein associated with mature virions. *AIDS Res Hum Retroviruses*. 1988;4(4):243–250.

63. Henderson LE, Sowder RC, Copeland TD, Benveniste RE, Oroszlan S. Isolation and characterization of a novel protein (X-ORF product) from SIV and HIV-2. *Science.* 1988;241(4862):199–201.
64. Kappes JC, Morrow CD, Lee SW, et al. Identification of a novel retroviral gene unique to human immunodeficiency virus type 2 and simian immunodeficiency virus SIVMAC. *J Virol.* 1988;62(9):3501–3505.
65. Beer BE, Foley BT, Kuiken CL, et al. Characterization of novel simian immunodeficiency viruses from red-capped mangabeys from Nigeria (SIVrcmNG409 and -NG411). *J Virol.* 2001;75(24):12014–12027.
66. Hu J, Switzer WM, Foley BT, et al. Characterization and comparison of recombinant simian immunodeficiency virus from drill (Mandrillus leucophaeus) and mandrill (Mandrillus sphinx) isolates. *J Virol.* 2003;77(8):4867–4880.
67. Tristem M, Marshall C, Karpas A, Petrik J, Hill F. Origin of vpx in lentiviruses. *Nature.* 1990;347(6291):341–342.
68. Sharp PM, Bailes E, Stevenson M, Emerman M, Hahn BH. Gene acquisition in HIV and SIV. *Nature.* 1996;383(6601):586–587.
69. Mahalingam S, Van Tine B, Santiago ML, Gao F, Shaw GM, Hahn BH. Functional analysis of the simian immunodeficiency virus Vpx protein: identification of packaging determinants and a novel nuclear targeting domain. *J Virol.* 2001;75(1):362–374.
70. Wu X, Conway JA, Kim J, Kappes JC. Localization of the Vpx packaging signal within the C terminus of the human immunodeficiency virus type 2 Gag precursor protein. *J Virol.* 1994;68(10):6161–6169.
71. Accola MA, Ohagen A, Gottlinger HG. Isolation of human immunodeficiency virus type 1 cores: retention of Vpr in the absence of p6(gag). *J Virol.* 2000;74(13):6198–6202.
72. Guyader M, Emerman M, Montagnier L, Peden K. VPX mutants of HIV-2 are infectious in established cell lines but display a severe defect in peripheral blood lymphocytes. *EMBO J.* 1989;8(4):1169–1175.
73. Yu XF, Yu QC, Essex M, Lee TH. The vpx gene of simian immunodeficiency virus facilitates efficient viral replication in fresh lymphocytes and macrophage. *J Virol.* 1991;65(9):5088–5091.
74. Kappes JC, Conway JA, Lee SW, Shaw GM, Hahn BH. Human immunodeficiency virus type 2 vpx protein augments viral infectivity. *Virology.* 1991;184(1):197–209.
75. Mangeot PE, Duperrier K, Negre D, et al. High levels of transduction of human dendritic cells with optimized SIV vectors. *Mol Ther.* 2002;5(3):283–290.
76. Goujon C, Jarrosson-Wuilleme L, Bernaud J, Rigal D, Darlix JL, Cimarelli A. With a little help from a friend: increasing HIV transduction of monocyte-derived dendritic cells with virion-like particles of SIV(MAC). *Gene Ther.* 2006;13(12):991–994.
77. Sharova N, Wu Y, Zhu X, et al. Primate lentiviral Vpx commandeers DDB1 to counteract a macrophage restriction. *PLoS Pathog.* 2008;4(5):e1000057.
78. Hrecka K, Hao C, Gierszewska M, et al. Vpx relieves inhibition of HIV-1 infection of macrophages mediated by the SAMHD1 protein. *Nature.* 2011;474(7353):658–661.
79. Laguette N, Sobhian B, Casartelli N, et al. SAMHD1 is the dendritic- and myeloid-cell-specific HIV-1 restriction factor counteracted by Vpx. *Nature.* 2011;474(7353):654–657.
80. Baldauf HM, Pan X, Erikson E, et al. SAMHD1 restricts HIV-1 infection in resting CD4(+) T cells. *Nat Med.* 2012;18(11):1682–1687.
81. Goldstone DC, Ennis-Adeniran V, Hedden JJ, et al. HIV-1 restriction factor SAMHD1 is a deoxynucleoside triphosphate triphosphohydrolase. *Nature.* 2011;480(7377):379–382.
82. Lahouassa H, Daddacha W, Hofmann H, et al. SAMHD1 restricts the replication of human immunodeficiency virus type 1 by depleting the intracellular pool of deoxynucleoside triphosphates. *Nat Immunol.* 2012;13(3):223–228.

83. Cribier A, Descours B, Valadao AL, Laguette N, Benkirane M. Phosphorylation of SAMHD1 by cyclin A2/CDK1 regulates its restriction activity toward HIV-1. *Cell Reports*. 2013;3(4):1036–1043.
84. White TE, Brandariz-Nunez A, Valle-Casuso JC, et al. The retroviral restriction ability of SAMHD1, but not its deoxynucleotide triphosphohydrolase activity, is regulated by phosphorylation. *Cell Host Microbe*. 2013;13(4):441–451.
85. Beloglazova N, Flick R, Tchigvintsev A, et al. Nuclease activity of the human SAMHD1 protein implicated in the Aicardi-Goutieres syndrome and HIV-1 restriction. *J Biol Chem*. 2013;288(12):8101–8110.
86. Le Rouzic E, Belaidouni N, Estrabaud E, et al. HIV1 Vpr arrests the cell cycle by recruiting DCAF1/VprBP, a receptor of the Cul4-DDB1 ubiquitin ligase. *Cell Cycle*. 2007;6(2):182–188.
87. Srivastava S, Swanson SK, Manel N, Florens L, Washburn MP, Skowronski J. Lentiviral Vpx accessory factor targets VprBP/DCAF1 substrate adaptor for cullin 4 E3 ubiquitin ligase to enable macrophage infection. *PLoS Pathog*. 2008;4(5):e1000059.
88. Ahn J, Hao C, Yan J, et al. HIV/simian immunodeficiency virus (SIV) accessory virulence factor Vpx loads the host cell restriction factor SAMHD1 onto the E3 ubiquitin ligase complex CRL4DCAF1. *J Biol Chem*. 2012;287(15):12550–12558.
89. Schwefel D, Groom HC, Boucherit VC, et al. Structural basis of lentiviral subversion of a cellular protein degradation pathway. *Nature*. 2014;505(7482):234–238.
90. Gramberg T, Kahle T, Bloch N, et al. Restriction of diverse retroviruses by SAMHD1. *Retrovirology*. 2013;10:26.
91. Pertel T, Reinhard C, Luban J. Vpx rescues HIV-1 transduction of dendritic cells from the antiviral state established by type 1 interferon. *Retrovirology*. 2011;8:49.
92. Goujon C, Arfi V, Pertel T, et al. Characterization of simian immunodeficiency virus SIVSM/human immunodeficiency virus type 2 Vpx function in human myeloid cells. *J Virol*. 2008;82(24):12335–12345.
93. Reinhard C, Bottinelli D, Kim B, Luban J. Vpx rescue of HIV-1 from the antiviral state in mature dendritic cells is independent of the intracellular deoxynucleotide concentration. *Retrovirology*. 2014;11:12.
94. Belshan M, Mahnke LA, Ratner L. Conserved amino acids of the human immunodeficiency virus type 2 Vpx nuclear localization signal are critical for nuclear targeting of the viral preintegration complex in non-dividing cells. *Virology*. 2006;346(1):118–126.
95. Pancio HA, Vander Heyden N, Ratner L. The C-terminal proline-rich tail of human immunodeficiency virus type 2 Vpx is necessary for nuclear localization of the viral preintegration complex in nondividing cells. *J Virol*. 2000;74(13):6162–6167.
96. Oberste MS, Gonda MA. Conservation of amino-acid sequence motifs in lentivirus Vif proteins. *Virus Genes*. 1992;6(1):95–102.
97. Lee TH, Coligan JE, Allan JS, McLane MF, Groopman JE, Essex M. A new HTLV-III/LAV protein encoded by a gene found in cytopathic retroviruses. *Science*. 1986;231(4745):1546–1549.
98. Strebel K, Daugherty D, Clouse K, Cohen D, Folks T, Martin MA. The HIV 'A' (sor) gene product is essential for virus infectivity. *Nature*. 1987;328(6132):728–730.
99. Gabuzda DH, Lawrence K, Langhoff E, et al. Role of vif in replication of human immunodeficiency virus type 1 in CD4+ T lymphocytes. *J Virol*. 1992;66(11):6489–6495.
100. Gabuzda DH, Li H, Lawrence K, Vasir BS, Crawford K, Langhoff E. Essential role of vif in establishing productive HIV-1 infection in peripheral blood T lymphocytes and monocyte/macrophages. *J Acquir Immune Defic Syndr*. 1994;7(9):908–915.
101. Simon JH, Gaddis NC, Fouchier RA, Malim MH. Evidence for a newly discovered cellular anti-HIV-1 phenotype. *Nat Med*. 1998;4(12):1397–1400.

102. Sheehy AM, Gaddis NC, Choi JD, Malim MH. Isolation of a human gene that inhibits HIV-1 infection and is suppressed by the viral Vif protein. *Nature*. 2002;418 (6898):646–650.
103. Harris RS, Bishop KN, Sheehy AM, et al. DNA deamination mediates innate immunity to retroviral infection. *Cell*. 2003;113(6):803–809.
104. Hultquist JF, Lengyel JA, Refsland EW, et al. Human and rhesus APOBEC3D, APOBEC3F, APOBEC3G, and APOBEC3H demonstrate a conserved capacity to restrict Vif-deficient HIV-1. *J Virol*. 2011;85(21):11220–11234.
105. Newman EN, Holmes RK, Craig HM, et al. Antiviral function of APOBEC3G can be dissociated from cytidine deaminase activity. *Curr Biol*. 2005;15(2):166–170.
106. Bishop KN, Verma M, Kim EY, Wolinsky SM, Malim MH. APOBEC3G inhibits elongation of HIV-1 reverse transcripts. *PLoS Pathog*. 2008;4(12):e1000231.
107. Iwatani Y, Chan DS, Wang F, et al. Deaminase-independent inhibition of HIV-1 reverse transcription by APOBEC3G. *Nucl Acids Res*. 2007;35(21):7096–7108.
108. Guo F, Cen S, Niu M, Saadatmand J, Kleiman L. Inhibition of formula-primed reverse transcription by human APOBEC3G during human immunodeficiency virus type 1 replication. *J Virol*. 2006;80(23):11710–11722.
109. Mbisa JL, Barr R, Thomas JA, et al. Human immunodeficiency virus type 1 cDNAs produced in the presence of APOBEC3G exhibit defects in plus-strand DNA transfer and integration. *J Virol*. 2007;81(13):7099–7110.
110. Esnault C, Heidmann O, Delebecque F, et al. APOBEC3G cytidine deaminase inhibits retrotransposition of endogenous retroviruses. *Nature*. 2005;433(7024):430–433.
111. Wiegand HL, Cullen BR. Inhibition of alpharetrovirus replication by a range of human APOBEC3 proteins. *J Virol*. 2007;81(24):13694–13699.
112. Okeoma CM, Lovsin N, Peterlin BM, Ross SR. APOBEC3 inhibits mouse mammary tumour virus replication in vivo. *Nature*. 2007;445(7130):927–930.
113. Sasada A, Takaori-Kondo A, Shirakawa K, et al. APOBEC3G targets human T-cell leukemia virus type 1. *Retrovirology*. 2005;2:32.
114. Doehle BP, Schafer A, Wiegand HL, Bogerd HP, Cullen BR. Differential sensitivity of murine leukemia virus to APOBEC3-mediated inhibition is governed by virion exclusion. *J Virol*. 2005;79(13):8201–8207.
115. Delebecque F, Suspene R, Calattini S, et al. Restriction of foamy viruses by APOBEC cytidine deaminases. *J Virol*. 2006;80(2):605–614.
116. Zennou V, Perez-Caballero D, Gottlinger H, Bieniasz PD. APOBEC3G incorporation into human immunodeficiency virus type 1 particles. *J Virol*. 2004;78(21):12058–12061.
117. Khan MA, Kao S, Miyagi E, et al. Viral RNA is required for the association of APOBEC3G with human immunodeficiency virus type 1 nucleoprotein complexes. *J Virol*. 2005;79(9):5870–5874.
118. Huthoff H, Autore F, Gallois-Montbrun S, Fraternali F, Malim MH. RNA-dependent oligomerization of APOBEC3G is required for restriction of HIV-1. *PLoS Pathog*. 2009;5(3):e1000330.
119. Sheehy AM, Gaddis NC, Malim MH. The antiretroviral enzyme APOBEC3G is degraded by the proteasome in response to HIV-1 Vif. *Nat Med*. 2003;9(11):1404–1407.
120. Yu X, Yu Y, Liu B, et al. Induction of APOBEC3G ubiquitination and degradation by an HIV-1 Vif-Cul5-SCF complex. *Science*. 2003;302(5647):1056–1060.
121. Yu Y, Xiao Z, Ehrlich ES, Yu X, Yu XF. Selective assembly of HIV-1 Vif-Cul5-ElonginB-ElonginC E3 ubiquitin ligase complex through a novel SOCS box and upstream cysteines. *Genes Dev*. 2004;18(23):2867–2872.
122. Mehle A, Strack B, Ancuta P, Zhang C, McPike M, Gabuzda D. Vif overcomes the innate antiviral activity of APOBEC3G by promoting its degradation in the ubiquitin-proteasome pathway. *J Biol Chem*. 2004;279(9):7792–7798.

123. Jager S, Kim DY, Hultquist JF, et al. Vif hijacks CBF-beta to degrade APOBEC3G and promote HIV-1 infection. *Nature.* 2012;481(7381):371–375.
124. Mercenne G, Bernacchi S, Richer D, et al. HIV-1 Vif binds to APOBEC3G mRNA and inhibits its translation. *Nucl Acids Res.* 2010;38(2):633–646.
125. Stopak K, de Noronha C, Yonemoto W, Greene WC. HIV-1 Vif blocks the antiviral activity of APOBEC3G by impairing both its translation and intracellular stability. *Mol Cell.* 2003;12(3):591–601.
126. LaRue RS, Jonsson SR, Silverstein KA, et al. The artiodactyl APOBEC3 innate immune repertoire shows evidence for a multi-functional domain organization that existed in the ancestor of placental mammals. *BMC Mol Biol.* 2008;9:104.
127. Russell RA, Wiegand HL, Moore MD, Schafer A, McClure MO, Cullen BR. Foamy virus Bet proteins function as novel inhibitors of the APOBEC3 family of innate antiretroviral defense factors. *J Virol.* 2005;79(14):8724–8731.
128. Lochelt M, Romen F, Bastone P, et al. The antiretroviral activity of APOBEC3 is inhibited by the foamy virus accessory Bet protein. *Proc Natl Acad Sci U S A.* 2005;102(22):7982–7987.
129. Jaguva Vasudevan AA, Perkovic M, Bulliard Y, et al. Prototype foamy virus Bet impairs the dimerization and cytosolic solubility of human APOBEC3G. *J Virol.* 2013;87(16):9030–9040.
130. Kolokithas A, Rosenke K, Malik F, et al. The glycosylated Gag protein of a murine leukemia virus inhibits the antiretroviral function of APOBEC3. *J Virol.* 2010;84(20):10933–10936.
131. Stavrou S, Nitta T, Kotla S, et al. Murine leukemia virus glycosylated Gag blocks apolipoprotein B editing complex 3 and cytosolic sensor access to the reverse transcription complex. *Proc Natl Acad Sci U S A.* 2013;110(22):9078–9083.
132. Rulli Jr SJ, Mirro J, Hill SA, et al. Interactions of murine APOBEC3 and human APOBEC3G with murine leukemia viruses. *J Virol.* 2008;82(13):6566–6575.
133. Buonocore L, Rose JK. Blockade of human immunodeficiency virus type 1 production in CD4+ T cells by an intracellular CD4 expressed under control of the viral long terminal repeat. *Proc Natl Acad Sci U S A.* 1993;90(7):2695–2699.
134. Crise B, Buonocore L, Rose JK. CD4 is retained in the endoplasmic reticulum by the human immunodeficiency virus type 1 glycoprotein precursor. *J Virol.* 1990; 64(11):5585–5593.
135. Jabbar MA, Nayak DP. Intracellular interaction of human immunodeficiency virus type 1 (ARV-2) envelope glycoprotein gp160 with CD4 blocks the movement and maturation of CD4 to the plasma membrane. *J Virol.* 1990;64(12):6297–6304.
136. Lama J, Mangasarian A, Trono D. Cell-surface expression of CD4 reduces HIV-1 infectivity by blocking Env incorporation in a Nef- and Vpu-inhibitable manner. *Curr Biol.* 1999;9(12):622–631.
137. Tanaka M, Ueno T, Nakahara T, Sasaki K, Ishimoto A, Sakai H. Downregulation of CD4 is required for maintenance of viral infectivity of HIV-1. *Virology.* 2003;311(2):316–325.
138. Lindwasser OW, Chaudhuri R, Bonifacino JS. Mechanisms of CD4 downregulation by the Nef and Vpu proteins of primate immunodeficiency viruses. *Curr Mol Med.* 2007;7(2):171–184.
139. Aiken C, Konner J, Landau NR, Lenburg ME, Trono D. Nef induces CD4 endocytosis: requirement for a critical dileucine motif in the membrane-proximal CD4 cytoplasmic domain. *Cell.* 1994;76(5):853–864.
140. Rhee SS, Marsh JW. Human immunodeficiency virus type 1 Nef-induced downmodulation of CD4 is due to rapid internalization and degradation of surface CD4. *J Virol.* 1994;68(8):5156–5163.

141. Chaudhuri R, Lindwasser OW, Smith WJ, Hurley JH, Bonifacino JS. Downregulation of CD4 by human immunodeficiency virus type 1 Nef is dependent on clathrin and involves direct interaction of Nef with the AP2 clathrin adaptor. *J Virol.* 2007;81(8):3877–3890.
142. daSilva LL, Sougrat R, Burgos PV, Janvier K, Mattera R, Bonifacino JS. Human immunodeficiency virus type 1 Nef protein targets CD4 to the multivesicular body pathway. *J Virol.* 2009;83(13):6578–6590.
143. Margottin F, Bour SP, Durand H, et al. A novel human WD protein, h-beta TrCp, that interacts with HIV-1 Vpu connects CD4 to the ER degradation pathway through an F-box motif. *Mol Cell.* 1998;1(4):565–574.
144. Dube M, Bego MG, Paquay C, Cohen EA. Modulation of HIV-1-host interaction: role of the Vpu accessory protein. *Retrovirology.* 2010;7:114.
145. Hong Y, Fink E, Hu QY, Kiosses WB, Elder JH. OrfA downregulates feline immunodeficiency virus primary receptor CD134 on the host cell surface and is important in viral infection. *J Virol.* 2010;84(14):7225–7232.
146. Arya SK, Gallo RC. Three novel genes of human T-lymphotropic virus type III: immune reactivity of their products with sera from acquired immune deficiency syndrome patients. *Proc Natl Acad Sci U S A.* 1986;83(7):2209–2213.
147. Franchini G, Robert-Guroff M, Wong-Staal F, et al. Expression of the protein encoded by the 3′ open reading frame of human T-cell lymphotropic virus type III in bacteria: demonstration of its immunoreactivity with human sera. *Proc Natl Acad Sci U S A.* 1986;83(14):5282–5285.
148. Luciw PA, Cheng-Mayer C, Levy JA. Mutational analysis of the human immunodeficiency virus: the orf-B region down-regulates virus replication. *Proc Natl Acad Sci U S A.* 1987;84(5):1434–1438.
149. Ahmad N, Venkatesan S. Nef protein of HIV-1 is a transcriptional repressor of HIV-1 LTR. *Science.* 1988;241(4872):1481–1485.
150. Niederman TM, Thielan BJ, Ratner L. Human immunodeficiency virus type 1 negative factor is a transcriptional silencer. *Proc Natl Acad Sci U S A.* 1989;86(4):1128–1132.
151. Hammes SR, Dixon EP, Malim MH, Cullen BR, Greene WC. Nef protein of human immunodeficiency virus type 1: evidence against its role as a transcriptional inhibitor. *Proc Natl Acad Sci U S A.* 1989;86(23):9549–9553.
152. Kim S, Ikeuchi K, Byrn R, Groopman J, Baltimore D. Lack of a negative influence on viral growth by the nef gene of human immunodeficiency virus type 1. *Proc Natl Acad Sci U S A.* 1989;86(23):9544–9548.
153. Kestler 3rd HW, Ringler DJ, Mori K, et al. Importance of the nef gene for maintenance of high virus loads and for development of AIDS. *Cell.* 1991;65(4):651–662.
154. Deacon NJ, Tsykin A, Solomon A, et al. Genomic structure of an attenuated quasi species of HIV-1 from a blood transfusion donor and recipients. *Science.* 1995; 270(5238):988–991.
155. Kirchhoff F, Greenough TC, Brettler DB, Sullivan JL, Desrosiers RC. Brief report: absence of intact nef sequences in a long-term survivor with nonprogressive HIV-1 infection. *N Engl J Med.* 1995;332(4):228–232.
156. Terwilliger EF, Langhoff E, Gabuzda D, Zazopoulos E, Haseltine WA. Allelic variation in the effects of the nef gene on replication of human immunodeficiency virus type 1. *Proc Natl Acad Sci U S A.* 1991;88(23):10971–10975.
157. de Ronde A, Klaver B, Keulen W, Smit L, Goudsmit J. Natural HIV-1 NEF accelerates virus replication in primary human lymphocytes. *Virology.* 1992;188(1):391–395.
158. Zazopoulos E, Haseltine WA. Effect of nef alleles on replication of human immunodeficiency virus type 1. *Virology.* 1993;194(1):20–27.

159. Miller MD, Warmerdam MT, Gaston I, Greene WC, Feinberg MB. The human immunodeficiency virus-1 nef gene product: a positive factor for viral infection and replication in primary lymphocytes and macrophages. *J Exp Med*. 1994;179(1):101–113.
160. Spina CA, Kwoh TJ, Chowers MY, Guatelli JC, Richman DD. The importance of nef in the induction of human immunodeficiency virus type 1 replication from primary quiescent CD4 lymphocytes. *J Exp Med*. 1994;179(1):115–123.
161. Arold S, Franken P, Strub MP, et al. The crystal structure of HIV-1 Nef protein bound to the Fyn kinase SH3 domain suggests a role for this complex in altered T cell receptor signaling. *Structure*. 1997;5(10):1361–1372.
162. Grzesiek S, Bax A, Hu JS, et al. Refined solution structure and backbone dynamics of HIV-1 Nef. *Protein Sci*. 1997;6(6):1248–1263.
163. Lee CH, Saksela K, Mirza UA, Chait BT, Kuriyan J. Crystal structure of the conserved core of HIV-1 Nef complexed with a Src family SH3 domain. *Cell*. 1996;85(6):931–942.
164. Grzesiek S, Bax A, Clore GM, et al. The solution structure of HIV-1 Nef reveals an unexpected fold and permits delineation of the binding surface for the SH3 domain of Hck tyrosine protein kinase. *Nat Struct Biol*. 1996;3(4):340–345.
165. Bentham M, Mazaleyrat S, Harris M. Role of myristoylation and N-terminal basic residues in membrane association of the human immunodeficiency virus type 1 Nef protein. *J Gen Virol*. 2006;87(Pt 3):563–571.
166. Kohleisen B, Neumann M, Herrmann R, et al. Cellular localization of Nef expressed in persistently HIV-1-infected low-producer astrocytes. *AIDS*. 1992;6(12):1427–1436.
167. Fujii Y, Otake K, Tashiro M, Adachi A. Soluble Nef antigen of HIV-1 is cytotoxic for human CD4+ T cells. *FEBS Lett*. 1996;393(1):93–96.
168. Greenberg ME, Bronson S, Lock M, Neumann M, Pavlakis GN, Skowronski J. Co-localization of HIV-1 Nef with the AP-2 adaptor protein complex correlates with Nef-induced CD4 down-regulation. *EMBO J*. 1997;16(23):6964–6976.
169. Pandori MW, Fitch NJ, Craig HM, Richman DD, Spina CA, Guatelli JC. Producer-cell modification of human immunodeficiency virus type 1: Nef is a virion protein. *J Virol*. 1996;70(7):4283–4290.
170. Bukovsky AA, Dorfman T, Weimann A, Gottlinger HG. Nef association with human immunodeficiency virus type 1 virions and cleavage by the viral protease. *J Virol*. 1997;71(2):1013–1018.
171. Welker R, Harris M, Cardel B, Krausslich HG. Virion incorporation of human immunodeficiency virus type 1 Nef is mediated by a bipartite membrane-targeting signal: analysis of its role in enhancement of viral infectivity. *J Virol*. 1998;72(11):8833–8840.
172. Welker R, Kottler H, Kalbitzer HR, Krausslich HG. Human immunodeficiency virus type 1 Nef protein is incorporated into virus particles and specifically cleaved by the viral proteinase. *Virology*. 1996;219(1):228–236.
173. Landi A, Iannucci V, Nuffel AV, Meuwissen P, Verhasselt B. One protein to rule them all: modulation of cell surface receptors and molecules by HIV Nef. *Curr HIV Res*. 2011;9(7):496–504.
174. Jia B, Serra-Moreno R, Neidermyer W, et al. Species-specific activity of SIV Nef and HIV-1 Vpu in overcoming restriction by tetherin/BST2. *PLoS Pathog*. 2009;5(5):e1000429.
175. Piguet V, Gu F, Foti M, et al. Nef-induced CD4 degradation: a diacidic-based motif in Nef functions as a lysosomal targeting signal through the binding of beta-COP in endosomes. *Cell*. 1999;97(1):63–73.
176. Piguet V, Wan L, Borel C, et al. HIV-1 Nef protein binds to the cellular protein PACS-1 to downregulate class I major histocompatibility complexes. *Nat Cell Biol*. 2000;2(3):163–167.

177. Lock M, Greenberg ME, Iafrate AJ, et al. Two elements target SIV Nef to the AP-2 clathrin adaptor complex, but only one is required for the induction of CD4 endocytosis. *EMBO J*. 1999;18(10):2722–2733.
178. Baur AS, Sawai ET, Dazin P, Fantl WJ, Cheng-Mayer C, Peterlin BM. HIV-1 Nef leads to inhibition or activation of T cells depending on its intracellular localization. *Immunity*. 1994;1(5):373–384.
179. Schrager JA, Marsh JW. HIV-1 Nef increases T cell activation in a stimulus-dependent manner. *Proc Natl Acad Sci U S A*. 1999;96(14):8167–8172.
180. Alexander L, Du Z, Rosenzweig M, Jung JU, Desrosiers RC. A role for natural simian immunodeficiency virus and human immunodeficiency virus type 1 nef alleles in lymphocyte activation. *J Virol*. 1997;71(8):6094–6099.
181. Simmons A, Aluvihare V, McMichael A. Nef triggers a transcriptional program in T cells imitating single-signal T cell activation and inducing HIV virulence mediators. *Immunity*. 2001;14(6):763–777.
182. Stolp B, Reichman-Fried M, Abraham L, et al. HIV-1 Nef interferes with host cell motility by deregulation of Cofilin. *Cell Host Microbe*. 2009;6(2):174–186.
183. Chowers MY, Spina CA, Kwoh TJ, Fitch NJ, Richman DD, Guatelli JC. Optimal infectivity in vitro of human immunodeficiency virus type 1 requires an intact nef gene. *J Virol*. 1994;68(5):2906–2914.
184. Miller MD, Warmerdam MT, Page KA, Feinberg MB, Greene WC. Expression of the human immunodeficiency virus type 1 (HIV-1) nef gene during HIV-1 production increases progeny particle infectivity independently of gp160 or viral entry. *J Virol*. 1995;69(1):579–584.
185. Goldsmith MA, Warmerdam MT, Atchison RE, Miller MD, Greene WC. Dissociation of the CD4 downregulation and viral infectivity enhancement functions of human immunodeficiency virus type 1 Nef. *J Virol*. 1995;69(7):4112–4121.
186. Aiken C, Trono D. Nef stimulates human immunodeficiency virus type 1 proviral DNA synthesis. *J Virol*. 1995;69(8):5048–5056.
187. Papkalla A, Munch J, Otto C, Kirchhoff F. Nef enhances human immunodeficiency virus type 1 infectivity and replication independently of viral coreceptor tropism. *J Virol*. 2002;76(16):8455–8459.
188. Khan M, Garcia-Barrio M, Powell MD. Restoration of wild-type infectivity to human immunodeficiency virus type 1 strains lacking nef by intravirion reverse transcription. *J Virol*. 2001;75(24):12081–12087.
189. Tobiume M, Tokunaga K, Kiyokawa E, et al. Requirement of nef for HIV-1 infectivity is biased by the expression levels of Env in the virus-producing cells and CD4 in the target cells. *Arch Virol*. 2001;146(9):1739–1751.
190. Tokunaga K, Kiyokawa E, Nakaya M, et al. Inhibition of human immunodeficiency virus type 1 virion entry by dominant-negative Hck. *J Virol*. 1998;72(7):6257–6259.
191. Pizzato M. MLV glycosylated-Gag is an infectivity factor that rescues Nef-deficient HIV-1. *Proc Natl Acad Sci U S A*. 2010;107(20):9364–9369.
192. Akari H, Arold S, Fukumori T, Okazaki T, Strebel K, Adachi A. Nef-induced major histocompatibility complex class I down-regulation is functionally dissociated from its virion incorporation, enhancement of viral infectivity, and CD4 down-regulation. *J Virol*. 2000;74(6):2907–2912.
193. Fackler OT, Moris A, Tibroni N, et al. Functional characterization of HIV-1 Nef mutants in the context of viral infection. *Virology*. 2006;351(2):322–339.
194. Cortes MJ, Wong-Staal F, Lama J. Cell surface CD4 interferes with the infectivity of HIV-1 particles released from T cells. *J Biol Chem*. 2002;277(3):1770–1779.

195. Arganaraz ER, Schindler M, Kirchhoff F, Cortes MJ, Lama J. Enhanced CD4 down-modulation by late stage HIV-1 nef alleles is associated with increased Env incorporation and viral replication. *J Biol Chem*. 2003;278(36):33912–33919.
196. Aiken C. Pseudotyping human immunodeficiency virus type 1 (HIV-1) by the glycoprotein of vesicular stomatitis virus targets HIV-1 entry to an endocytic pathway and suppresses both the requirement for Nef and the sensitivity to cyclosporin A. *J Virol*. 1997;71(8):5871–5877.
197. Pizzato M, Popova E, Gottlinger HG. Nef can enhance the infectivity of receptor-pseudotyped human immunodeficiency virus type 1 particles. *J Virol*. 2008;82(21): 10811–10819.
198. Munch J, Rajan D, Schindler M, et al. Nef-mediated enhancement of virion infectivity and stimulation of viral replication are fundamental properties of primate lentiviruses. *J Virol*. 2007;81(24):13852–13864.
199. Carl S, Greenough TC, Krumbiegel M, et al. Modulation of different human immunodeficiency virus type 1 Nef functions during progression to AIDS. *J Virol*. 2001;75(8):3657–3665.
200. Lai RP, Yan J, Heeney J, et al. Nef decreases HIV-1 sensitivity to neutralizing antibodies that target the membrane-proximal external region of TMgp41. *PLoS Pathog*. 2011;7(12):e1002442.
201. Schwartz O, Marechal V, Danos O, Heard JM. Human immunodeficiency virus type 1 Nef increases the efficiency of reverse transcription in the infected cell. *J Virol*. 1995;69(7):4053–4059.
202. Fournier C, Cortay JC, Carbonnelle C, Ehresmann C, Marquet R, Boulanger P. The HIV-1 Nef protein enhances the affinity of reverse transcriptase for RNA in vitro. *Virus Genes*. 2002;25(3):255–269.
203. Chazal N, Singer G, Aiken C, Hammarskjold ML, Rekosh D. Human immunodeficiency virus type 1 particles pseudotyped with envelope proteins that fuse at low pH no longer require Nef for optimal infectivity. *J Virol*. 2001;75(8):4014–4018.
204. Luo T, Douglas JL, Livingston RL, Garcia JV. Infectivity enhancement by HIV-1 Nef is dependent on the pathway of virus entry: implications for HIV-based gene transfer systems. *Virology*. 1998;241(2):224–233.
205. Usami Y, Gottlinger H. HIV-1 Nef responsiveness is determined by Env variable regions involved in trimer association and correlates with neutralization sensitivity. *Cell Reports*. 2013;5(3):802–812.
206. Schaeffer E, Geleziunas R, Greene WC. Human immunodeficiency virus type 1 Nef functions at the level of virus entry by enhancing cytoplasmic delivery of virions. *J Virol*. 2001;75(6):2993–3000.
207. Campbell EM, Nunez R, Hope TJ. Disruption of the actin cytoskeleton can complement the ability of Nef to enhance human immunodeficiency virus type 1 infectivity. *J Virol*. 2004;78(11):5745–5755.
208. Cavrois M, Neidleman J, Yonemoto W, Fenard D, Greene WC. HIV-1 virion fusion assay: uncoating not required and no effect of Nef on fusion. *Virology*. 2004; 328(1):36–44.
209. Tobiume M, Lineberger JE, Lundquist CA, Miller MD, Aiken C. Nef does not affect the efficiency of human immunodeficiency virus type 1 fusion with target cells. *J Virol*. 2003;77(19):10645–10650.
210. Miyauchi K, Kim Y, Latinovic O, Morozov V, Melikyan GB. HIV enters cells via endocytosis and dynamin-dependent fusion with endosomes. *Cell*. 2009;137(3):433–444.
211. Qi M, Aiken C. Selective restriction of Nef-defective human immunodeficiency virus type 1 by a proteasome-dependent mechanism. *J Virol*. 2007;81(3):1534–1536.

212. Schwartz O, Marechal V, Friguet B, Arenzana-Seisdedos F, Heard JM. Antiviral activity of the proteasome on incoming human immunodeficiency virus type 1. *J Virol.* 1998;72(5):3845–3850.
213. Chen YL, Trono D, Camaur D. The proteolytic cleavage of human immunodeficiency virus type 1 Nef does not correlate with its ability to stimulate virion infectivity. *J Virol.* 1998;72(4):3178–3184.
214. Miller MD, Warmerdam MT, Ferrell SS, Benitez R, Greene WC. Intravirion generation of the C-terminal core domain of HIV-1 Nef by the HIV-1 protease is insufficient to enhance viral infectivity. *Virology.* 1997;234(2):215–225.
215. Laguette N, Benichou S, Basmaciogullari S. HIV-1 Nef incorporation into virions does not increase infectivity. *J Virol.* 2009;83(2):1093–1104.
216. Mendonca LM, Poeys SC, Abreu CM, Tanuri A, Costa LJ. HIV-1 Nef inhibits protease activity and its absence alters protein content of mature viral particles. *PLoS One.* 2014;9(4):e95352.
217. Schiavoni I, Trapp S, Santarcangelo AC, et al. HIV-1 Nef enhances both membrane expression and virion incorporation of Env products. A model for the Nef-dependent increase of HIV-1 infectivity. *J Biol Chem.* 2004;279(22):22996–23006.
218. Bregnard C, Zamborlini A, Leduc M, et al. Comparative proteomic analysis of HIV-1 particles reveals a role for ezrin and EHD4 in the Nef-dependent increase of virus infectivity. *J Virol.* 2013;87(7):3729–3740.
219. Zheng YH, Plemenitas A, Linnemann T, Fackler OT, Peterlin BM. Nef increases infectivity of HIV via lipid rafts. *Curr Biol.* 2001;11(11):875–879.
220. Zheng YH, Plemenitas A, Fielding CJ, Peterlin BM. Nef increases the synthesis of and transports cholesterol to lipid rafts and HIV-1 progeny virions. *Proc Natl Acad Sci U S A.* 2003;100(14):8460–8465.
221. Brugger B, Krautkramer E, Tibroni N, et al. Human immunodeficiency virus type 1 Nef protein modulates the lipid composition of virions and host cell membrane microdomains. *Retrovirology.* 2007;4:70.
222. Edwards SA, Fan H. gag-Related polyproteins of Moloney murine leukemia virus: evidence for independent synthesis of glycosylated and unglycosylated forms. *J Virol.* 1979;30(2):551–563.
223. Evans LH, Dresler S, Kabat D. Synthesis and glycosylation of polyprotein precursors to the internal core proteins of Friend murine leukemia virus. *J Virol.* 1977;24(3):865–874.
224. Neil JC, Smart JE, Hayman MJ, Jarrett O. Polypeptides of feline leukemia virus: a glycosylated gag-related protein is released into culture fluids. *Virology.* 1980; 105(1):250–253.
225. Prats AC, De Billy G, Wang P, Darlix JL. CUG initiation codon used for the synthesis of a cell surface antigen coded by the murine leukemia virus. *J Mol Biol.* 1989;205(2):363–372.
226. Goff SP, Lobel LI. Mutants of murine leukemia viruses and retroviral replication. *Biochim Biophys Acta.* 1987;907(2):93–123.
227. Chun R, Fan H. Recovery of glycosylated gag virus from mice infected with a glycosylated gag-negative mutant of moloney murine leukemia virus. *J Biomed Sci.* 1994;1(4):218–223.
228. Fan H, Chute H, Chao E, Feuerman M. Construction and characterization of Moloney murine leukemia virus mutants unable to synthesize glycosylated gag polyprotein. *Proc Natl Acad Sci U S A.* 1983;80(19):5965–5969.
229. Corbin A, Prats AC, Darlix JL, Sitbon M. A nonstructural gag-encoded glycoprotein precursor is necessary for efficient spreading and pathogenesis of murine leukemia viruses. *J Virol.* 1994;68(6):3857–3867.
230. Fujisawa R, McAtee FJ, Wehrly K, Portis JL. The neuroinvasiveness of a murine retrovirus is influenced by a dileucine-containing sequence in the cytoplasmic tail of glycosylated Gag. *J Virol.* 1998;72(7):5619–5625.

231. Munk C, Prassolov V, Rodenburg M, Kalinin V, Lohler J, Stocking C. 10A1-MuLV but not the related amphotropic 4070A MuLV is highly neurovirulent: importance of sequences upstream of the structural Gag coding region. *Virology*. 2003;313(1):44–55.
232. Usami Y, Popov S, Gottlinger HG. The Nef-like effect of murine leukemia virus glycosylated gag on HIV-1 infectivity is mediated by its cytoplasmic domain and depends on the AP-2 adaptor complex. *J Virol*. 2014;88(6):3443–3454.
233. Pizzato M, Helander A, Popova E, et al. Dynamin 2 is required for the enhancement of HIV-1 infectivity by Nef. *Proc Natl Acad Sci U S A*. 2007;104(16):6812–6817.
234. Tsukahara T, Ratner L. Substitution of HIV type 1 Nef with HTLV-1 p12. *AIDS Res Hum Retroviruses*. 2004;20(9):938–943.
235. Gemeniano MC, Sawai ET, Leutenegger CM, Sparger EE. Feline immunodeficiency virus ORF-Ais required for virus particle formation and virus infectivity. *J Virol*. 2003;77(16):8819–8830.
236. London N, Raveh B, Schueler-Furman O. Druggable protein-protein interactions—from hot spots to hot segments. *Curr Opin Chem Biol*. 2013;17(6):952–959.
237. Basmaciogullari S, Bouchet J, Chrobak P, et al. Inhibition of the Nef regulatory protein of HIV-1 by a single-domain antibody. *Blood*. 2011;117(13):3559–3568.
238. Breuer S, Schievink SI, Schulte A, Blankenfeldt W, Fackler OT, Geyer M. Molecular design, functional characterization and structural basis of a protein inhibitor against the HIV-1 pathogenicity factor Nef. *PLoS One*. 2011;6(5):e20033.
239. Smith JL, Bu W, Burdick RC, Pathak VK. Multiple ways of targeting APOBEC3-virion infectivity factor interactions for antiHIV-1 drug development. *Trends Pharmacol Sci*. 2009;30(12):638–646.

CHAPTER NINE

The Cytoplasmic Tail of Retroviral Envelope Glycoproteins

Philip R. Tedbury, Eric O. Freed[1]
Virus-Cell Interaction Section, HIV Drug Resistance Program, Center for Cancer Research, National Cancer Institute, Frederick, Maryland, USA
[1]Corresponding author: e-mail address: efreed@nih.gov

Contents

Abstract

Retroviruses comprise a large, diverse group that infects a broad range of host organisms. Pathogenicity varies widely; the human immunodeficiency virus is the causative agent of acquired immunodeficiency syndrome, one of the world's leading infectious causes of death, while many nonhuman retroviruses cause cancer in the host. Retroviruses have been studied intensively, and great strides have been made in understanding aspects of retroviral biology. While the principal functions of the viral structural proteins are well understood, there remain many incompletely characterized domains. One of these is the cytoplasmic tail (CT) of the envelope glycoprotein. Several functions of the CT are highly conserved, whereas other properties are unique to a specific retrovirus. For example, the lentiviruses encode envelope glycoproteins with particularly large cytoplasmic domains. The functions of the long lentiviral envelope CT are still being deciphered. The reported functions of retroviral envelope CTs are discussed in this chapter.

Progress in Molecular Biology and Translational Science, Volume 129
ISSN 1877-1173
http://dx.doi.org/10.1016/bs.pmbts.2014.10.009

1. INTRODUCTION

Retroviruses are ancient, with the genomes of all vertebrates showing evidence for extensive retroviral activity throughout their evolution, even in species that are not known to harbor active retroviruses at present.[1] Retroviruses belong to a larger family of retro-elements that includes long terminal repeat (LTR) retrotransposons, characterized by the presence of the enzymes reverse transcriptase (RT) and integrase (IN), the capsid (CA) protein Gag, and the eponymous LTRs at the 5′ and 3′ ends of the transposon.[2] These retrotransposons are transcribed and translated by host cell machinery, then assemble intracellular particles containing the RNA genome, RT and IN, within a core of Gag. They propagate by reverse-transcribing their RNA genome to DNA and re-entering the nucleus where IN inserts the retrotransposon DNA genome back into the host cell genome. By this "copy and paste" mechanism, retrotransposons have accumulated to high numbers in the genomes of many species. The human genome, for example, comprises 42% retrotransposon sequences.[3] Retro-elements have therefore contributed significantly to the evolution of species, by continually enlarging and shuffling the genomes in which they reside.

Retroviruses can be viewed as a class of retrotransposon that has acquired both the ability to produce extracellular particles and an envelope (*env*) gene that encodes a surface-exposed Env glycoprotein to mediate viral entry into new cells.[4–6] These *env* genes have a variety of potential sources; in many cases, they may have been acquired from other viruses by recombination of the retrotransposon RNA with the RNA of the incoming virus. Whatever their origin, retroviral Env glycoproteins share certain general properties (Fig. 1).[8] They possess an extracellular domain, which is exposed on the outside of the viral particle. This external domain is responsible for binding to receptor proteins on target cells. Receptor binding triggers conformational changes in the Env glycoprotein to bring a second – fusion – domain into play. The fusion apparatus is principally composed of a hydrophobic fusion peptide and two heptad repeats (HR) that fold into a six-helix bundle during the fusion process.[9,10] The concerted action of the fusion peptide and the HRs mediates the fusion of the viral membrane with target cell membrane [typically plasma membrane (PM) or endosomal membrane] during virus entry. The domains that mediate receptor binding and membrane fusion are generally translated as a single polyprotein precursor that is cleaved during trafficking through the Golgi apparatus. The extracellular Env (referred to as the

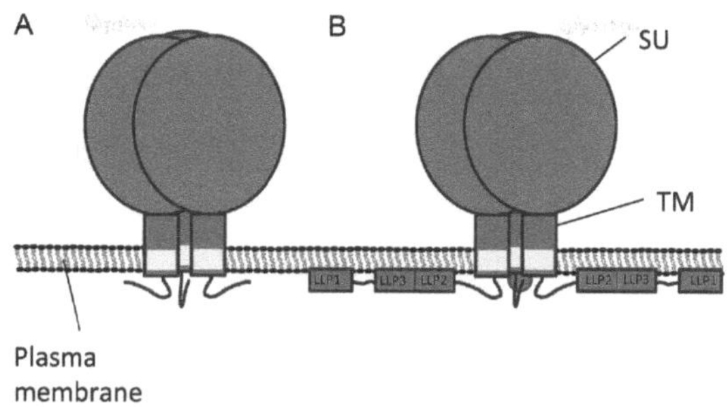

Figure 1 Schematic representation of retroviral Env glycoproteins. (A) Retroviruses express Env glycoproteins with a large, glycosylated soluble subunit [blue (gray in the print version)], oriented to the outside of the cell/virus, and a smaller transmembrane protein [green (gray in the print version)]. In most retroviral families, the transmembrane domain [yellow (light gray in the print version)] is followed by a short cytoplasmic tail of unknown structure. (B) Lentiviral Env glycoproteins have arrangement of their ectodomains as found in other families, but in most cases possess a larger cytoplasmic tail. Shown here is the cytoplasmic tail for HIV-1 Env, with three alpha-helical lentiviral lytic peptides, although their organization *in vivo* is unknown. Other lentiviral cytoplasmic tails may contain similar features, but have not been subject to detailed examination. SU, surface unit; TM, transmembrane subunit; LLP, lentiviral lytic peptide. *Figure adapted with permission of Cell Press from Ref. 7.*

surface, or SU, subunit) and transmembrane (TM) subunit remain covalently or noncovalently associated as heterodimers, which in turn form trimeric Env glycoprotein spikes, in their mature forms. Env is a type I membrane protein, anchored to the membrane by a single transmembrane domain (TMD).[11,12] In addition to functioning as an anchor, studies in HIV-1 and SIV indicate that the TMD influences the function of Env. In particular, a conserved arginine residue and a Gly-X-X-X-Gly (where X is any amino acid) motif are required for efficient membrane fusion[11,13]; the positioning of these features is also important, as the arginine residue is thought to interact with the phosphate head-groups of the phospholipid membrane and the Gly-X-X-X-Gly motif likely mediates interactions between gp41 monomers.[14–16] The final major structural and functional region of the Env glycoprotein is the cytoplasmic tail (CT). Retroviral CTs range in size from short peptides of 20–30 amino acids found in most retroviral genera, to the large CT of the lentiviruses, which can be over 200 amino acids in length (Fig. 2). These domains are typically dispensable for receptor binding and fusion activities, but nevertheless play essential roles in viral replication. In particular, the presence of a CT allows communication with the interior of the host cell and, postrelease, the virus particle, and coordination of functions during assembly, release, and infection of a new cell.

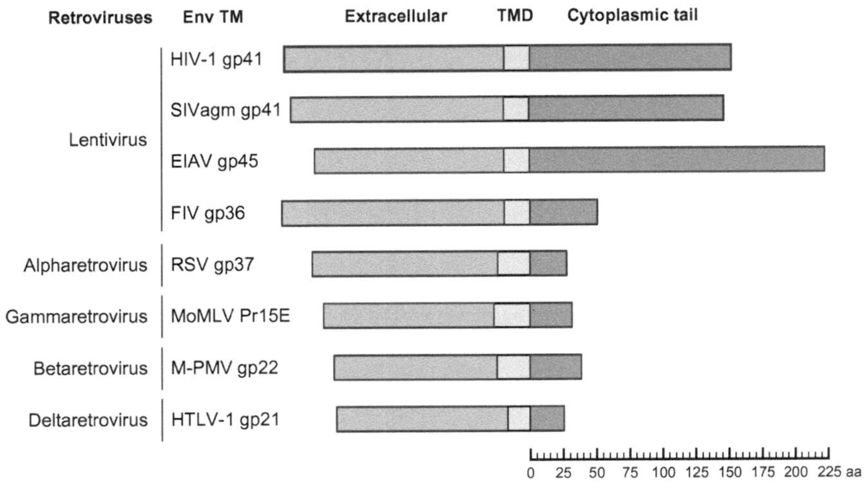

Figure 2 Schematic representation of retroviral TM subunits, indicating the lengths [in amino acids (aa)] of their respective cytoplasmic tails. As illustrated, lentiviral TM CTs tend to be significantly longer than those of other retroviruses, with the exception of feline immunodeficiency virus (FIV). HIV-1, human immunodeficiency virus 1; SIVagm, simian immunodeficiency virus from African green monkey; EIAV, equine infectious anemia virus; RSV, Rous sarcoma virus; MoMLV, Moloney-MLV; M-PMV, Mason-Pfizer monkey virus; HTLV-1, human T-cell lymphotropic virus type 1. *Figure reproduced with permission of Elsevier from Ref. 17.*

The acquisition of an Env glycoprotein and consequent addition of a transmission stage to the viral replication cycle dramatically alters the lifestyle of the virus, as it is no longer confined to the cell in which it was produced or even to the individual organism. The spread of retroviruses can occur through cell-free or cell–cell mechanisms; the latter is often referred to as transmission via a viral synapse.[18] In many cases, cell–cell transmission is far more efficient than cell-free transmission, and some viruses, such as human T-lymphotropic virus (HTLV), appear to spread exclusively through this route.[19–21] The freedom of the virus to spread from the host favors the development of pathogenic viruses, whose replication takes place at the expense of the host cell or organism. It is not surprising that those viruses that adversely affect the health of their host have been the favored subjects for research, particularly when humans are the host species.

A great many retroviruses are now known, infecting a wide variety of species (for detailed listings see Ref. 22 and http://www.ncbi.nlm.nih.gov/retroviruses/). The retrovirus family contains two subfamilies, the orthoretrovirinae and the spumavirinae. The spumavirinae contain a single genus, the spumaviruses, while the orthoretrovirinae contain six genera, the alpha-, beta-, gamma-, delta-, and episilon-retroviruses, and the lentiviruses (Table 1). While all share the basic *gag, pol,* and *env* genomic arrangement, there is considerable variability in the accessory and regulatory proteins, and in replication strategies and pathogenicity.

This review will focus primarily on the functions of the CT of the human immunodeficiency virus (HIV), the causative agent of acquired immunodeficiency syndrome (AIDS), and its close relatives, the simian immunodeficiency viruses (SIVs). Other retroviruses will also be discussed, albeit in less detail, in an attempt to convey the diversity of functions attributed to the CT of retroviral Env proteins.

Table 1 Summary of retroviral taxonomy, including the viruses discussed in this review and other commonly studied retroviruses

Family	Genus	Species
Orthoretrovirinae	Alpharetrovirus	Avian sarcoma leukosis virus Rous sarcoma virus
	Betaretrovirus	Enzootic nasal tumor virus of goats Jaagsiekte sheep retrovirus Mason-Pfizer monkey virus Mouse mammary tumor virus Simian retrovirus 4
	Gammaretrovirus	Feline leukemia virus Gibbon ape leukemia virus Murine leukemia virus Spleen focus-forming virus Xenotropic murine leukemia virus-related virus
	Deltaretrovirus	Bovine leukemia virus Human T-lymphotropic virus 1, 2, 3, and 4
	Epsilonretrovirus	Walleye dermal sarcoma virus
	Lentivirus	Bovine immunodeficiency virus Equine infectious anemia virus Feline immunodeficiency virus Human immunodeficiency virus 1 and 2 Simian immunodeficiency virus
Spumaretrovirinae	Spumavirus	Bovine foamy virus Feline foamy virus Simian foamy virus

2. RETROVIRAL ASSEMBLY

Retroviral particles are composed of either an RNA (orthoretroviruses) or DNA (spumaviruses) genome, contained in a protein core.[23] This core is surrounded in turn by a lipid membrane, embedded in which are the trimeric Env glycoprotein spikes. The Gag polyprotein is responsible for the formation of the protein core.[24] Domain organization varies, but common features are: (a) a membrane-targeting matrix (MA); (b) a multimerizing CA; (c) a nucleic acid-binding nucleocapsid (NC) in orthoretroviruses or an arginine-rich region in foamy viruses[25]; and (d) one or multiple motifs that recruit the endosomal sorting complex required for transport (ESCRT) proteins to promote virus release. In the pathway exploited by lentiviruses, such as HIV-1, the Gag precursor is translated in the cytoplasm, then forms low-order oligomers through CA–CA interactions. The MA domain mediates Gag binding to the PM.[26] In the case of HIV, membrane targeting involves a highly basic patch of residues in MA that bind specifically to phosphatidylinositol-4,5-bisphosphate, and subsequent exposure of an amino-terminal myristic acid moiety that inserts into the lipid bilayer, anchoring Gag to the membrane.[27–29] The RNA-binding domain of Gag, typically NC, possesses general nucleic acid-binding properties, but preferentially binds the viral genome.[30] In addition to ensuring packaging of the genome within the nascent particle, NC also promotes higher-order multimerization of Gag, which forms an extensive lattice containing several thousand Gag molecules.[31] The central feature of the Gag lattice is the hexameric arrangement of CA.[32–34] Release of the particle from the host cell is achieved through the recruitment of the ESCRT complex by late domains in Gag.[35] Three forms of these domains are known: Pro-[Thr/Ser]-Ala-Pro binds TSG101, a component of ESCRT-I, which is responsible for binding cargo[36–39]; Tyr-Pro-X_n-Leu (where X_n is a variable number of amino acids) binds to the V-domain of ALIX, which in turn recruits ESCRT-III directly[40]; Pro-Pro-Pro-Tyr recruits NEDD4-like ubiquitin ligases, ubiquitination is one of the principal signals used by ESCRT to identify cargo.[41,42]

Following, or concurrent with, release of the virus from the host cell, retroviruses undergo a maturation step wherein the structural proteins of the virus undergo processing by the viral protease (PR), causing the breakdown of the immature Gag lattice and the formation of a mature infectious particle.[8,43–45] In addition to the dramatic reorganization of Gag (from a shell of the complete Gag protein located immediately beneath the lipid membrane, to a smaller core composed of CA and containing the viral genome), maturation can also influence the function of the Env protein, via interactions with the CT of the TM Env subunit, as discussed below.

The viral particle also contains the viral enzymes PR, RT, and IN.[24] Additional accessory proteins may be present, depending on the specific virus. HIV-1 particles, for example, contain Vif and Vpr, and SIV and HIV-2 particles additionally contain Vpx. These accessory proteins often have roles in enhancing the infectivity of the virus by countering host cell defenses; Vif is known to counter the antiviral activity of APOBEC3 proteins,[46–48] and Vpx triggers the degradation of the cellular dNTPase SAMHD1.[49–51]

3. SYNTHESIS AND FUNCTION OF ENV

Retroviral Env glycoproteins are type I membrane proteins, with an extracellular N-terminus and a cytoplasmic C-terminus[52]; they are synthesized as a precursor at the endoplasmic reticulum (ER), and cotranslationally inserted across the membrane. Following synthesis, the precursor organizes as a trimer and undergoes initial glycosylation steps at the ER. The level of glycosylation is highly variable; for example, mouse mammary tumor virus possesses only 4 putative sites for N-linked glycosylation, while HIV-1 has over 30.[8,53,54] Env then traffics through the Golgi apparatus where the mature glycans are added and the precursor is cleaved into the mature forms, the extracellular SU subunit and TM subunit. These mature proteins remain noncovalently associated as heterodimers, which in turn form the homotrimeric Env spikes. Trimerization of Env is driven primarily by motifs in the extracellular portion of the protein; in avian sarcoma leukosis virus (ASLV),[55] murine leukemia virus (MLV),[56] and HIV,[57] truncated, soluble forms of the Env glycoproteins have been shown to form trimers, in the absence of TM-SU cleavage. To obtain structures of a processed, soluble HIV-1 Env trimer, modifications to enhance stability were necessary. Specifically, a disulphide bridge was introduced between residues 501 (in SU) and 605 (in TM), and an Ile to Pro mutation was introduced at residue 559 to improve TM:TM interactions.[58,59] All numbering is based on the HXB2 isolate.

The mature Env subunits have distinct functions, with SU being exposed on the exterior of the virion and possessing the receptor binding functions. The glycosylation of the SU is thought to play roles in avoiding antibody binding,[60,61] promoting correct Env folding,[62] and receptor binding.[63] The binding of receptor (and potentially coreceptor) triggers conformational changes that expose the fusion machinery of TM. The fusion function is conferred by the ecto- and transmembrane-domains of TM, and follows a common mechanism. In HIV-1, -2, and SIV, the critical "fusion peptide" is located at the N-terminus of gp41 (the TM), with two amphipathic HR (HR1 and HR2) C-terminal to the fusion peptide. HR1 and HR2 pack as an antiparallel six-helix bundle within the Env trimer. The structural similarities between gp41 and the influenza fusion protein hemagglutinin 2 suggest a similar spring-loaded mechanism, where the fusion peptide inserts into the PM or endosomal membrane of the target cell, and the energy to promote fusion is derived from the structural changes in the six-helix bundle. Peptides mimicking HR1 can bind to and disrupt the function of the six-helix bundle; this approach yielded the fusion inhibitor enfurvitide. Fusion mediated by retroviral Env glycoproteins often occurs at the PM,[64,65] or following endocytosis and low-pH triggering in the endosome.[66,67] Some retroviruses, like ASLV, fuse in a pH-dependent fashion at the endosomal membrane after receptor priming at the PM.[68]

The most C-terminal domain of Env is the CT of TM. This region of the Env is not required for the primary Env functions of receptor binding and membrane fusion, and some retroviruses with deletions of the CT have been shown to replicate in culture[69–72]; however, the CT has been reported to play a wide variety of regulatory functions in viral biology and in many viruses is required in culture and animal models.[69,70,73–75] The CT is the most variable region of Env in terms of its size, presumably reflecting the diverse roles it plays in different viruses. In most retroviruses, the CT is relatively short, 25–35 amino acids, in keeping with the CT of most known viral envelope proteins.[17] By contrast, the majority of lentiviral CT domains are large, well over 100 amino acids in length. One exception to this rule is the CT of feline immunodeficiency virus (FIV); at around 50 amino acids it is substantially shorter than those of other lentiviruses, although still longer than most other retroviruses. These larger CT domains have greater potential for function and specific structures. The CT of HIV-1 contains three lentiviral lytic peptides (LLP1–3), highly alpha-helical regions that have been shown to interact with membranes.[76–78] The remainder of the CT, approximately 70 amino acids, is of unknown structure. The length of the lentiviral CT also opens the possibility of a more complex topology than the simple cytoplasmic orientation of the short-tailed retroviral Env. A motif in the HIV-1 gp41 CT known as the Kennedy sequence immediately follows the TMD and has been suggested to be exposed outside the membrane, based primarily on detection and neutralization by antibodies.[79–81] Such a topology would be in conflict with studies that reported a single membrane-spanning domain.[82] Two resolutions of this conflict have been proposed. It is possible that the surface exposure of the Kennedy sequence is context dependent, and these Env molecules are found on infected cells and potentially in fusion intermediates, but are not incorporated into or found on cell-free virions[79,80]; alternatively, the detection of surface-exposed regions of the CT may be an artifact of the experimental conditions, as regions have been found to be exposed to antibody detection that should be cytoplasmic in both topology models.[83] It is noteworthy that viral isolates have been described that have acquired PR cleavage sites in the gp41 CT near the Kennedy sequence[84,85] (see below). These cleavage sites would have to be inside the membrane to be accessed by PR, arguing against the three-membrane-pass model for HIV-1 Env topology.

4. FUNCTION OF THE RETROVIRAL ENV CT

4.1. Anterograde trafficking

One function for a cytoplasmic region in any viral Env glycoprotein is to aid in regulating the trafficking of the protein. The vesicular stomatitis virus glycoprotein (VSV-G) has been used extensively as a model protein for ER-to-PM trafficking. The retroviral Env glycoproteins likewise follow an ER-to-PM pathway, and display some specificity of trafficking. Work with several retroviruses has shown that Env traffics preferentially to the basolateral membrane in polarized cells.[86–88] Furthermore, HIV-1 Env appears to be able to direct Gag to the site of virus budding in polarized epithelial cells, in a CT- and MA-dependent manner.[89,90]

The CT also plays a role in determining the site of the PM to which Env localizes. With a few exceptions, such as Rous sarcoma virus (RSV),[91] retroviral Env proteins typically demonstrate a tendency to localize at lipid rafts.[92–96] In MLV, this function has been shown to require palmitylation of conserved cysteine residues in the CT.[93] In contrast, studies of HIV-1 and SIV found that although the CT is palmitylated, this posttranslational modification is not required for raft localization.[94,97,98] There are conflicting reports on the role of palmitylation in the HIV-1 Env CT. The existence of HIV-1 isolates that lack cysteine residues in their CT, and efficient replication of palmitylation-deficient mutants, suggest this modification may not be absolutely required[98,99]; however, it does not rule out a potential contribution to Env function in some isolates.[100] The accumulation of HIV-1 Env in lipid rafts may also be influenced by Gag and other motifs of Env, such as a cholesterol recognition amino acid consensus motif between the membrane-proximal external region and the TMD.[96,101]

In the case of viruses that exploit raft localization of Gag and Env proteins, this colocalization may represent a mechanism to enrich Env at sites of Gag assembly. In HIV-1, a further aspect of this colocalization has been reported; an interaction between the gp41 CT and Rab11-family interacting protein 1c (FIP1c).[102] FIP1c appears to mediate an interaction between HIV-1 and the Rab14 trafficking pathway, which targets proteins toward the PM.[103,104] It is likely that Env is additionally targeted to specific sites on the PM, as deletion of the CT does not prevent Env from trafficking to the PM, but does prevent it from localizing at sites of viral assembly in physiologically relevant cell types.[102,105]

4.2. Retrograde trafficking

The CT contains motifs that regulate endocytosis of Env. HIV and SIV are exposed to the host immune system in part through Env on the PM; Env can be recognized by circulating antibodies and triggers antibody-dependent cell-mediated cell killing. To avoid this scenario, the CTs of HIV and SIV Env contain endocytic signals that promote the rapid internalization of Env, limiting its accumulation at the PM.[106–109] The primary endocytic signals in the HIV and SIV Env CTs are Tyr-X-X-Φ (Tyr712 in HIV-1 and Tyr721 in SIV, Φ is any hydrophobic amino acid) and the C-terminal dileucine.[110–112] Intriguingly, the deletion of Tyr721 and the preceding Gly720 in SIV generates a virus that retains efficient replication in animals, but does not deplete the host CD4+ cells.[113] Passage of the virus yielded a compensatory mutation of Ser727 to proline, which restored the ability to deplete host CD4+ cells.[114]

A Tyr-X-X-Φ motif has also been found to regulate endocytosis of HTLV Env, and its removal was shown to increase both levels of cell-surface Env and Env incorporation into VLPs.[115,116] The endocytic motifs recruit subunits of the clathrin adaptor protein complexes AP1–3, with AP1 and AP2 playing the dominant roles in cells.[111,112,115,117] A second Tyr-X-X-Φ motif has been identified in the HIV-1 CT, based on Tyr763, and able to interact with AP1 and AP3[111]; however, this motif appears to be less efficient at recruiting endocytic machinery than the motif based on Tyr712. The rapid endocytosis of HIV-1 Env may only apply to Env molecules outside sites of assembly, as coexpression of Env with high levels of HIV-1 Gag has been reported to reduce the rate of Env internalization.[109] The presence of the Gag lattice may prevent the endocytic machinery from interacting with the Env CT at sites of particle assembly. If the Gag lattice indeed blocks AP subunits from accessing the endoctyic signals in the Env CT, then the mechanism could be broadly applicable to other viruses.

The HTLV-1 Env CT contains a second trafficking motif, close to Tyr-X-X-Φ: a PDZ-binding motif that appears to play a role in countering the effects of Env endocytosis.[116] Deletion or mutation of the PDZ-binding motif dramatically reduces Env incorporation into VLPs, an effect that can be alleviated by the silencing of AP2 components that interact with the Tyr-X-X-Φ motif. At this time, the identity of the interaction partner(s) for the PDZ-binding motif in the HTLV Env CT is unknown and the precise mechanistic role of the motif is thus unclear. A comparable motif may also exist in the Env CT of RSV.[91] It is likely that the interplay of the two trafficking motifs is important in regulating the abundance of Env at the cell surface, with consequences for incorporation into particles, cell–cell fusion and immune recognition.

Deletion of the SIV Env CT leads to enhanced accumulation of Env on the PM; this increase is further associated with elevated incorporation of Env into viral particles.[106] When cultured in human cells, SIV isolates will rapidly lose their CT to enhance replication.[72] However, the CT is retained during culture in the native host, suggesting that the CT plays a critical role in the natural setting. Indeed, it has been reported that mutation of the Tyr-X-X-Φ motif in SIV did not prevent replication, but did render the virus more susceptible to the host immune response.[113]

The motifs regulating HIV-1 Env endocytosis may additionally play roles in mediating optimal virus fusion and infectivity.[118,119] It has been shown that LLP2 can influence the physical properties of the membrane in the proximity of the CT, which may influence membrane fusion[120]; however, the specific mechanisms and functional relevance of these findings remain unexplained.

4.3. Env packaging into particles

In addition to the trafficking pathways that direct retroviral Env toward particle assembly sites, there is clear evidence that viral Envs cluster with Gag at sites of assembly. Clustering has been observed by scanning electron microscopy and by super-resolution microscopy.[105,121,122] This clustering is seen with a wide variety of Env glycoproteins, including non-retroviral Envs, such as VSV-G and Ebola glycoprotein, suggesting a general property of retroviral assembly sites that is recognized by the Env glycoproteins.[122] These data agree with the general observation that many glycoproteins can pseudotype retroviral particles, nevertheless, there is a degree of specificity, as MLV Env will preferentially package into MLV particles over HIV particles when both are assembling in the same cell.[74,123] Surprisingly, the specificity of particle selectivity does not appear to be driven by an interaction between the Env CT and MA, as MLV Env continues to select MLV particles containing an HIV-1 MA, over HIV-1 particles with MLV MA[123]; instead selection appears to be driven by CA and NC domains.

In the case of HIV-1, the clustering of Env at assembly sites requires the Env CT[105,121]; CT-truncated Env is still observed at the PM, but does not colocalize with Gag. In the 293T cells used for these experiments, CT-truncated Env is still incorporated into particles, despite the lack of colocalization with Gag. However, it is likely that the clustering of Env at sites of assembly (which is also seen in T cells that do require CT for efficient Env packaging) reflects a necessary component of the Env packaging mechanism under physiological conditions. The presence of the CT renders the Env immobile at assembly sites, while the CT-truncated Env retains the ability to diffuse in the membrane, despite the presence of Gag.[121] It is not known whether this phenotype is due to a direct interaction between Gag and Env, or a more indirect trapping, where the large CT is unable to diffuse through the densely packed Gag lattice. The clustering of MLV Env appears to occur through a slightly different mechanism, as the truncation of the CT does not prevent clustering of Env at viral assembly sites.[74] The MLV Env CT is required, however, for the preferential packaging of MLV Env into MLV particles over HIV-1 particles.[74]

While trafficking of Env contributes to its incorporation into particles, there are clearly additional factors that influence Env levels in virions, particularly in the long-tailed lentiviruses whose Env glycoproteins do not efficiently package into the particles of other viruses. In the case of HIV-1, point mutations in MA or small deletions in the gp41 CT, which do not affect known trafficking motifs, can inhibit Env incorporation. Furthermore, these Env incorporation defects can be corrected by modifications to the reciprocal domain; that is, a deletion in the CT can be rescued by a point mutation in MA (observed in culture- and in patient-derived samples), and point mutations in MA can be rescued by deletion of the entire gp41 CT.[124–127] In a similar vein, it has long been observed that although HIV particles can package the short-CT Env of MLV, MLV particles are unable to package HIV Env, unless the CT is removed.[128] A similar phenotype is seen in ASLV, where HIV-1 particles pseudotyped with a CT-truncated Env are far more infectious than those bearing full-length Env.[129] In the case of HIV-1, the reciprocal rescue of MA/Env mutations has been interpreted as genetic evidence for an interaction between these protein domains that is required for Env incorporation. Any form of interaction would likely drive coevolution of the Env CT and MA. Accordingly, incorporation of Env from primary isolates into laboratory strain-derived Gag particles can be relatively inefficient relative to incorporation of the cognate Env.[130] Also, during infection, coevolution of MA has been observed as compensation for a mutant Env CT.[127] However, these phenomena have not been thoroughly explained mechanistically. More recent data additionally point to the importance of the structure of the MA lattice in rescuing or permitting Env incorporation.[131] The long-range structure of MA may be important either to allow a MA–Env interaction through avidity effects, compensating for a hypothetically low-affinity interaction between the two proteins; alternatively, the structure of the MA lattice may be an adaptation to the large CT of lentiviruses, potentially accommodating the tail by relieving steric hindrance between MA and the CT.[7]

A potential role for Env in nucleating assembly, rather than simply packaging into particles has also been reported. In RSV, a correlation was observed between the number of Env spikes on a particle and the completed assembly of the viral core.[132] The mechanism by which Env would achieve this nucleation is unknown; an attractive model would involve the Env CT interacting with Gag to promote a region of high Gag concentration, which in turn would promote core formation.[132] If such a mechanism is at work in RSV, it would likely be found in other retroviruses, although it would perhaps be less likely in those viruses with low virion Env content, such as HIV-1.[133]

4.4. Signaling

The CT domain of HIV-1 gp41, in addition to containing motifs involved in regulating its own trafficking, has been shown to impact other cellular functions through signaling motifs. The most persuasive evidence that the CT influences signaling derives from the finding that the CT activates NF-κB through the canonical pathway, via an interaction with the kinase

TAK1.[134] The same study showed that the conserved CT motif responsible for the interaction in HIV-1 gp41 is also present and functional in SIVmac239 gp41. The critical residues of the signaling motif are also conserved in HIV-2 and a variety of other SIV strains.[134] As the HIV-1 5′LTR contains NF-κB responsive elements, this activation of NF-κB may enhance viral gene expression. Activation of NF-κB may also contribute indirectly to enhancing viral gene expression through activation of T cells harboring the provirus.

A potential NF-κB activating domain has also been identified in the Env CT of bovine leukosis virus (BLV), a deltaretrovirus closely related to HTLV.[135] This domain contains two motifs of Tyr-X-X-Leu/Ile. BLV Env induces activation of B and T cells, as assayed by induction of IL-2 production.[135] The signaling pathway was not investigated in this study, but the similarity between BLV and HIV-1 motifs suggests that the NF-κB pathway may again be involved. In contrast, the CTs of HTLV-1 and simian retrovirus 3 Env are unable to induce activation of NF-κB, suggesting that such signaling is common, but not ubiquitous, in retroviruses.[135]

The betaretroviruses enzootic nasal tumor virus (ENTV-1) and Jaagsiekte sheep retrovirus (JSRV) disrupt cell signaling through perturbation of phosphatidylinositide 3-kinase (PI3K) signaling, a distinct mechanism from that of BLV or HIV-1. JSRV causes lung cancer in infected animals,[136] and induces tumor formation in infected cells.[137] In contrast to most transforming retroviruses that carry an activated oncogene or induce tumors via insertional mutagenesis, the JSRV Env glycoprotein is responsible for malignant transformation, with an SH2-binding motif (Tyr-X-X-Met) in the CT of the TM protein playing a central role.[138,139] ENTV predominantly infects goats, and shares the Env-mediated ability to transform cells in culture.[140] The CTs of both ENTV-1 and JRSV Env contain the Tyr-X-X-Met motif that contributes to the activation of cAkt via PI3K signaling,[141,142] and inhibitors of PI3K signaling have been shown to inhibit transformation by ENTV-1 and JSRV.[143,144] The specific mechanism by which Env drives transformation remains unclear; in a conventional PI3K docking site, the tyrosine residue would be phosphorylated, however, no evidence for phosphorylation or direct interaction with PI3K has been obtained for JSRV Env.[142] In addition, the importance of the CT in relation to other regions of Env varies depending on cell type.[142,145–147]

4.5. Role of the CT of retroviral TM proteins in regulating Env conformation

A number of studies have demonstrated that the CT of retroviral Env proteins influences the conformation of Env sequences on the exterior of the membrane (for review see Ref. 148). This so-called "inside-out" regulation can affect Env-mediated fusion, reactivity with anti-Env antibodies, and sensitivity to antibody-mediated neutralization. To activate the fusion activity of the MLV Env complex, the MLV PR cleaves the Env CT to remove the C-terminal 16 amino acids from the C-terminus of Env. The removal of the C-terminal so-called "R peptide" produces a glycoprotein with greatly increased fusogenic capacity.[149] It should be noted, however, that truncation of the MLV Env CT beyond the R peptide inhibits the fusogenicity of MLV Env; this result was also seen in xenotropic murine leukemia virus-related virus (XMRV).[74,150] The truncated MLV Env glycoprotein displays greater reactivity with antibodies to the SU subunit of the Env complex, consistent with an altered conformation.[151] Early studies with SIV demonstrated that Env truncation likewise altered antibody reactivity to epitopes in the ectodomain of the TM glycoprotein[152] and later work showed that HIV-1 gp41 truncation increased exposure of a CD4-induced epitope in gp120.[153,154] Consistent with a connection between the gp41 CT and conformational changes at the gp120 CD4-binding site, truncation of the tail was found to contribute to CD4 independence of an HIV-1 isolate.[155,156] Point mutations in LLP2 that did not significantly affect virus infectivity were shown to render mutant virions resistant to neutralizing antibodies[157]; truncation of the gp41 CT also increased sensitivity to antibody-mediated neutralization in a cell–cell transfer model.[158]

A role for the Env CT in regulating the conformation and function of the external portions of the glycoprotein complex extends to the betaretrovirus JSRV. Truncation of the JSRV Env CT led to markedly enhanced cell–cell fusion and an increase in the amount of shed SU glycoprotein, suggesting an effect of the truncations on SU/TM interactions.[159] This may be related to the observation in HIV-1, that immature particles are more resistant to SU shedding when treated with soluble CD4.[160] Both observations are likely reflections of the influence of Gag on Env, mediated via the TM CT.

The take-home message from these studies is that mutations (both deletions and point mutations) in the CT of retroviral Env proteins can induce changes in the conformation and activity of the ectodomain of TM and in SU; the mechanism by which this occurs remains to be clarified, but could in some cases relate to altered interaction of the CT with Gag or with the inner leaflet of the cellular/viral membrane.

4.6. Role of Gag and virion maturation in regulating Env function

In multiple retroviral systems, the fusogenicity of the viral Env glycoprotein complex is linked to the state of virion maturation. This regulation, which is presumably in place to limit premature membrane fusion during particle production, is achieved via two main mechanisms: (1) cleavage of the Env CT by the viral PR concomitant with particle release and maturation, and (2) suppression of Env fusogenicity by unprocessed Gag in the immature virion. Cleavage of the CT by PR has been observed with a gammaretrovirus [MLV[73,161,162]]; a betaretrovirus [Mason-Pfizer monkey virus (M-PMV)[163]]; and a lentivirus [equine infectious anemia virus (EIAV)[164]]. This strategy allows the C-terminal end of the CT to be present during assembly to facilitate Env incorporation into virions, and then be removed to activate fusion. The mechanism by which Env CT truncation enhances fusion is not fully understood, but, as mentioned above, a number of studies have demonstrated that such truncations alter the conformation of the ectodomain of the TM protein (e.g., Refs. 151–154).

In the case of HIV-1, which does not normally undergo PR-mediated processing of the gp41 CT, two initial reports observed that the fusogenic activity of the Env complex was markedly lower in immature virions (containing uncleaved Gag) than in mature virions.[165,166] In studies of cell–cell transfer, it was similarly observed that although immature particles can transfer into target cells, they do not fuse until maturation has occurred.[167]

This suppression of Env fusion activity could be overcome by deletion of the gp41 CT, suggesting that an interaction between the long CT and uncleaved Gag was responsible for the fusion defect, perhaps by limiting the mobility of the Env complex in the viral membrane or by constraining the ability of Env to undergo conformational changes necessary to catalyze membrane fusion. Indeed, super-resolution microscopy analysis revealed that Env clustered on the surface of HIV-1 virions in a manner that depended upon the gp41 CT.[168] Env on immature and mature virions was found to react differentially with several antibodies specific for the gp41 ectodomain[169]; this effect was partially alleviated by gp41 CT truncation.

As mentioned above, the HIV-1 Env CT does not typically undergo PR-mediated processing. However, propagation of HIV-1 in the presence of a cholesterol-binding entry inhibitor [amphotericin B methyl ester (AME)] led to the selection of resistant viruses that had acquired mutations in the gp41 CT that resulted in PR-mediated cleavage at the site of the mutations.[84,85] These HIV-1 isolates thus behaved like MLV, M-PMV, and EIAV in that PR cleaved their CTs within the virus particle. This strategy for resistance allowed the CT to be present during assembly to promote Env incorporation into virions, and then be removed to activate fusion activity and confer AME resistance. When selection for AME resistance was performed with SIVmac, which tolerates gp41 CT truncations when virus is propagated in human cells, the virus acquired a stop codon in the gp41 CT rather than a site for PR-mediate CT cleavage.[84] There is also evidence that mutations in the HIV-1 gp41 CT can contribute to PR-inhibitor resistance; however, the mechanism for this resistance is not clear.[170] Because interactions between uncleaved Gag and the gp41 CT suppress Env fusion activity,[165,166] PR inhibitors act at both the level of virus maturation and Env function during entry.[170] One could therefore speculate that mutations in the CT that modulate interactions with Gag could allow escape from the Env-dependent component of PR-inhibitor activity.

The hypothesis that uncleaved Gag regulates Env mobility in the membrane gains support from several recent studies that used super-resolution microscopy to measure the localization of Env at budding sites and on virions. Env was found to cluster at sites of Gag assembly; deletion of the gp41 CT, or mutations in MA that block Env incorporation, resulted in a loss of Env/Gag coclustering.[105] Similarly, the mobility of Env on the cell surface was constrained by Gag in a gp41 CT-dependent manner.[121] The hypothesis that the gp41 CT interacts directly with uncleaved Gag in immature virions is supported by the following observations: removal of the viral membrane of mature virions by detergent treatment results in the loss of Env from the detergent-stripped virions; however, when the same experiment is performed with immature virions (composed of unprocessed Gag) Env is retained following removal of membrane. This retention of Env in the detergent-stripped particles is lost if the gp41 CT is truncated.[171]

4.7. HIV-1 versus SIV gp41 CT

Although, there have been occasional reports of naturally occurring isolates of HIV-1 that bear truncated gp41s,[172,173] the loss of the gp41 CT is most commonly observed with SIV propagated in human cells.[72,174,175] Intriguingly, such SIV isolates lacking their gp41 CTs reacquire full-length CTs when passaged in simian cells *in vitro*[72,174] and in monkeys.[75] These differences may reflect use of alternative trafficking pathways by HIV-1 and SIV Env, and the interaction motifs for these putative anterograde trafficking cofactors may reside in different parts of the CT; however, the basis for this species-specific preference for long versus short gp41 CT in simian versus human cells, respectively, has not been elucidated.

4.8. Role of Env in counteracting BST-2

BST-2 (also known as CD317, HM1.24, and tetherin) is an interferon-induced viral restriction factor, with activity against a wide range of enveloped viruses.[176–179] It is expressed on the PM and causes the accumulation of viral particles, which are trapped at the cell surface after completion of the budding process.[180] The primary mechanism of viral restriction by BST-2 *in vivo* is unclear, but may include direct inhibition of release through tethering, enhanced exposure of virions to the immune system, and signaling through NF-κB.[178,179,181–185] Whatever the mechanism, the potency of BST-2 restriction *in vivo* is illustrated by the fact that a number of viruses have evolved strategies to counteract BST-2.[186] HIV-1 Vpu and SIV Nef proteins have been shown to counter BST-2[187]; however, in the case of HIV-2 and FIV, this function is associated with the Env protein. The ability of HIV-2 Env to enhance virus release in HeLa and T-cell lines was identified before the antiviral role of BST-2 was discovered[188,189]; that activity was mapped to the CT of the HIV-2 Env.[190,191] HIV-2 Env is able to interact with and down-regulate BST-2 from the PM, sequestering it in the *trans*-Golgi network.[192,193] Two specific regions of HIV-2 Env are involved in counteracting BST-2. The membrane-proximal endocytic signaling motif, Gly-Tyr-X-X-Φ, is required for the down-regulation of surface BST-2; however, neither this motif nor any other part of the CT confers specificity for human BST-2.[193] Interestingly, when *nef*-deficient SIV was passaged in a tetherin-expressing cell line, it acquired the ability to counteract BST-2 through mutations in the gp41 CT, mimicking the activity of HIV-2 gp41.[194]

The BST-2-countering function of FIV is less well understood than that of the primate lentiviruses, although like HIV-2, the Env protein confers this activity.[195,196] Unlike HIV-2 Env, FIV Env must be packaged into particles to counter restriction by BST-2.[195] It is not clear which domains of FIV Env are required to counteract BST-2, although it appears that the CT is dispensable.[195] EIAV is also restricted by equine BST-2, when replicating in physiologically relevant cells, and the EIAV Env protein has been shown to interact with equine BST-2 and redistribute it from the PM to intracellular sites.[197] As is the case for FIV, the specific domain of Env responsible for this activity is unknown; however, the mechanism by which EIAV counteracts BST-2 more closely resembles that of HIV-2 than FIV. Like HIV-2 Env, EIAV Env can counter BST-2 without being incorporated into particles, and does not cause BST-2 degradation.[197] It is, therefore, probable that the EIAV Env CT is involved in counteracting BST-2-mediated restriction.

4.9. Env CT as a therapeutic target?

As detailed above, the CT of many retroviruses performs a variety of functions critical to the virus replication cycle; these functions include the "classical" functions of Env (packaging into viral particles and mediating fusion with target cells) as well as functions in regulating the cellular environment through signaling or interaction with cellular factors. The Env CT could then be amenable to targeting for antiviral therapy in conditions such as HIV-AIDS. Unfortunately, at the present time, there is a lack of information regarding the structure of any retroviral CT, particularly in the context of their interactions with the membrane and other proteins. As more information becomes available, it may become possible to target the Env CT to inhibit viral replication and pathogenicity, providing a novel approach to alleviating the problems of viral drug resistance.

5. CONCLUSIONS

All retroviral Env glycoproteins share a common structural arrangement, with the CT being the most variable region. It seems likely that all CTs play a role in the trafficking and incorporation of Env into particles. In addition, for many retroviruses, the Env CT functions in the regulation of Env fusogenicity. In some cases, the CT influences signaling within the cell, typically stimulating pathways that contribute to activating cells that may otherwise remain in a resting, inactive state unfavorable for viral gene expression. In the case of lentiviruses, the large CT poses particular questions. Most viruses, including most retroviruses, have only a small CT; organisms as genetically efficient as viruses do not encode large non-functional domains. Additional functions can be attributed to the large lentiviral CT that are not shared by their smaller-CT relatives, such as the countering of BST-2 by the HIV-2 Env CT. In many cases, however, the activities ascribed to the large lentiviral CT are functions that in theory could also be performed by a small CT. It is unclear whether the additional size is required for specific functions or represents a domain with a greater number of functions. The reduction in replication efficiency by CT-deleted retroviruses in their authentic hosts clearly demonstrates that this region of the virus plays important roles, including some that no doubt have yet to be described.

ACKNOWLEDGMENTS

We thank the members of the Freed laboratory for productive discussions. Work in the Freed laboratory is supported by the Intramural Research Program of the Center for Cancer Research, National Cancer Institute, NIH, and by the Intramural AIDS Targeted Antiviral Program.

REFERENCES

1. Eickbush TH, Jamburuthugoda VK. The diversity of retrotransposons and the properties of their reverse transcriptases. *Virus Res*. 2008;134(1–2):221–234. http://dx.doi.org/10.1016/j.virusres.2007.12.010.
2. Cordaux R, Batzer MA. The impact of retrotransposons on human genome evolution. *Nat Rev Genet*. 2009;10(10):691–703. http://dx.doi.org/10.1038/nrg2640.
3. Lander ES, Linton LM, Birren B, et al. Initial sequencing and analysis of the human genome. *Nature*. 2001;409(6822):860–921. http://dx.doi.org/10.1038/35057062.
4. Lerat E, Capy P. Retrotransposons and retroviruses: analysis of the envelope gene. *Mol Biol Evol*. 1999;16(9):1198–1207.
5. McClure MA. Evolution of retroposons by acquisition or deletion of retrovirus-like genes. *Mol Biol Evol*. 1991;8(6):835–856.
6. Kim FJ, Battini J-L, Manel N, Sitbon M. Emergence of vertebrate retroviruses and envelope capture. *Virology*. 2004;318(1):183–191. http://dx.doi.org/10.1016/j.virol.2003.09.026.
7. Tedbury PR, Freed EO. The role of matrix in HIV-1 envelope glycoprotein incorporation. *Trends Microbiol*. 2014;22(7):372–378. http://dx.doi.org/10.1016/j.tim.2014.04.012.
8. Vogt VM. Virion proteins. In: Coffin JM, Hughes SH, Varmus HE, eds. *Retroviruses*. Cold Spring Harbor, NY: Cold Spring Harbor Laboratory Press; 1997.
9. Hunter E. Membrane fusion. In: Coffin JM, Hughes SH, Varmus HE, eds. *Retroviruses*. Cold Spring Harbor, NY: Cold Spring Harbor Laboratory Press; 1997.
10. Blumenthal R, Durell S, Viard M. HIV entry and envelope glycoprotein-mediated fusion. *J Biol Chem*. 2012;287(49):40841–40849. http://dx.doi.org/10.1074/jbc.R112.406272.
11. Helseth E, Olshevsky U, Gabuzda D, Ardman B, Haseltine W, Sodroski J. Changes in the transmembrane region of the human immunodeficiency virus type 1 gp41 envelope glycoprotein affect membrane fusion. *J Virol*. 1990;64(12):6314–6318.
12. Gabuzda D, Olshevsky U, Bertani P, Haseltine WA, Sodroski J. Identification of membrane anchorage domains of the HIV-1 gp160 envelope glycoprotein precursor. *J Acquir Immune Defic Syndr*. 1991;4(1):34–40.
13. Shang L, Yue L, Hunter E. Role of the membrane-spanning domain of human immunodeficiency virus type 1 envelope glycoprotein in cell–cell fusion and virus infection. *J Virol*. 2008;82(11):5417–5428. http://dx.doi.org/10.1128/JVI.02666-07.

14. Senes A, Engel DE, DeGrado WF. Folding of helical membrane proteins: the role of polar, GxxxG-like and proline motifs. *Curr Opin Struct Biol.* 2004;14(4):465–479. http://dx.doi.org/10.1016/j.sbi.2004.07.007.
15. Yue L, Shang L, Hunter E. Truncation of the membrane-spanning domain of human immunodeficiency virus type 1 envelope glycoprotein defines elements required for fusion, incorporation, and infectivity. *J Virol.* 2009;83(22):11588–11598. http://dx.doi.org/10.1128/JVI.00914-09.
16. Miyauchi K, Curran AR, Long Y, et al. The membrane-spanning domain of gp41 plays a critical role in intracellular trafficking of the HIV envelope protein. *Retrovirology.* 2010;7:95. http://dx.doi.org/10.1186/1742-4690-7-95.
17. Checkley MA, Luttge BG, Freed EO. HIV-1 envelope glycoprotein biosynthesis, trafficking, and incorporation. *J Mol Biol.* 2011;410(4):582–608. http://dx.doi.org/10.1016/j.jmb.2011.04.042.
18. Jolly C, Sattentau QJ. Retroviral spread by induction of virological synapses. *Traffic.* 2004;5(9):643–650. http://dx.doi.org/10.1111/j.1600-0854.2004.00209.x.
19. Bangham CRM. The immune control and cell-to-cell spread of human T-lymphotropic virus type 1. *J Gen Virol.* 2003;84(Pt 12):3177–3189.
20. Majorovits E, Nejmeddine M, Tanaka Y, Taylor GP, Fuller SD, Bangham CRM. Human T-lymphotropic virus-1 visualized at the virological synapse by electron tomography. *PLoS One.* 2008;3(5):e2251. http://dx.doi.org/10.1371/journal.pone.0002251.
21. Yamamoto N, Okada M, Koyanagi Y, Kannagi M, Hinuma Y. Transformation of human leukocytes by cocultivation with an adult T cell leukemia virus producer cell line. *Science.* 1982;217(4561):737–739. http://dx.doi.org/10.1126/science.6980467.
22. Petropoulos C. Retroviral taxonomy, protein structures, sequences, and genetic maps. In: Coffin JM, Hughes SH, Varmus HE, eds. *Retroviruses.* Cold Spring Harbor, NY: Cold Spring Harbor Laboratory Press; 1997.
23. Freed EO, Martin MA. HIVs and their replication. In: Knipe DM, Howley PM, eds. *Field's Virology.* 6th ed. Philadelphia, PA: Lippincott, Williams, and Wilkins; 2013:1502–1560.
24. Sundquist WI, Kräusslich H-G. HIV-1 assembly, budding, and maturation. *Cold Spring Harb Perspect Med.* 2012;2(7):a006924. http://dx.doi.org/10.1101/cshperspect.a006924.
25. Müllers E, Uhlig T, Stirnnagel K, Fiebig U, Zentgraf H, Lindemann D. Novel functions of prototype foamy virus Gag glycine- arginine-rich boxes in reverse transcription and particle morphogenesis. *J Virol.* 2011;85(4):1452–1463. http://dx.doi.org/10.1128/JVI.01731-10.
26. Ono A, Orenstein JM, Freed EO. Role of the Gag matrix domain in targeting human immunodeficiency virus type 1 assembly. *J Virol.* 2000;74(6):2855–2866.
27. Ono A, Ablan SD, Lockett SJ, Nagashima K, Freed EO. Phosphatidylinositol (4,5) bisphosphate regulates HIV-1 Gag targeting to the plasma membrane. *Proc Natl Acad Sci U S A.* 2004;101(41):14889–14894.
28. Saad JS, Miller J, Tai J, Kim A, Ghanam RH, Summers MF. Structural basis for targeting HIV-1 Gag proteins to the plasma membrane for virus assembly. *Proc Natl Acad Sci U S A.* 2006;103(30):11364–11369. http://dx.doi.org/10.1073/pnas.0602818103.
29. Saad JS, Loeliger E, Luncsford P, et al. Point mutations in the HIV-1 matrix protein turn off the myristyl switch. *J Mol Biol.* 2007;366(2):574–585. http://dx.doi.org/10.1016/j.jmb.2006.11.068.
30. D'Souza V, Summers MF. How retroviruses select their genomes. *Nat Rev Microbiol.* 2005;3(8):643–655. http://dx.doi.org/10.1038/nrmicro1210.
31. Rein A, Datta SAK, Jones CP, Musier-Forsyth K. Diverse interactions of retroviral Gag proteins with RNAs. *Trends Biochem Sci.* 2011;36(7):373–380. http://dx.doi.org/10.1016/j.tibs.2011.04.001.

32. Huseby D, Barklis RL, Alfadhli A, Barklis E. Assembly of human immunodeficiency virus precursor gag proteins. *J Biol Chem*. 2005;280(18):17664–17670. http://dx.doi.org/10.1074/jbc.M412325200.
33. Briggs JAG, Kräusslich H-G. The molecular architecture of HIV. *J Mol Biol*. 2011;410(4):491–500. http://dx.doi.org/10.1016/j.jmb.2011.04.021.
34. Pornillos O, Ganser-Pornillos BK, Banumathi S, Hua Y, Yeager M. Disulfide bond stabilization of the hexameric capsomer of human immunodeficiency virus. *J Mol Biol*. 2010;401(5):985–995. http://dx.doi.org/10.1016/j.jmb.2010.06.042.
35. Votteler J, Sundquist WI. Virus budding and the ESCRT pathway. *Cell Host Microbe*. 2013;14(3):232–241. http://dx.doi.org/10.1016/j.chom.2013.08.012.
36. Garrus JE, von Schwedler UK, Pornillos OW, et al. Tsg101 and the vacuolar protein sorting pathway are essential for HIV-1 budding. *Cell*. 2001;107(1):55–65.
37. Martin-Serrano J, Zang T, Bieniasz PD. HIV-1 and Ebola virus encode small peptide motifs that recruit Tsg101 to sites of particle assembly to facilitate egress. *Nat Med*. 2001;7(12):1313–1319. http://dx.doi.org/10.1038/nm1201-1313.
38. VerPlank L, Bouamr F, LaGrassa TJ, et al. Tsg101, a homologue of ubiquitin-conjugating (E2) enzymes, binds the L domain in HIV type 1 Pr55(Gag). *Proc Natl Acad Sci U S A*. 2001;98(14):7724–7729. http://dx.doi.org/10.1073/pnas.131059198.
39. Demirov DG, Ono A, Orenstein JM, Freed EO. Overexpression of the N-terminal domain of TSG101 inhibits HIV-1 budding by blocking late domain function. *Proc Natl Acad Sci U S A*. 2002;99(2):955–960. http://dx.doi.org/10.1073/pnas.032511899.
40. Strack B, Calistri A, Craig S, Popova E, Göttlinger HG. AIP1/ALIX is a binding partner for HIV-1 p6 and EIAV p9 functioning in virus budding. *Cell*. 2003;114(6):689–699.
41. Kikonyogo A, Bouamr F, Vana ML, et al. Proteins related to the Nedd4 family of ubiquitin protein ligases interact with the L domain of Rous sarcoma virus and are required for gag budding from cells. *Proc Natl Acad Sci U S A*. 2001;98(20): 11199–11204. http://dx.doi.org/10.1073/pnas.201268998.
42. Strack B, Calistri A, Accola MA, Palu G, Gottlinger HG. A role for ubiquitin ligase recruitment in retrovirus release. *Proc Natl Acad Sci U S A*. 2000;97(24): 13063–13068. http://dx.doi.org/10.1073/pnas.97.24.13063.
43. Bell NM, Lever AML. HIV Gag polyprotein: processing and early viral particle assembly. *Trends Microbiol*. 2013;21(3):136–144. http://dx.doi.org/10.1016/j.tim.2012.11.006.
44. Müllers E. The foamy virus Gag proteins: what makes them different? *Viruses*. 2013;5(4):1023–1041. http://dx.doi.org/10.3390/v5041023.
45. Keller PW, Huang RK, England MR, et al. A two-pronged structural analysis of retroviral maturation indicates that core formation proceeds by a disassembly-reassembly pathway rather than a displacive transition. *J Virol*. 2013;87(24):13655–13664. http://dx.doi.org/10.1128/JVI.01408-13.
46. Marin M, Rose KM, Kozak SL, Kabat D. HIV-1 Vif protein binds the editing enzyme APOBEC3G and induces its degradation. *Nat Med*. 2003;9(11):1398–1403. http://dx.doi.org/10.1038/nm946.
47. Mangeat B, Turelli P, Caron G, Friedli M, Perrin L, Trono D. Broad antiretroviral defence by human APOBEC3G through lethal editing of nascent reverse transcripts. *Nature*. 2003;424(6944):99–103. http://dx.doi.org/10.1038/nature01709.
48. Goila-Gaur R, Strebel K. HIV-1 Vif, APOBEC, and intrinsic immunity. *Retrovirology*. 2008;5(1):51. http://dx.doi.org/10.1186/1742-4690-5-51.
49. Laguette N, Sobhian B, Casartelli N, et al. SAMHD1 is the dendritic- and myeloid-cell-specific HIV-1 restriction factor counteracted by Vpx. *Nature*. 2011;474(7353): 654–657. http://dx.doi.org/10.1038/nature10117.

50. Hrecka K, Hao C, Gierszewska M, et al. Vpx relieves inhibition of HIV-1 infection of macrophages mediated by the SAMHD1 protein. *Nature.* 2011;474(7353):658–661. http://dx.doi.org/10.1038/nature10195.
51. Schaller T, Bauby H, Hué S, Malim MH, Goujon C. New insights into an X-traordinary viral protein. *Front Microbiol.* 2014;5:126. http://dx.doi.org/10.3389/fmicb.2014.00126.
52. Haffar OK, Dowbenko DJ, Berman PW. Topogenic analysis of the human immunodeficiency virus type 1 envelope glycoprotein, gp160, in microsomal membranes. *J Cell Biol.* 1988;107(5):1677–1687.
53. Felkner RH, Roth MJ. Mutational analysis of the N-linked glycosylation sites of the SU envelope protein of Moloney murine leukemia virus. *J Virol.* 1992;66(7):4258–4264.
54. Dedera DA, Gu RL, Ratner L. Role of asparagine-linked glycosylation in human immunodeficiency virus type 1 transmembrane envelope function. *Virology.* 1992;187(1):377–382.
55. Einfeld D, Hunter E. Oligomeric structure of a prototype retrovirus glycoprotein. *Proc Natl Acad Sci U S A.* 1988;85(22):8688–8692.
56. Tucker SP, Srinivas RV, Compans RW. Molecular domains involved in oligomerization of the Friend murine leukemia virus envelope glycoprotein. *Virology.* 1991;185(2):710–720.
57. Earl PL, Moss B. Mutational analysis of the assembly domain of the HIV-1 envelope glycoprotein. *AIDS Res Hum Retroviruses.* 1993;9(7):589–594.
58. Julien J-P, Cupo A, Sok D, et al. Crystal structure of a soluble cleaved HIV-1 envelope trimer. *Science.* 2013;342(6165):1477–1483. http://dx.doi.org/10.1126/science.1245625.
59. Lyumkis D, Julien J-P, de Val N, et al. Cryo-EM structure of a fully glycosylated soluble cleaved HIV-1 envelope trimer. *Science.* 2013;342(6165):1484–1490. http://dx.doi.org/10.1126/science.1245627.
60. Montefiori DC, Robinson WE, Mitchell WM. Role of protein N-glycosylation in pathogenesis of human immunodeficiency virus type 1. *Proc Natl Acad Sci U S A.* 1988;85(23):9248–9252.
61. Raska M, Takahashi K, Czernekova L, et al. Glycosylation patterns of HIV-1 gp120 depend on the type of expressing cells and affect antibody recognition. *J Biol Chem.* 2010;285(27):20860–20869. http://dx.doi.org/10.1074/jbc.M109.085472.
62. Li H, Chien PC, Tuen M, et al. Identification of an N-linked glycosylation in the C4 region of HIV-1 envelope gp120 that is critical for recognition of neighboring CD4 T cell epitopes. *J Immunol.* 2008;180(6):4011–4021.
63. Raska M, Novak J. Involvement of envelope-glycoprotein glycans in HIV-1 biology and infection. *Arch Immunol Ther Exp (Warsz).* 2010;58(3):191–208. http://dx.doi.org/10.1007/s00005-010-0072-3.
64. Felts RL, Narayan K, Estes JD, et al. 3D visualization of HIV transfer at the virological synapse between dendritic cells and T cells. *Proc Natl Acad Sci U S A.* 2010;107(30):13336–13341. http://dx.doi.org/10.1073/pnas.1003040107.
65. Clavel F, Charneau P. Fusion from without directed by human immunodeficiency virus particles. *J Virol.* 1994;68(2):1179–1185.
66. Miyauchi K, Kim Y, Latinovic O, Morozov V, Melikyan GB. HIV enters cells via endocytosis and dynamin-dependent fusion with endosomes. *Cell.* 2009;137(3):433–444. http://dx.doi.org/10.1016/j.cell.2009.02.046.
67. Sloan RD, Kuhl BD, Mesplède T, Münch J, Donahue DA, Wainberg MA. Productive entry of HIV-1 during cell-to-cell transmission via dynamin-dependent endocytosis. *J Virol.* 2013;87(14):8110–8123. http://dx.doi.org/10.1128/JVI.00815-13.

68. Mothes W, Boerger AL, Narayan S, Cunningham JM, Young JA. Retroviral entry mediated by receptor priming and low pH triggering of an envelope glycoprotein. *Cell.* 2000;103(4):679–689.
69. Murakami T, Freed EO. The long cytoplasmic tail of gp41 is required in a cell type-dependent manner for HIV-1 envelope glycoprotein incorporation into virions. *Proc Natl Acad Sci U S A.* 2000;97(1):343–348.
70. Akari H, Fukumori T, Adachi A. Cell-dependent requirement of human immunodeficiency virus type 1 gp41 cytoplasmic tail for Env incorporation into virions. *J Virol.* 2000;74(10):4891–4893.
71. Perez LG, Davis GL, Hunter E. Mutants of the Rous sarcoma virus envelope glycoprotein that lack the transmembrane anchor and cytoplasmic domains: analysis of intracellular transport and assembly into virions. *J Virol.* 1987;61(10):2981–2988.
72. Hirsch VM, Edmondson P, Murphey-Corb M, Arbeille B, Johnson PR, Mullins JI. SIV adaptation to human cells. *Nature.* 1989;341(6243):573–574. http://dx.doi.org/10.1038/341573a0.
73. Rein A, Mirro J, Haynes JG, Ernst SM, Nagashima K. Function of the cytoplasmic domain of a retroviral transmembrane protein: p15E-p2E cleavage activates the membrane fusion capability of the murine leukemia virus Env protein. *J Virol.* 1994;68(3):1773–1781.
74. Lucas TM, Lyddon TD, Grosse SA, Johnson MC. Two distinct mechanisms regulate recruitment of murine leukemia virus envelope protein to retroviral assembly sites. *Virology.* 2010;405(2):548–555. http://dx.doi.org/10.1016/j.virol.2010.06.017.
75. Shacklett BL, Weber CJ, Shaw KE, et al. The intracytoplasmic domain of the Env transmembrane protein is a locus for attenuation of simian immunodeficiency virus SIVmac in rhesus macaques. *J Virol.* 2000;74(13):5836–5844.
76. Viard M, Ablan SD, Zhou M, et al. Photoinduced reactivity of the HIV-1 envelope glycoprotein with a membrane-embedded probe reveals insertion of portions of the HIV-1 Gp41 cytoplasmic tail into the viral membrane. *Biochemistry.* 2008;47(7):1977–1983. http://dx.doi.org/10.1021/bi701920f.
77. Eisenberg D, Wesson M. The most highly amphiphilic alpha-helices include two amino acid segments in human immunodeficiency virus glycoprotein 41. *Biopolymers.* 1990;29(1):171–177. http://dx.doi.org/10.1002/bip.360290122.
78. Srinivas S, Srinivas R, Anantharamaiah G, Segrest J, Compans R. Membrane interactions of synthetic peptides corresponding to amphipathic helical segments of the human immunodeficiency virus type-1 envelope glycoprotein. *J Biol Chem.* 1992; 267(10):7121–7127.
79. Steckbeck JD, Sun C, Sturgeon TJ, Montelaro RC. Topology of the C-terminal tail of HIV-1 gp41: differential exposure of the Kennedy epitope on cell and viral membranes. *PLoS One.* 2010;5(12):e15261. http://dx.doi.org/10.1371/journal.pone.0015261.
80. Steckbeck JD, Sun C, Sturgeon TJ, Montelaro RC. Detailed topology mapping reveals substantial exposure of the "cytoplasmic" C-terminal tail (CTT) sequences in HIV-1 Env proteins at the cell surface. *PLoS One.* 2013;8(5):e65220. http://dx.doi.org/10.1371/journal.pone.0065220.
81. Kennedy RC, Henkel RD, Pauletti D, et al. Antiserum to a synthetic peptide recognizes the HTLV-III envelope glycoprotein. *Science.* 1986;231(4745):1556–1559.
82. Liu S, Kondo N, Long Y, Xiao D, Iwamoto A, Matsuda Z. Membrane topology analysis of HIV-1 envelope glycoprotein gp41. *Retrovirology.* 2010;7:100. http://dx.doi.org/10.1186/1742-4690-7-100.
83. Postler TS, Martinez-Navio JM, Yuste E, Desrosiers RC. Evidence against extracellular exposure of a highly immunogenic region in the C-terminal domain of the SIVmac gp41 transmembrane protein. *J Virol.* 2012;86(2):1145–1157. http://dx.doi.org/10.1128/JVI.06463-11.

84. Waheed AA, Ablan SD, Roser JD, et al. HIV-1 escape from the entry-inhibiting effects of a cholesterol-binding compound via cleavage of gp41 by the viral protease. *Proc Natl Acad Sci U S A*. 2007;104(20):8467–8471. http://dx.doi.org/10.1073/pnas.0701443104.
85. Waheed AA, Ablan SD, Sowder RC, et al. Effect of mutations in the human immunodeficiency virus type 1 protease on cleavage of the gp41 cytoplasmic tail. *J Virol*. 2010;84(6):3121–3126. http://dx.doi.org/10.1128/JVI.02002-09.
86. Ball JM, Mulligan MJ, Compans RW. Basolateral sorting of the HIV type 2 and SIV envelope glycoproteins in polarized epithelial cells: role of the cytoplasmic domain. *AIDS Res Hum Retroviruses*. 1997;13(8):665–675.
87. Lodge R, Delamarre L, Lalonde JP, et al. Two distinct oncornaviruses harbor an intracytoplasmic tyrosine-based basolateral targeting signal in their viral envelope glycoprotein. *J Virol*. 1997;71(7):5696–5702.
88. Owens RJ, Compans RW. Expression of the human immunodeficiency virus envelope glycoprotein is restricted to basolateral surfaces of polarized epithelial cells. *J Virol*. 1989;63(2):978–982.
89. Owens RJ, Dubay JW, Hunter E, Compans RW. Human immunodeficiency virus envelope protein determines the site of virus release in polarized epithelial cells. *Proc Natl Acad Sci U S A*. 1991;88(9):3987–3991.
90. Lodge R, Göttlinger H, Gabuzda D, Cohen EA, Lemay G. The intracytoplasmic domain of gp41 mediates polarized budding of human immunodeficiency virus type 1 in MDCK cells. *J Virol*. 1994;68(8):4857–4861.
91. Ochsenbauer C, Dubay SR, Hunter E. The Rous sarcoma virus Env glycoprotein contains a highly conserved motif homologous to tyrosine-based endocytosis signals and displays an unusual internalization phenotype. *Mol Cell Biol*. 2000;20(1):249–260. http://dx.doi.org/10.1128/MCB.20.1.249-260.2000.
92. Beer C, Pedersen L, Wirth M. Amphotropic murine leukaemia virus envelope protein is associated with cholesterol-rich microdomains. *Virol J*. 2005;2:36. http://dx.doi.org/10.1186/1743-422X-2-36.
93. Li M, Yang C, Tong S, Weidmann A, Compans RW. Palmitoylation of the murine leukemia virus envelope protein is critical for lipid raft association and surface expression. *J Virol*. 2002;76(23):11845–11852.
94. Vzorov AN, Weidmann A, Kozyr NL, Khaoustov V, Yoffe B, Compans RW. Role of the long cytoplasmic domain of the SIV Env glycoprotein in early and late stages of infection. *Retrovirology*. 2007;4:94. http://dx.doi.org/10.1186/1742-4690-4-94.
95. Nguyen DH, Hildreth JE. Evidence for budding of human immunodeficiency virus type 1 selectively from glycolipid-enriched membrane lipid rafts. *J Virol*. 2000;74(7):3264–3272.
96. Schwarzer R, Levental I, Gramatica A, et al. The cholesterol-binding motif of the HIV-1 glycoprotein gp41 regulates lateral sorting and oligomerization. *Cell Microbiol*. 2014;16(10):1565–1581. http://dx.doi.org/10.1111/cmi.12314.
97. Yang P, Ai L-S, Huang S-C, et al. The cytoplasmic domain of human immunodeficiency virus type 1 transmembrane protein gp41 harbors lipid raft association determinants. *J Virol*. 2010;84(1):59–75. http://dx.doi.org/10.1128/JVI.00899-09.
98. Chan W-E, Lin H-H, Chen SS-L. Wild-type-like viral replication potential of human immunodeficiency virus type 1 envelope mutants lacking palmitoylation signals. *J Virol*. 2005;79(13):8374–8387. http://dx.doi.org/10.1128/JVI.79.13.8374-8387.2005.
99. Bhattacharya J, Peters PJ, Clapham PR. Human immunodeficiency virus type 1 envelope glycoproteins that lack cytoplasmic domain cysteines: impact on association with membrane lipid rafts and incorporation onto budding virus particles. *J Virol*. 2004;78(10):5500–5506.

100. Rousso I, Mixon MB, Chen BK, Kim PS. Palmitoylation of the HIV-1 envelope glycoprotein is critical for viral infectivity. *Proc Natl Acad Sci U S A.* 2000;97(25):13523–13525. http://dx.doi.org/10.1073/pnas.240459697.
101. Bhattacharya J, Repik A, Clapham PR. Gag regulates association of human immunodeficiency virus type 1 envelope with detergent-resistant membranes. *J Virol.* 2006;80(11):5292–5300. http://dx.doi.org/10.1128/JVI.01469-05.
102. Qi M, Williams JA, Chu H, et al. Rab11-FIP1C and Rab14 direct plasma membrane sorting and particle incorporation of the HIV-1 envelope glycoprotein complex. *PLoS Pathog.* 2013;9(4):e1003278. http://dx.doi.org/10.1371/journal.ppat.1003278, Swanstrom R, ed.
103. Linford A, Yoshimura S, Nunes Bastos R, et al. Rab14 and its exchange factor FAM116 link endocytic recycling and adherens junction stability in migrating cells. *Dev Cell.* 2012;22(5):952–966. http://dx.doi.org/10.1016/j.devcel.2012.04.010.
104. Junutula JR, De Maziére AM, Peden AA, et al. Rab14 is involved in membrane trafficking between the Golgi complex and endosomes. *Mol Biol Cell.* 2004;15(5):2218–2229. http://dx.doi.org/10.1091/mbc.E03-10-0777.
105. Muranyi W, Malkusch S, Müller B, Heilemann M, Kräusslich H-G. Super-resolution microscopy reveals specific recruitment of HIV-1 envelope proteins to viral assembly sites dependent on the envelope C-terminal tail. *PLoS Pathog.* 2013;9(2):e1003198. http://dx.doi.org/10.1371/journal.ppat.1003198.
106. LaBranche CC, Sauter MM, Haggarty BS, et al. A single amino acid change in the cytoplasmic domain of the simian immunodeficiency virus transmembrane molecule increases envelope glycoprotein expression on infected cells. *J Virol.* 1995;69(9):5217–5227.
107. Polydefkis M, Koenig S, Flexner C, et al. Anchor sequence-dependent endogenous processing of human immunodeficiency virus 1 envelope glycoprotein gp160 for CD4+ T cell recognition. *J Exp Med.* 1990;171(3):875–887.
108. Callahan KM, Rowell JF, Soloski MJ, Machamer CE, Siliciano RF. HIV-1 envelope protein is expressed on the surface of infected cells before its processing and presentation to class II-restricted T lymphocytes. *J Immunol.* 1993;151(6):2928–2942.
109. Egan MA, Carruth LM, Rowell JF, Yu X, Siliciano RF. Human immunodeficiency virus type 1 envelope protein endocytosis mediated by a highly conserved intrinsic internalization signal in the cytoplasmic domain of gp41 is suppressed in the presence of the Pr55gag precursor protein. *J Virol.* 1996;70(10):6547–6556.
110. Rowell JF, Stanhope PE, Siliciano RF. Endocytosis of endogenously synthesized HIV-1 envelope protein. Mechanism and role in processing for association with class II MHC. *J Immunol.* 1995;155(1):473–488.
111. Ohno H, Aguilar RC, Fournier MC, Hennecke S, Cosson P, Bonifacino JS. Interaction of endocytic signals from the HIV-1 envelope glycoprotein complex with members of the adaptor medium chain family. *Virology.* 1997;238(2):305–315. http://dx.doi.org/10.1006/viro.1997.8839.
112. Wyss S, Berlioz-Torrent C, Boge M, et al. The highly conserved C-terminal dileucine motif in the cytosolic domain of the human immunodeficiency virus type 1 envelope glycoprotein is critical for its association with the AP-1 clathrin adaptor [correction of adapter]. *J Virol.* 2001;75(6):2982–2992. http://dx.doi.org/10.1128/JVI.75.6.2982-2992.2001.
113. Breed MW, Jordan APO, Aye PP, et al. Loss of a tyrosine-dependent trafficking motif in the simian immunodeficiency virus envelope cytoplasmic tail spares mucosal CD4 cells but does not prevent disease progression. *J Virol.* 2013;87(3):1528–1543. http://dx.doi.org/10.1128/JVI.01928-12.
114. Breed MW, Jordan APO, Aye PP, et al. A single amino acid mutation in the envelope cytoplasmic tail restores the ability of an attenuated simian immunodeficiency virus

mutant to deplete mucosal CD4 + T cells. *J Virol*. 2013;87(23):13048–13052. http://dx.doi.org/10.1128/JVI.02126-13.
115. Berlioz-Torrent C, Shacklett BL, Erdtmann L, et al. Interactions of the cytoplasmic domains of human and simian retroviral transmembrane proteins with components of the clathrin adaptor complexes modulate intracellular and cell surface expression of envelope glycoproteins. *J Virol*. 1999;73(2):1350–1361.
116. Ilinskaya A, Heidecker G, Derse D. Opposing effects of a tyrosine-based sorting motif and a PDZ-binding motif regulate human T-lymphotropic virus type 1 envelope trafficking. *J Virol*. 2010;84(14):6995–7004. http://dx.doi.org/10.1128/JVI.01853-09.
117. Byland R, Vance PJ, Hoxie JA, Marsh M. A conserved dileucine motif mediates clathrin and AP-2-dependent endocytosis of the HIV-1 envelope protein. *Mol Biol Cell*. 2007;18(2):414–425. http://dx.doi.org/10.1091/mbc.E06-06-0535.
118. Day JR, Münk C, Guatelli JC. The membrane-proximal tyrosine-based sorting signal of human immunodeficiency virus type 1 gp41 is required for optimal viral infectivity. *J Virol*. 2004;78(3):1069–1079.
119. Bhakta SJ, Shang L, Prince JL, Claiborne DT, Hunter E. Mutagenesis of tyrosine and di-leucine motifs in the HIV-1 envelope cytoplasmic domain results in a loss of Env-mediated fusion and infectivity. *Retrovirology*. 2011;8(1):37. http://dx.doi.org/10.1186/1742-4690-8-37.
120. Boscia AL, Akabori K, Benamram Z, et al. Membrane structure correlates to function of LLP2 on the cytoplasmic tail of HIV-1 gp41 protein. *Biophys J*. 2013;105(3):657–666. http://dx.doi.org/10.1016/j.bpj.2013.06.042.
121. Roy NH, Chan J, Lambele M, Thali M. Clustering and mobility of HIV-1 Env at viral assembly sites predict its propensity to induce cell–cell fusion. *J Virol*. 2013;87(13):7516–7525. http://dx.doi.org/10.1128/JVI.00790-13.
122. Jorgenson RL, Vogt VM, Johnson MC. Foreign glycoproteins can be actively recruited to virus assembly sites during pseudotyping. *J Virol*. 2009;83(9):4060–4067. http://dx.doi.org/10.1128/JVI.02425-08.
123. Gregory DA, Lyddon TD, Johnson MC. Multiple Gag domains contribute to selective MLV Env recruitment to MLV virions. *J Virol*. 2013;87(3). http://dx.doi.org/10.1128/JVI.02604-12.
124. Murakami T, Freed EO. Genetic evidence for an interaction between human immunodeficiency virus type 1 matrix and alpha-helix 2 of the gp41 cytoplasmic tail. *J Virol*. 2000;74(8):3548–3554.
125. Freed EO, Martin MA. Virion incorporation of envelope glycoproteins with long but not short cytoplasmic tails is blocked by specific, single amino acid substitutions in the human immunodeficiency virus type 1 matrix. *J Virol*. 1995;69(3):1984–1989.
126. Mammano F, Kondo E, Sodroski J, Bukovsky A, Göttlinger HG. Rescue of human immunodeficiency virus type 1 matrix protein mutants by envelope glycoproteins with short cytoplasmic domains. *J Virol*. 1995;69(6):3824–3830.
127. Beaumont E, Vendrame D, Verrier B, et al. Matrix and envelope coevolution revealed in a patient monitored since primary infection with human immunodeficiency virus type 1. *J Virol*. 2009;83(19):9875–9889. http://dx.doi.org/10.1128/JVI.01213-09.
128. Nack U, Schnierle BS. Replacement of the murine leukemia virus (MLV) envelope gene with a truncated HIV envelope gene in MLV generates a virus with impaired replication capacity. *Virology*. 2003;315(1):209–216. http://dx.doi.org/10.1016/S0042-6822(03)00519-1.
129. Lewis BC, Chinnasamy N, Morgan RA, Varmus HE. Development of an avian leukosis-sarcoma virus subgroup A pseudotyped lentiviral vector. *J Virol*. 2001;75(19):9339–9344. http://dx.doi.org/10.1128/JVI.75.19.9339-9344.2001.
130. Lambelé M, Labrosse B, Roch E, et al. Impact of natural polymorphism within the gp41 cytoplasmic tail of human immunodeficiency virus type 1 on the intracellular

distribution of envelope glycoproteins and viral assembly. *J Virol.* 2007;81(1):125–140. http://dx.doi.org/10.1128/JVI.01659-06.

131. Tedbury PR, Ablan SD, Freed EO. Global rescue of defects in HIV-1 envelope glycoprotein incorporation: implications for matrix structure. *PLoS Pathog.* 2013;9(11): e1003739. http://dx.doi.org/10.1371/journal.ppat.1003739.
132. Butan C, Winkler DC, Heymann JB, Craven RC, Steven AC. RSV capsid polymorphism correlates with polymerization efficiency and envelope glycoprotein content: implications that nucleation controls morphogenesis. *J Mol Biol.* 2008;376(4): 1168–1181. http://dx.doi.org/10.1016/j.jmb.2007.12.003.
133. Zhu P, Chertova E, Bess J, et al. Electron tomography analysis of envelope glycoprotein trimers on HIV and simian immunodeficiency virus virions. *Proc Natl Acad Sci U S A.* 2003;100(26):15812–15817. http://dx.doi.org/10.1073/pnas.2634931100.
134. Postler TS, Desrosiers RC. The cytoplasmic domain of the HIV-1 glycoprotein gp41 induces NF-κB activation through TGF-β-activated kinase 1. *Cell Host Microbe.* 2012;11(2):181–193. http://dx.doi.org/10.1016/j.chom.2011.12.005.
135. Beaufils P, Choquet D, Mamoun RZ, Malissen B. The (YXXL/I)2 signalling motif found in the cytoplasmic segments of the bovine leukaemia virus envelope protein and Epstein-Barr virus latent membrane protein 2A can elicit early and late lymphocyte activation events. *EMBO J.* 1993;12(13):5105–5112.
136. Hofacre A, Fan H. Jaagsiekte sheep retrovirus biology and oncogenesis. *Viruses.* 2010;2(12):2618–2648. http://dx.doi.org/10.3390/v2122618.
137. Maeda N, Palmarini M, Murgia C, Fan H. Direct transformation of rodent fibroblasts by jaagsiekte sheep retrovirus DNA. *Proc Natl Acad Sci U S A.* 2001;98(8):4449–4454. http://dx.doi.org/10.1073/pnas.071547598.
138. Liu S-L, Miller AD. Oncogenic transformation by the jaagsiekte sheep retrovirus envelope protein. *Oncogene.* 2007;26(6):789–801. http://dx.doi.org/10.1038/sj.onc.1209850.
139. Cousens C, Maeda N, Murgia C, Dagleish MP, Palmarini M, Fan H. In vivo tumorigenesis by Jaagsiekte sheep retrovirus (JSRV) requires Y590 in Env TM, but not full-length orfX open reading frame. *Virology.* 2007;367(2):413–421. http://dx.doi.org/10.1016/j.virol.2007.06.004.
140. Alberti A, Murgia C, Liu S-L, et al. Envelope-induced cell transformation by ovine betaretroviruses. *J Virol.* 2002;76(11):5387–5394.
141. Palmarini M, Maeda N, Murgia C, De-Fraja C, Hofacre A, Fan H. A phosphatidylinositol 3-kinase docking site in the cytoplasmic tail of the Jaagsiekte sheep retrovirus transmembrane protein is essential for envelope-induced transformation of NIH 3T3 cells. *J Virol.* 2001;75(22):11002–11009. http://dx.doi.org/10.1128/JVI.75.22.11002-11009.2001.
142. Liu S-L, Lerman MI, Miller AD. Putative phosphatidylinositol 3-kinase (PI3K) binding motifs in ovine betaretrovirus Env proteins are not essential for rodent fibroblast transformation and PI3K/Akt activation. *J Virol.* 2003;77(14):7924–7935.
143. Maeda N, Fu W, Ortin A, de las Heras M, Fan H. Roles of the Ras-MEK-mitogen-activated protein kinase and phosphatidylinositol 3-kinase-Akt-mTOR pathways in Jaagsiekte sheep retrovirus-induced transformation of rodent fibroblast and epithelial cell lines. *J Virol.* 2005;79(7):4440–4450. http://dx.doi.org/10.1128/JVI.79.7.4440-4450.2005.
144. Maeda N, Fan H. Signal transduction pathways utilized by enzootic nasal tumor virus (ENTV-1) envelope protein in transformation of rat epithelial cells resemble those used by jaagsiekte sheep retrovirus. *Virus Genes.* 2008;36(1):147–155. http://dx.doi.org/10.1007/s11262-007-0193-x.
145. Allen TE, Sherrill KJ, Crispell SM, Perrott MR, Carlson JO, DeMartini JC. The jaagsiekte sheep retrovirus envelope gene induces transformation of the avian fibroblast

cell line DF-1 but does not require a conserved SH2 binding domain. *J Gen Virol.* 2002;83(Pt 11):2733–2742.

146. Hofacre A, Fan H. Multiple domains of the Jaagsiekte sheep retrovirus envelope protein are required for transformation of rodent fibroblasts. *J Virol.* 2004;78(19):10479–10489. http://dx.doi.org/10.1128/JVI.78.19.10479-10489.2004.
147. Hull S, Lim J, Hamil A, Nitta T, Fan H. Analysis of jaagsiekte sheep retrovirus (JSRV) envelope protein domains in transformation. *Virus Genes.* 2012;45(3):508–517. http://dx.doi.org/10.1007/s11262-012-0793-y.
148. Steckbeck JD, Kuhlmann A-S, Montelaro RC. Structural and functional comparisons of retroviral envelope protein C-terminal domains: still much to learn. *Viruses.* 2014;6(1):284–300. http://dx.doi.org/10.3390/v6010284.
149. Zhao Y, Zhu L, Benedict CA, Chen D, Anderson WF, Cannon PM. Functional domains in the retroviral transmembrane protein. *J Virol.* 1998;72(7):5392–5398.
150. Côté M, Zheng Y-M, Liu S-L. Membrane fusion and cell entry of XMRV are pH-independent and modulated by the envelope glycoprotein's cytoplasmic tail. *PLoS One.* 2012;7(3):e33734. http://dx.doi.org/10.1371/journal.pone.0033734.
151. Aguilar HC, Anderson WF, Cannon PM. Cytoplasmic tail of Moloney murine leukemia virus envelope protein influences the conformation of the extracellular domain: implications for mechanism of action of the R peptide. *J Virol.* 2003;77(2):1281–1291.
152. Spies CP, Ritter GD, Mulligan MJ, Compans RW. Truncation of the cytoplasmic domain of the simian immunodeficiency virus envelope glycoprotein alters the conformation of the external domain. *J Virol.* 1994;68(2):585–591.
153. Edwards TG, Wyss S, Reeves JD, et al. Truncation of the cytoplasmic domain induces exposure of conserved regions in the ectodomain of human immunodeficiency virus type 1 envelope protein. *J Virol.* 2002;76(6):2683–2691.
154. Wyss S, Dimitrov AS, Baribaud F, Edwards TG, Blumenthal R, Hoxie JA. Regulation of human immunodeficiency virus type 1 envelope glycoprotein fusion by a membrane-interactive domain in the gp41 cytoplasmic tail. *J Virol.* 2005;79(19):12231–12241. http://dx.doi.org/10.1128/JVI.79.19.12231-12241.2005.
155. LaBranche CC, Hoffman TL, Romano J, et al. Determinants of CD4 independence for a human immunodeficiency virus type 1 variant map outside regions required for coreceptor specificity. *J Virol.* 1999;73(12):10310–10319.
156. Edwards TG, Hoffman TL, Baribaud F, et al. Relationships between CD4 independence, neutralization sensitivity, and exposure of a CD4-induced epitope in a human immunodeficiency virus type 1 envelope protein. *J Virol.* 2001;75(11):5230–5239. http://dx.doi.org/10.1128/JVI.75.11.5230-5239.2001.
157. Kalia V, Sarkar S, Gupta P, Montelaro RC. Antibody neutralization escape mediated by point mutations in the intracytoplasmic tail of human immunodeficiency virus type 1 gp41. *J Virol.* 2005;79(4):2097–2107. http://dx.doi.org/10.1128/JVI.79.4.2097-2107.2005.
158. Durham ND, Yewdall AW, Chen P, et al. Neutralization resistance of virological synapse-mediated HIV-1 Infection is regulated by the gp41 cytoplasmic tail. *J Virol.* 2012;86(14):7484–7495. http://dx.doi.org/10.1128/JVI.00230-12.
159. Côté M, Zheng Y-M, Albritton LM, Liu S-L. Fusogenicity of Jaagsiekte sheep retrovirus envelope protein is dependent on low pH and is enhanced by cytoplasmic tail truncations. *J Virol.* 2008;82(5):2543–2554. http://dx.doi.org/10.1128/JVI.01852-07.
160. Hammonds J, Chen X, Ding L, et al. Gp120 stability on HIV-1 virions and Gag-Env pseudovirions is enhanced by an uncleaved Gag core. *Virology.* 2003;314(2):636–649. http://dx.doi.org/10.1016/S0042-6822(03)00467-7.
161. Green N, Shinnick TM, Witte O, Ponticelli A, Sutcliffe JG, Lerner RA. Sequence-specific antibodies show that maturation of Moloney leukemia virus envelope

polyprotein involves removal of a COOH-terminal peptide. *Proc Natl Acad Sci U S A.* 1981;78(10):6023–6027.
162. Henderson LE, Sowder R, Copeland TD, Smythers G, Oroszlan S. Quantitative separation of murine leukemia virus proteins by reversed-phase high-pressure liquid chromatography reveals newly described gag and env cleavage products. *J Virol.* 1984;52(2):492–500.
163. Brody BA, Rhee SS, Sommerfelt MA, Hunter E. A viral protease-mediated cleavage of the transmembrane glycoprotein of Mason-Pfizer monkey virus can be suppressed by mutations within the matrix protein. *Proc Natl Acad Sci U S A.* 1992;89(8):3443–3447.
164. Rice NR, Henderson LE, Sowder RC, Copeland TD, Oroszlan S, Edwards JF. Synthesis and processing of the transmembrane envelope protein of equine infectious anemia virus. *J Virol.* 1990;64(8):3770–3778.
165. Murakami T, Ablan S, Freed E, Tanaka Y. Regulation of human immunodeficiency virus type 1 Env-mediated membrane fusion by viral protease activity. *J Virol.* 2004;78(2):1026–1031. http://dx.doi.org/10.1128/JVI.78.2.1026.
166. Wyma DJ, Jiang J, Shi J, et al. Coupling of human immunodeficiency virus type 1 fusion to virion maturation: a novel role of the gp41 cytoplasmic tail. *J Virol.* 2004;78(7):3429–3435. http://dx.doi.org/10.1128/JVI.78.7.3429-3435.2004.
167. Dale BM, McNerney GP, Thompson DL, et al. Cell-to-cell transfer of HIV-1 via virological synapses leads to endosomal virion maturation that activates viral membrane fusion. *Cell Host Microbe.* 2011;10(6):551–562. http://dx.doi.org/10.1016/j.chom.2011.10.015.
168. Chojnacki J, Staudt T, Glass B, et al. Maturation-dependent HIV-1 surface protein redistribution revealed by fluorescence nanoscopy. *Science.* 2012;338(6106):524–528. http://dx.doi.org/10.1126/science.1226359.
169. Joyner AS, Willis JR, Crowe JE, Aiken C. Maturation-induced cloaking of neutralization epitopes on HIV-1 particles. *PLoS Pathog.* 2011;7(9):e1002234. http://dx.doi.org/10.1371/journal.ppat.1002234.
170. Rabi SA, Laird GM, Durand CM, et al. Multi-step inhibition explains HIV-1 protease inhibitor pharmacodynamics and resistance. *J Clin Invest.* 2013;123(9):3848–3860. http://dx.doi.org/10.1172/JCI67399.
171. Wyma DJ, Kotov A, Aiken C. Evidence for a stable interaction of gp41 with Pr55(Gag) in immature human immunodeficiency virus type 1 particles. *J Virol.* 2000; 74(20):9381–9387.
172. Jones DR, Suzuki K, Piller SC. A 100-amino acid truncation in the cytoplasmic tail of glycoprotein 41 in the reference HIV type 1 strain RF. *AIDS Res Hum Retroviruses.* 2002;18(7):513–517. http://dx.doi.org/10.1089/088922202317406664.
173. Shimizu H, Hasebe F, Tsuchie H, Morikawa S, Ushijima H, Kitamura T. Analysis of a human immunodeficiency virus type 1 isolate carrying a truncated transmembrane glycoprotein. *Virology.* 1992;189(2):534–546.
174. Kodama T, Wooley DP, Naidu YM, et al. Significance of premature stop codons in env of simian immunodeficiency virus. *J Virol.* 1989;63(11):4709–4714.
175. Chakrabarti L, Emerman M, Tiollais P, Sonigo P. The cytoplasmic domain of simian immunodeficiency virus transmembrane protein modulates infectivity. *J Virol.* 1989;63(10):4395–4403.
176. Jouvenet N, Neil SJD, Zhadina M, et al. Broad-spectrum inhibition of retroviral and filoviral particle release by tetherin. *J Virol.* 2009;83(4):1837–1844. http://dx.doi.org/10.1128/JVI.02211-08.
177. Yondola MA, Fernandes F, Belicha-Villanueva A, et al. Budding capability of the influenza virus neuraminidase can be modulated by tetherin. *J Virol.* 2011;85(6):2480–2491. http://dx.doi.org/10.1128/JVI.02188-10.

178. Neil SJD, Zang T, Bieniasz PD. Tetherin inhibits retrovirus release and is antagonized by HIV-1 Vpu. *Nature*. 2008;451(7177):425–430. http://dx.doi.org/10.1038/nature06553.
179. Van Damme N, Goff D, Katsura C, et al. The interferon-induced protein BST-2 restricts HIV-1 release and is downregulated from the cell surface by the viral Vpu protein. *Cell Host Microbe*. 2008;3(4):245–252. http://dx.doi.org/10.1016/j.chom.2008.03.001.
180. Perez-Caballero D, Zang T, Ebrahimi A, et al. Tetherin inhibits HIV-1 release by directly tethering virions to cells. *Cell*. 2009;139(3):499–511. http://dx.doi.org/10.1016/j.cell.2009.08.039.
181. Veillette M, Désormeaux A, Medjahed H, et al. Interaction with cellular CD4 exposes HIV-1 envelope epitopes targeted by antibody-dependent cell-mediated cytotoxicity. *J Virol*. 2014;88(5):2633–2644. http://dx.doi.org/10.1128/JVI.03230-13.
182. Arias JF, Heyer LN, von Bredow B, et al. Tetherin antagonism by Vpu protects HIV-infected cells from antibody-dependent cell-mediated cytotoxicity. *Proc Natl Acad Sci U S A*. 2014;111(17):6425–6430. http://dx.doi.org/10.1073/pnas.1321507111.
183. Pham TN, Lukhele S, Hajjar F, Routy J-P, Cohen EA. HIV Nef and Vpu protect HIV-infected CD4+ T cells from antibody-mediated cell lysis through down-modulation of CD4 and BST2. *Retrovirology*. 2014;11(1):15. http://dx.doi.org/10.1186/1742-4690-11-15.
184. Galão RP, Le Tortorec A, Pickering S, Kueck T, Neil SJD. Innate sensing of HIV-1 assembly by tetherin induces NFκB-dependent proinflammatory responses. *Cell Host Microbe*. 2012;12(5):633–644. http://dx.doi.org/10.1016/j.chom.2012.10.007.
185. Tokarev A, Suarez M, Kwan W, Fitzpatrick K, Singh R, Guatelli J. Stimulation of NF-κB activity by the HIV restriction factor BST2. *J Virol*. 2013;87(4):2046–2057. http://dx.doi.org/10.1128/JVI.02272-12.
186. Douglas JL, Gustin JK, Viswanathan K, Mansouri M, Moses AV, Früh K. The great escape: viral strategies to counter BST-2/tetherin. *PLoS Pathog*. 2010;6(5):e1000913. http://dx.doi.org/10.1371/journal.ppat.1000913.
187. Zhang F, Wilson SJ, Landford WC, et al. Nef proteins from simian immunodeficiency viruses are tetherin antagonists. *Cell Host Microbe*. 2009;6(1):54–67. http://dx.doi.org/10.1016/j.chom.2009.05.008.
188. Bour S, Schubert U, Peden K, Strebel K. The envelope glycoprotein of human immunodeficiency virus type 2 enhances viral particle release: a Vpu-like factor? *J Virol*. 1996;70(2):820–829.
189. Bour S, Strebel K. The human immunodeficiency virus (HIV) type 2 envelope protein is a functional complement to HIV type 1 Vpu that enhances particle release of heterologous retroviruses. *J Virol*. 1996;70(12):8285–8300.
190. Bour SP, Aberham C, Perrin C, Strebel K. Lack of effect of cytoplasmic tail truncations on human immunodeficiency virus type 2 ROD env particle release activity. *J Virol*. 1999;73(1):778–782.
191. Bour S, Akari H, Miyagi E, Strebel K. Naturally occurring amino acid substitutions in the HIV-2 ROD envelope glycoprotein regulate its ability to augment viral particle release. *Virology*. 2003;309(1):85–98.
192. Hauser H, Lopez LA, Yang SJ, et al. HIV-1 Vpu and HIV-2 Env counteract BST-2/tetherin by sequestration in a perinuclear compartment. *Retrovirology*. 2010;7:51. http://dx.doi.org/10.1186/1742-4690-7-51.
193. Le Tortorec A, Neil SJD. Antagonism to and intracellular sequestration of human tetherin by the human immunodeficiency virus type 2 envelope glycoprotein. *J Virol*. 2009;83(22):11966–11978. http://dx.doi.org/10.1128/JVI.01515-09.
194. Serra-Moreno R, Jia B, Breed M, Alvarez X, Evans DT. Compensatory changes in the cytoplasmic tail of gp41 confer resistance to tetherin/BST-2 in a pathogenic nef-

deleted SIV. *Cell Host Microbe.* 2011;9(1):46–57. http://dx.doi.org/10.1016/j.chom.2010.12.005.
195. Morrison JH, Guevara RB, Marcano AC, et al. Feline immunodeficiency virus envelope glycoproteins antagonize tetherin through a distinctive mechanism that requires virion incorporation. *J Virol.* 2014;88(6):3255–3272. http://dx.doi.org/10.1128/JVI.03814-13.
196. Celestino M, Calistri A, Del Vecchio C, et al. Feline tetherin is characterized by a short N-terminal region and is counteracted by the feline immunodeficiency virus envelope glycoprotein. *J Virol.* 2012;86(12):6688–6700. http://dx.doi.org/10.1128/JVI.07037-11.
197. Yin X, Hu Z, Gu Q, et al. Equine tetherin blocks retrovirus release and its activity is antagonized by equine infectious anemia virus envelope protein. *J Virol.* 2014;88(2):1259–1270. http://dx.doi.org/10.1128/JVI.03148-13.

CHAPTER TEN

Molecular Determinants of the Ratio of Inert to Infectious Virus Particles

P.J. Klasse[1]
Department of Microbiology and Immunology, Weill Cornell Medical College, Cornell University, New York, USA
[1]Corresponding author: e-mail address: pek2003@med.cornell.edu

Contents

Abstract

The ratio of virus particles to infectious units is a classic measurement in virology and ranges widely from several million to below 10 for different viruses. Much evidence suggests a distinction be made between infectious and infecting particles or virions: out of many potentially infectious virions, few infect under regular experimental conditions, largely because of diffusion barriers. Still, some virions are inert from the start; others become defective through decay. And with increasing cell- and molecular-biological knowledge of each step in the replicative cycle for different viruses, it emerges that many processes entail considerable losses of potential viral infectivity. Furthermore, all-or-nothing assumptions about virion infectivity are flawed and should be replaced by descriptions that allow for spectra of infectious propensities. A more realistic understanding of the infectivity of individual virions has both practical and theoretical implications for virus neutralization, vaccine research, antiviral therapy, and the use of viral vectors.

Progress in Molecular Biology and Translational Science, Volume 129
ISSN 1877-1173
http://dx.doi.org/10.1016/bs.pmbts.2014.10.012

1. INTRODUCTION: A WIDE RANGE OF PARTICLE-TO-INFECTIOUS-UNIT RATIO

One mode of virus infection is mediated by virus particles, or virions, that diffuse in the extracellular fluid and encounter susceptible cells that they infect. How do we assess how infectious those particles are? A classic approach is to determine the ratio of total virions to infectious units.

The number of virus particles or virions per volume in, e.g., medium harvested from virus-producing cells can be determined by electron or confocal microscopy,[1] or by a number of new bio-physical techniques, some of which stem from the rapid development of nanotechnology.[2] Or when the number of molecules of a structural protein incorporated into each virion is known, provided all of that protein is virion-associated, then the number of virions per volume can be calculated from the concentration of the detergent-solubilized protein measured by, e.g., immunochemical detection.[3,4]

The infectious titer of a suspension of virions can be determined in a plaque- or focus-forming assay and expressed as infectious units per volume. Alternatively, the virus suspension can be titrated out to a point where it gives infection in half of the tissue-culture wells; then there would theoretically be $-\ln(0.5) \approx 0.69$ infectious units per well.

By dividing the number of particles per volume by the number of infectious units per volume one obtains the ratio of noninfectious or inert virus particles per infectious unit (P/IU). That ratio is the subject of this review.

Wide ranges of P/IU ratios have been described both within and among different virus species: poliovirus, 30–1000; adenovirus, 20–100; papillomavirus 10^4 (Ref. 5). The P/IU ratio for varicella-zoster virus of the Herpes virus family is high, $\sim 4 \times 10^4$, contrasting with that for herpes simplex virus, 50–200 (Refs. 5,6); and the P/IU ratio for dengue virus, a flavivirus, can range from 3×10^3 to 7×10^4, contrasting strikingly with that of the distantly related Semliki Forest virus, an alpha virus, which sets the record for lowest recorded ratios, 1–2 (Refs. 5,7). This low ratio for Semliki Forest virus has greatly facilitated pioneering studies on the entry mechanisms of that virus.[8,9] HIV-1, a retrovirus, has been reported to have P/IU ratios in an even wider range: 1–10^2 (Ref. 10); 10^2–10^4 (Ref. 4); 10^3–10^4 (Refs. 3,11); 10^2–10^4 (Ref. 12); $\sim 10^5$ (Ref. 13); and 10^4–10^7 (Ref. 14).

For strong reasons that will emerge, the wide ranges of ratios for individual species should not be taken to signify mere experimental uncertainty.

Rather, variants of the same virus can display divergent P/IU ratios. And some of the clearest ratio differences among virus species, sometimes between closely related species, probably reflect real molecularly determined variations in replicative capacity that have evolved under selection pressure.

2. INFECTIOUS OR INFECTING?

Some distinctions need to be made about the virions in the numerator of the P/IU ratio, i.e., the usually much larger number, approximately equal to the virions that do not infect. For distinct purposes the investigator may be interested in different degrees of completeness of replication by the virus. For example, in the context of gene therapy or the use of viral vectors for vaccination the recombinant virus under study may be known to be defective and what counts as a successful infectious event may be the expression of a gene carried by the viral vector. In contrast, virological studies aiming to understand viral pathogenesis or the inhibition of infection, for example, by neutralizing antibodies may define an infectious event more rigorously, *viz.* as ending with the production of infectious progeny (which of course may have a P/IU ratio that differs from that of the inoculum). Thus, in the latter case, virions that have genomic defects, failing to encode viral proteins that can properly assemble into infectious virions, would count as noninfective. The P/IU ratio is nearly as old as virology itself, but it was partly through the recent intense studies aiming towards gene therapy that the concept gained new currency and indeed was elucidated with greater clarity.

These studies, conducted largely on adenovirus and retroviruses, have challenged a prevalent albeit tacit assumption about the virions that do not infect. Now, several lines of evidence impel the conclusion that the virions that do not infect are not all necessarily noninfectious, i.e., defective or inert. Hence, the distinction between noninfectious and -infecting virions is fundamental. If the potentially infecting virions that have not had the chance to infect under the experimental conditions are included in the denominator, many of the high P/IU ratios cited above would shrink substantially. It would be demanding to determine what proportion of virions are potentially infectious, but crucial experiments have indicated which factors prevent them from infecting. One such factor is their slow movement by diffusion.

Virions are colloidal particles. Their density, i.e., mass per volume, depends on the relative contents of proteins and nucleic acids, whether the virus is enveloped or naked, and how tightly packed the proteins are. Retroviral particles, which are enveloped, are around 100 nm in diameter and have densities similar to that of culture medium (but slightly higher, which allows them to be pelleted by ultracentrifugation): ~1.2 g/ml.[15] Colloidal particles suspended in an aqueous solution undergo Brownian motion and their displacement (l) over time (t) was quantitatively related to the diffusion of the smaller molecules by Albert Einstein[16]: $l=(2Dt)^{1/2}$, in which D, the diffusion coefficient, is described by the Stokes–Einstein relation: $D=kT/(6\pi\eta r)$, where k is Boltzmann's constant, T the absolute temperature, η the viscosity of the medium, and r the radius of the particle. The diffusion constant D for retroviral virions can be estimated to $\sim 6.5\times 10^{-8}\ \text{cm}^2/\text{s}$.[15]

In order to infect, the virion must traverse the distance to the target cell. In an adherent-cell culture, this means that there is a zone closest to the cells that becomes depleted of virions as they adsorb to cells. Under these conditions Fick's laws of diffusion and equations for the number of virus-cell hits can be derived.[17]

As the time for the trajectory through the medium extends, more and more virions would by chance reach susceptible cells. But retroviruses, for example, usually have short infectivity half-lives at physiological temperatures: 5–8 h for recombinant murine amphotropic retrovisuses[15] and 6–18 h for different strains of HIV-1.[14,18] Therefore the maximum infectivity will be reduced by the decline in infectivity that occurs before virions reach susceptible cells. Thus virions with the lower range of half-life will only travel on average 500 μm during a half-life, and since the depth of medium in adherent-cell culture is several mm, a small fraction of the virions would reach the cells within one half-life.

Adsorption to cells would approximately follow the law of mass action, but we generally do not know the affinity of whole virions for their receptors, let alone constellations of receptors and ancillary attachment factors on the cell surface; furthermore, the amounts of binding sites on cells may not be negligible in relation to the virions on a molar basis. The local concentration of virions will therefore decline with adsorption, as suggested above. The total concentration of virions, however, is unlikely to be as high as the dissociation constant for their binding to the cells,[19] and the depleted zone of virions will reequilibrate with the rest of the suspension.[17] Even if equilibrium were reached, only some viral particles would thus bind onto the cells while others would remain unligated in suspension.[20] These factors will conspire to make the density of cells and receptors strong influences on the frequency of virion attachment to the cells and therefore on the degree of infection.[15,17,19–21]

The concept of saturation of infection, meaning that the infectivity asymptotically has approached a maximum through increases in cell or receptor density, is important in studies on viral ligands as entry inhibitors: unsaturated assays must be used if the inhibitory concentrations of the ligands are to reflect their degree of uncompeted binding and hence affinity.[14,21–24]

Cultures of cells in suspension, such as primary lymphocytes or lymphocytic cell lines, differ principally from those of adherent cells. The distances that the virions must diffuse in order to infect are shorter. But another factor may also promote infection in varying degrees: infection can occur through cell-to-cell contact, sometimes organized as a specific intercellular interaction zone called the virological synapse.[25–28] Cell-to-cell transfer is more effective than virion-based infection.[29,30] Indeed, coculture of producer and target cells can vastly increase the degree of infection compared with transfer of supernatants containing virions.[31,32] Although mere proximity between virions and target cells, which can be achieved by culturing the latter on grids through which the virion-containing medium is passed, enhances infection,[15] that is not the only factor at work in mixed cultures of infected and uninfected cells. Thus, gently shaking infected suspension cultures reduces the infection although it would increase the chances of virion encounters with susceptible cells; instead the shaking disrupts synapse formation and that influence dominates.[33]

A study that modeled the kinetics of HIV-1 replication in a T-cell line culture mathematically came to compatible conclusions, while conceiving the problems of virion- and cell-mediated infection differently, before the discovery of the virological synapse.[29] In culture the kinetic rate of infection was modeled to be proportional to the number of infectious progeny virions per cycle and inversely proportional to the duration of the cycle. This time period was estimated to 3–4 days for different cultures, thus considerably longer than the intracellular portion of one cycle, which was 24 h.[34] The medium in the infected culture contained virions with P/IU ratios of 10^3–10^4. Approximately 10^3 physical particles were released per infected lymphocytic cell. But according to the kinetic modeling, every cell produced ~10^2 infectious virions. Therefore, the P/IU ratio within the ongoing culture was 10^2–10^3 times lower than when supernatant was collected and reassayed on uninfected cells. This discrepancy might be accounted for by the chance of earlier encounters with a susceptible cell in the ongoing culture and the opportunities for cell-to-cell contact and thereby enhanced transfer of the freshly budded virions.

A simple but ingenious type of experiment demonstrates the tangible difference between the infecting and infectious fractions of virions: cells, most conveniently but not necessarily adherent cells, are incubated with virus-containing medium for a certain time to allow virus adsorption (Fig. 1).[17,35–39] Then the entire medium is aspirated and transferred to an identical layer of target cells; the procedure is repeated several times. The outcome is usually that the series of cultures show similar stepwise reductions in infectivity as controls for nonspecific loss of virus through adsorption to nonsusceptible cells or plastic. The conclusion is that the infecting proportion of virus is negligible compared with the infectious one. By only measuring the infectivity in the first culture, we would underestimate the total infectivity 20- to 100-fold and hence overestimate the corresponding P/IU ratios to the same extent.[36,40]

The hypothesis explaining these results implies a potential for increasing viral infections by a number of techniques, which have all yielded corroborating results. The already mentioned flow of virions through cell cultures on grids,[15] the conjugation of virions to magnetic beads combined with the application of magnetic fields over the medium in order to attract the virions to the cells,[37,41] and the centrifugation of virions onto target cells, so-called spinoculation,[37,42] have all been shown to enhance infection substantially. Regarding spinoculation, it should be noted, though, that it also affects the dynamic interplay in the cortical cytoskeleton of actin and cofilin, and this effect may also favor infection.[43] Thus, spinoculation reduces the HIV-1-blocking potency in T cells of Jasplakinolide, a net enhancer of actin polymerization.[43]

As further support for the hypothesis of diffusion-limited infection, elegant and simple experiments have shown that it is not the amount of virus in a medium overlay in a culture well that determines the infectivity: it is the concentration (Fig. 2). Thus, increasing the volume and keeping the concentration constant adds only marginally to the infection, whereas keeping the amount of virus constant and increasing the volume reduces infection proportionately, as illustrated both for retrovirus and adenovirus.[15,17,39,40] The slight deviations from these sketched ideal relationships, i.e. strict concentration dependence and volume independence might be explained by a weak tendency for virions to sediment, i.e., the same tendency that is vastly augmented by spinoculation; as mentioned, the density of retrovirus particles is only marginally above that of the medium,[15] but denser viruses might show somewhat greater influence of gravitation on the degree of infection.

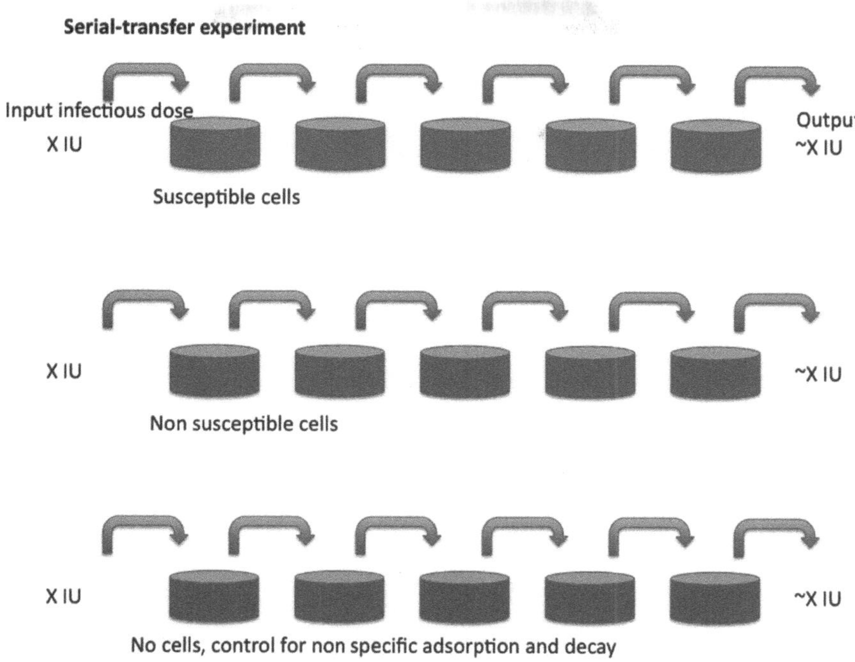

Figure 1 *Serial transfer.* A nominal infectious dose (X IU) of virus is added to the tissue-culture dishes in the left-hand column. After incubation for a fixed number of hours and at a constant temperature, the medium is aspirated from the second dish, to which all of the medium is transferred from the first dish, which is replenished with medium without virus. The procedure is repeated until the right-most dish has been incubated with supernatant for the same time as the others. The top row of dishes (orange (gray in the print version)) have susceptible cells forming a nonconfluent cell matt. The middle row (green (gray in the print version)) have nonsusceptible but otherwise similar cells (for example, lacking only a crucial receptor) growing at the same confluency. The bottom row has only tissue-culture dishes without cells. The outcome of the classical experiment as outlined is usually that infectious virus is transferred from the left to the right at nearly constant levels as detected in each dish; or there may be nonspecific losses, for example, some virus is lost for each step also in the middle row by binding to cell-surface glycosaminoglycans, equally abundant on nonsusceptible cells; or the nonspecific losses through binding only to plastic (bottom row) may be equally great; or the infectivity declines significantly in the medium within the time-frame of the experiment (not shown). The key finding is that the losses attributable to infection are negligible and that is because the diffusion time from the top to the bottom of the medium in the tissue-culture dish is too long for more than a small fraction of the virions to encounter target cells through diffusion.

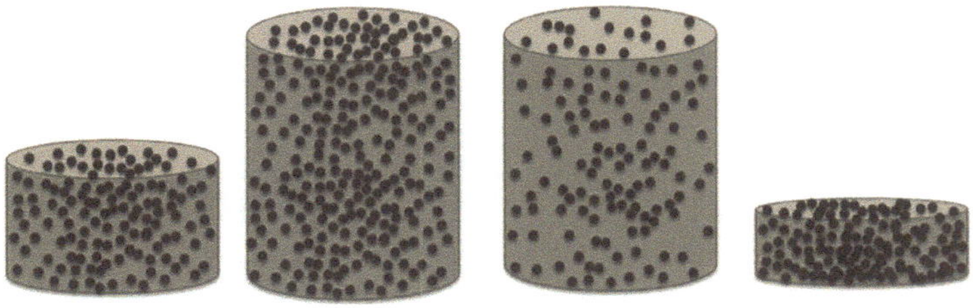

Figure 2 *Volume versus concentration.* Four tissue-culture wells are shown: the first, third, and fourth from the left are given the same amount of input virus; the second is given twice as much. In the two tissue-culture wells to the left, medium containing the same concentration of infectious virions is added. In the third one, the concentration is half of that for the two to the left but in the fourth it is double that (the volume is half of that of first one to the left). Experiments of this sort tend to show that the infectivity readout in the two wells to the left is similar; it is twice as low in the third one and twice as high in the fourth. Measured P/IU would have varied accordingly although the virus preparation was the same and hence the real P/IU was constant, as well as lower than all those based on measurements determined by the slow diffusion.

To correct for the underestimation of the infectious content of viral stocks, mathematical models have been introduced, some designed to improve upon previous ones.[17,20,40] But even improved estimates of the total infectious content in a suspension does not predict how many cells will be infected by how many virions. That relationship requires a more complex analysis, which also has a long history.

The concept of the *multiplicity of infection* (MOI) was introduced in the study of phage infection of bacteria.[44] If there is no impediment to a second, third, etc., infection of the same cell after the first infectious event, and if infections occur rarely but randomly in a culture, then the average number of virions infecting each cell can be described by the Poisson distribution. Of course, viral interference, and to some extent constitutive variation in receptor density among cells, would invalidate a strict adherence to Poisson analysis. Still, it may approximately hold up. According to the Poisson distribution, the multiplicity of infection, MOI = − ln(fraction of uninfected cells).

The concept of MOI has, however, been severely criticized; indeed, it has been dismissed as useless and unscientific (Refs. 17,40,45; cf., "L'État, c'est MOI!," attributed to Louis XIV). But I suggest that there are two distinct, indeed incompatible, prevalent uses of MOI; and that is how the concept acquired its ill repute. Some investigators have adopted the habit of simply dividing the measured number of infectious units (often the IU derived from an endpoint dilution) by the number of cells and calling that the MOI. Regardless of whether the IU measurement is orders of magnitude off the actual infectious content or not, that ratio will not agree with MOI = − ln(fraction of uninfected cells). For example, as we have seen, the same number of infectious units in different volumes will give vastly different degrees of infection; per definition they would have the same IU/cell; but they would not yield the same MOI = − ln(fraction of uninfected cells).

Since this discussion has been taking place in the context of gene therapy, and it may not be feasible to determine MOI = − ln(fraction of uninfected cells) in the target tissue, it is rational to look for more practicable approaches that would give reproducible results. IU/cell would most emphatically not pass muster. It would be laudable if virologists in general avoided calling IU/cells "MOI," while remembering that even the theoretically more justifiable quantity −ln(fraction of uninfected cells) is treacherous to interpret when infectious events do not occur independently and at random. As a good reminder of the complexity of the influences on how infectious events can be distributed *in vivo*, among splenic CD4-positive lymphocytes, a small minority of cells were infected by HIV-1, but the infected ones were multiply infected, a strong deviation from Poisson distribution for the whole population of cells.[46] In contrast, HIV-1-infected peripheral blood CD4-positive T cells mostly contain single proviral copies.[47] These and many other studies suggest a third application of the concept of MOI: in retrovirology, for example, various techniques are available for determining the number of proviral copies per infected cell. That quantity comes close to what the concept of MOI is theoretically aiming at. Then such measurements of the actual number of infectious events that the cells have on average undergone can be compared with −ln(fraction of uninfected cells) and IU/cell; and explanations can be sought for the discrepancies. There is nothing wrong with the concept: it has a place in virological theory and is not beyond the reach of experimental measurements. But practical use of viruses for gene transduction will benefit from other measurements, such as the number of active virions that will adsorb per cell and the extrapolated total infectious content.[17,20,40] New sophisticated techniques on the horizon may change our outlook further.

It has been aptly asked: "What could be a better way to study virus trafficking than 'miniaturizing oneself' and 'taking a ride with the virus particle' on its journey into the cell?".[48] What indeed? Single-virion tracking comes close to fulfilling this dream.[2,48–52] The approach has been made possible by new techniques for labeling viral components, either by conjugation with dyes or the creation of fusion proteins incorporating fluorescent moieties.

When dyes are incorporated at high densities in viral envelopes, fusion can be detected as a dequenching of the fluorescence. The fluorescence can also be sensitive to pH and thus be used for the monitoring of entry into endosomal compartments. Single-virion tracking has revealed actin-dependent influenza virion trajectories on the cell surface before endocytosis[48,53,54]; it has elucidated how vesicular stomatitis virus and retroviruses, including HIV-1, depend on the dynamics of the cortical cytoskeleton and myosin II for sliding along microvilli and filopodia towards sites conducive to productive internalization and entry.[48,55]

Furthermore, single-virion tracking was instrumental in advancing the case for productive HIV-1 entry through clathrin-dependent endocytosis,[56] thereby refuting a long held view of entry through cell-surface fusion by that virus. Previous evidence that HIV-1 depends on endocytosis for entry, although strong, was restricted to reductions in reporter gene expression and in viral antigen production when target cells expressed dominant-negative components of the endocytic machinery.[57,58]

The entry of the picornavirus coxsackievirus at tight junctions between epithelial cells has been elucidated through single-virion tracking and found to involve binding to the decay-accelarating factor and Abl- and Fyn-kinase effects on Rac that remold the actin network, directing the virion to the permissive spot.[59] Perhaps even the longstanding problem of how another picornavirus, *viz.*, poliovirus, enters has been solved through variations of these imaging techniques. Labeling of both the viral capsid and the genome helped demonstrate that the poliovirus particle is internalized by a distinct endocytic pathway that depends on tyrosine kinases and actin but not on clathrin, caveolin, flotillin, or microtubules. These findings would explain why dominant-negative mutants of dynamin do not block infection; the genome is rapidly injected into the cytoplasm from vesicles or tightly sealed membrane invaginations that juxtapose the cortical cytoskeleton, the last mode coming closest to a previously favored scenario of genome injection directly through the plasma membrane.[49]

Single-virion tracking has been instrumental in depicting how murine polyoma virus moves on the cell-surface, first diffusing randomly, then honing in on sites of potential entry. The technique has also revealed how a related virus, simian polyoma virus SV40, productively enters through uptake into caveolae.[48,60–63]

The corresponding techniques of tracking single core or capsid particles after entry have revealed the intricacies of kinesin- and dynein-mediated cytoplasmic transport towards the microtubule organizing center for several viruses: the enveloped viruses influenza virus, HIV-1, and herpes simplex virus,[48,64,65] but also the naked viruses reovirus (which includes the medically highly important rotavirus) and adenovirus.[48,66]

Still, these elegant techniques do not solve the conundrum of high P/IU ratios. The interpretations of results obtained by these techniques will, however, be facilitated by the drastic reductions in the relevant ratios discussed above: only high total-to-potentially infectious ratios, not high total-to-actually infecting ratios, would pose problems. But any ratio above one must be taken into account. Many safeguards have been proposed to insure valid interpretations in this area and the logic of the accompanying problems may border on the risk of circularity.

The proposed steps towards validating single-particle tracking data and interpretations are as follows. The behaviors of virions observed should be classified. It is then suggested that "Among these viruses, it should be possible to focus on the group of virus particles that are most likely to cause a productive infection."[48] This may be difficult if nothing is known about the entry mechanism and may be biased if something is erroneously thought to be known, as it was in the case of HIV and poliovirus. Nevertheless, an enveloped virus must presumably fuse, so that nonfusogenic behaviors might reasonably be categorized as unproductive. And when both the capsid and genome of naked viruses are labeled, these components would, at least hypothetically, be seen to separate at some point during or after productive entry.

It might seem ironic, however, that what would ultimately strengthen the relevance to productive infection of a trajectory observed by single-particle tracking would be a parallel regular infectivity assay. But it is not that assay, detecting bulk events, that gives ultimate evidence: it is the combination of the two. If the same intervention abolishes the signal in the infectivity assay and stops the particle in its track, the case is strong that the virion has been caught *in flagrante delicto*, as it were red-handed, whether fluorescing red or green or both.

3. DEFECTIVE FROM THE START

A virion that has failed to incorporate functional versions of all the components necessary for infection is inert, i.e., noninfectious, and this defectiveness can be absolute. But it would manifest itself differently depending on which infectivity assay is used and which step in the replicative cycle is knocked out. Thus, a mutant surface protein may be unable to interact with receptors, to penetrate membranes, or mediate fusion. These deficiencies would manifest themselves as a lack of signal in all infectivity assays. That would be so even in assays that involve the early to middle replicative events, such as activation of a *trans* gene in the target cell by a newly synthesized viral product. For example, the mutant would not raise luciferase or beta-galactosidase expression under the control of promoters activated by early viral products in engineered cell lines. But many deficiencies can affect steps between transcription and the formation of new virions as well: target cells could be infected in the first cycle but progeny virions would fail to form or be released. Or defective genomes can be incorporated into particles with functional proteins. Depending on what the genomic defect entails, different later steps would be affected. Progeny virions might form but would be defective.

Both replication-competent and -defective HIV-1 viral genomes were detected in uncultured brain tissue from a patient with AIDS-related dementia. Among 10 circular unintegrated proviruses were four full-length genomes with one or two long terminal repeats, two rearranged genomes, and five genomes that were truncated or had internal deletions. Only one genome, however, proved replication-competent. Thus mixtures of functional and defective proviruses, in integrated and circular unintegrated forms coexist *in vivo*; the defects may arise at reverse transcription in the newly infected cell, but defective genomes may also be transduced by virions that can only mediate incomplete replication and dead-end, abortive infections.[67]

Another study recorded that of eight HIV-1 virions initiating reverse transcription only one formed a provirus (intergrated DNA transcript), whereas one in 20 of RNA genomes in the suspension of virions was reverse transcribed. Hence, in this case the high P/IU ratios were largely not determined by absolute defects in the reverse transcriptase.[37]

In the preceding discussion, the straightforward possibilities of absolute or complete and relative or partial defectiveness of virions were outlined. But there is a more intricate kind of defectiveness with a long history of elegant studies in virology, that of defective interfering particles (DIPs). DIPs incorporate truncated genomes that arise both *in vitro* and *in vivo*; they can lack other essential structural components and be identified by EM as smaller than replicating viruses. Although they cannot replicate autonomously they can coinfect with functional virus and skew the replicative activity towards making more DIPs, eventually interfering with functional virus production—hence their name.

The phenomenon occurs with several DNA and RNA viruses; influenza virus, poliovirus, and the rhabdovirus vesicular stomatitis virus (VSV) are particularly well-studied prototypes.[68–71] It should be noted that defective viruses do not always interfere and are not necessarily truncated: defective retroviruses, for example, do not typically suppress their obligate helper viruses and the defects are often caused by recombinations with cellular genes.[72] But, whereas gammaretroviruses readily do form DIPs, lentiviruses do not; among the latter, natural DIPs have not been detected, although isolates of HIV and simian immunodeficiency virus (SIV) strains from infected organisms have been extensively sequenced.[73] Engineered mutations in Gag of HIV-1 that produce DIPs have, however, been described. The interference occurs at a late stage in the viral replicative cycle, when the progeny capsids are assembled.[74] It has been suggested that such interference could form the basis of intracellular immunization against HIV-1, more feasibly as therapy for the already infected than as prevention.[75]

In many natural infections DIPs appear to be acting as molecular parasites of the functional viruses. Why have natural DIPs not been detected among lentiviruses as well as among gammaretroviruses? The different modes of replication and spread within the organism for these viruses may explain the discrepancy. HIV-1 is more cytopathogenic than the gammaretroviruses. HIV-1 also depends strongly on horizontal replication within a host, whereas gammaretroviruses largely replicate through vertical transmission, i.e., through division of infected cells[73]; deltaretroviruses are the most restricted to the latter mode.[76,77]

Another possible explanation is that, although many cells infected with HIV-1 contain inactive proviruses, if the vast majority of them harbor a single proviral copy, the conditions for parasitic DIPs to prevail would be lacking.[67,73] At least in CD4-positive T cells in the spleen though, infection by more than one HIV-1 variant appears to be common, and this could explain the prevalence of recombination: assembling virions will incorporate two distinct genomes; and in the next round of replication these two will recombine, generating genetic mosaicism among the viral strains.[46] Then, however, impediments to DIP activity would arise for a rapidly replicating virus with a high error rate and a prodigious evolutionary plasticity: the virus would be under selective pressure to evolve resistance from such interference.[73] But if the DIPs evinced similar plasticity, an arms race in fitness competition would ensue.[73]

How would this coevolution play out? What would it mean for the prospects of using DIPs therapeutically? In this regard, it is important to define the potential mechanisms of interference by defective particles. There are two principal ones—two kinds of theft: first, *cis* stealing would occur at the step when the two viral genome copies are united; a DIP RNA molecule would compete with a full-length molecule to interact with another full-length RNA genome; second, in *trans* stealing, full-length and DIP RNA genomes would compete for viral capsids. Can any solutions to the problem of predicting the effects of DIPs be found—with a view to using DIPs to combat infections? Does the modeling of the coevolution explain the absence of natural lentiviral DIPs?

Already von Magnus noted that the prevalence of interference by defective virus depends on the MOI (strictly defined).[71] A strong dependence on initial conditions in every infected cell, and other sources of variation, may result in deterministic chaos and long-term intrinsic unpredictability.[78,79] How strong the interference is, whether it is complete or not, influences the predictability. When wildtype virions are produced only from singly infected cells, the interference is perfect. Artificial lentiviral DIPs interfere more weakly than that[80–82]; the more imperfect the interference, the stronger is the negative selection pressure against the DIPs.[73] This limit to the interference would militate against the tendency for chaotic dynamics of the DIP–HIV-1 interplay.[73]

In addition, the two kinds of parasitic stealing are predicted to have opposite dynamic tendencies. Genome theft is prone to be eliminated; capsid theft is more likely to be preserved: the fitness cost to HIV-1 for escaping through reduced packaging efficiency is greater than the harm done by the mutations to the DIPs. Even if HIV-1 evolves resistance through mutations that disproportionately favor packaging of functional over DIP genomes, the DIPs have an advantage. For while such an evolutionary strategy on the part of the functional virus necessitates combined *cis* and *trans* mutations in the capsid and the packaging signal of the viral genome, respectively, the countermove by the DIP could be a mere *cis* mutation in its packaging signal.[73] The logic of the asymmetry seems to predispose for a Darwinian checkmate against the wild-type virus. At least, while *cis* theft will predictably be selected out by the functional virus, *trans* theft might establish itself. Although the conditions favoring DIPs apparently do not prevail in natural HIV-1 infection, engineered DIPs might be evolutionarily stable, provided they fulfill highly demanding and specific requirements.

A recent study made use of deep sequencing to study the quantitative emergence of VSV DIPs.[83] The depth in sequencing refers to the number of reads of a nucleotide. To identify small minority populations in a pool of genome molecules, e.g., rare single-nucleotide polymorphisms, and to distinguish them from majority sequences, deep sequencing is required. VSV spontaneously produces virions of different sizes, with different deletions in essential genes and varying degrees of infectivity. Some variants are completely dependent on coinfection with functional virus for replication, but when they coinfect they will increase in relative frequency to the detriment of total viral replication. Deep sequencing identified the emergence of two differently truncated genomes; in parallel, transmission-electron microscopy detected an increasing frequency of smaller particles; the regular-size virions declined more in frequency than did the total count of virions.[83] Genetic analysis at that refined level may also contribute to better predictions of DIP behavior, at least for a few cycles.

The dynamism of defective-functional interactions contrasts with the static dichotomy of inert and infectious particles, which is still pervasive. When it comes to DIPs, in a sense, the defective is not dead; it is not even defective, except on its own. Direct interference with infection through the blocking of receptors by defective particles would require high concentrations of DIPs in relation to the relevant dissociation constants, K_ds, and the abundance of receptors.[84] One method for addressing the effect of virions that completely lack replicative capacity is based on the use UV-irradiated particles. Such virions of the paramyxovirus New Castle Disease virus seemingly interfered with the binding of infectious virus to receptors on the cell surface.[85–87] Since paramyxovirus virions on their surfaces have hemagglutinin-neuraminidase molecules, which can digest the sialic-acid moieties of the receptors, this is a special case of interference. It would not require as high particle concentrations as would direct receptor blocking. Receptor interference by infecting viruses, again, can be more efficient than direct block by inert particles through multiple mechanisms of receptor down-modulation.

Extreme examples of DIPs are genetically defective in all *trans* elements: a defective genome gets incorporated into particles, all components of which are provided by the functional helper virus. Another extreme situation also occurs: virus-like particles assemble but fail to incorporate any genome. The frequency with which this happens is determined by the molar synthesis ratio of capsid over genome, a malleable quantity that is central to the coevolution of DIPs and virus.[73] In the absence of detected DIPs, the proportion of genome-carrying HIV-1 cores was only 20% *in vivo*,[88,89] but *in vitro* it was 90% for virions expressed from transfected 293T human embryonic kidney cells.[90] Whether the empty particles interfere with receptor binding again depends on their concentrations, the relevant K_ds, and the abundance of receptors. But virions lacking genomes will obviously be an absolute contributor to elevated P/IU ratios. It is therefore significant that their prevalence varies among replication conditions: their abundance will set a lower limit for the P/IU ratio.

A newly budded HIV-1 virion is not yet infectious. The immature uncleaved form of the Gag protein drives the assembly and budding of virions, but after the virion release, Gag must be cleaved by the protease for the virions to become infectious. Hence protease inhibitors block infection by preventing this maturation. The proteolytic processing of Gag leads to a reorganization of the viral core to its characteristic conical mature shape.[91–93] In a sense the virions are stillborn but revived by the internal protease activity. Thus, if the P/IU ratio was assessed very early, it would be high but decreasing. We shall see later that other processes will dominate after a few hours and instead raise the ratio.

Gag processing has intriguing implications for cell-to-cell transfer of virus. As mentioned, one important mode of HIV-1 transmission is through the formation of virological synapses; amongst other features of the virological synapse, one effect is that the cell-to-cell bridge increases the efficacy of infection by reducing the diffusion distances that the virions must cover in order to reach a susceptible cell. Internalization of virions directly after budding entails proteolytic maturation within the acceptor-cell endosomes, a process which occurs over several hours. This is a mode of infection mediated by virions but during which infectious virions never appear in the open extracellular space; they traverse the semi-sealed synaptic cleft in an immature, noninfectious form.[94]

The maturation of Gag affects the functionality of Env: only after Gag reorganization does Env become fusogenic.[58,94] A new super-resolution fluorescence-microscopy technique, stimulated emission depletion (STED), showed that the Env-spike distribution on the virion surface changed upon Gag maturation: concomitant clustering of the spikes depended on the Gag-interacting Env tail and correlated with enhanced efficacy of viral entry. These events, coupling the viral interior and exterior and rearranging the capsid lattice to allow linked reorganization of the Env spikes externally, would thus explain why the Gag processing activates the Env-mediated fusion and entry events.[95] Env trimers have been observed to cluster[96]; it has previously been suggested that the sparse trimers on the virion surface come together to form entry claws[97]; and how the heterogeneity of the trimer distribution over the virion sphere would affect infectivity and neutralization of the virus has been mathematically modeled.[58,98,99] These intricate interactions and rearrangements will influence the P/IU ratio over time.

Other chapters in this volume discuss in greater detail how the cytoplasmic tail of Env interacts with Gag and how virally encoded accessory proteins enhance infectivity of the HIV-1 virions ("Retroviral factors promoting infectivity" by Cucurullo *et al.* and "The cytoplasmic tail of retroviral envelope glycoproteins" by Tedbury and Freed).

A clone of murine cells transformed by murine sarcoma virus and then super-infected with Moloney murine leukemia virus (MuLV-M) was found to produce high levels of gammaretrovirus (previously known as type C) particles but with very high P/IU ratios for both viruses, particularly MuLV. Uninfected mouse cells exposed to supernatants from the cell clone eventually produced fully infectious progeny after an eclipse in culture of several weeks. The ensuing virus was indistinguishable from MuLV-M. How could the regained replicative competence be explained? Either a genetic revertant or recombination with endogenous retroviruses in the murine cells had corrected the defect, which manifested itself at a step after entry.[100] The defective virus in the cell clone could also be rescued by coinfection with an amphotropic MuLV.[101]

A lymphoma cell line from the AKR strain of mice was likewise found to produce mostly defective virus; nonmalignant cells from the same mice produced larger quantities of replication-competent virus with lower P/IU ratios.[102] Although defectiveness is a dead end for the evolution of the virus, the defective viral phenotype can be maintained at the cellular level of selection and may be linked to pathogenesis.

An example of a mechanism of how defective MuLV can be rescued involves the Gag protein p12, which contains the PPPY motif and is instrumental in viral assembly and release. Deletion of the entire p12-encoding part manifested itself in the formation of tube-like Gag structures, whereas deletion of only the PPPY motif yielded chains of linked assembling particles.[103] If the culture was probed for virus in the supernatant at that stage, the overall particle count rather than merely the infectious unit content would be very low. Deletion mutants lacking the PPPY motif could be rescued by having the PPPY motif reintroduced in ectopic positions. Then the production of virus particles was restored.

Other Gag mutations can cause defects that affect both early and late events in the intracellular part of the replicative cycle. The matrix protein (MA) of the Moloney murine leukemia virus (M-MuLV) interacts with the Ras GTPase-activating-like protein IQGAP1, which regulates cytoskeletal dynamics. When MA was mutated such that IQGAP1 binding was affected in various degrees, IQGAP1 binding and replication correlated strongly. Revertant viruses restored the IQGAP1 interaction. The changes in cytoskeletal interactions resulting from different degrees of IQGAP1 binding apparently affect events both at the early post-entry and the later assembly stages of the cycle. It was suggested that IQGAPs link the capsid to the cytoskeleton and facilitate both afferent and efferent trafficking in the cell.[104]

The demands on Gag for the formation of infectious particles have been studied with HIV-1 as a model. Mutations affecting different aspects of Gag functions were introduced: thus Gag mutants were defective in binding to the cytoplasmic membrane, in dimerizing, or in binding to genomic RNA. Surprisingly, a single one of these defects still allowed core assembly and the formation of spherical particles, whereas a combination of any two completely blocked assembly and budding. The interesting interpretation was that the virus has evolved redundant functions: these aspects of how Gag functions, although qualitatively different, may quantitatively have similar effects by increasing the local Gag concentration at the assembly site; hence they could cooperate but also be redundant, thereby providing safety nets.[105]

The presence of host-cell proteins in virions can also modulate virus infectivity, although their absence may not cause virions to be completely inert. The two major sources of infectious HIV-1 virions *in vivo* are CD4-positive T lymphocytes and macrophages. A number of differences in the host-cell proteins taken up by virions produced by these cells have been identified, some with definite effects on infectivity.[106] For example, annexin II, which is abundant in macrophages but absent from peripheral blood T cells, normally regulates exocytosis. But in infected macrophages it interacts with the Gag precursor, p55, and when incorporated enhances the infectivity of the virions, specifically in relation to macrophages as target cells.[106]

Generally, enveloped virions incorporate cellular passenger proteins, both membrane proteins into the envelope and cytoplasmic proteins into the interior. The prevalence of these passenger molecules varies with their amounts in the cells from which the virions derive. For example, primary T lymphocytes stimulated to undergo mitosis express B7.1, B7.2, MHC class II (HLA-DR), CD28, CD40, CD40L, CD86, ICAM-1, and LFA-1; but HEK 293T cells, often used for transfection and virus or pseudovirus production, do not. And these molecules can enhance infectivity by interacting with their receptors or ligands on the target cells.[107–127] In addition, the lipids of the envelope vary in composition according to host cell and site of viral budding.[106] Clathrin, thioltransferase, and heat shock protein 70 are abundantly taken up into the interior of HIV-1 virions that bud from T-cell lines but less so when the virus is produced in HEK 293T cells. These proteins enhance viral infectivity, and hence lower the P/IU ratio, possibly by regulating the activity of the viral protease activity and by promoting the correct folding of the proteins derived from the Pol-precursor.[106] This host-cell-dependent epigenetic variation mainly molds the degree of functionality of the viral replicative machinery and thus the infectious propensity of the virion, rather than being all-or-nothing determinants.

How viral proteins, such as Vpr, Vpu, Vif, and Nef promote viral infectivity is dealt with elsewhere in the volume ("Retroviral factors promoting infectivity" by Cucurullo *et al.*).

Enveloped viruses are studded with regular or irregular arrays of surface glycoproteins, some or all of which mediate receptor binding and membrane fusion. Some forms of these proteins that get incorporated into the envelope are defective from the start. As discussed in the next section, they may also decay or disassemble. The glycans on viral surface envelope glycoproteins vary with the host cells in which the virus replicates. If functional and nonfunctional envelope glycoprotein oligomers are incorporated into the budding virions at random, a certain small fraction of virions will get fewer functional copies than what is required for infection; or the few functional copies may not be distributed on the virion surface in a constellation that is conducive to infectivity—the adequate proximity of the viral proteins for the formation of an entry complex may not be achieved.[58,95,98,99] The relative prevalence of the defective protein, the total number of copies of incorporated oligomers, and the minimum number of functional units required for infection will determine how large such a random fraction of completely inert virions is.[58,98,99,128] If the total number of incorporated envelope glycoprotein molecules also varies, the virions with fewest copies, even in the absence of defective protein, could then be inert; perhaps there are even some virions, known as bald particles, that completely lack the requisite surface protein.

The presence of both functional and nonfunctional envelope glycoprotein molecules becomes apparent from results obtained by virion capture assays.[129–133] Clearly, both neutralizing and nonneutralizing antibodies directed to the HIV-1 envelope glycoprotein can capture virions. Their capacity to do so correlates poorly with their capacity to neutralize. How can these findings be explained? The best interpretation comes from the well-corroborated hypothesis that it is necessary and sufficient for antibodies to bind to functional Env in order to neutralize. If the postulate is added that the Env spikes on the virions are a mixture of functional and nonfunctional forms, the capture would be explained: the nonneutralizing antibodies would capture by binding exclusively to nonfunctional Env and the neutralizing ones by binding to either form of Env, for some particular neutralizing antibodies only to the functional form. Then some intricacies arise. Although nonneutralizing antibodies as Fabs cannot block neutralization, which supports their lack of binding to functional Env, they do block capture by neutralizing antibodies.[133,134] This raises the question why the binding by the neutralizing antibodies to the functional Env spikes is not sufficient for capture. One explanation would be that the particular neutralizing antibodies used have a higher affinity for nonfunctional than functional Env and the higher affinity is required for the capture. Such affinity differences have been described.[135] Another explanation would be that the binding of the neutralizing antibodies to the functional spikes induces the dissociation of the outer subunit, gp120, from the transmembrane protein, gp41, and thus, while effectively inactivating the receptor-binding capacity of the protein, counteracts capture. The latter explanation may seem plausible but does not agree well with a more recent study that found that neutralizing antibodies preferentially capture infectious virions, whereas nonneutralizing antibodies favor the noninfectious ones.[131] Neutralizing antibodies that bind little to nonnative forms of Env, and by implication also little to nonfunctional Env, are now under intense study and these antibodies preferentially capture infectious virions too.[131] Possibly antibodies with that specificity do not induce shedding of the outer subunit, whereas others that also bind to nonfunctional forms of Env do. Why those binding to the nonfunctional forms would not induce shedding requires other explanations. A plausible one might have been that they bind to Env that is not proteolytically cleaved, representing precursor molecules, but those forms are found to be rarely incorporated into virions.[136] Remaining questions notwithstanding, for the current purposes, the capture experiments and analyses of virion-associated Env teach us a basis for the varying propensities of virions to infect. The results suggest that virions with the highest ratio of nonfunctional-to-functional Env, or simply in absolute terms those with the fewest functional molecules, are completely or relatively inert.

What defective forms of Env apart from uncleaved precursor molecules exist on the surface of HIV-1 particles? One study identified primarily two forms: gp120–gp41 protomers and trimeric gp41 stumps.[137] Antibodies to a region of gp41 that is occluded in the proteolytically processed, native trimers, but exposed on the stumps, do indeed capture virions effectively.[138]

Two further complications pertain to virion capture. Nonspecific capture can be substantial: one study found that several HIV-1-neutralizing antibodies captured enveloped particles lacking Env protein altogether, an artifact that can be obviated by letting the antibodies bind to the virions in suspension and removing unbound antibody before capturing the complex, rather than coating with primary antibody and capturing uncomplexed virions.[139] Building on the improved capture technique, this study identified differences between two viral isolates, obtained from the same patient, in the degree of trimerization of Env on virions. Thus genetic variation *in vivo* molds the degree of functionality of virion-incorporated Env and by implication the P/IU ratio, all other things being equal. Furthermore, in addition to the protomers and gp41 stumps identified in the previous study, virions were found to carry exogenous gp41 or parts of it as well as exogenous uncleaved precursor molecules; these unanchored molecules and fragments were more prevalent in preparations of pseudovirus than of infectious molecular clones. They must derive from the transfected cells or their excreted vesicles and attach to virions at or after budding without proper incorporation.[139]

One approach that may hold promise for the study of virions in suspension as they interact with antibodies, neutralizing or not, is fluorescence correlation spectroscopy. But even that method generated some anomalies in that partly trimer-specific HIV-1-neutralizing antibodies bound well to the virions but did not always neutralize all strains studied;[140] the particular antibody used (PG9) is, however, known to recognize nonneutralization-relevant forms of Env to some extent and to leave large nonneutralized fractions of infectivity.[135]

The number of Env trimers incorporated into virions is partly regulated by the interplay between the cytoplasmic tail of the protein and the underlying matrix protein, which forms a meshwork after Gag processing by the viral protease ("The cytoplasmic tail of retroviral envelope glycoproteins" by Tedbury and Freed). But the outer part of Env can also affect the degree of Env incorporation as well as the intrinsic fusogenicity of the individual trimer, both effects having an impact on infectivity. As demonstrated for an HIV-1 isolate of Clade C, the shorter the highly glycospylated and variable V1V2 region, the greater was the Env incorporation and also the fusogenicity. Since a shorter V1V2 is regularly found in recently transmitted, founder viruses, the higher infectivity that this feature confers may plausibly aid transmission. Demands on Env for high degrees of incorporation[141] or vigorous fusogenicity may represent a selective bottleneck in transmission: usually a single or very few variants out of the vast numbers harbored by the transmitting organism establish infection in the new host.[138,142] The longer V1V2 regions would then evolve in each infected organism under the selective pressure of neutralizing antibodies at some cost in specific infectivity, which would be reflected in raised P/IU ratios.[143] Indeed, the bottleneck in transmission may entail an augmented infectivity, i.e., a minimum in P/IU.

The *env* gene of HIV-1 is notoriously variable, and after few passages swarms of new variants with different sensitivities to neutralizing antibodies arise. Therefore, in order to have Env proteins of known sequence for neutralization, the test virus is often produced by cotransfecting the *env* gene and a viral genome that is *env*-defective, yielding pseudovirus, or by transfecting an infectious molecular clone. In one study, the neutralization sensitivities of virus produced in these different manners were found to be similar, but when the replication-competent virus was passaged once in PBMCs it became more neutralization resistant while also incorporating more Env,[144] which would explain the resistance.[145] A greater infectious propensity because of more Env per particle would translate into a reduced P/IU ratio after the PBMC passage. But it is important to note that both the pseudovirus and the infectious molecular clone, the latter being obtained by inserting *env* gene cassettes into a proviral backbone, involve a mismatch of Env and Gag.[144] Some *env* constructs used for making pseudovirus are truncated in the cytoplasmic tail. Possibly the mismatch and the truncation affect both the degree of Env incorporation and the functionality of the protein. The importance of the cytoplasmic tail of Env is discussed elsewhere in this volume ("The cytoplasmic tail of retroviral envelope glycoproteins" by Tedbury and Freed).

Another study compared the virion and infectious unit content, the genetic diversity, the degree of Env incorporation, and the cytokine and chemokine content of stocks of SIV (simian immunodeficiency virus) produced either directly by transfection or after passage of infectious virus. The stocks differed in genetic variability: as expected mutations had arisen through passaging, whereas the transfection-produced virus was clonal. The infectious virus and particle contents were both higher in the transfection-produced than the passaged virus, yielding similar P/IU ratios for the two kinds of preparations. As in the previous study, the Env content was higher for the passaged virus; it was surprisingly low for the highly infectious transfection-produced virus, less than one trimer per virion, which means that many particles were bald and thereby lacked infectivity.[146] Thus, the infectivity per amount of Env was considerably higher in the transfection-produced virus. A possible factor that might explain this was the higher contents of potentially inhibiting chemokines that were detected in the passaged virus. Another factor would be a higher degree of decay of the Env molecules in the passage culture. This kind of decay of entry-mediating proteins is the topic of the next section.

4. DECAY IN SUSPENSION

Particularly when under attack by the immune system *in vivo*, virions lead lives that can be solitary, poor, nasty, brutish, and short. The half-life of HIV-1 virions in the plasma of infected persons was estimated to 5 min.[147] It is longer *in vitro*. The infectivity half-life of virions derived from an infectious molecular clone (HxB3) of a T-cell line-adapted strain of HIV-1 was estimated to 36 h at 37 °C when measured on a T-cell line (CEM-SS) and 53 h on peripheral blood mononuclear cells. These values were compared with the half-life of the retention of the outer envelope glycoprotein, gp120, on the virions, which was 30 h, as well as with the half-lives of the reverse transcriptase activity, the physical integrity of the virions, the lipid envelope, and the core proteins, which were around 100 h.[14] Thus, the decline in Env integrity dominated kinetically and would be the prime causal determinant of the loss of infectivity. Furthermore, upon prolonged incubation, the loss of infectivity accelerated, which was attributed to a minimal threshold of intact Env oligomers per virion required for infectivity.[14]

A more recent study measured markedly shorter infectivity half-lives of pseudoviruses and molecular clones carrying Env derived from primary HIV-1 isolates of Clades A, B, or C in an assay based on reporter-gene activation in Tzm-bl cells. The half-lives ranged from 6 to 18 h with an average of 12 h and they correlated well with the temperatures required for inactivation of 90% of the infectivity in 1 h, the T(90) values, which varied between 40 and 49 °C. Biochemical analyses of Env indicated two stages of inactivation: first a perturbation of the Env trimer structure and then complete dissociation.[18]

Since the pseudovirions are identical except for the Env protein from different strains, differences in inactivation could be attributed to Env; and further evidence suggested that Env stability was the limiting factor determining infectivity.

Blue-native PAGE analyses indicated that somewhat higher temperatures were required for disintegration of Env than those registered as T(90) values. One explanation may be that the Env present in virions includes nonfunctional forms of Env, such as precursor molecules that have failed to be proteolytically cleaved, and that these forms are more resistant to heat-induced deoligomerization.[18]

Uncleaved Env precursors have, however, been reported to be excluded from virions.[136] But they are more efficiently excluded form viral stocks of infectious molecular clones than from the corresponding pseudoviruses.[139] And in the blue-native PAGE analyses mostly virions derived from infectious molecular clones were used. Therefore the gap between functional inactivation and deoligomerization is probably not explained by the prevalence of precursor forms of Env.

Another effect on Env of raised temperatures and chaotropic agents is the shedding of the outer Env subunit, gp120, from the transmembrane protein, gp41. This dissociation can be detected both by blue-native PAGE analyses and as an enhanced capacity of the virions to be captured by nonneutralizing antibodies to gp41. But loss of function also preceded gp120 shedding.[18] Therefore a more cogent explanation may be that as the temperature rises, and also during prolonged incubation at 37 °C, the Env trimers lose their function through conformational changes before they detectably disintegrate.

This more recent study,[18] in agreement with the earlier one,[14] found the reverse transcriptase to be more resistant to raised temperatures than Env. And even higher temperatures would be required to destabilize the lipid envelope. As a result of these differential stabilities, Env becomes the limiting factor in infection. Still, the stabilities of Env and the backbone components (the non-*env*-encoded parts in pseudoviruses and infectious molecular clones) of the virions will vary relative to each other; and the half-life of the individual the virion preparation may therefore be best described by the harmonic mean of these respective half-lives.[18]

Instability of the functional Env glycoprotein is a major problem in vaccine development, because after disintegration Env components are capable of inducing mainly nonneutralizing antibodies. In attempts to increase Env stability, HIV-1 virions were subjected to destabilization before they were allowed to initiate new replicative cycles. Since the HIV-1 genome is highly variable, particularly in *env*, it was possible to obtain more stable viral variants after a number of such iterations of directed evolution. The mutants showed prolonged half-lives and increased T(90) values.[148] Most of the crucial mutations were located in gp41, the rest in gp120. Furthermore, the Env proteins from virus of other strains and even other clades were also stabilized by some of the mutations.[148] During natural evolution of the virus, it is plausible that a balance is struck between more stable and looser structures by counteracting selective pressures.

The previously discussed studies applied both raised temperatures and chaotropic agents in order to destabilize Env.[18,148] In another study, reduced temperatures were instead used for iterative selection of variants of one particular strain of HIV-1 that was inactivated when incubated on ice. The mutation conferring resistance was mapped to the aminoterminal part of gp120, which is intimately involved in anchoring the subunit to gp41: it was a H66N substitution. The cold-sensitive phenotype was not attributable to protonation of the His residue since it was preserved at high pH. Cold inactivation resulted in an increased exposure of gp41 epitopes in the gp120–gp41 interface but not in complete gp120 shedding; rather, the resistance was associated with a less frequent sampling of the CD4-bound, open state.[149,150]

These different examples illustrate how the delicate balance between functionality and vulnerability makes the envelope glycoproteins a chief determinant of the infectious half-life of enveloped viruses.

5. ABORTIVE INFECTION

If some encounters of virions with susceptible cells have stochastic abortive outcomes, the apparent P/IU ratio for a population of virions should be proportionately corrected to reflect the real ratio by multiplying with a factor $b = (1 - a)$, where a is the proportion of abortive events for potentially infectious virions that reach the cell surface. We shall now examine what fates of virions at the cell surface and intracellularly can be abortive. The distinction between reversible and irreversible abortive events can be drawn immediately: if the virion dissociates from the cell surface, it might at least partly preserve its potential infectivity. In that scenario, a would have the weighted value of the probability of abortive infection through cumulative dissociations; in contrast, later in the replicative cycle, the virion or its remaining replication-mediating components may be degraded in lysosomes or by proteasomes in the cytoplasm (Fig. 3). The first interruption is temporary, the second definitive.

5.1. Abortive fates at the cell surface

Toyoshima and Vogt demonstrated that infection by avian sarcoma virus was inhibited by polyanions and enhanced by polycations.[151] These observations made it routine to include polycations such as Polybrene in cell cultures when aiming to isolate retroviruses. These ions affect viral contact with and attachment to target cells (Fig. 3). Thus, the convective flow of a virion suspension through a mesh on which the target cells grow was shown to supervene the requirement for Polybrene in achieving high degrees of infection.[15] What is then the precise mechanism of the polycation effect?

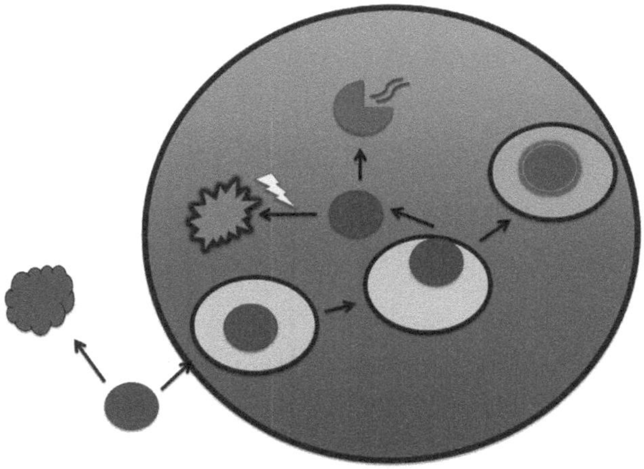

Figure 3 *Infectivity decay in suspension and abortive events during attachment and after entry.* While virions are diffusing at random, they may lose their infectivity and become inert particles (gray) because of the exponential decay of their components; for enveloped viruses the decay of the envelope glycoproteins may often determine the infectivity half-life. The virion (red) is depicted as getting endocytosed (the endosomal lumen is bright green) and as it later either fuses with or penetrates from the endosome. In case of excessive delay before cytoplasmic entry, the endosome continues to the lysosomal compartment and the virion gets degraded (gray). In case of successful entry of the capsid into the cytoplasm, a productive uncoating (orange) may lead to genome (green curved double lines) release for transcription or, in competition with this, degradation of the whole core by the proteasome (yellow flash). (See the color plate.)

How the enhancing effect of polycations on HIV-1 infection depends on both target cells and viral properties has been comprehensively dissected: not all strains of the virus are enhanced, although their usage of CCR5 or CXCR4 as a coreceptor does not influence the effect; T-cell line-adapted strains, which have multiple positively charged residues in the V3 region of their outer envelope glycoprotein subunit gp120, are not susceptible to the enhancement: on the contrary, infection by virus of one such clone, HxB2, was somewhat inhibited. Only infection by primary isolates is enhanced. Furthermore, the enhancement is contingent on prevalent glycosaminoglycans on the cell surface and probably acts by mitigating the electrostatic repulsion between these molecules and the negative polarity on the virions. The T-cell line-adapted strains instead, by virtue of their positive charges in V3, take advantage of the glycosaminoglycans as ancillary attachment factors; thereby such virions are enriched on the cell surface. These effects translate into several-fold increases and decreases of infectious units by the addition of polycations for primary isolates and T-cell line-adapted viruses, respectively.[152]

The same study, however, refuted an alleged role of surface-exposed cyclophilin A incorporated into the virions in enhancing infection by mediating attachment to target cells.[153] Otherwise such an effect of cyclophilin A would quite plausibly also have been counteracted by polycations. The effects of cyclophilin inside the virion on early post-entry steps are briefly described below.

The fate of virions that have adsorbed to target cells calibrates the P/IU ratios. A probe by immunofluorescence implied that the P/IU ratio for absorbed HIV-1 virions is lower than for virus in suspension.[154] Such an inference would agree with the previously described dominant diffusive barrier to infection. But the adsorbing minority of virions would also be expected to have higher Env content and therefore attach both to ancillary factors and specific receptors more avidly than would the bulk of virions. Cations prevented virions from dissociating from the cell surface, thereby enhancing infection 20- to 30-fold. On the resulting dynamic view of viral attachment, a postattachment race between entry and dissociation determines the degree of infection.[154] As noted though, the dissociated virions would still be infectious. And they would already be past the worst impediment, *viz.*, the long diffusion distance to the susceptible cell. Hence they would be likely to encounter susceptible cells again.

The cell-surface receptor T-cell immunoglobulin mucin domain 1 (TIM-1) has a specific and decisive role in the entry of filoviruses, including Ebola virus. Maybe that replicative step could be therapeutically targeted. In addition TIM-1 seems to promote infection by other enveloped viruses, including alpha viruses and baculoviruses, possibly independently of the viral envelope glycoproteins.[155–157] Effects of this sort, promoting infection by ancillary interactions at the cell surface, will determine the number of infectious events on specific cell types in proportion to the prevalence of the cell-surface molecules involved.

Another intriguing effect at the cell surface potentially affects the number of infectious events resulting from the attachment of virions. As a consequence of interactions between HIV-1 virions and its specific receptors, adenosine triphosphate (ATP) is released to the extracellular compartment through pannexin-1 channels and then activates purinergic receptors on the plasma membranes; further signaling events ensue and the membrane gets depolarized, which favors fusion.[158,159] But if HIV-1 mainly enters productively via the clathrin-dependent endocytic pathway and the site of complete fusion is the endosomal membrane, then somehow these events must be coordinated so that the signaling and depolarization occur in the right place and at the right time. Purinergic signaling pathways may be important also for entry of Hepatitis B, C, and D viruses.[160] A dynamic interplay of the cellular exterior and interior sets the stage for viral entry.

Inhibitory effects by the interferon-inducible transmembrane protein (IFITM) family are being elucidated against several medically important viruses, such as influenza A virus, dengue virus, hepatitis C virus, Ebola virus, HIV-1, and also naked viruses. These proteins are located in the cytoplasmic and endosomal membranes and block viral entry at these sites, possibly by changing the fluidity of the membranes, thereby interfering with fusion or penetration of different viruses.[161,162]

In general, these kinds of gatekeeper mechanisms emphasize the cellular-context dependence of P/IU measurements: whenever the various cellular variables are suboptimal or actively counter infection, the potential infectivity of the virus will be underestimated; the inflated P/IU ratio will say more about the target cells than about the state of the average virion.

5.2. Intracellular routes to abortive infection

An enveloped virus that enters via the endocytic route must fuse its envelope with the endocytic membrane and deliver the capsid proteins and genome into the cytoplasm for the replicative cycle to proceed. A substantial amount of intracellular HIV-1 Gag protein in cells that are exposed to infectious virus is found in endocytic vesicles.[163] The timing of fusion seems to be of the essence: if the endocytic trafficking continues too far the virion will be degraded in lysosomes before infection can occur. Indeed, if the lysosomal degradation is blocked, infectivity is enhanced several-fold.[164,165] Thus in natural infection a great number, sometimes the majority, of the potentially infectious internalized virions are lost through degradation. But those internalized virions represent a minority of the total population, the majority of which never made it to the susceptible cell (Fig. 3).

Any restriction factor that varies between cell type and the metabolic state or mitotic stage of a particular target cell adds to the context dependence of the P/IU ratio. For example, a number of intrinsic immunity factors restricting HIV-1 infection at post-entry steps have already been defined and more are being unraveled.[166,167] Viral accessory genes and the strategies by which HIV-1 overcomes these restrictions are reviewed elsewhere in this volume ("Retroviral factors promoting infectivity" by Cucurullo *et al.*). Suffice it to note here that most restriction, incompletely overcome, will increase the P/IU ratio. The IFITM proteins affecting entry have already been mentioned. APOBEC3G, which induces hypermutation of the proviral DNA, and TRIM5α, which interferes with uncoating of the viral core, must be understood as examples of post-entry restriction factors that also increase the P/IU ratio. Whether restriction factors affect the P/IU ratio depends on the replicative step they impede and the infectivity assay used. Tetherin, for example, acts by preventing the release of progeny HIV-1 virions. Hence it will not affect the outcome in common single-cycle infectivity assays; only when the infectivity read-out includes the capacity of progeny virions to initiate the next cycle would the tetherin effect and viral countermeasures affect the P/IU ratio.

Cyclophilin A is, as mentioned above, incorporated into HIV-1 virions and does enhance infectivity by promoting an early post-entry step in replication[168–171]: in human cells, specifically, it helps overcome the restriction factor Ref-1.[172] In nonhuman cells, in contrast, the interaction of cyclophilin with the Gag capsid protein is essential for imposing restriction. Thus, the usurpation of a nonviral protein mediates differential effects on infectivity depending on the host-cell species and thereby would correspondingly alter the P/IU ratios in a target-cell-dependent manner.

A virus that infects via the cell surface, whether it is an enveloped virus that fuses there or a naked virus that penetrates directly through the plasma membrane, would face a potential second barrier: the cortical cytoskeleton, particularly elaborate in polarized epithelial cells.[173] Some viral capsids are too large to diffuse through this actin meshwork. But by dint of endocytosis many viruses traverse the cortical cytoskeleton and travel to a locale, conducive to the subsequent replicative steps, whence they enter the cytoplasm. For example, Semliki Forest virus, an alpha virus, and vesicular stomatitis virus, a rhabdovirus, fail to infect when made to fuse at the artificial, ectopic site of the surface of CHO cells, which have extensive cortical cytoskeletal meshworks; they normally enter in a pH-dependent manner via endocytosis.[174]

HIV-1 interacts with the cytoskeleton through multiple and complex chains of events. The Env proteins of X4-tropic strains signal via CXCR4, thereby activating cofilin that depolymerizes the cortical cytoskeleton and promotes infection of resting T cells.[175] Why would this be so if penetration occurs from a vesicle that has already traversed the actin barrier by physiological mechanisms[56,176]? These effects will require further elucidation.

But the cortical actin seems to be modulated sequentially in different manners: initially Env–CD4 interactions at the cell surface induce the formation of a cytoskeletal cap, leading to local enrichment of Env-receptor complexes.[177] Then Env binding to the coreceptor elicits signaling through its linked Gα_q protein, activating kinase-dependent rearrangements of the actin cytoskeleton.[178–180] Tentatively, productive entry may thus require, first, a tightened cortical cytoskeletal meshwork and, thereafter, depolymerization and remodeling.[175] Recently, the actin modulator drebrin, which interacts with both CXCR4 and actin, has been implicated as a negative regulator of X4 HIV-1 entry.[181] An obligatory cytoskeletal role in the agglomeration of several Env-receptor complexes is compatible with the multioligomeric view of the expanding fusion pore.[176] Future research may elucidate how the endosomal sublocalization of the entry complex, differing from the cell surface in cytoskeletal or membrane conditions, permits progress from hemifusion to pore formation. The main point here is the evolutionary trade-off between the fitness value to the virus of overcoming obstacles and the cost of elaborate strategies for doing so: the usurpation of cytoskeletal functions by the virus is at least mildly imperfect; partial cytoskeletal barriers and negative regulators of productive entry and subsequent replicative steps remain. On this view, viral replication is suboptimal, suggesting some post-entry waste of potentially infectious virions.

We have seen how the slow extracellular diffusion of virions explains why many virions that could potentially infect do not; we have noted that the travel time exceeds several infectivity half-lives, so that often when the virion ultimately reaches the target cell it is defunct (Fig. 3). When viral components enter the cytoplasm, quite different diffusion-related problems arise: there, molecules larger than 500 kDa cannot diffuse freely. Strikingly diverse viruses, maybe most, including retroviruses, adenoviruses, parvoviruses, herpesviruses, poxviruses, baculoviruses, and even bacteriophages, have evolved mechanisms for dealing with the advantages and disadvantages that the cytoskeleton and the intracellular movement apparatus present. Some of these evolutionary strategies converge; others are apparently unique. In mammalian cells, interacting cellular partners for the virion components include actin, tubulin, dynein, and kinesins. Multiple evolutionary feats enable capsids to hitch-hike towards favorable sites for continued replication and to traverse cytoskeletal hindrances. These various cytoskeletal interactions begin before entry, during attachment, and continue to the ultimate release of infectious progeny.[182–188] Obviously the more successful an adaptation, the more it reduces the P/IU ratio, but the virus may sustain some losses at every step, adaptive imperfections continually militating against an approach of the P/IU ratio towards unity.

Once the viral capsid is in the cytoplasm it faces yet another threat to completing the replication: the proteasome. When proteasome activity is blocked, HIV-1 Gag accumulates in the cytoplasm, synthesis of proviral DNA is augmented, and the efficacy of HIV-1 infection is increased several–fold (Fig. 3).[165,189] Furthermore, physiological temperature is sub-optimal for a later step: in hyperthermia the viral regulatory protein Tat more effectively activates the viral promoter.[190]

6. CONCLUSIONS

The extremely high P/IU ratios that have been reported for many viruses should be understood as context- and assay-dependent phenomena that do not describe intrinsic properties of the virions. Rather, the high ratios are largely attributable to nonviral factors that prevent infectious virions from completing replicative cycles. Thus the distinction between infecting and infectious virions becomes crucial. The explanations for why infectious virions do not infect fall into two broad categories: first, the diffusion of virions in suspension is so slow that they do not reach any susceptible cells within the incubation time of standard experiments. Second, from the first contact with a susceptible cell onwards, every step carries some risk of abortive outcome: the reversion of attachment is a special case since it potentially allows a restart. The endocytic route offers several advantages: it avoids viral trapping in the apical cytoskeleton and delivers the genome to a more central location in the cell. But the degradation of internalized virions that fail to enter the cytoplasm before they reach the lysosomal compartment, just like the digestion of cores or naked virions by proteasomes in the cytoplasm, constitutes a definitive loss, a wastefulness accommodated for in evolutionary trade-offs.

The upshot of these considerations and potential corrections of the P/IU ratios is that at least two different ratios could be given: the phenomenological classical ratio, the total number of particles per infecting particle, with an emphasis on its dependence on many variables, and the—ever illusive—extrapolated ratio of (total virions)/(total virions – completely inert virions). Since the latter ratio in general would be substantially lower than the former, the enigma shifts from understanding why so few are infectious to explaining why some viruses, in particular Semliki Forest virus, presumably subject to the same diffusion barriers and some of the intracellular abortive mechanisms, can have so low ratios measured the classical way.

The specification that the inert virions be completely incapable of infecting is crucial. For another assumption underlying the classical usage of the ratios has been that infectivity is binary, all-or-nothing. This is partly unrealistic and clashes with many observations, old and new. While some virions do indubitably lack any infectivity, the rest may span a wide spectrum of infectious propensities. And that spectrum is not constant even in relation to a specific density of specific target cells: virions decay at different rates, usually exponentially; that decay is ultimately one source of completely inert particles, as necessary entry-mediating molecules or replicative enzymes cease to function.

The implications of refining the concept of virion infectivity are legion and both practically and theoretically consequential. An understanding of viral neutralization by antibodies requires a full account of the number of entry-mediating molecules that decorate the surface of a virion and how many of these are functional. It has been inferred from extremely high ratios of P/IU for HIV-1 that virions only have a single functional trimer, which would mean that any minimal occupancy on such antigens by antibody would neutralize.[191–193] When higher specific infectivities are postulated, as well as a spectrum of infectivity propensities, inferences to the best explanation suggest the participation in entry of several trimers per virion, which affects the demands on occupancy for neutralization.[58,98,99,145,194–197]

Understanding how neutralizing antibodies arise during infection *in vivo* will benefit from a realistic estimate of what fraction of virions are functional, because of the possible decoy effect of inert virions and decayed forms of entry-mediating proteins in eliciting humoral immune responses. Furthermore, in experimental infections *in vivo*, knowledge of the number of potentially infectious units in the inoculum will assist a dissection of how that amount translates into a number of infectious foci at the portal of entry. In all uses of viral vectors for vaccination and gene therapy, the improved quantitative understanding of infectivity and an avoidance of spurious MOI measurements will make for precision and accuracy in predicting the degree of *trans* gene expression *in vivo*.

As viruses evolved, redundancy and wastefulness were sometimes preserved as tolerably costly and sometimes indirectly advantageous to fitness. For example, the high error rate of viral RNA transcriptases allows enormous variability in viral targets for the immune system, thus enabling the viruses to escape attack. But the diffusion across an ocean of target-cell-free interstitial fluid, or worse, mucus, is unavoidably a voyage that entails both shipwreck and casualties. Indeed, the evolution of mechanisms for direct cell-to-cell transfer of viruses, the virological synapse,[25–27,198] and for replication with the infected cell as a proviral copy,[77] may represent adaptations that prevent excessive wastefulness. Even in the absence of direct cell-to-cell contact, though, diffusion distances in compartments *in vivo* with tightly packed cells may often be shorter than *in vitro*. Furthermore, potential intracellular restriction factors or barriers have been usurped and harnessed by viruses with the result of optimizing the replicative chances of each virion. Viruses have taken greater adaptive strides than classic measurements of extremely high P/IU ratios would suggest.

ACKNOWLEDGMENT

The author's work in this area is supported by the NIH grant R37 AI36082.

REFERENCES

1. Bhattacharya B, Weiss RA, Davis C, Holmes H, Hockley D, Fassati A. Detection and quantitation of human immunodeficiency virus type-1 particles by confocal microscopy. *J Virol Methods*. 2004;120:13–21.
2. Heider S, Metzner C. Quantitative real-time single particle analysis of virions. *Virology*. 2014;462-463C:199–206.
3. Kimpton J, Emerman M. Detection of replication-competent and pseudotyped human immunodeficiency virus with a sensitive cell line on the basis of activation of an integrated beta-galactosidase gene. *J Virol*. 1992;66:2232–2239.
4. Klasse PJ, McKeating JA. Soluble CD4 and CD4 immunoglobulin-selected HIV-1 variants: a phenotypic characterization. *AIDS Res Hum Retroviruses*. 1993;9:595–604.
5. Flint SJ, Enquist LW, Racaniello VR, Skalka AM. *Principles of Virology: Molecular Biology, Pathogenesis, and Control of Animal Viruses*. 2nd ed. Washington, DC: ASM Press; 2004.
6. Carpenter JE, Henderson EP, Grose C. Enumeration of an extremely high particle-to-PFU ratio for Varicella-zoster virus. *J Virol*. 2009;83:6917–6921.
7. van der Schaar HM, Rust MJ, Waarts BL, et al. Characterization of the early events in dengue virus cell entry by biochemical assays and single-virus tracking. *J Virol*. 2007;81:12019–12028.
8. Helenius A. Virus entry: what has pH got to do with it? *Nat Cell Biol*. 2013;15:125.
9. Helenius A, Moss B. Virus entry—an unwilling collaboration by the cell. *Curr Opin Virol*. 2013;3:1–2.
10. Bourinbaiar AS. The ratio of defective HIV-1 particles to replication-competent infectious virions. *Acta Virol*. 1994;38:59–61.
11. McDougal JS, Cort SP, Kennedy MS, et al. Immunoassay for the detection and quantitation of infectious human retrovirus, lymphadenopathy-associated virus (LAV). *J Immunol Methods*. 1985;76:171–183.
12. Rusert P, Fischer M, Joos B, et al. Quantification of infectious HIV-1 plasma viral load using a boosted in vitro infection protocol. *Virology*. 2004;326:113–129.
13. Piatak Jr M, Saag MS, Yang LC, et al. High levels of HIV-1 in plasma during all stages of infection determined by competitive PCR. *Science*. 1993;259:1749–1754.

14. Layne SP, Merges MJ, Dembo M, et al. Factors underlying spontaneous inactivation and susceptibility to neutralization of human immunodeficiency virus. *Virology*. 1992;189:695–714.
15. Chuck AS, Clarke MF, Palsson BO. Retroviral infection is limited by Brownian motion. *Hum Gene Ther*. 1996;7:1527–1534.
16. Einstein A. Über die von der molekularkinetischen Theorie der Wärme geforderte Bewegung von in ruhenden Flüssigkeiten suspendierten Teilchen. *Ann Phys*. 1905;17:549–560.
17. Nyberg-Hoffman C, Shabram P, Li W, Giroux D, Aguilar-Cordova E. Sensitivity and reproducibility in adenoviral infectious titer determination. *Nat Med*. 1997;3:808–811.
18. Agrawal N, Leaman DP, Rowcliffe E, et al. Functional stability of unliganded envelope glycoprotein spikes among isolates of human immunodeficiency virus type 1 (HIV-1). *PLoS One*. 2011;6:e21339.
19. Yu H, Soong N, Anderson WF. Binding kinetics of ecotropic (Moloney) murine leukemia retrovirus with NIH 3T3 cells. *J Virol*. 1995;69:6557–6562.
20. Kwon YJ, Hung G, Anderson WF, Peng CA, Yu H. Determination of infectious retrovirus concentration from colony-forming assay with quantitative analysis. *J Virol*. 2003;77:5712–5720.
21. Layne SP, Merges MJ, Spouge JL, Dembo M, Nara PL. Blocking of human immunodeficiency virus infection depends on cell density and viral stock age. *J Virol*. 1991;65:3293–3300.
22. Layne SP, Merges MJ, Dembo M, Spouge JL, Nara PL. HIV requires multiple gp120 molecules for CD4-mediated infection. *Nature*. 1990;346:277–279.
23. Spouge JI, Shrager RI, Dimitrov DS. HIV-1 infection kinetics in tissue cultures. *Math Biosci*. 1996;138:1–22.
24. Spouge JL, Layne SP, Dembo M. Analytic results for quantifying HIV infectivity. *Bull Math Biol*. 1989;51:715–730.
25. Feldmann J, Schwartz O. HIV-1 virological synapse: live imaging of transmission. *Viruses*. 2010;2:1666–1680.
26. Igakura T, Stinchcombe JC, Goon PK, et al. Spread of HTLV-I between lymphocytes by virus-induced polarization of the cytoskeleton. *Science*. 2003;299:1713–1716.
27. Nejmeddine M, Bangham CR. The HTLV-1 virological synapse. *Viruses*. 2010;2:1427–1447.
28. Piguet V, Sattentau Q. Dangerous liaisons at the virological synapse. *J Clin Invest*. 2004;114:605–610.
29. Dimitrov DS, Willey RL, Sato H, Chang LJ, Blumenthal R, Martin MA. Quantitation of human immunodeficiency virus type 1 infection kinetics. *J Virol*. 1993;67: 2182–2190.
30. Giese S, Marsh M. Tetherin can restrict cell-free and cell-cell transmission of HIV from primary macrophages to T cells. *PLoS Pathog*. 2014;10:e1004189.
31. Bodine DM, McDonagh KT, Seidel NE, Nienhuis AW. Survival and retrovirus infection of murine hematopoietic stem cells in vitro: effects of 5-FU and method of infection. *Exp Hematol*. 1991;19:206–212.
32. Hock RA, Miller AD. Retrovirus-mediated transfer and expression of drug resistance genes in human haematopoietic progenitor cells. *Nature*. 1986;320:275–277.
33. Sourisseau M, Sol-Foulon N, Porrot F, Blanchet F, Schwartz O. Inefficient human immunodeficiency virus replication in mobile lymphocytes. *J Virol*. 2007;81:1000–1012.
34. Kim SY, Byrn R, Groopman J, Baltimore D. Temporal aspects of DNA and RNA synthesis during human immunodeficiency virus infection: evidence for differential gene expression. *J Virol*. 1989;63:3708–3713.

35. Kabat D, Kozak SL, Wehrly K, Chesebro B. Differences in CD4 dependence for infectivity of laboratory-adapted and primary patient isolates of human immunodeficiency virus type 1. *J Virol.* 1994;68:2570–2577.
36. Tavoloni N. A simple procedure to determine the biological titer of recombinant retroviral vectors. *Gene Ther.* 1997;4:150–155.
37. Thomas JA, Ott DE, Gorelick RJ. Efficiency of human immunodeficiency virus type 1 postentry infection processes: evidence against disproportionate numbers of defective virions. *J Virol.* 2007;81:4367–4370.
38. Wang H, Paul R, Burgeson RE, Keene DR, Kabat D. Plasma membrane receptors for ecotropic murine retroviruses require a limiting accessory factor. *J Virol.* 1991;65:6468–6477.
39. Morgan JR, LeDoux JM, Snow RG, Tompkins RG, Yarmush ML. Retrovirus infection: effect of time and target cell number. *J Virol.* 1995;69:6994–7000.
40. Andreadis S, Lavery T, Davis HE, Le Doux JM, Yarmush ML, Morgan JR. Toward a more accurate quantitation of the activity of recombinant retroviruses: alternatives to titer and multiplicity of infection. *J Virol.* 2000;74:3431–3439.
41. Haim H, Steiner I, Panet A. Synchronized infection of cell cultures by magnetically controlled virus. *J Virol.* 2005;79:622–625.
42. O'Doherty U, Swiggard WJ, Malim MH. Human immunodeficiency virus type 1 spinoculation enhances infection through virus binding. *J Virol.* 2000;74:10074–10080.
43. Guo J, Wang W, Yu D, Wu Y. Spinoculation triggers dynamic actin and cofilin activity that facilitates HIV-1 infection of transformed and resting CD4 T cells. *J Virol.* 2011;85:9824–9833.
44. Ellis EL, Delbruck M. The growth of bacteriophage. *J Gen Physiol.* 1939;22:365–384.
45. Shabram P, Aguilar-Cordova E. Multiplicity of infection/multiplicity of confusion. *Mol Ther.* 2000;2:420–421.
46. Jung A, Maier R, Vartanian JP, et al. Recombination: multiply infected spleen cells in HIV patients. *Nature.* 2002;418:144.
47. Josefsson L, King MS, Makitalo B, et al. Majority of CD4+ T cells from peripheral blood of HIV-1-infected individuals contain only one HIV DNA molecule. *Proc Natl Acad Sci U S A.* 2011;108:11199–11204.
48. Brandenburg B, Zhuang X. Virus trafficking - learning from single-virus tracking. *Nat Rev Microbiol.* 2007;5:197–208.
49. Brandenburg B, Lee LY, Lakadamyali M, Rust MJ, Zhuang X, Hogle JM. Imaging poliovirus entry in live cells. *PLoS Biol.* 2007;5:e183.
50. Rust MJ, Lakadamyali M, Brandenburg B, Zhuang X. Single-virus tracking in live cells. *Cold Spring Harb Protoc.* 2011;2011.
51. Rust MJ, Lakadamyali M, Brandenburg B, Zhuang X. Single-particle virus tracking. *Cold Spring Harb Protoc.* 2011;2011.
52. Seisenberger G, Ried MU, Endress T, Buning H, Hallek M, Brauchle C. Real-time single-molecule imaging of the infection pathway of an adeno-associated virus. *Science.* 2001;294:1929–1932.
53. Lakadamyali M, Rust MJ, Babcock HP, Zhuang X. Visualizing infection of individual influenza viruses. *Proc Natl Acad Sci U S A.* 2003;100:9280–9285.
54. Rust MJ, Lakadamyali M, Zhang F, Zhuang X. Assembly of endocytic machinery around individual influenza viruses during viral entry. *Nat Struct Mol Biol.* 2004;11:567–573.
55. Lehmann MJ, Sherer NM, Marks CB, Pypaert M, Mothes W. Actin- and myosin-driven movement of viruses along filopodia precedes their entry into cells. *J Cell Biol.* 2005;170:317–325.

56. Miyauchi K, Kim Y, Latinovic O, Morozov V, Melikyan GB. HIV enters cells via endocytosis and dynamin-dependent fusion with endosomes. *Cell.* 2009;137:433–444.
57. Daecke J, Fackler OT, Dittmar MT, Krausslich HG. Involvement of clathrin-mediated endocytosis in human immunodeficiency virus type 1 entry. *J Virol.* 2005;79:1581–1594.
58. Klasse PJ. The molecular basis of HIV entry. *Cell Microbiol.* 2012;14:1183–1192.
59. Coyne CB, Bergelson JM. Virus-induced Abl and Fyn kinase signals permit coxsackievirus entry through epithelial tight junctions. *Cell.* 2006;124:119–131.
60. Ewers H, Smith AE, Sbalzarini IF, Lilie H, Koumoutsakos P, Helenius A. Single-particle tracking of murine polyoma virus-like particles on live cells and artificial membranes. *Proc Natl Acad Sci U S A.* 2005;102:15110–15115.
61. Pelkmans L, Burli T, Zerial M, Helenius A. Caveolin-stabilized membrane domains as multifunctional transport and sorting devices in endocytic membrane traffic. *Cell.* 2004;118:767–780.
62. Pelkmans L, Kartenbeck J, Helenius A. Caveolar endocytosis of simian virus 40 reveals a new two-step vesicular-transport pathway to the ER. *Nat Cell Biol.* 2001;3: 473–483.
63. Pelkmans L, Puntener D, Helenius A. Local actin polymerization and dynamin recruitment in SV40-induced internalization of caveolae. *Science.* 2002;296:535–539.
64. Dohner K, Wolfstein A, Prank U, et al. Function of dynein and dynactin in herpes simplex virus capsid transport. *Mol Biol Cell.* 2002;13:2795–2809.
65. McDonald D, Vodicka MA, Lucero G, et al. Visualization of the intracellular behavior of HIV in living cells. *J Cell Biol.* 2002;159:441–452.
66. Georgi A, Mottola-Hartshorn C, Warner A, Fields B, Chen LB. Detection of individual fluorescently labeled reovirions in living cells. *Proc Natl Acad Sci U S A.* 1990;87:6579–6583.
67. Li Y, Kappes JC, Conway JA, Price RW, Shaw GM, Hahn BH. Molecular characterization of human immunodeficiency virus type 1 cloned directly from uncultured human brain tissue: identification of replication-competent and -defective viral genomes. *J Virol.* 1991;65:3973–3985.
68. Huang AS, Baltimore D. Defective viral particles and viral disease processes. *Nature.* 1970;226:325–327.
69. Lazzarini RA, Keene JD, Schubert M. The origins of defective interfering particles of the negative-strand RNA viruses. *Cell.* 1981;26:145–154.
70. Roux L, Simon AE, Holland JJ. Effects of defective interfering viruses on virus replication and pathogenesis in vitro and in vivo. *Adv Virus Res.* 1991;40:181–211.
71. Von Magnus P. Incomplete forms of influenza virus. *Adv Virus Res.* 1954;2:59–79.
72. Weiss RA. AIDS. Defective viruses to blame? *Nature.* 1989;338:458.
73. Rouzine IM, Weinberger LS. Design requirements for interfering particles to maintain coadaptive stability with HIV-1. *J Virol.* 2013;87:2081–2093.
74. Trono D, Feinberg MB, Baltimore D. HIV-1 Gag mutants can dominantly interfere with the replication of the wild-type virus. *Cell.* 1989;59:113–120.
75. Baltimore D. Gene therapy intracellular immunization. *Nature.* 1988;335:395–396.
76. Bangham CR, Hall SE, Jeffery KJ, et al. Genetic control and dynamics of the cellular immune response to the human T-cell leukaemia virus, HTLV-I. *Philos Trans R Soc Lond B Biol Sci.* 1999;354:691–700.
77. Wodarz D, Nowak MA, Bangham CR. The dynamics of HTLV-I and the CTL response. *Immunol Today.* 1999;20:220–227.
78. Bangham CR, Kirkwood TB. Defective interfering particles and virus evolution. *Trends Microbiol.* 1993;1:260–264.
79. Kirkwood TB, Bangham CR. Cycles, chaos, and evolution in virus cultures: a model of defective interfering particles. *Proc Natl Acad Sci U S A.* 1994;91:8685–8689.

80. Bukovsky AA, Song JP, Naldini L. Interaction of human immunodeficiency virus-derived vectors with wild-type virus in transduced cells. *J Virol.* 1999;73:7087–7092.
81. Chen CJ, Banerjea AC, Harmison GG, Haglund K, Schubert M. Multitarget-ribozyme directed to cleave at up to nine highly conserved HIV-1 env RNA regions inhibits HIV-1 replication—potential effectiveness against most presently sequenced HIV-1 isolates. *Nucleic Acids Res.* 1992;20:4581–4589.
82. Evans JT, Garcia JV. Lentivirus vector mobilization and spread by human immunodeficiency virus. *Hum Gene Ther.* 2000;11:2331–2339.
83. Timm C, Akpinar F, Yin J. Quantitative characterization of defective virus emergence by deep sequencing. *J Virol.* 2014;88:2623–2632.
84. Klasse PJ, Moore JP. Is there enough gp120 in the body fluids of HIV-1-infected individuals to have biologically significant effects? *Virology.* 2004;323:1–8.
85. Baluda MA. Homologous interference by ultraviolet-inactivated Newcastle disease virus. *Virology.* 1957;4:72–96.
86. Baluda MA. Loss of viral receptors in homologous interference by ultraviolet-irradiated Newcastle disease virus. *Virology.* 1959;7:315–327.
87. Bratt MA, Rubin H. Specific interference among strains of Newcastle disease virus. II Comparison of interference by active and inactive virus. *Virology.* 1968;35:381–394.
88. Kuroda MJ, Schmitz JE, Charini WA, et al. Emergence of CTL coincides with clearance of virus during primary simian immunodeficiency virus infection in rhesus monkeys. *J Immunol.* 1999;162:5127–5133.
89. Letvin NL, Mascola JR, Sun Y, et al. Preserved CD4+ central memory T cells and survival in vaccinated SIV-challenged monkeys. *Science.* 2006;312:1530–1533.
90. Chen J, Nikolaitchik O, Singh J, et al. High efficiency of HIV-1 genomic RNA packaging and heterozygote formation revealed by single virion analysis. *Proc Natl Acad Sci U S A.* 2009;106:13535–13540.
91. Briggs JA, Krausslich HG. The molecular architecture of HIV. *J Mol Biol.* 2011;410:491–500.
92. Sundquist WI, Krausslich HG. HIV-1 assembly, budding, and maturation. *Cold Spring Harb Perspect Med.* 2012;2:a006924.
93. Zhao G, Perilla JR, Yufenyuy EL, et al. Mature HIV-1 capsid structure by cryo-electron microscopy and all-atom molecular dynamics. *Nature.* 2013;497:643–646.
94. Dale BM, McNerney GP, Thompson DL, et al. Cell-to-cell transfer of HIV-1 via virological synapses leads to endosomal virion maturation that activates viral membrane fusion. *Cell Host Microbe.* 2011;10:551–562.
95. Chojnacki J, Staudt T, Glass B, et al. Maturation-dependent HIV-1 surface protein redistribution revealed by fluorescence nanoscopy. *Science.* 2012;338:524–528.
96. Zhu P, Liu J, Bess Jr J, et al. Distribution and three-dimensional structure of AIDS virus envelope spikes. *Nature.* 2006;441:847–852.
97. Sougrat R, Bartesaghi A, Lifson JD, et al. Electron tomography of the contact between T cells and SIV/HIV-1: implications for viral entry. *PLoS Pathog.* 2007;3:e63.
98. Klasse PJ. Modeling how many envelope glycoprotein trimers per virion participate in human immunodeficiency virus infectivity and its neutralization by antibody. *Virology.* 2007;369:245–262.
99. Magnus C, Rusert P, Bonhoeffer S, Trkola A, Regoes RR. Estimating the stoichiometry of human immunodeficiency virus entry. *J Virol.* 2009;83:1523–1531.
100. Rein AL, Gerwin BI, Bassin RH, Schwarm L, Schidlovsky G. A replication-defective variant of Moloney murine leukemia virus. I. Biological characterization. *J Virol.* 1978;25:146–156.
101. Rein A, Benjers BM, Gerwin BI, Bassin RH, Slocum DR. Rescue and transmission of a replication-defective variant of moloney murine leukemia virus. *J Virol.* 1979;29:494–500.

102. Rein A, Athan E, Benjers BM, Bassin RH, Gerwin BI, Slocum DR. Isolation of a replication-defective murine leukaemia virus from cultured AKR leukaemia cells. *Nature*. 1979;282:753–754.
103. Yuan B, Campbell S, Bacharach E, Rein A, Goff SP. Infectivity of Moloney murine leukemia virus defective in late assembly events is restored by late assembly domains of other retroviruses. *J Virol*. 2000;74:7250–7260.
104. Leung J, Yueh A, Appah Jr FS, Yuan B, de los Santos K, Goff SP. Interaction of Moloney murine leukemia virus matrix protein with IQGAP. *EMBO J*. 2006;25:2155–2166.
105. O'Carroll IP, Crist RM, Mirro J, et al. Functional redundancy in HIV-1 viral particle assembly. *J Virol*. 2012;86:12991–12996.
106. Iordanskiy S, Santos S, Bukrinsky M. Nature, nurture and HIV: the effect of producer cell on viral physiology. *Virology*. 2013;443:208–213.
107. Arthur LO, Bess Jr JW, Sowder 2nd RC, et al. Cellular proteins bound to immunodeficiency viruses: implications for pathogenesis and vaccines. *Science*. 1992;258:1935–1938.
108. Bastiani L, Laal S, Kim M, Zolla-Pazner S. Host cell-dependent alterations in envelope components of human immunodeficiency virus type 1 virions. *J Virol*. 1997;71: 3444–3450.
109. Bounou S, Dumais N, Tremblay MJ. Attachment of human immunodeficiency virus-1 (HIV-1) particles bearing host-encoded B7-2 proteins leads to nuclear factor-kappa B- and nuclear factor of activated T cells-dependent activation of HIV-1 long terminal repeat transcription. *J Biol Chem*. 2001;276:6359–6369.
110. Bounou S, Giguere JF, Cantin R, et al. The importance of virus-associated host ICAM-1 in human immunodeficiency virus type 1 dissemination depends on the cellular context. *FASEB J*. 2004;18:1294–1296.
111. Bounou S, Leclerc JE, Tremblay MJ. Presence of host ICAM-1 in laboratory and clinical strains of human immunodeficiency virus type 1 increases virus infectivity and CD4(+)-T-cell depletion in human lymphoid tissue, a major site of replication in vivo. *J Virol*. 2002;76:1004–1014.
112. Cantin R, Fortin JF, Lamontagne G, Tremblay M. The acquisition of host-derived major histocompatibility complex class II glycoproteins by human immunodeficiency virus type 1 accelerates the process of virus entry and infection in human T-lymphoid cells. *Blood*. 1997;90:1091–1100.
113. Cantin R, Fortin JF, Lamontagne G, Tremblay M. The presence of host-derived HLA-DR1 on human immunodeficiency virus type 1 increases viral infectivity. *J Virol*. 1997;71:1922–1930.
114. Cantin R, Fortin JF, Tremblay M. The amount of host HLA-DR proteins acquired by HIV-1 is virus strain- and cell type-specific. *Virology*. 1996;218:372–381.
115. Esser MT, Graham DR, Coren LV, et al. Differential incorporation of CD45, CD80 (B7-1), CD86 (B7-2), and major histocompatibility complex class I and II molecules into human immunodeficiency virus type 1 virions and microvesicles: implications for viral pathogenesis and immune regulation. *J Virol*. 2001;75:6173–6182.
116. Fortin JF, Cantin R, Lamontagne G, Tremblay M. Host-derived ICAM-1 glycoproteins incorporated on human immunodeficiency virus type 1 are biologically active and enhance viral infectivity. *J Virol*. 1997;71:3588–3596.
117. Fortin JF, Cantin R, Tremblay MJ. T cells expressing activated LFA-1 are more susceptible to infection with human immunodeficiency virus type 1 particles bearing host-encoded ICAM-1. *J Virol*. 1998;72:2105–2112.
118. Giguere JF, Bounou S, Paquette JS, Madrenas J, Tremblay MJ. Insertion of host-derived costimulatory molecules CD80 (B7.1) and CD86 (B7.2) into human immunodeficiency virus type 1 affects the virus life cycle. *J Virol*. 2004;78:6222–6232.

119. Giguere JF, Paquette JS, Bounou S, Cantin R, Tremblay MJ. New insights into the functionality of a virion-anchored host cell membrane protein: CD28 versus HIV type 1. *J Immunol.* 2002;169:2762–2771.
120. Gomez MB, Hildreth JE. Antibody to adhesion molecule LFA-1 enhances plasma neutralization of human immunodeficiency virus type 1. *J Virol.* 1995;69:4628–4632.
121. Hioe CE, Chien Jr PC, Lu C, et al. LFA-1 expression on target cells promotes human immunodeficiency virus type 1 infection and transmission. *J Virol.* 2001;75:1077–1082.
122. Losier M, Fortin JF, Cantin R, Bergeron MG, Tremblay MJ. Virion-bound ICAM-1 and activated LFA-1: a combination of factors conferring resistance to neutralization by sera from human immunodeficiency virus type 1-infected individuals independently of the disease status and phase. *Clin Immunol.* 2003;108:111–118.
123. Martin G, Beausejour Y, Thibodeau J, Tremblay MJ. Envelope glycoproteins are dispensable for insertion of host HLA-DR molecules within nascent human immunodeficiency virus type 1 particles. *Virology.* 2005;335:286–290.
124. Martin G, Tremblay MJ. HLA-DR, ICAM-1, CD40, CD40L, and CD86 are incorporated to a similar degree into clinical human immunodeficiency virus type 1 variants expanded in natural reservoirs such as peripheral blood mononuclear cells and human lymphoid tissue cultured ex vivo. *Clin Immunol.* 2004;111:275–285.
125. Paquette JS, Fortin JF, Blanchard L, Tremblay MJ. Level of ICAM-1 surface expression on virus producer cells influences both the amount of virion-bound host ICAM-1 and human immunodeficiency virus type 1 infectivity. *J Virol.* 1998;72:9329–9336.
126. Rizzuto CD, Sodroski JG. Contribution of virion ICAM-1 to human immunodeficiency virus infectivity and sensitivity to neutralization. *J Virol.* 1997;71:4847–4851.
127. Tardif MR, Tremblay MJ. Presence of host ICAM-1 in human immunodeficiency virus type 1 virions increases productive infection of CD4 + T lymphocytes by favoring cytosolic delivery of viral material. *J Virol.* 2003;77:12299–12309.
128. Pang Y, Song H, Kim JH, Hou X, Cheng W. Optical trapping of individual human immunodeficiency viruses in culture fluid reveals heterogeneity with single-molecule resolution. *Nat Nanotechnol.* 2014;9:624–630.
129. Burrer R, Haessig-Einius S, Aubertin AM, Moog C. Neutralizing as well as non-neutralizing polyclonal immunoglobulin (Ig)G from infected patients capture HIV-1 via antibodies directed against the principal immunodominant domain of gp41. *Virology.* 2005;333:102–113.
130. Cavacini L, Posner M. Native HIV type 1 virion surface structures: relationships between antibody binding and neutralization or lessons from the viral capture assay. *AIDS Res Hum Retroviruses.* 2004;20:435–441.
131. Liu P, Williams LD, Shen X, et al. Capacity for infectious HIV-1 virion capture differs by envelope antibody specificity. *J Virol.* 2014;88:5165–5170.
132. Nyambi PN, Gorny MK, Bastiani L, van der Groen G, Williams C, Zolla-Pazner S. Mapping of epitopes exposed on intact human immunodeficiency virus type 1 (HIV-1) virions: a new strategy for studying the immunologic relatedness of HIV-1. *J Virol.* 1998;72:9384–9391.
133. Poignard P, Moulard M, Golez E, et al. Heterogeneity of envelope molecules expressed on primary human immunodeficiency virus type 1 particles as probed by the binding of neutralizing and nonneutralizing antibodies. *J Virol.* 2003;77:353–365.
134. Herrera C, Spenlehauer C, Fung MS, Burton DR, Beddows S, Moore JP. Nonneutralizing antibodies to the CD4-binding site on the gp120 subunit of human immunodeficiency virus type 1 do not interfere with the activity of a neutralizing antibody against the same site. *J Virol.* 2003;77:1084–1091.
135. Yasmeen A, Ringe R, Derking R, et al. Differential binding of neutralizing and nonneutralizing antibodies to native-like soluble HIV-1 Env trimers, uncleaved Env proteins, and monomeric subunits. *Retrovirology.* 2014;11:41.

136. McCune JM, Rabin LB, Feinberg MB, et al. Endoproteolytic cleavage of gp160 is required for the activation of human immunodeficiency virus. *Cell.* 1988;53:55–67.
137. Moore PL, Crooks ET, Porter L, et al. Nature of nonfunctional envelope proteins on the surface of human immunodeficiency virus type 1. *J Virol.* 2006;80:2515–2528.
138. Burton DR, Hessell AJ, Keele BF, et al. Limited or no protection by weakly or nonneutralizing antibodies against vaginal SHIV challenge of macaques compared with a strongly neutralizing antibody. *Proc Natl Acad Sci U S A.* 2011;108:11181–11186.
139. Leaman DP, Kinkead H, Zwick MB. In-solution virus capture assay helps deconstruct heterogeneous antibody recognition of human immunodeficiency virus type 1. *J Virol.* 2010;84:3382–3395.
140. Ray K, Mengistu M, Yu L, Lewis GK, Lakowicz JR, DeVico AL. Antigenic properties of the HIV envelope on virions in solution. *J Virol.* 2014;88:1795–1808.
141. Parrish NF, Gao F, Li H, et al. Phenotypic properties of transmitted founder HIV-1. *Proc Natl Acad Sci U S A.* 2013;110:6626–6633.
142. Keele BF, Li H, Learn GH, et al. Low-dose rectal inoculation of rhesus macaques by SIVsmE660 or SIVmac251 recapitulates human mucosal infection by HIV-1. *J Exp Med.* 2009;206:1117–1134.
143. Cavrois M, Neidleman J, Santiago ML, Derdeyn CA, Hunter E, Greene WC. Enhanced fusion and virion incorporation for HIV-1 subtype C envelope glycoproteins with compact V1/V2 domains. *J Virol.* 2014;88:2083–2094.
144. Louder MK, Sambor A, Chertova E, et al. HIV-1 envelope pseudotyped viral vectors and infectious molecular clones expressing the same envelope glycoprotein have a similar neutralization phenotype, but culture in peripheral blood mononuclear cells is associated with decreased neutralization sensitivity. *Virology.* 2005;339:226–238.
145. Klasse PJ, Moore JP. Quantitative model of antibody- and soluble CD4-mediated neutralization of primary isolates and T-cell line-adapted strains of human immunodeficiency virus type 1. *J Virol.* 1996;70:3668–3677.
146. Del Prete GQ, Scarlotta M, Newman L, et al. Comparative characterization of transfection- and infection-derived simian immunodeficiency virus challenge stocks for in vivo nonhuman primate studies. *J Virol.* 2013;87:4584–4595.
147. Murray JM, Kelleher AD, Cooper DA. Timing of the components of the HIV life cycle in productively infected CD4+ T cells in a population of HIV-infected individuals. *J Virol.* 2011;85:10798–10805.
148. Leaman DP, Zwick MB. Increased functional stability and homogeneity of viral envelope spikes through directed evolution. *PLoS Pathog.* 2013;9:e1003184.
149. Kassa A, Finzi A, Pancera M, Courter JR, Smith 3rd AB, Sodroski J. Identification of a human immunodeficiency virus type 1 envelope glycoprotein variant resistant to cold inactivation. *J Virol.* 2009;83:4476–4488.
150. Kassa A, Madani N, Schon A, et al. Transitions to and from the CD4-bound conformation are modulated by a single-residue change in the human immunodeficiency virus type 1 gp120 inner domain. *J Virol.* 2009;83:8364–8378.
151. Toyoshima K, Vogt PK. Enhancement and inhibition of avian sarcoma viruses by polycations and polyanions. *Virology.* 1969;38:414–426.
152. Zhang YJ, Hatziioannou T, Zang T, et al. Envelope-dependent, cyclophilin-independent effects of glycosaminoglycans on human immunodeficiency virus type 1 attachment and infection. *J Virol.* 2002;76:6332–6343.
153. Saphire AC, Bobardt MD, Gallay PA. Host cyclophilin A mediates HIV-1 attachment to target cells via heparans. *EMBO J.* 1999;18:6771–6785.
154. Platt EJ, Kozak SL, Durnin JP, Hope TJ, Kabat D. Rapid dissociation of HIV-1 from cultured cells severely limits infectivity assays, causes the inactivation ascribed to entry inhibitors, and masks the inherently high level of infectivity of virions. *J Virol.* 2010;84:3106–3110.

155. Kondratowicz AS, Lennemann NJ, Sinn PL, et al. T-cell immunoglobulin and mucin domain 1 (TIM-1) is a receptor for Zaire Ebolavirus and Lake Victoria Marburgvirus. *Proc Natl Acad Sci U S A*. 2011;108:8426–8431.
156. Moller-Tank S, Albritton LM, Rennert PD, Maury W. Characterizing functional domains for TIM-mediated enveloped virus entry. *J Virol*. 2014;88:6702–6713.
157. Moller-Tank S, Kondratowicz AS, Davey RA, Rennert PD, Maury W. Role of the phosphatidylserine receptor TIM-1 in enveloped-virus entry. *J Virol*. 2013;87: 8327–8341.
158. Paoletti A, Raza SQ, Voisin L, et al. Editorial: pannexin-1—the hidden gatekeeper for HIV-1. *J Leukoc Biol*. 2013;94:390–392.
159. Seror C, Melki MT, Subra F, et al. Extracellular ATP acts on P2Y2 purinergic receptors to facilitate HIV-1 infection. *J Exp Med*. 2011;208:1823–1834.
160. Paoletti A, Raza SQ, Voisin L, et al. Multifaceted roles of purinergic receptors in viral infection. *Microbes Infect*. 2012;14:1278–1283.
161. Perreira JM, Chin CR, Feeley EM, Brass AL. IFITMs restrict the replication of multiple pathogenic viruses. *J Mol Biol*. 2013;425:4937–4955.
162. Smith S, Weston S, Kellam P, Marsh M. IFITM proteins-cellular inhibitors of viral entry. *Curr Opin Virol*. 2014;4:71–77.
163. Marechal V, Clavel F, Heard JM, Schwartz O. Cytosolic Gag p24 as an index of productive entry of human immunodeficiency virus type 1. *J Virol*. 1998;72:2208–2212.
164. Fredericksen BL, Wei BL, Yao J, Luo T, Garcia JV. Inhibition of endosomal/lysosomal degradation increases the infectivity of human immunodeficiency virus. *J Virol*. 2002;76:11440–11446.
165. Wei BL, Denton PW, O'Neill E, Luo T, Foster JL, Garcia JV. Inhibition of lysosome and proteasome function enhances human immunodeficiency virus type 1 infection. *J Virol*. 2005;79:5705–5712.
166. Roesch F, Schwartz O. The SAMHD1 knockout mouse model: in vivo veritas? *EMBO J*. 2013;32:2427–2429.
167. Harris RS, Hultquist JF, Evans DT. The restriction factors of human immunodeficiency virus. *J Biol Chem*. 2012;287:40875–40883.
168. Braaten D, Franke EK, Luban J. Cyclophilin A is required for an early step in the life cycle of human immunodeficiency virus type 1 before the initiation of reverse transcription. *J Virol*. 1996;70:3551–3560.
169. Braaten D, Luban J. Cyclophilin A regulates HIV-1 infectivity, as demonstrated by gene targeting in human T cells. *EMBO J*. 2001;20:1300–1309.
170. Klasse PJ, Schulz TF, Willison KR. HIV. Cyclophilins unfold the Gag? *Nature*. 1993;365:395–396.
171. Luban J, Bossolt KL, Franke EK, Kalpana GV, Goff SP. Human immunodeficiency virus type 1 Gag protein binds to cyclophilins A and B. *Cell*. 1993;73:1067–1078.
172. Towers GJ, Hatziioannou T, Cowan S, Goff SP, Luban J, Bieniasz PD. Cyclophilin A modulates the sensitivity of HIV-1 to host restriction factors. *Nat Med*. 2003;9:1138–1143.
173. Delorme-Axford E, Coyne CB. The actin cytoskeleton as a barrier to virus infection of polarized epithelial cells. *Viruses*. 2011;3:2462–2477.
174. Marsh M, Bron R. SFV infection in CHO cells: cell-type specific restrictions to productive virus entry at the cell surface. *J Cell Sci*. 1997;110(Pt 1):95–103.
175. Yoder A, Yu D, Dong L, et al. HIV envelope-CXCR4 signaling activates cofilin to overcome cortical actin restriction in resting CD4 T cells. *Cell*. 2008;134:782–792.
176. Melikyan GB. Common principles and intermediates of viral protein-mediated fusion: the HIV-1 paradigm. *Retrovirology*. 2008;5:111.
177. Liu Y, Belkina NV, Shaw S. HIV infection of T cells: actin-in and actin-out. *Sci Signal*. 2009;2:pe23.

178. Harmon B, Campbell N, Ratner L. Role of Abl kinase and the Wave2 signaling complex in HIV-1 entry at a post-hemifusion step. *PLoS Pathog.* 2010;6:e1000956.
179. Spear M, Guo J, Wu Y. The trinity of the cortical actin in the initiation of HIV-1 infection. *Retrovirology.* 2012;9:45.
180. Vorster PJ, Guo J, Yoder A, et al. LIM kinase 1 modulates cortical actin and CXCR4 cycling and is activated by HIV-1 to initiate viral infection. *J Biol Chem.* 2011;286:12554–12564.
181. Gordon-Alonso M, Rocha-Perugini V, Alvarez S, et al. Actin-binding protein drebrin regulates HIV-1-triggered actin polymerization and viral infection. *J Biol Chem.* 2013;288:28382–28397.
182. Erb ML, Pogliano J. Cytoskeletal proteins participate in conserved viral strategies across kingdoms of life. *Curr Opin Microbiol.* 2013;16:786–789.
183. Spear M, Wu Y. Viral exploitation of actin: force-generation and scaffolding functions in viral infection. *Virol Sin.* 2014;29:139–147.
184. Taylor MP, Koyuncu OO, Enquist LW. Subversion of the actin cytoskeleton during viral infection. *Nat Rev Microbiol.* 2011;9:427–439.
185. Dohner K, Nagel CH, Sodeik B. Viral stop-and-go along microtubules: taking a ride with dynein and kinesins. *Trends Microbiol.* 2005;13:320–327.
186. Dohner K, Sodeik B. The role of the cytoskeleton during viral infection. *Curr Top Microbiol Immunol.* 2005;285:67–108.
187. Radtke K, Dohner K, Sodeik B. Viral interactions with the cytoskeleton: a hitchhiker's guide to the cell. *Cell Microbiol.* 2006;8:387–400.
188. Sodeik B. Unchain my heart, baby let me go—the entry and intracellular transport of HIV. *J Cell Biol.* 2002;159:393–395.
189. Schwartz O, Marechal V, Friguet B, Arenzana-Seisdedos F, Heard JM. Antiviral activity of the proteasome on incoming human immunodeficiency virus type 1. *J Virol.* 1998;72:3845–3850.
190. Roesch F, Meziane O, Kula A, et al. Hyperthermia stimulates HIV-1 replication. *PLoS Pathog.* 2012;8:e1002792.
191. Yang X, Kurteva S, Lee S, Sodroski J. Stoichiometry of antibody neutralization of human immunodeficiency virus type 1. *J Virol.* 2005;79:3500–3508.
192. Yang X, Kurteva S, Ren X, Lee S, Sodroski J. Stoichiometry of envelope glycoprotein trimers in the entry of human immunodeficiency virus type 1. *J Virol.* 2005;79:12132–12147.
193. Yang X, Kurteva S, Ren X, Lee S, Sodroski J. Subunit stoichiometry of human immunodeficiency virus type 1 envelope glycoprotein trimers during virus entry into host cells. *J Virol.* 2006;80:4388–4395.
194. Magnus C, Brandenberg OF, Rusert P, Trkola A, Regoes RR. Mathematical models: a key to understanding HIV envelope interactions? *J Immunol Methods.* 2013;398–399:1–18.
195. Magnus C, Regoes RR. Estimating the stoichiometry of HIV neutralization. *PLoS Comput Biol.* 2010;6:e1000713.
196. Magnus C, Regoes RR. Restricted occupancy models for neutralization of HIV virions and populations. *J Theor Biol.* 2011;283:192–202.
197. Magnus C, Regoes RR. Analysis of the subunit stoichiometries in viral entry. *PLoS One.* 2012;7:e33441.
198. Sattentau Q. Avoiding the void: cell-to-cell spread of human viruses. *Nat Rev Microbiol.* 2008;6:815–826.

CHAPTER ELEVEN

The Role of Chance in Primate Lentiviral Infectivity: From Protomer to Host Organism

Roland R. Regoes*,[1], Carsten Magnus†
*Institute of Integrative Biology, ETH Zurich, Zurich, Switzerland
†Institute of Medical Virology, University of Zurich, Zurich, Switzerland
[1]Corresponding author: e-mail address: roland.regoes@env.ethz.ch

Contents

Abstract

Infection is best described as a stochastic process. Whether a host becomes infected upon exposure has a strong random element. The same applies to cells exposed to virions. In this review, we show how the mathematical formalism for stochastic processes has been used to describe and understand the infection by the Human and Simian Immunodeficiency Virus on different levels.

We survey quantitative studies on the establishment of infection in the host (the organismal level) and on the infection of target cells (the cellular and molecular level). We then discuss how a synthesis of the approaches across these levels could give rise to a predictive framework for assessing the efficacy of microbicides and vaccines.

1. INTRODUCTION

While Mendel is often called the first geneticist, he could also be referred to as the first statistical biologist (Stanislav Leibler in the Institute for Theoretical Studies Science Colloquium on May 2, 2014). In Mendelian

Progress in Molecular Biology and Translational Science, Volume 129
ISSN 1877-1173
http://dx.doi.org/10.1016/bs.pmbts.2014.10.013

inheritance, the offspring inherits one of two alleles from each parent randomly. While this process can have many outcomes if only one offspring is observed, the frequencies of outcomes in large ensembles reveal the underlying process probabilities according to the law of large numbers.

The probabilistic logic of Mendelian inheritance can nowadays be found in many fields of biology. In particular, in population genetics,[1] mathematical epidemiology,[2] and studies on gene expression within cells[3] probabilistic approaches, combinatorial arguments, and stochastic process play a central role. In large parts of the microbiology and immunology communities, however, randomness is a marginal topic and probabilistic approaches are rare.

For a few virus infections, such as HIV, SIV, and HCV, the within-host population dynamics has been worked out in detail (see Alizon and Magnus[4]; Nowak and May[5]; Perelson[6]; Perelson and Ribeiro[7] for reviews). With a few exceptions, the modeling of these systems is deterministic, i.e., ignores the stochastic aspects of the dynamics. Especially in the early phases of the infection, however, when the fate of the virus population is uncertain because of small population sizes, stochastic approaches can be very insightful. Early phases are also most relevant for vaccination, and other interventions aiming at preventing the establishment of the virus in the host.

In this chapter, we give an overview of probabilistic and stochastic approaches that have been used to study the establishment of infection within hosts and the entry of viruses into cells. Rather than introducing the mathematical formalisms rigorously, we outline the conceptual underpinning. Our target audience is experimental immunologists and virologists with an interest in the quantitative aspects of HIV infection. We aim to illustrate how one can rationally treat random events in the context of infection biology, and how probabilistic approaches have been used to interpret experimental data.

We start with discussing the role of stochasticity in infection on the organismal level. We will introduce the stochastic approaches that can be used to estimate infection probabilities upon exposure to a certain amount of virus. We then show how the efficacy of a vaccine or other preventive treatments can be assessed and quantified within this stochastic framework. Lastly, we will illustrate that stochastic approaches describing the population dynamics within the host are qualitatively different from the deterministic approaches commonly used.

After that, we turn to cell entry and the probabilistic approaches that have been used to study the infection of cells and how it can be impaired by antibodies. In the last section, we combine the organismal and cellular levels to investigate pre-exposure prophylaxis with neutralizing antibodies or microbicides. In particular, we elaborate on how a combination of existing mathematical models can be used to predict the concentration of neutralizing antibodies needed to reduce the infection probability.

2. HOST ENTRY

2.1. Modeling acquisition

The rate at which HIV is acquired upon naturally occurring exposure is very low: estimates range from as low as 1/1000 (Ref. 8) to 1/10 per sex act.[9] The highest estimate to date, 1/4, stems from a study conducted in the adult film industry.[10] But even for the high estimates, establishment of infection is by no means a matter of certainty. Thus, on the crudest level, infection can be likened to a coin toss with an unbalanced coin.

In the context of clinical trials, the rarity and stochasticity of infection is accounted for in the trial design and analysis. On a very basic level, clinical trials involve the assignment of individuals to placebo and vaccine groups. After some time, the number of infected hosts between the two groups is compared. In such trials, infection is the result of stochastic exposure and stochastic acquisition. The probability of acquisition is modulated by the intrinsic susceptibility of the host. The combination of low natural exposure levels and low probability of acquisition of infection upon exposure makes large study populations necessary to assess whether the vaccine protects against infection (see Fig. 1). In a typical HIV vaccine trial, less than 1 in 10 of the individuals enrolled become infected, and the majority of study participants remains uninfected.[11]

While the most straightforward effect of a vaccine is the reduction of the acquisition probability, many alternative effects have been considered. A vaccine can also reduce the rate of progression to disease and the infectiousness of individuals who are infected. In addition, vaccination can confer indirect benefits to nonvaccinated individuals in a community through herd immunity. Sophisticated trial designs and methods of statistical inference have been developed to estimate the different effects vaccines can have.[12] However, because directly preventing infection confers the largest benefit to the vaccinated individual and the entire community, it remains the most desirable characteristic of a vaccine.

In preclinical trials conducted between the very beginning of the HIV pandemic to approximately 10 years ago, the stochasticity of the infection process was excluded by design: very high viral doses were used to challenge animals resulting in certain infection (for typical studies see Amara *et al.*[13]; Barouch *et al.*[14]; Rose *et al.*[15]; Shiver *et al.*[16]). The use of high challenge doses was justified with two arguments. First, by using high challenge doses

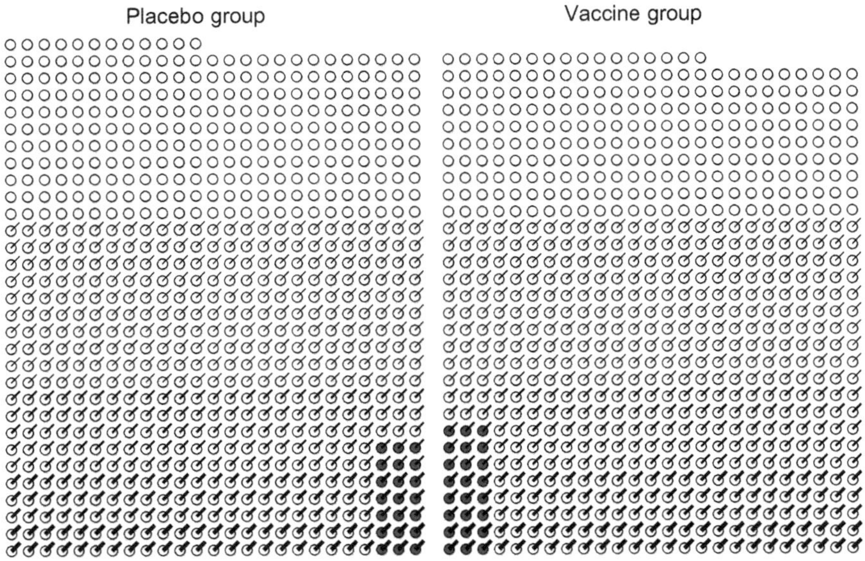

Figure 1 Schematic of a vaccine trial. The circles signify individuals enrolled in the study, red (gray in the print version) circles denote individuals who become infected. The numbers in this diagram are inspired by the STEP trial.[11] The short diagonal line denotes exposures. Their thickness corresponds to the number of exposures. Many of the individuals enrolled will not be exposed to HIV at all, even in a high-risk group, because the probability of exposure depends on the prevalence of HIV. Most infections will be found in individuals who are frequently exposed. However, some individuals with less frequent exposure are also expected to become infected. From the number of infected individuals in the two groups of this example, one can roughly estimate a vaccine efficacy against infection of $1-\frac{24/741}{21/762}=-0.18$. That this is negative means that the vaccine, if it has any effect at all, facilitates infection.

the bar for vaccine candidates was intentionally set high. Second, experimenters wanted to make sure that animals were infected. Uninfected animals, while providing essential information on infection probability, preclude further virological assessment and were thus seen as wasteful use of resources.

Over the past decade, a paradigm shift toward challenging with lower viral inocula has been taking place. One driver behind this shift was the realization that the failure to infect animals contains important information on the infection probability. Additionally, experimental designs were proposed that minimized the wasteful failure to infect animals: the repeated low-dose challenge design.[17,18] The idea is simply to challenge animals again if they have not been infected upon the preceding challenge, as depicted in Fig. 2. Through this procedure almost all animals become infected

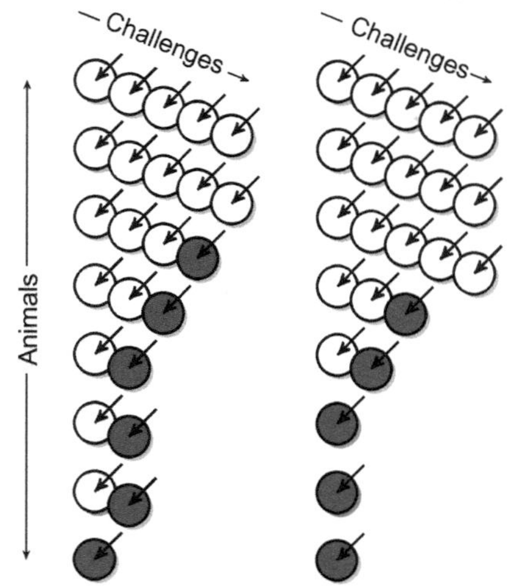

Figure 2 Schematic of a preclinical trial using repeated, low-dose challenges. Depicted are the data by Wilson *et al.*[19] The circles signify individual animals enrolled in the study, red (gray in the print version) circles denote animals that become infected. The small arrows denote challenges. Compared to clinical trials, this design is economical in that a large fraction of hosts become infected (over half). In this trial, approximately one in five challenges resulted in infection. To quantify vaccine efficacy, one often assumes that the probability of infection is the same for each challenge. For the data shown, one can then obtain an efficacy of $1-\frac{5/23}{6/24}=0.13$. While this is positive, it is not significantly different from zero.[19]

eventually. Assuming that the probability of infection is independent of the challenge history, vaccine efficacies can be estimated.[20]

Challenging repeatedly with low doses has become the design of choice[19,21–37] initially mostly because it more realistically reflects natural exposure to HIV. But since this design allows to quantify the effect of vaccines on acquisition, correlates of protection are being identified.

Interesting patterns have emerged from comparing the vaccine effect on acquisition and on virus replication after infection.[38] While vaccines that did not include the viral envelope protein as an immunogen failed to reduce the probability of acquisition, they often resulted in lower peak or set-point virus loads.[19,25,37] Vaccines that include the viral envelope protein often result in protection from infection,[21–23,27,28,30,35] although there are exceptions to this rule.[39] Two recent studies[21,35] showed that there is no correlation between the effects on acquisition and virus load in the vaccinated host after infection—the primary end-point in the early preclinical vaccine trials.

These patterns are largely consistent with the outcomes of recent clinical HIV vaccine trials. Adenovirus 5-based vaccines that did not include the viral envelope failed to prevent infection.[11,40] In contrast, the vaccine tested in the RV144 trial[41] that used gp120 as an immunogen showed moderate protection. However, none of these vaccines had any effect on the virus load after infection, which—at least for the Adenovirus 5-based vaccines—is not consistent with preclinical trials. Overall, this strongly suggests to take preclinical repeated challenge data and what they reveal about the effects of the vaccine into serious consideration when choosing vaccine candidates for clinical testing.

In comparison with clinical trials, repeated low-dose challenge trials in nonhuman primates resolve the exposure history of individual hosts in much greater detail: the number of challenges each animal received is known and the challenge that gave rise to infection can be ascertained. This allows addressing questions that are almost impossible to elucidate in a clinical setting. One such question is the role of exposure history on infection probability: it is conceivable that noninfecting challenges lead to partial immunization, making future infection less likely. Another question is if there are susceptibility differences between animals. Both, immunization and susceptibility differences would manifest themselves as infection probabilities that become smaller and smaller with every challenge repeat. In the case of susceptibility differences, this pattern emerges as animals with high susceptibility are, on average, infected earlier in the challenge schedule, leaving predominantly animals with low susceptibility at later stages.

We made a first step into the direction of answering whether noninfecting challenges immunize or whether there are significant susceptibility differences between animals.[42] Analyzing repeated challenge data of the control groups in many experimental studies,[19,22,24,25,30,37,39] we found no evidence for immunization by noninfecting challenges. We also could not find any significant susceptibility differences in most studies. In our study, small sample sizes in the challenge experiments we analyzed are an issue. These analyses could, however, be conducted with data from more recent and substantially larger repeated challenge studies (e.g., Roederer *et al.*[35]), and tied into the design of future experiments.

2.2. Modeling initial viral replication and establishment

The conceptualization of infection presented in the previous subsection is very crude as it tries to capture many intricate processes with one parameter:

the probability of infection. Human and Simian Immunodeficiency Viruses (HIV and SIV), when transmitted sexually, have to cross a layer of epithelial cells.[43] If successful, the virus has to replicate in the lamina propria beneath the epithelium. The most abundant cell type in this compartment is the resting CD4+ T cell. Subsequently, the virus is disseminated into the lymphoid tissue compartment where the conditions for maintaining virus replication are much more favorable. Virus either spreads as free virus particles or is carried within infected cells or by other cells, which are resident in the lamina propria and express HIV-specific attachment factors.[44] In the lymphoid tissue, the fast replicating virus elicits humoral and cellular immune responses, which, however, are not strong enough to clear the infection.[43]

Some of these host entry processes are stochastic and have also been included into mathematical models of the virus dynamics.[45–50] In all these models, infection is conceived as an initial virus inoculum racing to infect cells before it is cleared. Figure 3 shows such a process, modeled in line with the arguments presented in a recent opinion article by Talbert-Slagle *et al.*[51]

One of the most important insights from these stochastic models is that vaccine induced immunity does not have to completely control virus replication to be effective.[50] We illustrate this important point with a simple stochastic model for virus replication within the host shortly after entry. Figure 3B shows simulations of this model and compares how many exposures lead to infection in unvaccinated and vaccinated hosts. In the simulations, we assume that the vaccine elicits immune responses that only reduce the replication rate of the virus, and have no effect on the inoculum. If the vaccine induces a level of immunity that reduces the average number of viral progeny from 6 to 1.5, we obtain a vaccine efficacy of 44% in terms of reducing the probability of establishing an infection. Thus, the vaccine is partly effective despite the fact that the virus population in vaccinated individuals is still able to grow. This effect is most pronounced if the average number of viral progeny is brought close to 1 by the vaccine. In deterministic models in contrast, a successful vaccine needs to control virus replication completely otherwise the virus is predicted to establish an infection with certainty.[52,53]

These simulations can also show how the initial dose affects the number of founder viruses. For a high dose, many viruses in the inoculum leave offspring, while at low initial inoculum sizes a single virus particle founds the population. This is in line with studies in humans, which established that in most HIV infections a single genotype founds the viral population using stochastic models of neutral virus evolution during acute infection.[54,55] It is also consistent with studies in nonhuman primates,[56] in which high challenge

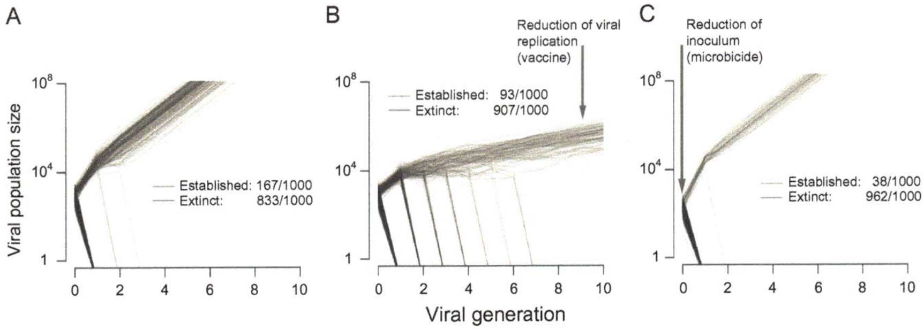

Figure 3 Simulation of the early virus dynamics. (A) Simulation in untreated hosts. We assume that V_0 virions enter the host. The number is drawn from a discretized Γ-distribution with mean 1000. Each virion has a large probability to be cleared before infecting a cell. Cell infection is assumed to occur with a probability 1/5000. Cell infection gives rise to *B* secondary virions. *B* is drawn from a discretized Γ distribution with mean 3×10^4. The process is iterated over many viral generations. In 833 of the 1000 simulations shown, the virus population goes extinct, while in 167 simulations the virus establishes within the host. (B) Simulation of the early virus dynamics in vaccinated hosts. Vaccination is assumed to leave the inoculum unchanged and to reduce the number of virions produced by a productively infected cell from an average of 3×10^4 to 0.75×10^4. This assumption is valid for vaccines that elicit late-acting, central immunity, rather than immune responses at the virus's port of entry. While the virus population still grows under the vaccine, a lower fraction of exposures is predicted to establish an infection: 93/1000 versus 167/1000 in (A). Comparing the simulations without and with vaccine, the vaccine efficacy can be calculated as $1 - \frac{93/1000}{167/1000} = 0.44$. (C) Simulation of the early virus dynamics in host treated prophylactically with neutralizing antibodies or an antiretroviral agent. For the sake of simplicity, prophylaxis is assumed to reduce only the initial number of virions, from an average of 1000 to 250, although an effect on viral replication cannot be ruled out. This results in a lower fraction of exposures establishing an infection: 38/1000 versus 167/1000 in (A). Comparing the simulations without and with prophylaxis, the vaccine efficacy can be calculated as $1 - \frac{38/1000}{167/1000} = 0.77$.

doses resulted in many viral founders, while low challenge doses mirrored the situation in humans. (We note that we use the term *founder virus* subtly different from Keele *et al.*[54] In our simulation, a founder virus is the individual virus particle, the progeny of which gives rise to the viral population, while Keele *et al.*'s[54] definition is genetical referring to a unique viral lineage that may consist of many genetically identical virus particles.)

The next frontier in modeling the early dynamics of HIV and SIV will be to take into account the anatomy of early infection. To-date, most models assume a well-mixed population dynamics within the infected host. There are some studies, which investigated the compartmentalization of the dynamics by analyzing viral sequence data obtained during the chronic phase of the infection.[57–61] Only the virus population in the brain and central nervous system was found to be genetically distinct from the population in other compartments, such as the blood or the lymphatic tissues.[58–60] However, the chronic stage, on which these studies focused, is not immediately relevant for the establishment of the infection. Anatomically resolved modeling of the processes unfolding shortly after virus entry into the host is needed to obtain a quantitative understanding of the establishment of HIV infection. This quantitative understanding could inform us how to optimally target vaccination and treatment strategies to specific tissue compartments to prevent infection.

3. CELL ENTRY

In this section, we switch from the level of infection of a host by a population of viruses to that of infection of a single cell by a virus particle. Hereby, we focus solely on HIV entry into its target cells, and how entry can be prevented by antibodies or inhibitors. Although the subsequent steps in the viral life cycle also display stochasticity and have been modeled mathematically,[62–64] their consideration are beyond the scope of the present article.

3.1. Infecting a cell

To establish an infection, the virions that have successfully entered a human host need to find and enter target cells. While there are no estimates of the probability of finding and infecting a target cell *in vivo*, *in vitro* studies show very low numbers with a high variability even within the studies ranging from 1 in 10^2 to 1 in 10^7 (Refs. 65–71).

Do the virions that fail to infect a cell look different from those that succeed, or were they simply unlucky? To discuss this question, we differentiate between *infecting* virions—those virions that end up infecting a cell—and *infectious* virions. Infectious virions are those virions that fulfill certain molecular requirements for cell infection but do not necessarily end up infecting a cell due to stochastic events, e.g., viral disintegration before finding a cell to infect or a too large distance to the next target cell. In infectivity assays, only infecting viruses can be measured. To further discuss the characteristics of infectious virions, we briefly recapitulate the molecular aspects of HIV entry and introduce the stoichiometric parameters, which are often used to describe the interaction between virions and target cells.

The first step in infecting a cell is attachment to the cellular surface. This contact is established via the viral spikes in which HIV particles express on their membrane.[72,73] These spikes are formed by three identical envelope glycoprotein heterodimers, comprising a gp120 subunit noncovalently linked to the transmembrane protein gp41 (Refs. 74–76). The trimers establish a first contact with CD4 receptors expressed on target cells.[77,78] After successful engagement of CD4, the envelope trimer undergoes conformational changes that allow a coreceptor, most commonly chemokine receptors CCR5 and CXCR4, to bind.[79] Binding to the coreceptor initiates a series of rearrangements in the viral envelope protein gp41, which upon insertion of the fusion peptide into the cellular membrane brings together viral and cellular membrane and triggers the fusion process. Finally, the viral core containing the RNA genome is released into the cytosol.

Many of these processes have a random element. First, the number of trimers that end up on the virions during budding varies from virion to virion. Zhu *et al.*[80] visualized 40 HIV-1 virions with cryoelectron microscopy, counted the trimers on each virion and found that 14 ± 7 trimers are expressed on the viral surface. Second, these trimers do not have predefined fixed positions but are randomly distributed on the viral surface. The probability distribution of the trimer locations on the virion surface has not been determined yet. It is also unknown if trimers can move on the virion's surface. However, clusters of trimers with at least seven spikes having a pairwise distance $\leq$30 nm are more frequently observed than expected by a uniform distribution.[80] In a later study, Sougrat *et al.*[81] identified areas on the viral surface of 40 nm diameter in which five to seven trimers are clustered together, referred to as *entry claw*. When looking at virions in the act of attaching and fusing with a cell, almost no trimers were observed outside the entry claw.

The existence of such an entry claw raises the question whether all trimers belonging to the entry claw are needed for membrane fusion. The numeric requirement, i.e., the number of trimers that are needed to engage with cell receptors for cell entry, is called *stoichiometry of entry*, T, with reference to numerical requirements in chemical reactions.[82] If this number is greater than one, several probabilistic questions arise: How high is the probability that T trimers engage with cellular receptors at the same time? If the trimers cannot move on the viral surface and thus cannot be recruited to the entry site, how likely is it that T trimers are close enough on the virion's surface to establish an entry claw?

The easiest way to study whether there is a threshold number of trimers that is required for viral entry would be to sort the virions according to their trimer numbers and test the infectivity of these viral stocks. Despite recent advances in virometry, i.e., optical trapping and analysis of HIV virions,[83] this technique does not allow to separate virions with, e.g., eight and nine trimers. By introducing a dominant negative mutation into one of the trimer's protomers, the whole trimer can be rendered nonfunctional. With this technique, the average number of functional trimers of a whole virion stock can be artificially lowered (see Fig. 4). These virus stocks were used in infectivity assays to study the stoichiometry of entry.[82,84–86] These assays do not directly lead to an estimate of the stoichiometry of entry. Due to the stochastic nature of these experiments, e.g., the randomness of trimer formation and

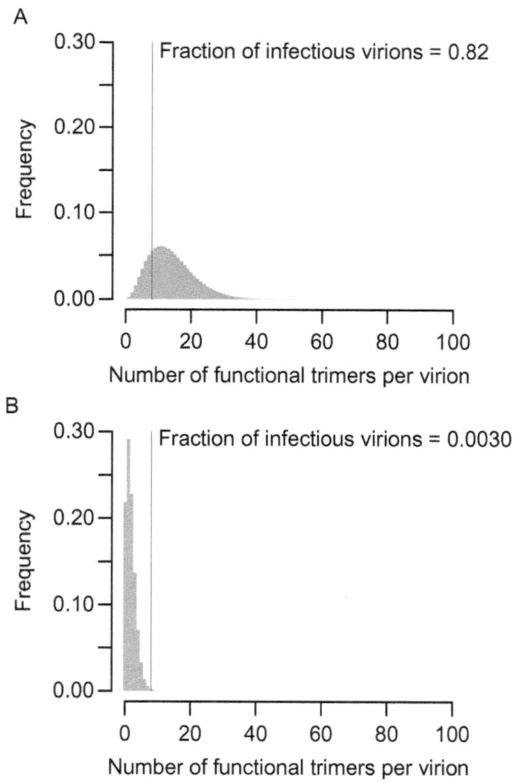

Figure 4 Examples of distributions of functional trimer numbers. (A) Discretized B-distribution with 14 ± 7 trimers, according to the experimentally found distribution in Zhu *et al.*[80] (B) By mixing wild type and dominant negative mutated envelope proteins in the relation 1:1, the distribution of functional trimers shifts toward low trimer numbers.

expression on the viral surface, a mathematical model is required to infer the stoichiometry of entry.

Several models have been proposed for this purpose.[82,84,86,87] The first models did not consider multiple trimers per virion[82,86] and estimated that one trimer is sufficient for cell entry. A model that assumed exactly nine trimers per virion[87] estimated this number to be five. We extended these models by a trimer number distribution with mean 14 and variance 49 according to the experimentally obtained trimer numbers[80] and estimated $T=8$. The models assume perfect transfection of virus producer cells, perfectly random assembly of trimers, and ignore the spatial locations of trimers on the surface of the virions. Relaxing these assumptions leads to estimates of the stoichiometry of entry, T, ranging from 2 to 19. Which of these models best describes reality cannot be determined on the basis of their goodness of fit to the data alone. A definitive estimate will require to quantitatively characterize the many processes involved in the production of pseudotyped virions.

Coming back to the problem of the fraction of infectious virions we raised at the beginning of this section, we can define one requirement for being infectious as having at least T functional trimers. In Fig. 4A, we show the trimer number distribution we used in[84] with mean 14 trimers and variance 49. If the stoichiometry of entry was eight, only 82% of virions would be infectious. This is a substantially higher fraction than the experimentally observed values for infecting virions that range from 1 in 10^2 to 1 in 10^7 (Refs. 65–71). According to the concept of stoichiometries, the fraction of infectious virions decreases when T increases or the average number of functional trimers on the virion surface decreases (the latter is depicted in Fig. 4B). However, whether an infectious virion ends up infecting a cell is not just determined by the number of trimers on its surface but also contingent on many other factors and processes, such as the location of the virion relative to target cells, its orientation, and its pattern of movement, which are themselves associated with randomness. Once the stoichiometry of entry is determined, we can attempt to disentangle the molecular from other factors and calculate the probability with which an infectious virion infects a cell.

3.1.1 Cell factors influencing viral entry and cell infection

In analogy to genetic and other factors affecting the host's susceptibility to infection there are intrinsic cellular factors influencing the probability that a virus can infect a cell. As described above, the virus needs to attach to CD4 receptors on the host cell followed by binding to a coreceptor. Johnston *et al.*[88] introduced a cellular system that allows to manipulate the density of CD4 and the coreceptor CCR5 as well as to test the cell lines' ability of being infected (for an extensive review of the experimental system and its use see Ref. 89). As a general rule, cells with a higher surface density of CD4 and coreceptors can be infected more easily than those cells with a lower receptor density. Mulampaka and Dixit[90] built a model to describe trimer binding to the coreceptor CCR5 in a cell–cell fusion system and estimated that the critical surface density for cell–cell fusion is approximately 20 μm^{-2}. Their model describes the binding reaction network with deterministic differential equations. Extending these approaches by describing the binding reactions stochastically and by taking into consideration varying trimer numbers across virions, might further our understanding of receptor densities required for cell infection.

3.2. Preventing the infection of a cell: The role of neutralizing antibodies

Why is it important to know how many trimers are needed to infect a cell and which practical implications does this threshold have? For HIV, it has been shown that binding of envelope-specific antibodies to the envelope trimer interferes with the entry process and may result in virus neutralization.[91,92] Possible targets for broadly neutralizing antibodies have been identified over the past decades and are more or less restricted to accessible conserved regions on the Env trimers.[93,94] Antibody binding to functional trimers is necessary for neutralization. Rendering a certain number of such trimers nonfunctional by simply binding to them may be sufficient for neutralization. Knowing this numerical requirement is crucial for estimating the antibody concentrations necessary to prevent infection at the primary sites of virus entry into the new host.

One early approach to determine numerical requirements for virion neutralization tackled this problem on the virion level borrowing concepts from chemical binding kinetics.[95,96] The virion was modeled as an entity with noncritical and $n \geq 1$ critical binding sites. As soon as an antibody binds to one of the critical sites, the virion is neutralized.[96] When the neutralization kinetics is of first order, Dulbecco *et al.*[96] hypothesized that, in combination with a linear relationship between the antibody concentration and rate of neutralization, binding of one single antibody is sufficient for virus neutralization (single-hit). A direct decline in the time-neutralization curve without a lag phase was interpreted as first-order kinetics. In contrast, a lag phase at the beginning of the time-neutralization curve was interpreted as a multihit mechanism. By applying these methods to the specific case of three rat-derived monoclonal HIV antibodies, McLain and Dimmock[97] concluded that one, two, or three antibodies are sufficient to neutralize a complete virion, respectively. Klasse and Sattentau[98] reviewed these low numbers critically and related the neutralizing occupancy by antibody on virions to the stoichiometry of entry. They showed that the number of required hits cannot be inferred from the order of the neutralization kinetics. Additionally, they pointed out that while a lag on a kinetic curve may refute single-hit neutralization, the absence of a lag does not confirm it.

The idea behind another early model of antibody binding is complete surface coverage by antibodies. Moran and Fazekas de St Groth[99] applied a model for how circular objects are independently and uniformly distributed over the surface of the sphere. Neutralization of a virion was reached, when the whole surface was covered by antibodies. Today we know that, in the case of HIV, only a few viral spikes need to be rendered nonfunctional for neutralization of the virion. Thus, these studies are mostly of theoretical value as a stochastic model of virus neutralization, but cannot be applied to HIV.

Instead, the requirements on the number of antibodies for virion neutralization was broken down into studying the stoichiometry of entry (as described in Section 3.1) and the number of antibodies needed to bind to one single trimer for neutralization. The latter is called *stoichiometry of (trimer) neutralization*, N (Refs. 82,85,87,100). The exact definition of the stoichiometry of trimer neutralization is the number of protomer–antibody interactions needed to render the trimer nonfunctional. For monovalent antibodies binding to one epitope located on one protomer only, the two definitions are the same. The experimental approach is similar to the one described in Section 3.1: The neutralization sensitivity of different stocks of pseudotyped virions expressing mixed trimers consisting of an antibody sensitive and an antibody resistant envelope protein in various ratios are tested in infectivity assays. Also the theoretical framework is similar to the one used for the analysis of the entry data. Different studies included different levels of complexity and different random effects.[82,85,87,100] Despite the different approaches, all these studies indicate that one antibody needs to bind to one trimer to render it nonfunctional. However, the possibility that more than one antibody is needed cannot be ruled out, especially for yet untested antibody-virus combinations.

According to the stoichiometrical framework, we can calculate the number of antibodies that need to bind to a virion for neutralization by combining the estimate of the number of antibodies needed to neutralize one trimer (stoichiometry of trimer neutralization), N, the number of trimers that are needed for cell entry (stoichiometry of entry), T, and the number of trimers per virion. Virions with less than T trimers are not infectious. As an example, let us assume a virion with four trimers, $T=2$ and $N=2$ (see Fig. 5A). At least three trimers need to be bound to at least two antibodies for neutralization. However, these six antibodies need to bind to exactly three different trimers and two antibodies to each of the trimers, as shown in Fig. 5A case I. If one trimer is bound to three antibodies, one to two and another one to one antibody, the virion is still infectious (Fig. 5A case II). If two trimers are bound to three antibodies and the remaining two trimers to one antibody each—that makes eight antibodies—the virion is still infectious. Binding of an additional antibody will inevitably lead to

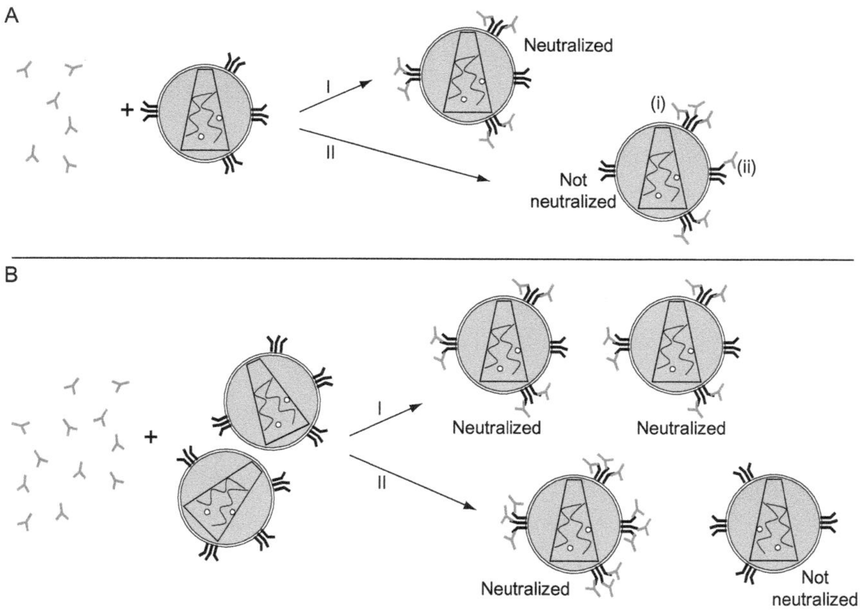

Figure 5 The location of antibodies binding to a virion or a population of virions plays a pivotal role for neutralization. For this scheme, we assume $T=2$ and $N=2$. (A) Six antibodies bind to one virion with four trimers. The virion is only neutralized when exactly three trimers are bound to two antibodies each (case I). If the antibodies do not bind in the right locations, the virion is not neutralized (case II). (B) Neutralization on the population level (here an example with two virions) adds another complexity: virions that are already neutralized can act as decoys for the remaining antibodies (case II).

neutralization. This example shows that there is a lower bound for the number of antibodies needed for virion neutralization but also an upper bound of the number of bound antibodies with and above which the virion is neutralized with certainty. For all antibody numbers between these bounds, there is a positive probability with which the virion is neutralized depending on the position of binding. By translating this problem into an urn model, Magnus and Regoes[101] could derive a mathematical expression for the average number of antibodies that is required for virion neutralization. The authors also scaled up this theory to calculate the average number of antibodies needed to bind to a whole virion population for complete neutralization. In this example, one can also see that antibodies can bind *unproductively* and do not add to neutralization. On the virion level this can happen, when an antibody binds to an already neutralized trimer (if $N<3$, shown in Fig. 5A case II(i)) or binds to a trimer that is not completely neutralized before a sufficient number of other trimers fulfill the neutralization condition (shown in Fig. 5A case II(ii)). On the population level, antibodies can bind to already neutralized virions.[101] An example for a population of two virions is shown in Fig. 5B.

The stoichiometrical framework is solely based on stochastic effects and only accounts for the number of antibodies bound. It neglects the affinity and concentration of antibodies as well as the kinetics of their binding. Magnus[102] combined the stoichiometrical framework with antibody binding kinetics and shows that this framework can also be used to estimate stoichiometrical parameters out of kinetic neutralization curves obtained earlier.[97] Hu *et al.*[103] combined reaction kinetics and stoichiometries to calculate the probability of antibodies directed to the membrane proximal external region to render the trimer nonfunctional after it has bound to the CD4 receptor. However, this approach does not allow estimating the number of antibodies required to render a trimer nonfunctional.

Stoichiometric parameters have not definitively been determined yet. However, the stochastic framework that captures the combinatorial aspects of antibody neutralization on different levels has proven to be useful in defining and estimating important parameters in antibody induced neutralization. Based on the experimental measurements available to-date, which constitute averages over large ensembles of virions and antibodies, it will not be possible to determine whether there is a threshold of trimers that have to engage with cellular receptors and whether it is sufficient to bind a high number of trimers to antibodies such that the whole virion is neutralized.

3.2.1 Analogies in antiviral drug treatment

Not only antibodies can interfere with the infection of a cell but also antiviral drugs. Different drug classes have been identified that act on different steps of the viral infection and virion production. Shen *et al.*[104] showed that stochastic effects also play a big role in HIV inhibition by drugs. The authors of this study introduce a critical subset model in which a number of target molecules within a virion needs to remain unbound for the virion to successfully produce offspring. The concept behind this model is the same as the stoichiometric argumentation for viral entry and antibody neutralization.[84–87,100] This especially becomes clear when comparing the expressions for the relative infectivity of a viral stock. The number of drug molecules per virion and the receptor occupation is a stochastic event and is therefore modeled with a binomial distribution. With this model the authors can explain the high antiretroviral activity of two different drug classes that are important for HIV treatment.

4. SYNTHESIS AND OUTLOOK

The two seemingly disparate topics presented in this article intersect in the context of pre-exposure prophylaxis with neutralizing antibodies and antiretroviral agents. Broadly neutralizing antibodies have been considered as prophylactic agents, and were shown to be efficacious in nonhuman primate models after intravenous[26,105–107] and intravaginal administration.[108] While passive immunization with broadly neutralizing antibodies has been applied to HIV infected humans,[109] this strategy has not been tested prophylactically yet. Also prophylactically administered antiretroviral drugs were shown to effectively reduce infection in animal models,[24,110] and some of these approaches have been tested in clinical trials (reviewed by Klasse *et al.*[111]).

From the perspective of the early stochastic dynamics of HIV infection, microbicides reduce mainly the inoculum, to which the host is exposed. In contrast, a vaccine will affect mostly the growth rate of the virus within the host, unless it also induces immune responses at the mucosal surfaces, where the virus enters the host. Intravenous administration of neutralizing antibodies will act similarly to a vaccine. For the broadly neutralizing antibodies, 2G12 (Ref. 105), b12 (Ref. 26), 2F5, 4E10 (Ref. 106), and PGT121 (Ref. 107) intravenous administration results also in protective antibody concentrations in the mucus.

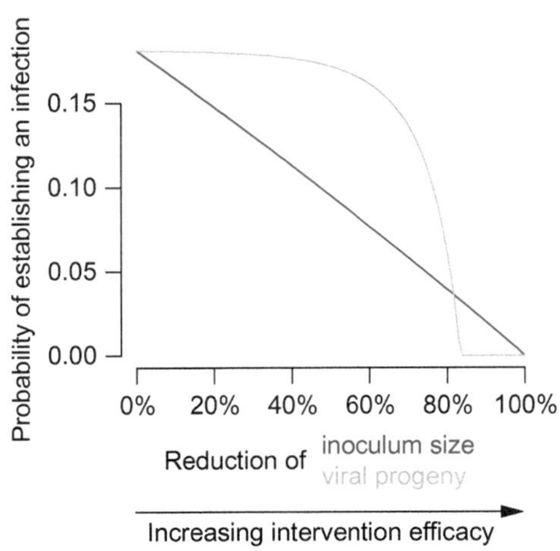

Figure 6 Predicted effect of reducing the inoculum (red (dark gray in the print version)) versus the viral growth rate (gray). Reducing the inoculum has a linear effect on the probability of HIV acquisition, In contrast, reducing the growth rate of the virus within the host results in a threshold-type effect. Over a wide range of reductions, reducing the inoculum is predicted to be more effective.

How do these two potential effects—reduction of inoculum size versus within-host replication—contribute to the efficacy of an intervention? To address this question, we use the stochastic simulation model shown in Fig. 3, and consider the whole conceivable range of effects of an intervention. Figure 6 shows how the probability of HIV acquisition changes in response to a given reduction of the inoculum size or of the within-host replication rate of the virus.

The probability of HIV acquisition decreases continuously with reduction of the inoculum size. Thus, an intervention that reduces the inoculum by, for example, a factor 4, will reduce the probability of establishment of an infection by approximately the same factor, unless the inoculum is very high and gives rise to infection with certainty. This is in line with the simulations in Fig. 3C, in which a fourfold reduction of the inoculum resulted in an approximately fourfold lower probability of acquisition (equivalent to a vaccine efficacy of approximately 3/4).

The relation between the within-host replication and the probability of HIV acquisition is more complex. As soon as the number of progeny virions cannot replenish the previous generation, extinction of the virus population becomes certain. In our simulations, we assume that, without intervention, one virion gives rise to six virions on average. This results in a threshold reduction of viral replication by a factor 6, or by 83%. For reductions less than this level, the probability of establishing an infection is mostly unaffected. Thus, over a wide range of efficacies, reducing the inoculum promises greater protection. This is even more likely if one considers that, in order to reduce viral replication, it is necessary to maintain the concentration high enough to durably reduce the number of viral progeny.

In the future, considerations as the ones presented here should be extended to predict quantitatively the efficacy of various interventions for preventing HIV infection. For such predictions, one would combine the conceptual, quantitative framework for neutralization with the stochastic approaches to describe the early virus dynamics. In particular, one could derive the local concentrations of neutralizing antibodies necessary to attain a certain level of protection from infection. Since Shen *et al.*[104] extended the stoichiometric framework to the action of antiretrovirals, an analogous logic can also be applied to prophylactic use of these agents.

From a mathematical modeling perspective, such predictions would require to better understand three particular aspects. First, the anatomical aspects of early virus dynamics need to be elucidated. At what rate does the virus replicate in different tissue compartments, and how does the virus population spread between relevant compartments. Second, the kinetics of antibodies or antiretroviral agents in the relevant tissues—i.e., how fast neutralizing antibodies or drugs are cleared in the mucosal tissues, and how they penetrate into neighboring anatomical compartments—needs to be quantitatively determined and incorporated into a predictive framework. These aspects are starting to be elucidated in experiments linking plasma and mucosal concentrations of antibodies in nonhuman primate models.[107] Lastly, it will have to be determined how antibodies and antiretroviral agents affect viral replication in these relevant tissue compartments *in vivo*. Such a combination of our quantitative understanding of early virus dynamics and antibody neutralization or pharmacodynamics will enable us to develop a predictive mathematical model that can guide the development and selection of antibodies and inhibitors for testing and use in humans.

ACKNOWLEDGMENTS

We would like to thank David Beauparlant, Thomas Liechti, and Lucia Reh for helpful discussions, as well as Frederic Bertels, Branislav Ivan, and Per Johan Klasse for comments on our manuscript. R. R. R. gratefully acknowledges the financial support of the Swiss National Science Foundation (grant number: 315230_130855).

REFERENCES

1. Hartl DL, Clark AG. *Priciples of Population Genetics*. 4th ed. Sunderland, MA: Sinauer; 2007.
2. Bailey NJT. *The Mathematical Theory of Infectious Diseases*. New York, NY: Macmillan; 1975.
3. Elowitz MB, Levine AJ, Siggia ED, Swain PS. Stochastic gene expression in a single cell. *Science*. 2002;297(5584):1183–1186.
4. Alizon S, Magnus C. Modelling the course of an HIV infection: insights from ecology and evolution. *Viruses*. 2012;4(10):1984–2013.
5. Nowak MA, May RM. *Virus Dynamics: Mathematical Principles of Immunology and Virology*. Oxford: Oxford University Press; 2000.
6. Perelson AS. Modelling viral and immune system dynamics. *Nat Rev Immunol*. 2002;2(1):28–36.
7. Perelson AS, Ribeiro RM. Modeling the within-host dynamics of HIV infection. *BMC Biol*. 2013;11:96.
8. Gray RH, Wawer MJ, Brookmeyer R, et al. Probability of HIV-1 transmission per coital act in monogamous, heterosexual, HIV-1-discordant couples in Rakai, Uganda. *Lancet*. 2001;357(9263):1149–1153.
9. Mastro TD, Kitayaporn D. HIV type 1 transmission probabilities: estimates from epidemiological studies. *AIDS Res Hum Retroviruses*. 1998;14:S223–S227.
10. Taylor MM, Rotblatt H, Brooks JT, et al. Epidemiologic investigation of a cluster of workplace HIV infections in the adult film industry: Los Angeles, California, 2004. *Clin Infect Dis*. 2007;44(2):301–305.
11. Buchbinder SP, Mehrotra DV, Duerr A, et al. Efficacy assessment of a cell-mediated immunity HIV-1 vaccine (the Step Study): a double-blind, randomised, placebo-controlled, test-of-concept trial. *Lancet*. 2008;372(9653):1881–1893.
12. Halloran E, Longini Jr Ira M, Struchiner C. *Design and Analysis of Vaccine Studies. Statistics for Biology and Health*. New York, NY: Springer; 2010, ISBN: 9780387403137.
13. Amara RR, Villinger F, Altman JD, et al. Control of a mucosal challenge and prevention of AIDS by a multiprotein DNA/MVA vaccine. *Science*. 2001;292(5514):69–74.
14. Barouch DH, Santra S, Schmitz JE, et al. Control of viremia and prevention of clinical AIDS in rhesus monkeys by cytokine-augmented DNA vaccination. *Science*. 2000;290(5491):486–492.
15. Rose NF, Marx PA, Luckay A, et al. An effective AIDS vaccine based on live attenuated vesicular stomatitis virus recombinants. *Cell*. 2001;106(5):539–549.
16. Shiver JW, Fu TM, Chen L, et al. Replication-incompetent adenoviral vaccine vector elicits effective anti-immunodeficiency-virus immunity. *Nature*. 2002;415(6869):331–335.
17. Hudgens MG, Gilbert PB, Mascola JR, Wu CD, Barouch DH, Self SG. Power to detect the effects of HIV vaccination in repeated low-dose challenge experiments. *J Infect Dis*. 2009;200(4):609–613.
18. Regoes RR, Longini IM, Feinberg MB, Staprans SI. Preclinical assessment of HIV vaccines and microbicides by repeated low-dose virus challenges. *PLoS Med*. 2005;2(8):e249.
19. Wilson NA, Keele BF, Reed JS, et al. Vaccine-induced cellular responses control simian immunodeficiency virus replication after heterologous challenge. *J Virol*. 2009;83(13):6508–6521.
20. Hudgens MG, Gilbert PB. Assessing vaccine effects in repeated low-dose challenge experiments. *Biometrics*. 2009;65(4):1223–1232.
21. Barouch DH, Liu J, Li H, et al. Vaccine protection against acquisition of neutralization-resistant SIV challenges in rhesus monkeys. *Nature*. 2012;482(7383):89–93.

22. Ellenberger D, Otten RA, Li B, et al. HIV-1 DNA/MVA vaccination reduces the per exposure probability of infection during repeated mucosal SHIV challenges. *Virology*. 2006;352(1):216–225.
23. Flatz L, Cheng C, Wang L, et al. Gene-based vaccination with a mismatched envelope protects against simian immunodeficiency virus infection in nonhuman primates. *J Virol.* 2012;86(15):7760–7770.
24. Garcia-Lerma JG, Otten RA, Qari SH, et al. Prevention of rectal SHIV transmission in macaques by daily or intermittent prophylaxis with emtricitabine and tenofovir. *PLoS Med.* 2008;5(2):e28.
25. Hansen SG, Vieville C, Whizin N, et al. Effector memory T cell responses are associated with protection of rhesus monkeys from mucosal simian immunodeficiency virus challenge. *Nat Med.* 2009;15(3):293–299.
26. Hessell AJ, Poignard P, Hunter M, et al. Effective, low-titer antibody protection against low-dose repeated mucosal SHIV challenge in macaques. *Nat Med.* 2009;15(8): 951–954.
27. Lai L, Kwa SF, Kozlowski PA, et al. SIVmac239 MVA vaccine with and without a DNA prime, similar prevention of infection by a repeated dose SIVsmE660 challenge despite different immune responses. *Vaccine.* 2012;30(9):1737–1745.
28. Lai L, Kwa S, Kozlowski PA, et al. Prevention of infection by a granulocyte-macrophage colony-stimulating factor co-expressing DNA/modified vaccinia Ankara simian immunodeficiency virus vaccine. *J Infect Dis.* 2011;204(1):164–173.
29. Letvin NL, Rao SS, Dang V, et al. No evidence for consistent virus-specific immunity in simian immunodeficiency virus-exposed, uninfected rhesus monkeys. *J Virol.* 2007;81(22):12368–12374.
30. Letvin NL, Rao SS, Montefiori DC, et al. Immune and genetic correlates of vaccine protection against mucosal infection by SIV in monkeys. *Sci Transl Med.* 2011;3(81):81ra36.
31. Ma ZM, Abel K, Rourke T, Wang Y, Miller CJ. A period of transient viremia and occult infection precedes persistent viremia and antiviral immune responses during multiple low-dose intravaginal simian immunodeficiency virus inoculations. *J Virol.* 2004;78(24):14048–14052.
32. McDermott AB, Mitchen J, Piaskowski S, et al. Repeated low-dose mucosal simian immunodeficiency virus SIVmac239 challenge results in the same viral and immunological kinetics as high-dose challenge: a model for the evaluation of vaccine efficacy in non-human primates. *J Virol.* 2004;78(6):3140–3144.
33. Otten RA, Adams DR, Kim CN, et al. Multiple vaginal exposures to low doses of R5 simian-human immunodeficiency virus: strategy to study HIV preclinical interventions in nonhuman primates. *J Infect Dis.* 2005;191(2):164–173.
34. Reynolds MR, Weiler AM, Piaskowski SM, et al. Macaques vaccinated with simian immunodeficiency virus SIVmac239Delta nef delay acquisition and control replication after repeated low-dose heterologous SIV challenge. *J Virol.* 2010;84(18): 9190–9199.
35. Roederer M, Keele BF, Schmidt SD, et al. Immunological and virological mechanisms of vaccine-mediated protection against SIV and HIV. *Nature.* 2014;505(7484): 502–508.
36. Van Rompay KK, Abel K, Lawson JR, et al. Attenuated poxvirus-based simian immunodeficiency virus (SIV) vaccines given in infancy partially protect infant and juvenile macaques against repeated oral challenge with virulent SIV. *J Acquir Immune Defic Syndr.* 2005;38(2):124–134.
37. Wilson NA, Reed J, Napoe GS, et al. Vaccine-induced cellular immune responses reduce plasma viral concentrations after repeated low-dose challenge with pathogenic simian immunodeficiency virus SIVmac239. *J Virol.* 2006;80(12):5875–5885.

38. Lynch RM, Yamamoto T, McDermott AB. HIV vaccine research and discovery in the nonhuman primates model: a unified theory in acquisition prevention and control of SIV infection. *Curr Opin HIV AIDS*. 2013;8(4):288–294.
39. Hansen SG, Ford JC, Lewis MS, et al. Profound early control of highly pathogenic SIV by an effector memory T-cell vaccine. *Nature*. 2011;473(7348):523–527.
40. Hammer SM, Sobieszczyk ME, Janes H, et al. Efficacy trial of a DNA/rAd5 HIV-1 preventive vaccine. *N Engl J Med*. 2013;369(22):2083–2092.
41. Rerks-Ngarm S, Pitisuttithum P, Nitayaphan S, et al. Vaccination with ALVAC and AIDSVAX to prevent HIV-1 infection in Thailand. *N Engl J Med*. 2009;361(23):2209–2220.
42. Regoes RR. The role of exposure history on hiv acquisition: insights from repeated low-dose challenge studies. *PLoS Comput Biol*. 2012;8(11):e1002767.
43. Haase AT. Perils at mucosal front lines for HIV and SIV and their hosts. *Nat Rev Immunol*. 2005;5(10):783–792.
44. Wu L, KewalRamani VN. Dendritic-cell interactions with HIV: infection and viral dissemination. *Nat Rev Immunol*. 2006;6(11):859–868.
45. Chakraborty H, Sen PK, Helms RW, et al. Viral burden in genital secretions determines male-to-female sexual transmission of HIV-1: a probabilistic empiric model. *AIDS*. 2001;15(5):621–627.
46. Merrill SJ. The stochastic dance of early hiv infection. *J Comput Appl Math*. 2005;184(1):242–257.
47. Tuckwell HC, Shipman PD, Perelson AS. The probability of HIV infection in a new host and its reduction with microbicides. *Math Biosci*. 2008;214(1–2):81–86.
48. Wick D, Self SG. Early HIV infection in vivo: branching-process model for studying timing of immune responses and drug therapy. *Math Biosci*. 2000;165(2):115–134.
49. Wick WD. On modeling the effects of T-cell vaccines on HIV acquisition and disease. *Stat Med*. 2008;27(23):4805–4816.
50. Wick WD, Gilbert PB, Self SG. On modeling HIV and T cells in vivo: assessing causal estimators in vaccine trials. *PLoS Comput Biol*. 2006;2(6):e64.
51. Talbert-Slagle K, Atkins KE, Yan KK, et al. Cellular superspreaders: an epidemiological perspective on HIV infection inside the body. *PLoS Pathog*. 2014;10(5):e1004092.
52. Davenport MP, Ribeiro RM, Zhang L, Wilson DP, Perelson AS. Understanding the mechanisms and limitations of immune control of HIV. *Immunol Rev*. 2007;216:164–175.
53. De Boer RJ. Understanding the failure of cd8+ t-cell vaccination against simian/human immunodeficiency virus. *J Virol*. 2007;81(6):2838–2848.
54. Keele BF, Giorgi EE, Salazar-Gonzalez JF, et al. Identification and characterization of transmitted and early founder virus envelopes in primary HIV-1 infection. *Proc Natl Acad Sci U S A*. 2008;105(21):7552–7557.
55. Lee HY, Giorgi EE, Keele BF, et al. Modeling sequence evolution in acute HIV-1 infection. *J Theor Biol*. 2009;261(2):341–360.
56. Keele BF, Li H, Learn GH, et al. Low-dose rectal inoculation of rhesus macaques by SIVsmE660 or SIVmac251 recapitulates human mucosal infection by HIV-1. *J Exp Med*. 2009;206(5):1117–1134.
57. Frost SD, Dumaurier MJ, Wain-Hobson S, Brown AJ. Genetic drift and within-host metapopulation dynamics of HIV-1 infection. *Proc Natl Acad Sci U S A*. 2001;98(12):6975–6980.
58. Pillai SK, Pond SLK, Liu Y, et al. Genetic attributes of cerebrospinal fluid-derived HIV-1 env. *Brain*. 2006;129(Pt 7):1872–1883.
59. Strain MC, Letendre S, Pillai SK, et al. Genetic composition of human immunodeficiency virus type 1 in cerebrospinal fluid and blood without treatment and during failing antiretroviral therapy. *J Virol*. 2005;79(3):1772–1788.

60. Wong JK, Hezareh M, Gunthard HF, et al. Recovery of replication-competent HIV despite prolonged suppression of plasma viremia. *Science*. 1997;278(5341):1291–1295.
61. Zárate S, Pond SLK, Shapshak P, Frost SDW. Comparative study of methods for detecting sequence compartmentalization in human immunodeficiency virus type 1. *J Virol*. 2007;81(12):6643–6651.
62. Althaus CL, De Boer RJ. Intracellular transactivation of HIV can account for the decelerating decay of virus load during drug therapy. *Mol Syst Biol*. 2010;6:348.
63. Reddy B, Yin J. Quantitative intracellular kinetics of HIV type 1. *AIDS Res Hum Retroviruses*. 1999;15(3):273–283.
64. Weinberger LS, Burnett JC, Toettcher JE, Arkin AP, Schaffer DV. Stochastic gene expression in a lentiviral positive-feedback loop: HIV-1 Tat fluctuations drive phenotypic diversity. *Cell*. 2005;122(2):169–182.
65. Bourinbaiar A. The ratio of defective HIV-1 particles to replication-competent infectious virions. *Acta Virol*. 1994;38(1):59–61.
66. Kimpton J, Emerman M. Detection of replication-competent and pseudotyped human-immunodeficiency-virus with a sensitive cell-line on the basis of activation of an integrated beta-galactosidase gene. *J Virol*. 1992;66(4):2232–2239.
67. Klasse PJ, McKeating J. Soluble Cd4 and Cd4 immunoglobulin-selected Hiv-1 variants - a phenotypic characterization. *AIDS Res Hum Retrovir*. 1993;9(7):595–604.
68. Layne SP, Merges MJ, Dembo M, et al. Factors underlying spontaneous inactivation and susceptibility to neutralization of human-immunodeficiency-virus. *Virology*. 1992;189(2):695–714.
69. McDougal J, Cort S, Kennedy M, et al. Immunoassay for the detection and quantitation of infectious human retrovirus, lymphadenopathy-associated virus (LAV). *J Immunol Methods*. 1984;76(1):171–183.
70. Piatak Jr M, Saag MS, Yang LC, et al. High levels of HIV-1 in plasma during all stages of infection determined by competitive PCR. *Science*. 1993;259:1749–1754.
71. Rusert P, Fischer M, Joos B, et al. Quantification of infectious HIV-1 plasma viral load using a boosted *in vitro* infection protocol. *Virology*. 2004;326(1):113–129.
72. Gallo SA, Finnegan CM, Viard M, et al. The HIV Envmediated fusion reaction. *Biochim Biophys Acta*. 2003;1614(1):36–50.
73. Wilen CB, Tilton JC, Doms RW. Molecular mechanisms of HIV entry. *Adv Exp Med Biol*. 2012;726:223–242.
74. Robey W, Safai B, Oroszlan S, et al. Characterization of envelope and core structural gene-products of HTLV-III with sera from AIDS patients. *Science*. 1985;228(4699):593–595.
75. Veronese F, Devico A, Copeland T, Oroszlan S, Gallo R, Sarngadharan M. Characterization of gp41 as the transmembrane protein coded by the HTLV-III/LAV envelope gene. *Science*. 1985;229(4720):1402–1405.
76. Wyatt R, Sodroski J. The HIV-1 envelope glycoproteins: fusogens, antigens, and immunogens. *Science*. 1998;280(5371):1884–1888.
77. Klatzmann D, Champagne E, Chamaret S, et al. T-lymphocyte T4 molecule behaves as the receptor for human retrovirus LAV. *Nature*. 1984;312(5996):767–768.
78. McDougal J, Kennedy M, Sligh J, Cort S, Mawle A, Nicholson J. Binding of HTLV-III/LAV to T4+ T-cells by a complex of the 110K viral protein and the T4 molecule. *Science*. 1986;231(4736):382–385.
79. Sattentau Q, Moore J. Conformational-changes induced in the human-immunodeficiency-virus envelope glycoprotein by soluble cd4 binding. *J Exp Med*. 1991;174(2):407–415.
80. Zhu P, Liu J, Bess Jr J, et al. Distribution and three-dimensional structure of AIDS virus envelope spikes. *Nature*. 2006;441(7095):847–852.

81. Sougrat R, Bartesaghi A, Lifson JD, et al. Electron tomography of the contact between T cells and SIV/HIV-1: implications for viral entry. *PLoS Pathog.* 2007;3(5):e63.
82. Schonning K, Lund O, Lund OS, Hansen JES. Stoichiometry of monoclonal antibody neutralization of T-cell line-adapted human immunodeficiency virus type 1. *J Virol.* 1999;73:8364–8370.
83. Pang Y, Song H, Kim JH, Hou X, Cheng W. Optical trapping of individual human immunodeficiency viruses in culture fluid revealsheterogeneity with single-molecule resolution. *Nat Nanotechnol.* 2014;9:1–7.
84. Magnus C, Rusert P, Bonhoeffer S, Trkola A, Regoes RR. Estimating the stoichiometry of human immunodeficiency virus entry. *J Virol.* 2009;83(3):1523–1531.
85. Yang X, Kurteva S, Lee S, Sodroski J. Stoichiometry of antibody neutralization of human immunodeficiency virus type 1. *J Virol.* 2005;79(6):3500–3508.
86. Yang X, Kurteva S, Ren X, Lee S, Sodroski J. Stoichiometry of envelope glycoprotein trimers in the entry of human immunodeficiency virus type 1. *J Virol.* 2005;79(19):12132–12147.
87. Klasse PJ. Modeling how many envelope glycoprotein trimers per virion participate in human immunodeficiency virus infectivity and its neutralization by antibody. *Virology.* 2007;369(2):245–262.
88. Johnston SH, Lobritz MA, Nguyen S, et al. A quantitative affinity-profiling system that reveals distinct CD4/CCR5 usage patterns among human immunodeficiency virus type 1 and simian immunodeficiency virus strains. *J Virol.* 2009;83(21):11016–11026.
89. Chikere K, Chou T, Gorry PR, Lee B. Affinofile profiling How efficiency of CD4/CCR5 usage impacts the biological and pathogenic phenotype of HIV. *Virology.* 2013;435(1):81–91.
90. Mulampaka SN, Dixit NM. Estimating the threshold surface density of Gp120-CCR5 complexes necessary for HIV-1 envelope-mediated cell-cell fusion. *PLoS One.* 2011;6(5):e19941.
91. Burton DR, Poignard P, Stanfield RL, Wilson IA. Broadly neutralizing antibodies present new prospects to counter highly. *Science.* 2012;337(6091):183–186.
92. Kwong PD, Mascola JR. Human antibodies that neutralize HIV-1: identification, structures, and B cell ontogenies. *Immunity.* 2012;37(3):412–425.
93. Burton DR, Desrosiers RC, Doms RW, et al. HIV vaccine design and the neutralizing antibody problem. *Nat Immunol.* 2004;5(3):233–236.
94. Hedestam GBK, Fouchier RAM, Phogat S, Burton DR, Sodroski J, Wyatt RT. The challenges of eliciting neutralizing antibodies to HIV-1 and to influenza virus. *Nat Rev Microbiol.* 2008;6(2):143–155.
95. Dulbecco R, Vogt M. Plaque formation and isolation of pure lines with poliomyelitis viruses. *J Exp Med.* 1954;99(2):167–182.
96. Dulbecco R, Vogt M, Strickland A. A study of the basic aspects of neutralization of 2 animal viruses, western equine encephalitis virus and poliomyelitis virus. *Virology.* 1956;2(2):162–205.
97. McLain L, Dimmock NJ. Single- and multi-hit kinetics of immunoglobulin G neutralization of human immunodeficiency virus type 1 by monoclonal antibodies. *J Gen Virol.* 1994;75(Pt 6):1457–1460.
98. Klasse PJ, Sattentau QJ. Occupancy and mechanism in antibody-mediated neutralization of animal viruses. *J Gen Virol.* 2002;83(Pt 9):2091–2108.
99. Moran P, S. Fazekas de St Groth. Random circles on a sphere. *Biometrika.* 1962;49:389–396.
100. Magnus C, Regoes RR. Estimating the stoichiometry of HIV neutralization. *PLoS Comput Biol.* 2010;6(3):e1000713.

101. Magnus C, Regoes R. Restricted occupancy models for neutralization of HIV virions and populations. *J Theor Biol.* 2011;283(1):192–202.
102. Magnus C. Virus neutralisation: new insights from kinetic neutralisation curves. *PLoS Comput Biol.* 2013;9(2):e1002900.
103. Hu B, Liao HX, Alam SM, Goldstein B. Estimating the probability of polyreactive antibodies 4E10 and 2F5 disabling a gp41 trimer after T cell-HIV adhesion. *PLoS Comput Biol.* 2014;10(1):e1003431.
104. Shen L, Rabi SA, Sedaghat AR, et al. A critical subset model provides a conceptual basis for the high antiviral activity of major HIV drugs. *Sci Transl Med.* 2011;3(91):91ra63.
105. Hessell AJ, Rakasz EG, Poignard P, et al. Broadly neutralizing human anti-HIV antibody 2G12 is effective in protection against mucosal SHIV challenge even at low serum neutralizing titers. *PLoS Pathog.* 2009;5(5):e1000433.
106. Hessell AJ, Rakasz EG, Tehrani DM, et al. Broadly neutralizing monoclonal antibodies 2F5 and 4E10 directed against the human immunodeficiency virus type 1 gp41 membrane-proximal external region protect against mucosal challenge by simian-human immunodeficiency virus SHIVBa-L. *J Virol.* 2010;84(3):1302–1313.
107. Moldt B, Rakasz EG, Schultz N, et al. Highly potent HIV-specific antibody neutralization in vitro translates into effective protection against mucosal SHIV challenge in vivo. *Proc Natl Acad Sci U S A.* 2012;109(46):18921–18925.
108. Veazey RS, Shattock RJ, Pope M, et al. Prevention of virus transmission to macaque monkeys by a vaginally applied monoclonal antibody to HIV-1 gp120. *Nat Med.* 2003;9(3):343–346.
109. Trkola A, Kuster H, Rusert P, et al. Delay of HIV-1 rebound after cessation of antiretroviral therapy through passive transfer of human neutralizing antibodies. *Nat Med.* 2005;11(6):615–622.
110. Veazey RS, Springer MS, Marx PA, Dufour J, Klasse PJ, Moore JP. Protection of macaques from vaginal SHIV challenge by an orally delivered CCR5 inhibitor. *Nat Med.* 2005;11(12):1293–1294.
111. Klasse PJ, Shattock R, Moore JP. Antiretroviral drug-based microbicides to prevent HIV-1 sexual transmission. *Annu Rev Med.* 2008;59:455–471.

CHAPTER TWELVE

Virus-Encoded 7 Transmembrane Receptors

Ann-Sofie Mølleskov-Jensen*,[1], Martha Trindade Oliveira†,[1], Helen Elizabeth Farrell†, Nick Davis-Poynter†,[2]

*Department of Neuroscience and Pharmacology, Laboratory for Molecular Pharmacology, University of Copenhagen, Copenhagen, Denmark

†Queensland Children's Medical Research Institute, Sir Albert Sakzewski Virus Research Centre, The University of Queensland & Royal Children's Hospital, Brisbane, Queensland, Australia

[1]These authors contributed equally to the work.

[2]Corresponding author: e-mail address: n.davispoynter@uq.edu.au

Contents

Abstract

Herpesviruses are an ancient group which have exploited gene capture of multiple cellular modulators of the immune response. Viral homologues of 7 transmembrane receptors (v7TMRs) are a consistent feature of beta- and gammaherpesviruses; the majority of the v7TMRs are homologous to cellular chemokine receptors (CKRs). Conserved families of v7TMRs distinguish between beta- versus gammaherpesviruses; furthermore, significant divisions within these subfamilies, such as between genera of the gammaherpesviruses or between the primate and rodent cytomegaloviruses, coincide with specific v7TMR gene families. Divergence of functional properties between the viral 7TMR and their cellular counterparts is likely, therefore, to reflect adaptation supporting various aspects of the viral lifecycle with concomitant effects upon viral pathogenesis.

Consistent with their long evolutionary history, the v7TMRs have acquired a range of distinctive characteristics. This chapter reviews key features of the v7TMRs which are likely to impact upon their functional roles: trafficking properties, ligand specificity, and signaling capacity. Rapid, constitutive endocytosis, reminiscent of cellular "scavenger"

Progress in Molecular Biology and Translational Science, Volume 129
ISSN 1877-1173
http://dx.doi.org/10.1016/bs.pmbts.2014.10.010

receptors, may provide a mechanism for immune evasion, or alternatively relate to virion assembly, including incorporation of v7TMRs within the virion envelope. Some v7TMRs display relatively broad chemokine-binding specificity, whereas others remain "orphan" and may be completely independent of ligand activation. Indeed, many of the v7TMRs have been shown to signal constitutively, associated in some cases with notable divergence of highly conserved regulatory elements such as the "DRY" motif of TMIII.

The availability of rodent models for v7TMR functional studies has provided evidence for important biological roles, including cellular transformation, tissue tropism, and viral persistence. Recent studies addressing signaling pathways critical to these phenotypes will be discussed, with reference to both beta- and gammaherpesviruses.

1. EVOLUTIONARY CONTEXT OF v7TMRs

Mammalian herpesviruses comprise three major subfamilies: the alpha-, beta-, and gammaherpesviruses.[1] These subfamilies are distinguished by their overall genome content and subfamily-specific features including several conserved gene families. Whereas homologues of chemokines have been found in members of all three subfamilies, viral homologues of 7 transmembrane receptors (v7TMRs) are restricted to the beta- and gammaherpesviruses. Furthermore, families of v7TMRs are apparent, on the basis of amino acid sequence homology and genomic position, which are characteristic of either the beta- or gamma-subfamilies. A phylogenetic tree of herpesvirus v7TMRs and selected cellular chemokine receptors (CKRs) is shown in Fig. 1, compiled using protein sequences listed in Table 1. The overall topology of the tree and positioning of the v7TMRs in relation to blocks of conserved viral genes within the respective genomes are consistent with several independent gene capture events from the host.

For the betaherpesviruses, three distinct families of v7TMRs are apparent, designated herein *beta 27/28*, *beta 33*, and *beta 78*. The *beta 33* and *beta 78* families, named after the prototype genes of human CMV (*UL33* and *UL78*, respectively), are conserved throughout the betaherpesviruses, being present in the four defined genera: *cytomegalovirus* (human and other primate CMVs), *muromegalovirus* (mouse and rat CMVs), *roseolovirus* (*human herpesvirus* (*HHV*)-*6* and -*7*), and *proboscivirus* (*elephantid herpesvirus-1*), plus unassigned betaherpesviruses of guinea pig, tupaia, bat, and pig (see Table 1). The *beta 27/28* family, named after the prototype genes of human CMV (*US27* and *US28*), is present only in primate CMVs. Presumably, the *beta 33* and *beta 78* v7TMRs derive from independent gene capture events

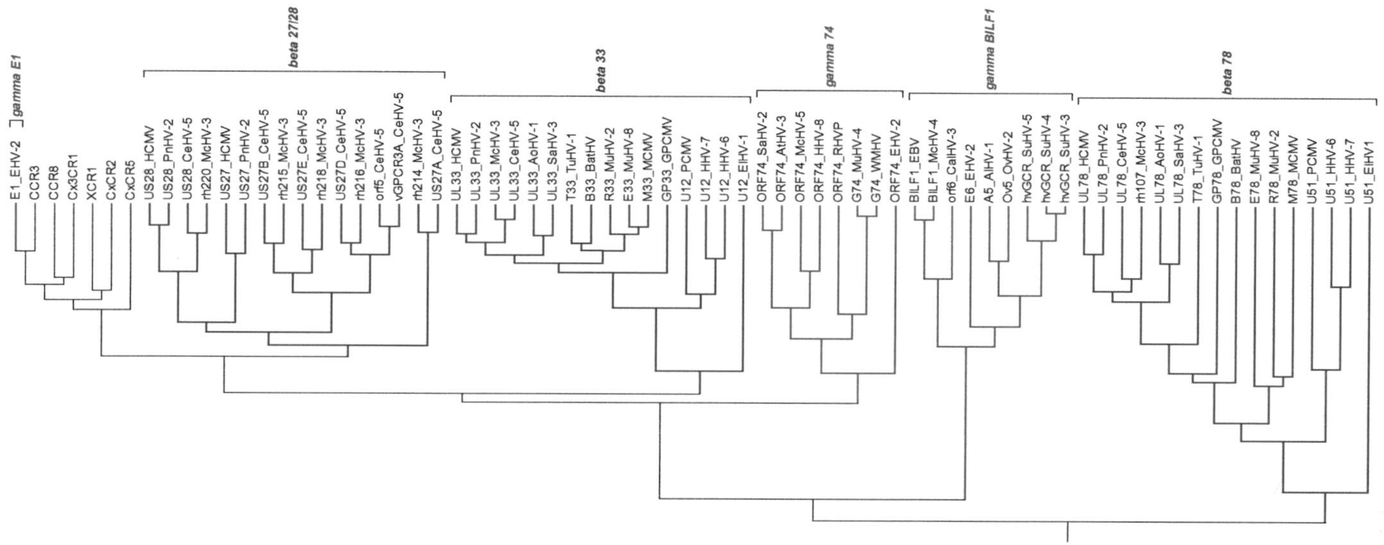

Figure 1 Evolutionary relationships of v7TMR taxa. Amino acid sequences were aligned using ClustalW (Geneious v 5.6.7), with manual adjustment of gaps to align likely transmembrane domains. The evolutionary history was inferred using the UPGMA method.[2] The optimal tree (with sum of branch length = 58.88) is drawn to scale, with branch lengths in the same units as those of the evolutionary distances used to infer the phylogenetic tree. The evolutionary distances were computed using the JTT matrix-based method[3] and are in the units of the number of amino acid substitutions per site. The analysis involved 72 amino acid sequences. All ambiguous positions were removed for each sequence pair, and there were a total of 676 positions in the final dataset. Evolutionary analyses were conducted in MEGA5.[4]

Table 1 Cellular and viral genes used for sequence alignments and phylogeny[a]

Identifier	Accession	Source (species name; common name)
*CCR1	NP_001286	*Homo sapiens*
*CCR2	NP_001828	
CCR3	NP_001828	
*CCR4	NP_005499	
*CCR5	NP_000570	
*CCR6	NP_004358	
*CCR7	NP_001829	
CCR8	NP_005192	
*CCR9	NP_112477	
*CCR10	NP_057686	
*CCR11	NP_057641	
CX3CR1	NP_001328	
XCR1	NP_005274	
*CXCR1	NP_000625	
CXCR2	NP_001548	
*CXCR3	NP_001495	
*CXCR4	NP_003458	
CxCR5	NP_001707	
*CXCR6	NP_006555	
Betaherpesviruses		
Cytomegalovirus		
US27_HCMV	YP_081611	Human cytomegalovirus (*Human herpesvirus-5*)
US28_HCMV	YP_081612	
UL33_HCMV	YP_081492	
UL78_HCMV	YP_081526	
US27_PnHV-2	NP_612799	*Panine herpesvirus-2* (chimpanzee cytomegalovirus)
US28_PnHV-2	NP_612800	
UL33_PnHV-2	NP_612677	
UL78_PnHV-2	NP_612711	

Table 1 Cellular and viral genes used for sequence alignments and phylogeny—cont'd

Identifier	Accession	Source (species name; common name)
US27A_CeHV-5	YP_004936139	*Cercopithecine herpesvirus-5* (Green monkey cytomegalovirus/Simian cytomegalovirus) isolate 2715
US27B_CeHV-5	YP_004936137	
US27D_CeHV-5	YP_004936138	
US27E_CeHV-5	YP_004936139	
US28_CeHV-5	YP_004936140	
UL33_CeHV-5	AEV80398	
UL78_CeHV-5	YP_004936044	
orf5_CeHV-5		*Cercopithecine herpesvirus-5* isolate 9610
vGPCR3A_CeHV-5	ACQ55268	*Cercopithecine herpesvirus-5* isolate 4915
rh214_McHV-3	AAP50735	*Macacine herpesvirus-3* (Rhesus cytomegalovirus) strain 68.1
rh220_McHV-3	YP_068307	
rh215_McHV-3	YP_068303	
rh107_McHV-3	YP_068200	
rh218_McHV-3	AAZ80736	*Macacine herpesvirus-3* strain 180.92
rh216_McHV-3	AAZ80734	
UL33_McHV-3	AAZ80554[b]	
UL33_SaHV-3	YP_004940205	*Saimirine herpesvirus-3* (squirrel monkey cytomegalovirus)
UL78_SaHV-3	YP_004940241	
UL33_AoHV-1	YP_004940059	*Aotine herpesvirus-1* (owl monkey cytomegalovirus)
UL78_AoHV-1	YP_004940095	
Muromegalovirus		
M33_MCMV	CAJ87071	Murine cytomegalovirus (*Murid herpesvirus-1*)
M78_MCMV	CAP08118	
R33_MuHV-2	NP_064138	*Murid herpesvirus-2* (rat cytomegalovirus isolate Maastricht)
R78_MuHV-2	AAC64046	
E33_MuHV-8	YP_007016437	*Murid herpesvirus-8* (rat cytomegalovirus isolate England)
E78_MuHV-8	YP_007016476	

Continued

Table 1 Cellular and viral genes used for sequence alignments and phylogeny—cont'd

Identifier	Accession	Source (species name; common name)
Roseolovirus		
U12_HHV-6	NP_042903	*Human herpesvirus-6A*
U51_HHV-6	NP_042944	
U12_HHV-7	P52381	*Human herpevirus-7*
U51_HHV-7	AGV28676	
Proboscivirus		
U12_ElHV-1	YP_007969825	*Elephantid herpesvirus-1*
U51_ElHV-1	AGG16090	
Unassigned betaherpesvirus		
B33_BatHV	AFK83852	Bat herpesvirus isolate B7D8
B78_BatHV	AFK83916	
T33_THV-1	NP_116383	*Tupaiid herpesvirus-1* (Tree shrew herpesvirus)
T78_THV-1	NP_116428	
GP33_GPCMV	AAK43591	Guinea pig cytomegalovirus (*Caviid herpesvirus-2*)
GP78_GPCMV	YP_007417844	
U12_PCMV	YP_008492955[b]	Porcine cytomegalovirus (*Suid herpesvirus-2*)
U51_PCMV	YP_008492988	
Gammaherpesviruses		
Rhadinovirus		
G74_WMHV	ACY41143	Wood mouse herpesvirus
G74_MuHV-4	NP_044914	*Murid herpesvirus-4* (Murine gammaherpesvirus 68)
ORF74_RHVP	YP_004207910	Rodent herpesvirus Peru
ORF74_SaHV-2	NP_040276	*Saimirine herpesvirus-2* (Herpesvirus saimiri)
ORF74_AtHV-3	NP_048046	*Ateline herpesvirus-3* (Herpesvirus ateles)
ORF74_McHV-5	NP_570822	*Macacine herpesvirus-5* (Rhesus rhadinovirus)
ORF74_HHV-8	YP_001129433	*Human herpesvirus-8* (Kaposi's sarcoma-associated herpesvirus)

Table 1 Cellular and viral genes used for sequence alignments and phylogeny—cont'd

Identifier	Accession	Source (species name; common name)
Percavirus		
E1_EHV-2	Q89609	*Equid herpesvirus-2*
E6_EHV-2	NP_042607	
ORF74_EHV-2	NP_042670	
Lymphocryptovirus		
BILF1_EBV	YP_401711	Epstein–Barr virus (*Human herpesvirus-4*)
BILF1_McHV-4	YP_068006	*Macacine herpesvirus-4* (Rhesus lymphocryptovirus)
orf6_CalHV-3	NP_733858	*Callitrichine herpesvirus-3* (Marmoset lymphocryptovirus)
Macavirus		
A5_AlHV-1	NP_065513	*Alcelaphine herpesvirus-5*
Ov5_OvHV-2	YP_438137	*Ovine herpesvirus-2*
hvGCR_SuHV-3	AAF16521	*Suid herpesvirus-3* (Porcine lymphotropic herpesvirus-1)
hvGCR_SuHV-4	AAO12283	*Suid herpesvirus-4* (Porcine lymphotropic herpesvirus-2)
hvGCR_SuHV-5	AAO12302	*Suid herpesvirus-5* (Porcine lymphotropic herpesvirus-3)

[a]A subset of cellular CKRs (indicated by *) were not included in the phylogenic analysis but were included in the derivation of sequence logos.
[b]Manual join of exons 1 and 2 predicted from genomic sequence (database sequence did not include exon 1).

predating the divergence of betaherpesviruses into the different genera. The *beta 78* family is more highly diverged than *beta 33* and less well conserved with cellular CKR, suggesting either a more ancient date of acquisition or reduced selective pressure to retain CKR features. It may be that the *beta 78* family do not derive from a cellular CKR, but an alternative G protein-coupled receptor (GPCR), although U12 of HHV-6 and HHV-7 has been confirmed as a functional CKR.[5,6] The *beta 27/28* family is an example of gene duplication; human and chimpanzee CMV encode *US27* and *US28* as a tandem repeat, whereas rhesus macaque and African green monkey CMVs encode five/six tandemly repeated homologues. Differences in the number of genes at this locus correspond to major divergence in the lineage of primates. The *beta 27/28* family is lacking in New World monkeys (Squirrel monkey, Owl monkey) and has multiple copies in the Old World monkeys and two copies in apes/humans. This is consistent with acquisition after divergence of New World from Old World monkeys (~40 My) and prior to divergence of Old World monkeys from apes (~25 My). Given the inferred much shorter evolutionary history of the *beta 27/28* family, the sequence diversity suggests a substantially higher rate of evolution than that of *beta 33*—it may be speculated that gene duplication, enabling functional degeneracy, may have enabled more rapid evolution.

A similar history of independent gene capture events may be inferred for the gammaherpesviruses, with three distinct v7TMR gene families, designated herein *gamma BILF1*, *gamma 74*, and *gamma E1*. The *gamma BILF1* family, named after the *BILF1* gene of Epstein–Barr virus (EBV), is present in several gammaherpesvirus genera: *lymphocryptovirus* (EBV and other primate gammaherpesviruses), *macavirus* (*alcelaphine herpesvirus-1* (cattle), *ovine herpesvirus-2* (sheep), *suid herpesvirus-3*, *-4*, and *-5* (pigs)), and *percavirus* (*equid herpesvirus-2* (EHV-2)). In contrast, the *gamma 74* family, named after the prototype gene *ORF74* of *saimirine herpesvirus-2* (*SaHV-2*, common name herpesvirus saimiri), is restricted to the *rhadinovirus* and *percavirus* genera. This suggests either that an ancestral gammaherpesvirus encoded progenitors of both the *BILF1* and *74* families, or each was acquired independently, subsequent to the divergence of the *rhadinovirus/percavirus* genera from the other lineages. The *gamma E1* v7TMR, named after *ORF E1* of EHV-2, is clearly the most recent evolutionary acquisition, on the basis of genomic location and high sequence similarity to cellular CKR, specifically CCR3. To date, *gamma E1* has been identified only in EHV-2, but other species in the *percavirus* genus have yet to be sequenced.

Predominant sequence motifs ("sequence logos") characteristic of each v7TMR family (except *gamma E1*, only identified in one species) for three regions important for CKR signaling, transmembrane (TM) domains II, III, and VII, are shown in Fig. 2. Likely functions of specific conserved residues are as follows: TMII (central Asp (D))—stabilization of the inactive receptor conformation; TMIII ("DRY" box)—"ionic lock" involved in stabilization of conformations between active and inactive states, with the R having a direct role in G protein coupling; TMVII (NPxxY motif)—"tyrosine toggle switch," involved in stabilization of the active conformation.[7] Consistent with the divergence apparent in the phylogenetic tree (Fig. 1), the *beta 78*

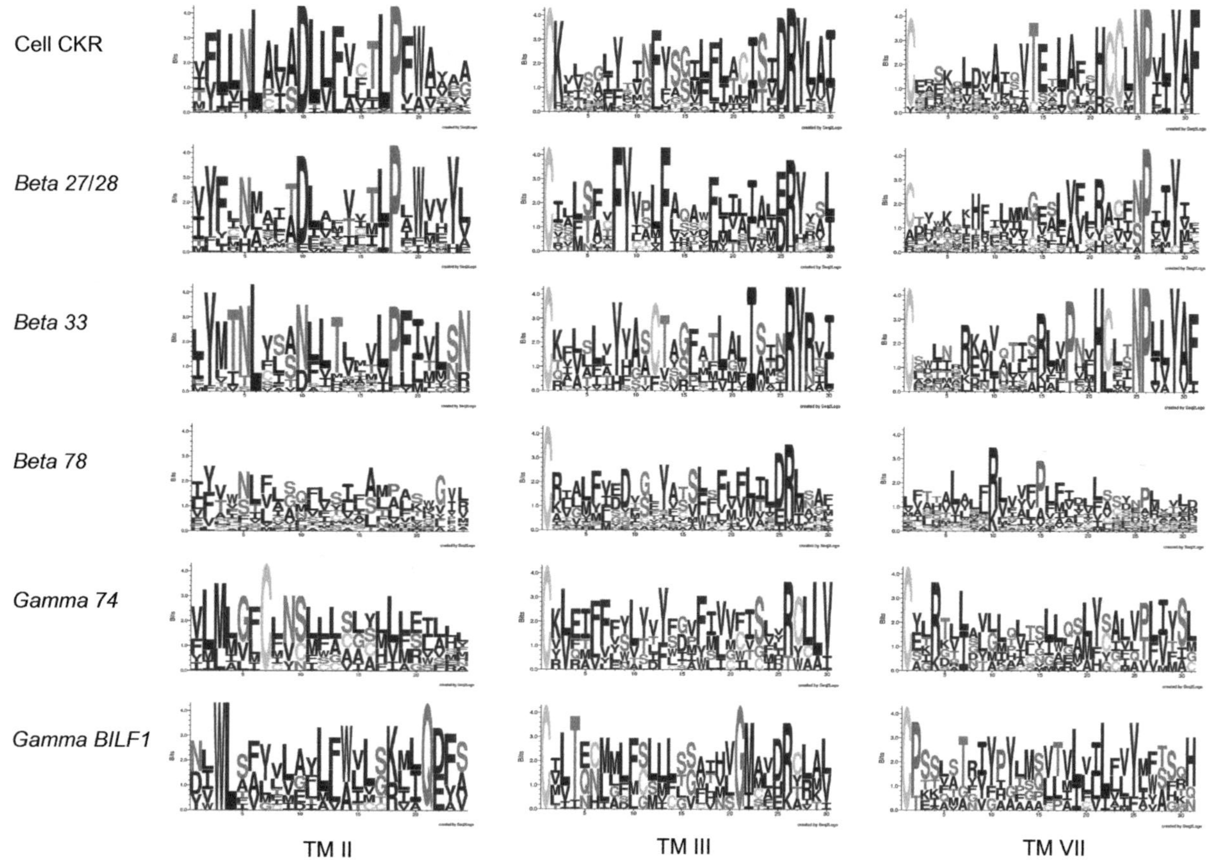

Figure 2 "Sequence logos" of selected transmembrane domains of cellular CKR and viral 7TMR. Sequence logos were derived using the Seq2logo-1.0 server (http://www.cbs.dtu.dk/biotools/Seq2Logo-1.0/) with the following options: logo type = Shannon, clustering method = none, weight on prior = 0. Custom colors (different gray shades in the print version) were selected to highlight amino acids.

family has the lowest overall conservation for each of the analyzed transmembrane domains, with the position of TMII and TMVII uncertain due to the lack of conserved "landmark" residues. In contrast, the *beta 33* family (same number of members and diverse host species as *beta 78*) has noticeable conservation within these transmembrane domains, likely to reflect functionally important motifs. For TMII, the Asp is only well conserved for *beta 27/28* and a subset of *beta 33* (U12 members). For TMIII, the Cys at the extracellular face of TMIII and the Arg of the DRY box are generally well conserved, but many of the v7TMRs have atypical variants of the DRY motif. For TMVII, the NPxxY motif is retained for *beta 33* and partly conserved for *beta 27/28*, but largely diverged for the other v7TMRs. For *gamma 74* and *gamma BILF1*, a number of conserved residues distinct from other v7TMRs and the cellular CKRs are apparent in each of the TM domains. Potentially, signature characteristic of each v7TMR family may reflect selective pressure relating to function; however, additional bioinformatic analyses or experimental data would be required to test that hypothesis.

2. FUNCTIONAL DIVERGENCE FROM CELLULAR CHEMOKINE RECEPTORS

2.1. Ligand specificity—Broadened repertoire/lack of (known) ligands

The cellular CKRs are divided into two major (CC and CXC) and two minor (XC and CX3C) subgroups, plus a group recently termed "atypical CKRs"; a comprehensive review of CKRs has recently been published by the International Union of Basic and Clinical Pharmacology.[8] Though the binding of chemokines to cellular CKRs is remarkably redundant, it is restricted to a single subgroup, i.e., CCR receptors only bind CC chemokines, whereas CXCR receptors only bind CXC ligands. The v7TMRs include members that are more promiscuous than the cellular CKRs (Table 2). US28 from HCMV binds both CX3CL1 and various CC chemokines[9–11]; ORF74 from KSHV binds chemokines of both the CC and CXC kinds[19,20]; U51 from HHV-6 binds XCL1 and several CC ligands.[5,14,15] In addition to the cellular chemokines, some v7TMRs also bind virally encoded ligands; for instance, US28 (HCMV, RhCMV), U51 (HHV-6), and ORF74 (HHV8) bind the HHV8-produced vCCL2.[13,14,27] In general, the pharmacology of the v7TMRs is relatively poorly investigated and mainly focused on v7TMRs encoded by the human

Table 2 Viral 7TMR ligand repertoire

Receptor family	Virus	v7TMR	Ligands		Ligand profile similarity		References
Beta 27/28	HCMV (HHV-5)	US28	CCL2, 3, 4, 5, 7, 11, 13, 26, and 28	CX3CL1	CCR3 (CCR1, CCR2, CCR10)	CX3CR1	9–12
	RhCMV (McHV-3)	US28	CCL2, 3, 4, 5, 7, 11, and 13	CX3CL1	CCR3 (CCR1, CCR2)	CX3CR1	13
Beta 33	HHV-6	U12	CCL2, 3, 4, and 5		CCR5, CCR3		5,14,15
	HHV-7	U12	CCL17, 19, 21, and 22		CCR4, CCR7		16
	MCMV (MuHV-1)	M33	mCCL5[a]		CCR1, CCR3[a]		17
Beta 78	HHV-6	U51	CCL2, 5, 7, 11, 13, 19, and 22	CX3CL1 and XCL1	CCR3 (CCR1, CCR2, CCR5)	CX3CR, XCR1	14,15,18
	HHV-7	U51	CCL17, 19, 21 & 22		CCR4, CCR7		16

Continued

Table 2 Viral 7TMR ligand repertoire—cont'd

Receptor family	Virus	v7TMR	Ligands		Ligand profile similarity		References
Gamma 74	KSHV (HHV-8)	ORF74	CCL1	CXCL1, 2, 3, 4, 5, 6, 7, 8, 10, and 12	(CCR8)	CXCR2 (CXCR3)	19–21
	HVS (SaHV-2)	ORF74		CXCL1, 7, and 8			22
	MHV-68 (MuHV-4)	ORF74		CXCL2 and 10		CXCR2	23
	RhRV (McHV-5)	ORF74		CXCL1[b]		CXCR2	24
	EHV-2	ORF74		CXCL6		CXCR1, CXCR2	25
Gamma E1	EHV-2	E1	CCL11		CCR3, CCR5		26

[a]Ligand interaction of M33 with mCCL5 has been inferred from modulation of chemotaxis, but has yet to be confirmed.
[b]Ligand interaction of RRV-ORF74 with CXCL1 was inferred from modulation (weak) of signaling, but has yet to be confirmed.

herpesviruses (HHVs), whereas v7TMRs from nonhuman viruses are less well studied.[28–31] A number of v7TMRs (human and nonhuman HVs) are still orphan, in some cases despite screening against panels of likely chemokine ligands. It should be noted that whereas some studies of v7TMRs have demonstrated ligand binding directly, others have relied solely on functional readouts, such as modulation of v7TMR-dependent signaling.

US28 from HCMV is the best-characterized v7TMR of the *beta 27/28* family binding a broad spectrum of both CC chemokines and the membrane-tethered CX3CL1.[9–11] US28 from rhesus CMV (rh220, rhUS28.5) has been shown to have a similar ligand repertoire.[13] The *beta 28* v7TMRs resemble cellular CCR1 and CCR3 with respect to their CC ligand binding profile. Unlike the broad repertoire of the *beta 28* v7TMRs, no ligands have been identified for the *beta 27* v7TMRs.

The v7TMRs of the *beta 33* family from HHV-6 and HHV-7 have closest homology to cellular CCR3 and CCR7, respectively, and bind some of the same chemokines as these cellular CKRs including CCL5 (HHV-6 U12) and CCL19 (HHV-7 U12). Additionally, they bind CC chemokines which classically bind other cellular CC CKRs.[5,6,16] So far, no ligands have been identified for the other *beta 33* v7TMRs. Murine CCL5 was reported to stimulate M33 (MCMV)-dependent smooth muscle cell (SMC) migration,[17] but direct demonstration of binding has not been reported and other studies have not demonstrated CCL5 modulation of M33 signaling[32] (Farrell HE, unpublished data); similarly, CCL5 and other chemokines tested to date have not been found to either bind to or modulate constitutive signaling activities of UL33 (HCMV) or R33 (RCMV)[33,34] (Farrell HE, unpublished data).

Most of the v7TMRs of the *beta 78* family are orphan. The few with known ligands have a very diverse binding profile with U51 from HHV-6 binding both CC and XC chemokines, with evidence for weaker association with CX3CL1.[14,15] Additionally, HHV-6 U51 has been reported to downregulate expression of one of its ligands, CCL5, via a mechanism involving transcriptional repression, in both transfected and virus-infected cells.[14,15,35] U51 from HHV-7 was reported to bind CC chemokines exclusively, resembling a combination of ligands of cellular CCR4 and CCR7.[16] The initial study by Tadagaki *et al.* reported very similar profiles of chemokine activity (Ca^{2+} flux) for HHV-7 U51 and U12 expressed in K562 cells,[16] whereas a subsequent report by Tadagaki *et al.* using transfected mouse L1.2 cells reported that U12- or U51-dependent Ca^{2+} flux in response to CCL19 or CCL22 required cotransfection of either

CCR4 or CCR7, suggesting that HHV-7 U12 and U51 work in concert with these cellular CKRs.[35] Thus, there is no general picture of the ligand profile of this family which could be used as a scaffold for searching for ligands for the orphan v7TMRs of the *beta 78* family.

The smallest v7TMR family of the gammaherpesviruses is *gamma E1* which currently includes only one member, E1 from EHV2. With closer homology to cellular CKRs than any other v7TMRs (>50% amino acid identity to *equine caballus* CCR3; unpublished data), of the CC and CXC chemokines tested, E1 was found to bind to a single chemokine, namely CCL11 (eotaxin).[26]

The v7TMRs of the *gamma 74* family have closest homology to the cellular CXCR2 and those with known ligands all bind CXC chemokines. The cellular CXC chemokines are divided into two subgroups depending on an ELR (Glu-Leu-Arg) motif prior to the first cysteine. ORF74 from SaHV-2 (HVS) only binds CXC chemokines of the ELR type,[22,23,36] whereas ORF74 from HHV8 and MHV68 binds both types. Interestingly, the two types of CXC chemokines have the opposite effect on ORF74 (HHV8 or MHV68) with the ELR-type CXC ligands acting as agonist or neutral ligands, while the non-ELR CXC chemokines act as antagonist.[20,37] In addition to CXC ligands, ORF74 from HHV8 also binds CC chemokines.[19,20] In contrast to these fairly broad ligand binging profiles, ORF74 of EHV2 was found to selectively bind CXCL6, although it should be noted that human rather than horse chemokines were tested in this study.[25]

A third family of gammaherpesvirus v7TMRs, initially identified on the basis of relatively weak sequence homology to GPCRs, includes the EBV-encoded *BILF1*.[38,39] BILF1 has been demonstrated to function as a GPCR and although no ligand has so far been identified for BILF1, it may indirectly inhibit the signaling of cellular chemokines by dimerization with CXCR4.[40–42] Broadly speaking, it seems like the ligand binding profiles of the *gamma* v7TMRs (with the exception of *gamma E1*) are more similar to cellular CXC CKRs, whereas the *beta* v7TMRs resemble CC CKRs.

2.2. Constitutive signaling activity

Though only some of the v7TMRs have been deorphanized, many show constitutive activity, stimulating host cell signaling pathways without activation by a ligand.[28–31] This independence of ligand activation may be one of the reasons why so many v7TMRs are still orphan. The receptors may have evolved from cellular CKRs losing ligand interaction along the way and gaining constitutive activity. For example, atypical "DRY" box sequences of TMIII (Fig. 2) may favor constitutive signaling. It may be conjectured that in the context of viral infection, the vGPCRs are no longer constrained by the requirement (of most host GPCRs) to be tightly regulated via ligand interactions to avoid pathological consequences of being constitutively active; loss of ligand binding and gain of constitutive activity may therefore have occurred via lack of negative selection, rather than the presence of positive selection. Alternatively, it may be hypothesized that the receptors have evolved a more efficient system via positive selection, whereby they can utilize host cell signaling directly and independently compared with a two-component system in which the virally induced signaling would depend upon the access and availability of (host) ligands. The ability to still bind and in some cases modify signaling by association with host chemokines may enable dual exploitation of constitutive signaling with the ability to respond to the local environment (e.g., tissue type, inflammation, etc., associated with specific chemokine signals). Additionally, the v7TMRs are more promiscuous in their G protein coupling. Whereas most cellular CKRs primarily couple to $G_{\alpha i}$, several v7TMRs couple to two or more G proteins ($G_{\alpha s}$, $G_{\alpha i}$, $G_{\alpha q}$, $G_{\alpha 12/13}$).

US28 (HCMV) of the *beta 27/28* family was one of the first v7TMRs for which constitutive activity was shown.[43] It is able to stimulate a variety of cellular signaling pathways (PLC, NFkB, CREB, ERK) by coupling to several G proteins including both α and $\beta\gamma$ subunits.[43–46] Moreover, the signaling pathways which are activated constitutively and/or by ligand stimulation seem to be cell type specific.[47] US27 (HCMV) has recently been implicated in upregulation of proliferative responses, but signaling pathways involved have yet to be determined.[48] The v7TMRs of the *beta 33* family have shown similar activity as US28.[44,49] Several of the *beta 33* family members display constitutive activity.[32–34,44] Though they influence the same pathways (PLC, CREB), the individual *beta 33* v7TMRs can have opposite effects on a particular pathway; for instance R33 (MuHV-2, common name rat CMV [Maastricht]) constitutively inhibits CRE, whereas this is constitutively activated by UL33 (HCMV) and M33 (MCMV).[33,44] Ligand-dependent (but not constitutive) signaling activity has been reported for U12 (HHV-6/HHV-7).[5,6]

Constitutive activity has only been reported for one member of the *beta 78* family, namely U51 from HHV-6, which stimulates PLC and inhibits CRE.[50] Additionally, signaling by this v7TMR is modulated by cellular chemokines. The activity of U51 (HHV 6) has been implicated in viral replication,[51] and although no constitutive activity has been reported for the other members of the *beta 78* family, they may serve similar functions since UL78, R78, and M78 (HCMV, RCMV, and MCMV) have been shown to be important for replication and/or pathogenesis of the virus.[52–54]

The first report of a virally encoded receptor with constitutive activity featured ORF74 from HHV-8 (Kaposi's sarcoma (KS)-associated herpesvirus) which stimulates PLC.[21] Other v7TMRs of the *gamma 74* family have also been shown to constitutively activate cellular signaling pathways; for instance ORF74 from McHV-5 (rhesus rhadinovirus) stimulates ERK and VEGF secretion[24] and ORF74 (ECRF3) from SaHV-2 stimulates transcription factors through various G proteins.[22] Interestingly, signaling by ORF74 from SaHV-2 seems to be more ligand dependent than ORF74 from HHV-8 as only some of several signaling pathways are constitutively activated by SaHV-2 ORF74.[22] The constitutive activity of ORF74 from HHV-8 can be modified by chemokine binding, where angiogenic chemokines further stimulate the v7TMRs and angiostatic chemokines inhibit it.[55,56] ORF74 of MuHV-4 (mouse gammaherpesvirus-68) has not directly shown constitutive activity; however, ORF74 (MuHV-4) is expressed on the cell surface and is able to transform cells *in vitro* similar to ORF74 (HHV-8).[23,57]

The v7TMRs of the *gamma BILF1* family such as BILF1 from human and rhesus EBV show constitutive activity through the $G_{\alpha i}$ pathway, inhibiting forskolin-induced CREB activity in HEK or COS cells, but activating CREB in EBV-transformed B cells.[40,41] Furthermore, BILF1 from EBV has been shown to dimerize with CXCR4 causing inhibition of CXCL12-induced signaling by the cellular CKR—a feature linked to the constitutive activity of the v7TMR as a signaling-deficient receptor did not inhibit cellular signaling of CXCR4.[42]

2.3. Constitutive endocytosis

In addition to being constitutively active, many of the v7TMRs are constitutively endocytosed and recycled to the host cell membrane. In this way, the v7TMRs that bind chemokines may potentially act as scavengers, removing inflammatory chemokines from the surroundings and thus modulating detection by the host immune system. Classically, endocytosis is seen as a regulatory process in which ligand-activated GPCRs are internalized to stop signaling. However, endocytosis may trigger G protein-independent signaling pathways.[58,59] Moreover, constitutive endocytosis may localize v7TMR to sites of virion assembly/maturation, thereby promoting incorporation of v7TMR into the virion envelope.

US28 from HCMV has a predominant intracellular rather than cell surface distribution, primarily due to constitutive endocytosis. It has been shown by several groups that ligand-bound US28 (HCMV) is rapidly internalized together with the ligand thus removing chemokines from the extracellular environment as a potential immune evasion mechanism. The v7TMR (without ligand) is then recycled to the membrane where the process can be repeated.[12,60–63] The endocytic properties of US28 (HCMV) have been linked to the C-terminal tail of the v7TMR and the cellular distribution pattern was shown to be transferable when the tail was substituted onto nonconstitutively internalized GPCRs of both viral and cellular origin.[64] Likewise, the C-terminal tail of US27 (HCMV) was shown to determine cellular localization and promote constitutive endocytosis.[65] Similar to US28, US27 is primarily localized intracellularly.[66] The v7TMRs of the *beta 33* family are also found both on the cell surface and localized within endocytic vesicles inside the cell.[66,67] Similarly, v7TMRs of the *beta 78* family have been found to be rapidly endocytosed as shown for M78 from MCMV[54] and UL78 from HCMV.[68]

The v7TMRs of the *gamma BILF1* family are unique in that they not only interfere with cellular chemokines but interact with other components related to the host immune system. As shown for EBV, BILF1 associates with MHC (major histocompatibility complex) class I molecules downregulating their cell surface expression and thus avoiding presentation before the immune system. Furthermore, BILF1 (EBV) targets newly synthesized MHC class I molecules inhibiting trafficking to the host cell surface for immune surveillance.[69,70] Structurally, the C-terminal tail of BILF1 is involved in the endocytosis and lysosomal degradation of MHC class I molecules.[69] The ability to downregulate cell surface expression of MHC class I molecules is shared by BILF1 from rhesus EBV, but not BILF1 from New World marmoset lymphocryptovirus.[70]

2.4. Homo- and heterodimerization of v7TMRs

Dimerization of GPCRs is a common physiological event; however, the significance of this for many GPCRs is still poorly understood. Many reasons for dimerization of these receptors have been proposed (activation of the receptor, proper interaction with G proteins, quality control during synthesis), with a variety of postulated requisites for dimerization (similarity between TM domains, chaperone proteins, and ER colocalization). However, as a rule, most of the cellular GPCRs form homodimers.[71]

It is not surprising that v7TMRs have been found to form dimers. Detection of homodimers has been reported for members of different families of v7TMRs (UL78, M78, US28, BILF, U51) and is likely that most of them present this phenotype.[14,54,68,72,73] Moreover, v7TMRs seem to have explored this physiological pathway in a more promiscuous manner, forming heterodimers with diverse receptors. EBV BILF1 receptor can directly interact with CCR9, CCR10, CXCR3, and CXCR4. The functional consequences of all these interactions are not clear, but it could alter the sensitivity of B cells to chemokines and thus viral dissemination and pathogenesis.[73] In fact, BILF1 interaction with CxCR4 seems to prevent CXCL12 binding to the cellular receptor, reducing G protein signaling and CREB activation.[42] US28 was shown to interact with the three other HCMV v7TMRs. While no functional effect was attributed to dimerization with US27, dimerization with UL33 and UL78 resulted in significant repression of NF-κB activation; thus, interaction between receptors could be another way to regulate virus-induced signaling.[68,72] UL33 and UL78 are also able to dimerize with the cellular receptors CCR5 and CXCR4, thereby modulating host cell responses. Despite keeping the same binding profile, the interaction with the v7TMR induced negative effects upon the regular phenotype of the cellular receptors: reduced cell surface expression, reduced cellular migration, impaired coreceptor activity (to HIV entry), and ligand-induced internalization. Signaling changes upon ligand binding were also modified: in the presence of CCL5, expression of UL78 increased calcium mobilization by CCR5, while expression of UL33 decreased it; in the presence of CXCL12, inositol phosphate induction was reduced by half by CXCR4 in cells expressing either v7TMR.[74] US27, on the other hand, may potentiate CxCR4 signaling. Expression of US27 increased calcium mobilization (dependent on DRY motif and C-terminal region of US27) and enhanced chemotaxis in the presence of CXCL12. Furthermore, US27 also increased CXCR4 expression and translation. However, while colocalization has been reported, heterodimerization between US27 and CXCR4 has not yet been confirmed.[75] It is important to note that all these studies were performed as transient or stable expression of each viral receptor and it is essential to confirm these findings in the context of virus infection. In this regard, a recent report by Niemann *et al.* of tagged US27 and UL78 expressed by recombinant HCMV found that the subcellular trafficking of these receptors changed during the course of infection and did not observe colocalization, in contrast to observations in transfected cells.[76]

3. BIOLOGICAL ROLES OF VIRAL CKRs

Tissue culture studies of v7TMRs have revealed considerable information regarding subcellular trafficking, ligand binding, and signaling properties. Determination of biological roles has been somewhat more elusive. Generally, the v7TMRs have been found to have little or no effect upon virus replication in tissue culture, although some cell type-specific attenuation has been described (see below); in marked contrast, substantial phenotypes have been observed in animal infection studies. The species specificity of the beta- and gammaherpesviruses has restricted direct assessment of v7TMR function in the context of human herpesvirus infection *in vivo*, but roles may be inferred from studies of related animal viruses and from "humanized" animal models. Studies to date have implicated v7TMR in three main aspects of the viral lifecycle and pathogenesis: oncogenesis/angiogenesis/vascular disease, tissue-specific tropism, and persistence. A summary of signaling pathways activated by v7TMRs and potential functional effects is given in Table 3.

3.1. Oncogenesis, angiogenesis, and vascular disease

The oncogenic potential of gammaherpesviruses is well established, for example, EBV (Burkitt's lymphoma, nasopharyngeal carcinoma, Hodgkin's lymphoma) and HHV-8 (KS). Association of betaherpesviruses with neoplasia is uncertain. Both HCMV and HHV-6 have been proposed as potential contributors to neoplasia, either as primary agents of transformation or as oncomodulatory agents, with several studies supporting an association between virus infection and a variety of cancers.[108–110] However, this is an area of controversy. Although initial studies (mainly based upon PCR detection rates) suggested that HHV-6 infection may be associated with a number of hematological malignancies, confirmation that viral DNA/antigen is present within the neoplastic cells has been elusive.[111] In the case of HCMV, current debate is focused upon a potential role in glioblastoma.[112–114] The initial report by Cobbs and colleagues of high-frequency detection of HCMV antigen and DNA in glioblastoma, but not control tissues, prompted considerable interest[115]; this group also reported HCMV detection in colorectal, prostatic, and breast neoplastic tissues.[116–118] Whereas some independent studies have confirmed a high prevalence of

Table 3 Signaling pathways and proposed functions of viral 7TMR

	Activation	Pathway	Downstream effectors[a]	Proposed functions	References
US28 (HCMV)	$G_{q/11}$ $G_{12/13}$ $[G_i, G_{16}]$[b] GRK/β-arrestin	PLC-β/PKC ERK, JNK, p38 MAPK Calcineurin RhoA/ROCK Src-FAK[L]	NF-κB, CREB, SRF, NFAT β-Catenin *STAT3* (via *IL-6R/JAK1*) Survivin, cyclin-D1, c-myc VEGF, IL-6, COX-2Ca^{2+} flux[L]	Transformation Angiogenesis Inflammation Cell migration Latency/reactivation Virus replication (cell–cell spread)	10,43–47,49,77–90
UL33 (HCMV)	$G_{q/11}$ G_s $G_{i/0}$	PLC-β/PKC Rho/p38 MAPK	CREB (CREB downregulated via $G_{i/0}$)	Tissue tropism Latency/reactivation Vascular sclerosis	33,44,49,67
M33 (MCMV)	$G_{q/11}$ GRK/β-arrestin	PLC-β/PKC ERK, p38 MAPK Rac1	CREB, NFAT, NF-κB[c]		17,32,44,49,91,92
R33 (RCMV)	$G_{q/11}$ $G_{i/0}$	PLC-β/PKC	NF-κB CREB down		33,34,93,94
ORF74 (HHV-8)	$G_{q/11}$ $G_{i/0}$ $G_{12/13}$	PLC-β/PKC ERK, JNK, p38 MAPK PI3Kγ/Akt/mTOR Rac1 Src-FAK Calcineurin	NF-κB, CREB, NFAT, SRF, AP1, HIF IL-8, VEGF, angiopoietin-2, angiopoietin-like-4, angiogenin VE-cadherin	Transformation Angiogenesis Inflammation Cell migration Vascular permeability Virus replication Latency/reactivation	22,84,95–107

ORF74 (HVS)	$G_{q/11^L}$ $G_{i/0}$ $G_{12/13}$	PLC-β/PKC Rho/ROCK	CREB, NFAT, SRF $NFkB^L$		22
ORF74 (MuHV-4)	$G_{i/0^L}$	ERK^L, Akt^L PLC-β^L	CREB $down^L$ $NFkB^L$		23
BILF1 (EBV)	$G_{i/0}$		CREB (up or down, depending on cell type) NFkB	Transformation Modulation of cellular CKR	40–42,69

[a]Unless noted otherwise, effectors are upregulated.
[b]Most studies suggest that these G proteins are minor contributors to signaling.
[c]NF-κB activation is not a consistent finding.

PLC, phospholipase C; PKC, protein kinase C; ERK, extracellular signal-regulated kinase; JNK, c-JUN N terminal kinase; MAPK, mitogen-activated protein kinase; ROCK, Rho-associated protein kinase; FAK, focal adhesion kinase; Rac1, Ras-related C3 botulinum toxin substrate 1; PI3Kγ, phosphoinositide-3-kinase-γ polypeptide; Akt, AKT8 virus oncogene homologue; mTOR, mammalian target of rapamycin; NF-κB, nuclear factor-κB; CREB, cAMP response element binding protein; NFAT, nuclear factor of activated T cells; SRF, serum response factor; HIF, hypoxia-inducible factor; AP1, activator protein-1; STAT3, signal transducer and activator of transcription 3; VEGF, vascular endothelial growth factor; COX-2, cyclooxygenase-2.

L, ligand dependent, but note constitutive signaling may also be ligand modulated.

HCMV in glioblastoma,[119–122] others have found no such association.[123–128] Preliminary clinical trials of adjunct anti-CMV chemotherapy or immunotherapy for glioblastoma treatment have been reported,[129,130] but an interpretation of treatment benefit in the chemotherapy trial has been questioned.[131–133] Although a definite causal relationship between any betaherpesvirus and cancer has yet to be established, a number of viral proteins, including the v7TMRs, have been suggested as potential transforming or oncomodulatory factors.[110]

The v7TMRs have attracted attention as potential oncogenes, due to their capacity to activate a range of gene expression pathways associated with cellular transformation and angiogenesis. Often such studies have been based upon isolated v7TMR expression in tissue culture or transgenic mouse models; demonstration of important functional roles in the context of virus-induced tumors is less well studied. In particular, assessment of the relative importance of v7TMRs compared with other viral oncoproteins may be complicated by the availability of physiologically relevant model systems. Nevertheless, accumulating evidence supports the involvement of v7TMRs in gammaherpesvirus neoplastic disease and potentially also for betaherpesviruses.

3.1.1 Beta 27/28 *and* 33

The potential for US28 to promote oncogenesis was demonstrated by Maussang and colleagues, whereby US28 induced transformation of NIH-3T3 cells, associated with upregulated cyclin D and VEGF expression, with vascularized tumor formation in nude mice.[77] Constitutive G protein-mediated signaling was shown to be important via disruption of the DRY motif (US28-R128A), and US28 was shown to be required for VEGF promoter activation in the context of HCMV infection of a glioblastoma cell line. These results contrast with earlier studies demonstrating US28-induced apoptosis in HEK293, COS7, and HeLa cell lines,[134] which may reflect cell-type-specific differences in signaling or other pathways related to transformation. Studies of US28-mediated transformation of NIH-3T3 cells have highlighted NF-κB signaling as a key inducer of oncogenic pathways. Microarray analysis identified upregulation of a number of potential oncogenic mediators, including COX-2, which is implicated in the etiology of several cancers and associated with VEGF upregulation.[135] Subsequent studies highlighted IL-6/STAT3 signaling as a key pathway.[78] Inhibition of either COX-2 or STAT3 was found to inhibit US28-mediated transformation, and US28 was demonstrated to upregulate COX-2 expression in virus-infected cells. Transgenic mice with expression of US28 in intestinal epithelial cells had spontaneous development of adenoma/adenocarcinoma and enhanced susceptibility to inflammation-driven tumor development.[79] The tumors of the transgenic model were associated with enhanced activation of β-catenin and Wnt-target genes, known mediators of oncogenesis; studies in transfected cells confirmed activation of β-catenin which was dependent upon US28 signaling.[79,80] The detection of HCMV infection/US28 expression in a significant proportion of glioblastomas and medulloblastomas, associated with upregulation of VEGF/STAT3 and COX-2, respectively, supported the potential contribution of US28 as an oncomodulator.[136,137] Recently, US27 has been reported to stimulate cellular proliferation in transfected HEK293, HeLa, and COS cells, with stably transfected HEK293 cells showing upregulated expression (1.8–3-fold RNA levels) of several cellular genes including transcription factor AP-1 (jun, fos), antiapoptotic Bcl-x, and the IL-6-related cytokine oncostatin M, suggesting potential complementary activities of US28 and US27.[48]

For several years, HCMV has been a candidate cofactor for atherosclerosis and other vascular diseases; association between CMV and vascular disease is found in some, but not all, studies and remains an open question, with evidence stronger in the transplant recipient setting than the wider population.[138,139] SMCs have been implicated in the early stages of atherosclerosis, with increased proliferation and migration associated with intimal thickening. It has been demonstrated that M33 or R33 was required for induction of vascular SMC migration following infection by mouse or rat CMV, respectively[17,93]; this phenotype is conserved with US28, which was found to be required for HCMV-induced SMC migration.[81,82] Furthermore, US28-dependent migration was found to be cell-type specific, with SMC migration being stimulated via $G_{\alpha 12}$ signaling by CCL5 (RANTES) and antagonized by CX3CL (Fractalkine); in contrast, macrophage migration was stimulated via $G_{\alpha q}$ signaling by CX3CL and antagonized by CCL5.[83] These observations suggest that the *beta28/33* v7TMR may be contributory factors for the observed clinical association between CMV infection and vascular pathologies such as atherosclerosis, restenosis, and transplant vascular sclerosis. However, whether HCMV-induced SMC migration directly contributes to vascular disorders *in vivo* still needs solid confirmation. It has been suggested that in allografts, the overall increase of alloimmune response caused by HCMV infection would have a more active influence in the arteriosclerosis process, evidenced in an animal model by the low number of RCMV-infected cells found in aortic grafts of rats.[140] It should also be noted that there are multiple virus-encoded and virus-induced inflammatory mediators, in addition to CMV-induced immune responses, with a potential role in vascular disease, aside from a direct impact of the v7TMR upon SMC migration.[141]

3.1.2 Gamma 74

The first member of the *gamma74* family of v7TMRs was identified in herpesvirus saimiri (HVS, SaHV-2) and was termed *ECRF3* subsequently renamed *ORF74*.[142] At that time, it was postulated that ORF74 and a neighboring viral cyclin homologue may play roles in cellular transformation. HVS is a T-cell lymphotropic virus which, in common with many herpesviruses, is not usually associated with overt clinical signs in the natural host (squirrel monkey), but which can cause serious disease in other New World primates, inducing fatal T-cell lymphoma.[143] ORF74 of HVS was identified as a viral CKR with closest homology to the IL-8 receptor (CXCR2), and while agonist effects of IL-8 and other CXC chemokines have been confirmed, the function of ORF74 in HVS pathogenesis has not been determined. It is not required for cellular transformation *in vitro*.[143] The transforming potential of ORF74 of the related rhesus rhadinovirus (RhRV, McHV-5) has been demonstrated in an NIH 3T3/nude mouse model, associated with moderate constitutive activation of ERK1/2 (p44/42 MAPK) and secretion of vascular endothelial growth factor (VEGF), features similar to ORF74 of HHV-8 (see below).[24] Similarly, ORF74 of mouse gammaherpesvirus (MuHV-4, MHV-68) was reported to have transforming activity in an NIH 3T3/nude mouse model.[57] In contrast to HHV-8 and RRV ORF74, however, MuHV-4 ORF74 signaling was not constitutive, but dependent upon stimulation by ELR+ CXC chemokines.[23]

The identification of *ORF74* in HHV-8 prompted considerable interest in its pathogenic role. HHV-8 *ORF74* is located within a cluster of genes, most of which are conserved with other rhadinoviruses, which may be considered a "pathogenicity island" due to potential roles in controlling virus replication or cellular responses to infection. These are *ORF71* (vFLIP), a viral FLICE inhibitor protein with antiapoptotic and IL-6-inducing activity; *ORF72* (v-cyclin), which modulates cell-cycle regulation; *ORF73* (LANA-1), a transcriptional regulator expressed in latently infected cells which has been implicated in oncogenesis; and *K14* (vOX-2/CD200), which stimulates inflammatory cytokine production in myeloid cells.[144,145] Early studies demonstrated constitutive signaling activity of ORF74 (inositol phosphate accumulation, activation of JNK and p38 MAPK), with stimulation of cellular proliferation and upregulation of VEGF.[21,146] A potential role of ORF74 in oncogenesis was supported by detection of ORF74 in KS lesions and other HHV-8-associated neoplasia (primary effusion lymphoma (PEL) and multicentric Castleman's disease), although in a minority of cells comprising the tumor.[147–150] Transfection of primary human endothelial cells with ORF74 was found to induce transformation and modulation of cytokines and angiogenic factors, with morphological changes similar to "spindle cells," a characteristic feature of KS lesions.[151,152] Transgenic mouse models have confirmed that ORF74 is sufficient to induce neoplastic lesions resembling KS and that ORF74 signaling activity is required.[153–156] Recently, IKKε activation mediated by ORF74, with concomitant upregulation of NFkB transactivator activity, was reported to be a critical component of oncogenesis in a mouse model.[157] Other studies have indicated important roles in oncogenesis for TGF beta-activating kinase (TAK1) with activation of the NFkB pathway[158] and for CXCL1 agonist activity for ORF74 in activation of the Akt pathway.[159] Significantly in these models, ORF74 expression in a minority of cells within the lesion was sufficient to induce "spindle cell" formation and promote angiogenesis, consistent with ORF74 exerting paracrine effects and its expression during the "lytic" cycle (i.e., in cells likely to be killed during the course of virus replication). A number of studies have supported an important role for VEGF in ORF74-induced neoplasia, whereby ORF74 signaling induces VEGF upregulation with concomitant immortalization of endothelial cells via the VEGF receptor and activation of Akt/protein kinase B signaling.[95,151,160,161] A number of additional factors that are mediators of angiogenesis/oncogenesis have been found to be induced by ORF74, including angiogenin, angiopoietin-2, and angiopoietin-like-4.[96–98] Taken together, these studies provide strong evidence that ORF74 is likely to be at least a contributory factor to HHV-8 oncogenesis and hence a potential target for chemotherapy.

3.1.3 Gamma BILF1

Similar to *gamma 74*, two of the *gamma BILF1* v7TMRs have been found to signal constitutively via $G_{\alpha i}$, namely EBV and AlHV-1.[41,53,162] Furthermore, EBV BILF1 has been found to induce transformation, with evidence for a role of constitutive signaling, whereby disruption of $G_{\alpha i}$ reduced tumor rates from 100% to 40% (NIH 3T3/nude mouse model).[163] However, there is to date no direct evidence that BILF1 contributes to EBV-mediated transformation and oncogenesis. AlHV-1 causes malignant catarrhal fever (MCF), an often fatal lymphoproliferative disease in domestic cattle. By analogy with HHV-8 ORF74, the AlHV-1 BILF1 homologue (A5) may be a candidate contributor to MCF. However, in a rabbit model, knockout of A5 was not found to affect induction of MCF; in contrast, knockout of the major latency-expressed protein LANA-1 (encoded by ORF73) abolished MCF.[162,164]

3.2. Tissue-specific tropism and virus persistence/latency

None of the v7TMRs have been found to be essential for virus replication in tissue culture. However, more subtle effects upon replication in specific cell types and results from *in vivo* pathogenesis have highlighted their likely contribution to virus infection of specific tissue sites. In addition, a number of studies have implicated v7TMRs as modulators of virus persistence or reactivation from latency.

3.2.1 Beta 27/28

HCMV US28 was found to upregulate expression of the major immediate early promoter, suggesting potential to enhance HCMV gene expression/replication.[165,166] It has been postulated that US28 incorporation into the virion envelope may enable delivery to the host cell immediately after viral entry, with potential to activate MAPK, NF-κB, NFAT, and CREB pathways,[43,44,60,84] altering the cellular environment to favor expression of viral genes, which use these transcriptional factors as transactivators. However, the impact of expression changes in the host cell by US28 is not clear. While many studies show that US28 is capable of inducing cell proliferation or expression of genes involved with this process,[48,77,80,83] Hertel and Mocarski found that deletion of US28 did not influence the induction of a promitosis environment during late HCMV infection of fibroblasts.[167] It could be argued that this result may be related to the cell type used in the study, since US28 seems to have a more profound effect in nonfibroblast cell types (epithelial cells, monocytes, endothelial cells). Furthermore, the Hertel study employed a common laboratory strain (AD169), which is now known to be heavily adapted to fibroblast replication with substantial genomic and cell tropism differences to clinical isolates. The differential activities of US28, comparing motility of SMC with macrophages in response to either CCL5 or CXCL3 (see above), highlight potential for US28 modulation of virus dissemination and that other functional effects of the v7TMR are likely to be cell-type specific. US28 has been shown to increase cell-to-cell fusion mediated by other viral proteins, and a role of US28 in viral entry was conjectured.[168] Recently, US28 was shown to facilitate cell-to-cell spread of virus infection, demonstrated by a marked growth deficiency of US28 null HCMV (moderate passage clinical isolate TB40/E) in epithelial cells, where infection strongly depends on cell-to-cell virus spread, and by increased production of extracellular progeny in fibroblasts.[85] Similarly, recent studies of US27-deficient viruses (derived from moderate passage clinical isolates FIX and TB40/E) demonstrated a 10-fold reduction in extracellular progeny in fibroblasts and growth defect in endothelial cells, maintaining, however, normal virus infectivity via direct cell-to-cell spread.[169] These results suggest that US27, which accumulates at sites of virus assembly, may assist virus dissemination by the extracellular route.

A potential role for US28 during latency was suggested by *in vitro* studies demonstrating expression of US28 in a model of latency (HCMV-infected THP-1 monocytes); moreover, US28 expression has been detected in both experimentally and naturally infected peripheral blood mononuclear cells (PBMC), including CD34+ hematopoietic progenitor cells, a site of HCMV latency.[170–172] Expression of US28 could then potentially improve viral dissemination by latently infected monocytes "sensing" tissues that are permissive for HCMV infection via engagement of chemokine ligands for US28—such as endothelial cells, intestinal epithelial cells, and Langerhans cells.

3.2.2 Beta 33

The importance of *beta33* v7TMR in CMV pathogenesis was initially demonstrated by independent studies of mouse and rat CMV. Disruption of M33 or R33 resulted in a profound defect (>4 logs reduction in titer) in replication in salivary glands of mice or rats respectively.[173,174] This strong *in vivo* phenotype was in contrast to minimal effects upon virus replication in tissue culture (mouse fibroblasts or endothelial cells), consistent with earlier demonstration that UL33 knockout did not have a major effect HCMV replication in fibroblasts.[175] Subsequent studies have demonstrated the importance of G protein-coupled signaling to the salivary gland replication phenotype of M33 in mice. Analysis of a panel of MCMV recombinants with mutations of M33 resulting in different effects upon signaling demonstrated a correlation between constitutive signaling ($G_{\alpha q}$/PLC/PKC pathway with CREB and NFAT upregulation) and replication in salivary glands; $G_{\alpha q}$ signaling-null mutants did not replicate to detectable levels in salivary glands, whereas a mutant with intermediate signaling was partially attenuated.[67,91] The M33 phenotype was tissue-specific, with M33 knockout (premature stop codon) or $G_{\alpha q}$ signaling-null virus replicating similarly to wild type in spleen, liver, lung, and pancreas (the original M33 knockout virus, with a lacZ insertion within M33, showed additional attenuation (spleen and pancreas)).[49,176] Recent studies using a severe immunodeficient mouse model (NOD-SCID-gamma mice) suggested that M33-null MCMV has a specific defect in salivary gland replication within the acinar epithelial cells.[92]

M33 knockout or G_q signaling-null viruses also showed reduced *ex vivo* efficiency of reactivation from latency following tissue explant.[49,176] Substitution of M33 by either HCMV UL33 or US28 demonstrated a degree of functional complementation; recombinant viruses expressing either UL33 or US28 replicated (to relatively low levels) in salivary glands and efficiency of reactivation from latency was restored either fully (US28) or partially (UL33).[49,67] Further studies of US28 complementation, comparing mutations affecting US28-mediated signaling, demonstrated that G_q signaling was a major determinant of functional rescue, with G protein-independent, MAPK-mediated signaling also playing a role.[45] The particular cell types within which M33 functions to promote dissemination to/replication in salivary glands or establishment/reactivation from latency are not currently known. It has recently been shown that M33 knockout results in reduced replication efficiency in primary bone marrow macrophage culture, a phenotype rescued by US28.[45] However, the G_q signaling-null mutants of M33 or US28 were not attenuated, so the significance of this observation to the *in vivo* phenotypes is uncertain.

3.2.3 Beta 78

The *beta 78* receptors have an important role in virus replication in a cell type-dependent manner. R78 was found to be important for efficient replication of rat CMV in fibroblast and SMC, but not necessary in monocytes/macrophages and endothelial cells.[53] Deletion of M78 resulted in reduced mouse CMV replication in fibroblasts and macrophages.[177] Initial studies of HCMV (strain AD169) found that UL78 did not affect replication in fibroblasts and renal artery culture[178]; however, subsequent studies (strain TB40/E) demonstrated that although UL78 did not affect replication in fibroblasts, it was required for efficient replication in epithelial and endothelial cells.[52] It has been suggested that UL78 and M78 are incorporated into the virion envelope and these diverse phenotypes may be related to different modes of viral entry in these cells.[52,54,177] HCMV enters fibroblast cells by membrane fusion at the cell surface, while epithelial cell infection is mediated by endocytosis.[179–181] The available data suggest that for UL78, the receptor takes part in the endocytic process, precisely in a step after binding and before entry itself. UL78 defective viruses, besides a delayed entry profile, also present a holdup in the transport of tegument proteins and DNA to the cellular nucleus. UL78 was also shown to be necessary for efficient accumulation of viral proteins and RNA prior to accumulation or viral IE RNAs in both epithelial and endothelial cells.[52] Both M78 and R78 are determinants for virus pathogenesis. Null virus for both M78 and R78 induced a lower mortality rate *in vivo*, with M78-null mouse CMV having lower viral titers in spleen, liver, and salivary glands.[53,177] Disruption of the R78 C-tail region alone was sufficient to reduce rat CMV titers *in vitro* and even though it reduced titers in the spleen, it did not affect infection of salivary glands *in vivo*.[53,182]

The HHV-6 U51 has been shown to play a direct part in viral replication and dissemination. Silencing of U51 in an infected T lymphoblast cell line inhibited both viral DNA replication and syncytium formation. The latter suggests that the v7TMRs may also enhance cell-to-cell fusion, thus assisting with virus dissemination, since HHV-6 is a highly cell associated virus that relies strongly upon cell-to-cell spread.[51]

3.2.4 Gamma 74

The observation that HHV-8 ORF74 transactivated expression of several viral gene promoters suggested a possible regulatory role during virus replication.[183] Studies of HHV-8, either in the context of viral BAC transfected HEK293 cells, or latently infected B cells (PEL) or endothelial cells (TIME), suggested a role for ORF74 in promoting viral replication. Thus, ORF74 null (BAC) or ORF74 knockdown via shRNA (PEL, TIME) reduced HHV8 genome copy number expansion in response to induction via TPA or (HEK cells only) expression of the viral transactivator protein RTA (encoded by *ORF50*).[184] Using the HEK/BAC system, ORF74 mutations were introduced to disrupt either $G_{\alpha i}$ (PI3-kinase/NFkB) or $G_{\alpha q}$ (PKC/MAPK) signaling; disruption of $G_{\alpha q}$ coupling attenuated genome replication, whereas disruption of $G_{\alpha i}$ coupling had little effect, suggesting that ORF74 signaling via the PLC/PKC/MAPK pathway, which is associated with VEGF upregulation, promoted virus replication.[99,184] The involvement of ORF74 during early stages of the switch from the latent to lytic cycle is consistent with the observations that the *ORF74* promoter is responsive to RTA transactivation and is 1 of 19 sites in the HHV-8 genome that bind RTA.[185,186] Furthermore, ORF74 has been found to stimulate expression of RTA, via upregulation of Sp1/Sp3 transactivation, suggesting a potential positive feedback mechanism for the lytic switch upon reactivation; consistent with the replication studies, transactivation was inhibited by disruption of $G_{\alpha q}$-(PLC/PKC pathway) signaling, but not affected by $G_{\alpha i}$ (PI3K/NFkB) disruption.[187] Studies of mouse gammaherpesvirus (MuHV-4) ORF74 have also suggested a role in promoting viral replication and reactivation from latency. Virus replication in tissue culture (fibroblasts: NIH 3T3) was found to be enhanced in the presence of ORF74 agonists (ELR + chemokines CXCL1 and CXCL2), an effect abolished for an ORF74-null virus.[188] Independent studies of effects of ORF74 knockout suggested a contribution to reactivation from latency, using either *in vitro* (mouse splenocytes) or *ex vivo* reactivation models[188,189]; it should be noted that the *in vivo* studies demonstrated no attenuation in replication or pathogenesis for the ORF74 knockout.

4. CONCLUDING REMARKS

A striking feature of all beta- and gammaherpesviruses is the encoding of one or more 7TMR homologues. In contrast, 7TMR homologues have yet to be identified for any alphaherpesvirus. It is reasonable to speculate that the acquisition of v7TMR by beta- and gammaherpesviruses is related to shared features of their biology that contrast to alphaherpesviruses. A potential factor may be the key involvement of immune effector cells (myeloid or lymphocyte lineages) in the pathobiology of beta- and gammaherpesviruses. Whether this distinction reflects enhanced opportunity to capture 7TMR genes from host immune effector cells, improved fitness of viruses able to subvert normal 7TMR-mediated regulation of such cells, or other factors is not known. However, given the plethora of physiological processes demonstrated or suspected to be influenced by the v7TMR, they present potential targets for development of antiviral therapies.

REFERENCES

1. Davison AJ. Herpesvirus systematics. *Vet Microbiol*. 2010;143(1):52–69.
2. Sneath PHA, Sokal RR. *Numerical Taxonomy; the Principles and Practice of Numerical Classification*. San Francisco: W. H. Freeman; 1973.
3. Jones DT, Taylor WR, Thornton JM. The rapid generation of mutation data matrices from protein sequences. *Comput Appl Biosci*. 1992;8(3):275–282.

4. Tamura K, Peterson D, Peterson N, Stecher G, Nei M, Kumar S. MEGA5: molecular evolutionary genetics analysis using maximum likelihood, evolutionary distance, and maximum parsimony methods. *Mol Biol Evol.* 2011;28(10):2731–2739.
5. Isegawa Y, Ping Z, Nakano K, Sugimoto N, Yamanishi K. Human herpesvirus 6 open reading frame U12 encodes a functional beta-chemokine receptor. *J Virol.* 1998;72(7):6104–6112.
6. Nakano K, Tadagaki K, Isegawa Y, Aye MM, Zou P, Yamanishi K. Human herpesvirus 7 open reading frame U12 encodes a functional beta-chemokine receptor. *J Virol.* 2003;77(14):8108–8115.
7. Trzaskowski B, Latek D, Yuan S, Ghoshdastider U, Debinski A, Filipek S. Action of molecular switches in GPCRs—theoretical and experimental studies. *Curr Med Chem.* 2012;19(8):1090–1109.
8. Bachelerie F, Ben-Baruch A, Burkhardt AM, et al. International Union of Basic and Clinical Pharmacology. [corrected]. LXXXIX. Update on the extended family of chemokine receptors and introducing a new nomenclature for atypical chemokine receptors. *Pharmacol Rev.* 2014;66(1):1–79.
9. Kledal TN, Rosenkilde MM, Schwartz TW. Selective recognition of the membrane-bound CX3C chemokine, fractalkine, by the human cytomegalovirus-encoded broad-spectrum receptor US28. *FEBS Lett.* 1998;441(2):209–214.
10. Billstrom MA, Johnson GL, Avdi NJ, Worthen GS. Intracellular signaling by the chemokine receptor US28 during human cytomegalovirus infection. *J Virol.* 1998;72(7):5535–5544.
11. Gao JL, Murphy PM. Human cytomegalovirus open reading frame US28 encodes a functional beta chemokine receptor. *J Biol Chem.* 1994;269(46):28539–28542.
12. Bodaghi B, Jones TR, Zipeto D, et al. Chemokine sequestration by viral chemoreceptors as a novel viral escape strategy: withdrawal of chemokines from the environment of cytomegalovirus-infected cells. *J Exp Med.* 1998;188(5):855–866.
13. Penfold ME, Schmidt TL, Dairaghi DJ, Barry PA, Schall TJ. Characterization of the rhesus cytomegalovirus US28 locus. *J Virol.* 2003;77(19):10404–10413.
14. Milne RS, Mattick C, Nicholson L, Devaraj P, Alcami A, Gompels UA. RANTES binding and down-regulation by a novel human herpesvirus-6 beta chemokine receptor. *J Immunol.* 2000;164(5):2396–2404.
15. Catusse J, Spinks J, Mattick C, et al. Immunomodulation by herpesvirus U51A chemokine receptor via CCL5 and FOG-2 down-regulation plus XCR1 and CCR7 mimicry in human leukocytes. *Eur J Immunol.* 2008;38(3):763–777. http://dx.doi.org/10.1002/eji.200737618.
16. Tadagaki K, Nakano K, Yamanishi K. Human herpesvirus 7 open reading frames U12 and U51 encode functional beta-chemokine receptors. *J Virol.* 2005;79(11):7068–7076.
17. Melnychuk RM, Smith P, Kreklywich CN, et al. Mouse cytomegalovirus M33 is necessary and sufficient in virus-induced vascular smooth muscle cell migration. *J Virol.* 2005;79(16):10788–10795.
18. Caruso A, Favilli F, Rotola A, et al. Human herpesvirus-6 modulates RANTES production in primary human endothelial cell cultures. *J Med Virol.* 2003;70(3):451–458.
19. Gershengorn MC, Geras-Raaka E, Varma A, Clark-Lewis I. Chemokines activate Kaposi's sarcoma-associated herpesvirus G protein-coupled receptor in mammalian cells in culture. *J Clin Invest.* 1998;102(8):1469–1472.
20. Rosenkilde MM, Kledal TN, Brauner-Osborne H, Schwartz TW. Agonists and inverse agonists for the herpesvirus 8-encoded constitutively active seven-transmembrane oncogene product, ORF-74. *J Biol Chem.* 1999;274(2):956–961.
21. Arvanitakis L, Geras-Raaka E, Varma A, Gershengorn MC, Cesarman E. Human herpesvirus KSHV encodes a constitutively active G-protein-coupled receptor linked to cell proliferation. *Nature.* 1997;385(6614):347–350.

22. Rosenkilde MM, McLean KA, Holst PJ, Schwartz TW. The CXC chemokine receptor encoded by herpesvirus saimiri, ECRF3, shows ligand-regulated signaling through Gi, Gq, and G12/13 proteins but constitutive signaling only through Gi and G12/13 proteins. *J Biol Chem*. 2004;279(31):32524–32533, Epub 32004 May 32520.
23. Verzijl D, Fitzsimons CP, Van Dijk M, et al. Differential activation of murine herpesvirus 68- and Kaposi's sarcoma-associated herpesvirus-encoded ORF74 G protein-coupled receptors by human and murine chemokines. *J Virol*. 2004;78(7):3343–3351.
24. Estep RD, Axthelm MK, Wong SW. A G protein-coupled receptor encoded by rhesus rhadinovirus is similar to ORF74 of Kaposi's sarcoma-associated herpesvirus. *J Virol*. 2003;77(3):1738–1746.
25. Rosenkilde MM, Kledal TN, Schwartz TW. High constitutive activity of a virus-encoded seven transmembrane receptor in the absence of the conserved DRY motif (Asp-Arg-Tyr) in transmembrane helix 3. *Mol Pharmacol*. 2005;68(1):11–19.
26. Camarda G, Spinetti G, Bernardini G, et al. The equine herpesvirus 2 E1 open reading frame encodes a functional chemokine receptor. *J Virol*. 1999;73(12):9843–9848.
27. Kledal TN, Rosenkilde MM, Coulin F, et al. A broad-spectrum chemokine antagonist encoded by Kaposi's sarcoma-associated herpesvirus. *Science*. 1997; 277(5332):1656–1659.
28. Montaner S, Kufareva I, Abagyan R, Gutkind JS. Molecular mechanisms deployed by virally encoded G protein-coupled receptors in human diseases. *Annu Rev Pharmacol Toxicol*. 2013;53:331–354.
29. Slinger E, Langemeijer E, Siderius M, Vischer HF, Smit MJ. Herpesvirus-encoded GPCRs rewire cellular signaling. *Mol Cell Endocrinol*. 2011;331(2):179–184.
30. Vischer HF, Siderius M, Leurs R, Smit MJ. Herpesvirus-encoded GPCRs: neglected players in inflammatory and proliferative diseases? *Nat Rev Drug Discov*. 2014;13(2): 123–139.
31. Rosenkilde MM, Kledal TN. Targeting herpesvirus reliance of the chemokine system. *Curr Drug Targets*. 2006;7(1):103–118.
32. Sherrill JD, Miller WE. G protein-coupled receptor (GPCR) kinase 2 regulates agonist-independent Gq/11 signaling from the mouse cytomegalovirus GPCR M33. *J Biol Chem*. 2006;281(52):39796–39805, Epub 32006 Nov 39796.
33. Casarosa P, Gruijthuijsen YK, Michel D, et al. Constitutive signaling of the human cytomegalovirus-encoded receptor UL33 differs from that of its rat cytomegalovirus homolog R33 by promiscuous activation of G proteins of the Gq, Gi, and Gs classes. *J Biol Chem*. 2003;278(50):50010–50023, Epub 52003 Sep 50030.
34. Gruijthuijsen YK, Casarosa P, Kaptein SJ, et al. The rat cytomegalovirus R33-encoded G protein-coupled receptor signals in a constitutive fashion. *J Virol*. 2002;76(3): 1328–1338.
35. Tadagaki K, Yamanishi K, Mori Y. Reciprocal roles of cellular chemokine receptors and human herpesvirus 7-encoded chemokine receptors, U12 and U51. *J Gen Virol*. 2007;88(Pt 5):1423–1428.
36. Ahuja SK, Murphy PM. Molecular piracy of mammalian interleukin-8 receptor type B by herpesvirus saimiri. *J Biol Chem*. 1993;268(28):20691–20694.
37. Rosenkilde MM, Schwartz TW. Potency of ligands correlates with affinity measured against agonist and inverse agonists but not against neutral ligand in constitutively active chemokine receptor. *Mol Pharmacol*. 2000;57(3):602–609.
38. Davis-Poynter NJ, Farrell HE. Masters of deception: a review of herpesvirus immune evasion strategies. *Immunol Cell Biol*. 1996;74(6):513–522.
39. Telford EA, Watson MS, Aird HC, Perry J, Davison AJ. The DNA sequence of equine herpesvirus 2. *J Mol Biol*. 1995;249(3):520–528.
40. Beisser PS, Verzijl D, Gruijthuijsen YK, et al. The Epstein-Barr virus BILF1 gene encodes a G protein-coupled receptor that inhibits phosphorylation of RNA-dependent protein kinase. *J Virol*. 2005;79(1):441–449.

41. Paulsen SJ, Rosenkilde MM, Eugen-Olsen J, Kledal TN. Epstein-Barr virus-encoded BILF1 is a constitutively active G protein-coupled receptor. *J Virol.* 2005;79(1):536–546.
42. Nijmeijer S, Leurs R, Smit MJ, Vischer HF. The Epstein-Barr virus-encoded G protein-coupled receptor BILF1 hetero-oligomerizes with human CXCR4, scavenges Galphai proteins, and constitutively impairs CXCR4 functioning. *J Biol Chem.* 2010;285(38):29632–29641. http://dx.doi.org/10.1074/jbc.M110.115618, Epub 2010 Jul 9.
43. Casarosa P, Bakker RA, Verzijl D, et al. Constitutive signaling of the human cytomegalovirus-encoded chemokine receptor US28. *J Biol Chem.* 2001;276(2): 1133–1137.
44. Waldhoer M, Kledal TN, Farrell H, Schwartz TW. Murine cytomegalovirus (CMV) M33 and human CMV US28 receptors exhibit similar constitutive signaling activities. *J Virol.* 2002;76(16):8161–8168.
45. Farrell HE, Abraham AM, Cardin RD, Molleskov-Jensen AS, Rosenkilde MM, Davis-Poynter N. Identification of common mechanisms by which human and mouse cytomegalovirus seven-transmembrane receptor homologues contribute to in vivo phenotypes in a mouse model. *J Virol.* 2013;87(7):4112–4117. http://dx.doi.org/10.1128/JVI.03406-12, Epub 2013 Jan 23.
46. Minisini R, Tulone C, Luske A, et al. Constitutive inositol phosphate formation in cytomegalovirus-infected human fibroblasts is due to expression of the chemokine receptor homologue pUS28. *J Virol.* 2003;77(8):4489–4501.
47. Miller WE, Zagorski WA, Brenneman JD, Avery D, Miller JL, O'Connor CM. US28 is a potent activator of phospholipase C during HCMV infection of clinically relevant target cells. *PLoS One.* 2012;7(11):e50524. http://dx.doi.org/10.51371/journal.pone.0050524, Epub 2012 Nov 29.
48. Lares AP, Tu CC, Spencer JV. The human cytomegalovirus US27 gene product enhances cell proliferation and alters cellular gene expression. *Virus Res.* 2013;176(1–2):312–320.
49. Farrell HE, Abraham AM, Cardin RD, et al. Partial functional complementation between human and mouse cytomegalovirus chemokine receptor homologues. *J Virol.* 2011;85(12):6091–6095. http://dx.doi.org/10.1128/JVI.02113-10, Epub 2011 Apr 13.
50. Fitzsimons CP, Gompels UA, Verzijl D, et al. Chemokine-directed trafficking of receptor stimulus to different g proteins: selective inducible and constitutive signaling by human herpesvirus 6-encoded chemokine receptor U51. *Mol Pharmacol.* 2006;69(3):888–898, Epub 2005 Dec 6.
51. Zhen Z, Bradel-Tretheway B, Sumagin S, Bidlack JM, Dewhurst S. The human herpesvirus 6 G protein-coupled receptor homolog U51 positively regulates virus replication and enhances cell-cell fusion in vitro. *J Virol.* 2005;79(18):11914–11924.
52. O'Connor CM, Shenk T. Human cytomegalovirus pUL78 G protein-coupled receptor homologue is required for timely cell entry in epithelial cells but not fibroblasts. *J Virol.* 2012;86(21):11425–11433.
53. Beisser PS, Grauls G, Bruggeman CA, Vink C. Deletion of the R78 G protein-coupled receptor gene from rat cytomegalovirus results in an attenuated, syncytium-inducing mutant strain. *J Virol.* 1999;73(9):7218–7230.
54. Sharp EL, Davis-Poynter NJ, Farrell HE. Analysis of the subcellular trafficking properties of murine cytomegalovirus M78, a 7 transmembrane receptor homologue. *J Gen Virol.* 2009;90(Pt 1):59–68. http://dx.doi.org/10.1099/vir.0.004853-0.
55. Rosenkilde MM, Kledal TN, Holst PJ, Schwartz TW. Selective elimination of high constitutive activity or chemokine binding in the human herpesvirus 8 encoded seven transmembrane oncogene ORF74. *J Biol Chem.* 2000;275(34):26309–26315.
56. Geras-Raaka E, Arvanitakis L, Bais C, Cesarman E, Mesri EA, Gershengorn MC. Inhibition of constitutive signaling of Kaposi's sarcoma-associated herpesvirus

G protein-coupled receptor by protein kinases in mammalian cells in culture. *J Exp Med.* 1998;187(5):801–806.
57. Wakeling MN, Roy DJ, Nash AA, Stewart JP. Characterization of the murine gammaherpesvirus 68 ORF74 product: a novel oncogenic G protein-coupled receptor. *J Gen Virol.* 2001;82(Pt 5):1187–1197.
58. Irannejad R, von Zastrow M. GPCR signaling along the endocytic pathway. *Curr Opin Cell Biol.* 2014;27:109–116.
59. Jean-Alphonse F, Hanyaloglu AC. Regulation of GPCR signal networks via membrane trafficking. *Mol Cell Endocrinol.* 2011;331(2):205–214.
60. Fraile-Ramos A, Kledal TN, Pelchen-Matthews A, Bowers K, Schwartz TW, Marsh M. The human cytomegalovirus US28 protein is located in endocytic vesicles and undergoes constitutive endocytosis and recycling. *Mol Biol Cell.* 2001;12(6): 1737–1749.
61. Tschische P, Moser E, Thompson D, et al. The G-protein coupled receptor associated sorting protein GASP-1 regulates the signalling and trafficking of the viral chemokine receptor US28. *Traffic.* 2010;11(5):660–674. http://dx.doi.org/10.1111/j.1600-0854.2010.1045.x.
62. Randolph-Habecker JR, Rahill B, Torok-Storb B, et al. The expression of the cytomegalovirus chemokine receptor homolog US28 sequesters biologically active CC chemokines and alters IL-8 production. *Cytokine.* 2002;19(1):37–46.
63. Billstrom MA, Lehman LA, Scott Worthen G. Depletion of extracellular RANTES during human cytomegalovirus infection of endothelial cells. *Am J Respir Cell Mol Biol.* 1999;21(2):163–167.
64. Waldhoer M, Casarosa P, Rosenkilde MM, et al. The carboxyl terminus of human cytomegalovirus-encoded 7 transmembrane receptor US28 camouflages agonism by mediating constitutive endocytosis. *J Biol Chem.* 2003;278(21):19473–19482, Epub 12003 Mar 19418.
65. Stapleton LK, Arnolds KL, Lares AP, Devito TM, Spencer JV. Receptor chimeras demonstrate that the C-terminal domain of the human cytomegalovirus US27 gene product is necessary and sufficient for intracellular receptor localization. *Virol J.* 2012;9:42. http://dx.doi.org/10.1186/1743-1422X-9-42.
66. Fraile-Ramos A, Pelchen-Matthews A, Kledal TN, Browne H, Schwartz TW, Marsh M. Localization of HCMV UL33 and US27 in endocytic compartments and viral membranes. *Traffic.* 2002;3(3):218–232.
67. Case R, Sharp E, Benned-Jensen T, Rosenkilde MM, Davis-Poynter N, Farrell HE. Functional analysis of the murine cytomegalovirus chemokine receptor homologue M33: ablation of constitutive signaling is associated with an attenuated phenotype in vivo. *J Virol.* 2008;82(4):1884–1898, Epub 2007 Dec 1885.
68. Wagner S, Arnold F, Wu Z, et al. The 7-transmembrane protein homologue UL78 of the human cytomegalovirus forms oligomers and traffics between the plasma membrane and different intracellular compartments. *Arch Virol.* 2012;157(5):935–949. http://dx.doi.org/10.1007/s00705-012-1246-6, Epub 2012 Feb 12.
69. Zuo J, Quinn LL, Tamblyn J, et al. The Epstein-Barr virus-encoded BILF1 protein modulates immune recognition of endogenously processed antigen by targeting major histocompatibility complex class I molecules trafficking on both the exocytic and endocytic pathways. *J Virol.* 2011;85(4):1604–1614. http://dx.doi.org/10.1128/JVI.01608-10, Epub 2010 Dec 1.
70. Griffin BD, Gram AM, Mulder A, et al. EBV BILF1 evolved to downregulate cell surface display of a wide range of HLA class I molecules through their cytoplasmic tail. *J Immunol.* 2013;190(4):1672–1684. http://dx.doi.org/10.4049/jimmunol.1102462, Epub 2013 Jan 11.

71. Bulenger S, Marullo S, Bouvier M. Emerging role of homo- and heterodimerization in G-protein-coupled receptor biosynthesis and maturation. *Trends Pharmacol Sci.* 2005;26(3):131–137.
72. Tschische P, Tadagaki K, Kamal M, Jockers R, Waldhoer M. Heteromerization of human cytomegalovirus encoded chemokine receptors. *Biochem Pharmacol.* 2011;82(6):610–619.
73. Vischer HF, Nijmeijer S, Smit MJ, Leurs R. Viral hijacking of human receptors through heterodimerization. *Biochem Biophys Res Commun.* 2008;377(1):93–97.
74. Tadagaki K, Tudor D, Gbahou F, et al. Human cytomegalovirus-encoded UL33 and UL78 heteromerize with host CCR5 and CXCR4 impairing their HIV coreceptor activity. *Blood.* 2012;119(21):4908–4918.
75. Arnolds KL, Lares AP, Spencer JV. The US27 gene product of human cytomegalovirus enhances signaling of host chemokine receptor CXCR4. *Virology.* 2013; 439(2):122–131.
76. Niemann I, Reichel A, Stamminger T. Intracellular trafficking of the human cytomegalovirus-encoded 7-trans-membrane protein homologs pUS27 and pUL78 during viral infection: a comparative analysis. *Viruses.* 2014;6(2):661–682.
77. Maussang D, Verzijl D, van Walsum M, et al. Human cytomegalovirus-encoded chemokine receptor US28 promotes tumorigenesis. *Proc Natl Acad Sci USA.* 2006;103(35):13068–13073.
78. Slinger E, Maussang D, Schreiber A, et al. HCMV-encoded chemokine receptor US28 mediates proliferative signaling through the IL-6-STAT3 axis. *Sci Signal.* 2010;3(133): ra58.
79. Bongers G, Maussang D, Muniz LR, et al. The cytomegalovirus-encoded chemokine receptor US28 promotes intestinal neoplasia in transgenic mice. *J Clin Invest.* 2010;120(11):3969–3978.
80. Langemeijer EV, Slinger E, de Munnik S, et al. Constitutive beta-catenin signaling by the viral chemokine receptor US28. *PLoS One.* 2012;7(11):e48935.
81. Streblow DN, Soderberg-Naucler C, Vieira J, et al. The human cytomegalovirus chemokine receptor US28 mediates vascular smooth muscle cell migration. *Cell.* 1999;99(5):511–520.
82. Streblow DN, Vomaske J, Smith P, et al. Human cytomegalovirus chemokine receptor US28-induced smooth muscle cell migration is mediated by focal adhesion kinase and Src. *J Biol Chem.* 2003;278(50):50456–50465.
83. Vomaske J, Nelson JA, Streblow DN. Human Cytomegalovirus US28: a functionally selective chemokine binding receptor. *Infect Disord Drug Targets.* 2009;9(5):548–556.
84. McLean KA, Holst PJ, Martini L, Schwartz TW, Rosenkilde MM. Similar activation of signal transduction pathways by the herpesvirus-encoded chemokine receptors US28 and ORF74. *Virology.* 2004;325(2):241–251.
85. Noriega VM, Gardner TJ, Redmann V, Bongers G, Lira SA, Tortorella D. Human cytomegalovirus US28 facilitates cell-to-cell viral dissemination. *Viruses.* 2014;6(3):1202–1218.
86. Miller WE, Houtz DA, Nelson CD, Kolattukudy PE, Lefkowitz RJ. G-protein-coupled receptor (GPCR) kinase phosphorylation and beta-arrestin recruitment regulate the constitutive signaling activity of the human cytomegalovirus US28 GPCR. *J Biol Chem.* 2003;278(24):21663–21671.
87. Moepps B, Tulone C, Kern C, et al. Constitutive serum response factor activation by the viral chemokine receptor homologue pUS28 is differentially regulated by Galpha(q/11) and Galpha(16). *Cell Signal.* 2008;20(8):1528–1537.
88. Stropes MP, Schneider OD, Zagorski WA, Miller JL, Miller WE. The carboxy-terminal tail of human cytomegalovirus (HCMV) US28 regulates both

chemokine-independent and chemokine-dependent signaling in HCMV-infected cells. *J Virol.* 2009;83(19):10016–10027.
89. Hjorto GM, Kiilerich-Pedersen K, Selmeczi D, Kledal TN, Larsen NB. Human cytomegalovirus chemokine receptor US28 induces migration of cells on a CX3CL1-presenting surface. *J Gen Virol.* 2013;94(Pt 5):1111–1120.
90. Melnychuk RM, Streblow DN, Smith PP, Hirsch AJ, Pancheva D, Nelson JA. Human cytomegalovirus-encoded G protein-coupled receptor US28 mediates smooth muscle cell migration through Galpha12. *J Virol.* 2004;78(15):8382–8391.
91. Sherrill JD, Stropes MP, Schneider OD, et al. Activation of intracellular signaling pathways by the murine cytomegalovirus G protein-coupled receptor M33 occurs via PLC-{beta}/PKC-dependent and -independent mechanisms. *J Virol.* 2009;83(16):8141–8152.
92. Bittencourt F, Wu SE, Bridges JP, Miller WE. The M33 GPCR encoded by MCMV is dispensable for hematogenous dissemination but is required for growth within the salivary gland. *J Virol.* 2014;88(20):11811–11824.
93. Streblow DN, Kreklywich CN, Smith P, et al. Rat cytomegalovirus-accelerated transplant vascular sclerosis is reduced with mutation of the chemokine-receptor R33. *Am J Transplant.* 2005;5(3):436–442.
94. Gruijthuijsen YK, Beuken EV, Smit MJ, Leurs R, Bruggeman CA, Vink C. Mutational analysis of the R33-encoded G protein-coupled receptor of rat cytomegalovirus: identification of amino acid residues critical for cellular localization and ligand-independent signalling. *J Gen Virol.* 2004;85(Pt 4):897–909.
95. Sodhi A, Montaner S, Patel V, et al. The Kaposi's sarcoma-associated herpes virus G protein-coupled receptor up-regulates vascular endothelial growth factor expression and secretion through mitogen-activated protein kinase and p38 pathways acting on hypoxia-inducible factor 1alpha. *Cancer Res.* 2000;60(17):4873–4880.
96. Ma T, Jham BC, Hu J, et al. Viral G protein-coupled receptor up-regulates Angiopoietin-like 4 promoting angiogenesis and vascular permeability in Kaposi's sarcoma. *Proc Natl Acad Sci USA.* 2010;107(32):14363–14368.
97. Sadagopan S, Sharma-Walia N, Veettil MV, et al. Kaposi's sarcoma-associated herpesvirus upregulates angiogenin during infection of human dermal microvascular endothelial cells, which induces 45S rRNA synthesis, antiapoptosis, cell proliferation, migration, and angiogenesis. *J Virol.* 2009;83(7):3342–3364.
98. Vart RJ, Nikitenko LL, Lagos D, et al. Kaposi's sarcoma-associated herpesvirus-encoded interleukin-6 and G-protein-coupled receptor regulate angiopoietin-2 expression in lymphatic endothelial cells. *Cancer Res.* 2007;67(9):4042–4051.
99. Liu C, Sandford G, Fei G, Nicholas J. Galpha protein selectivity determinant specified by a viral chemokine receptor-conserved region in the C tail of the human herpesvirus 8 g protein-coupled receptor. *J Virol.* 2004;78(5):2460–2471.
100. Cannon ML, Cesarman E. The KSHV G protein-coupled receptor signals via multiple pathways to induce transcription factor activation in primary effusion lymphoma cells. *Oncogene.* 2004;23(2):514–523.
101. Couty JP, Geras-Raaka E, Weksler BB, Gershengorn MC. Kaposi's sarcoma-associated herpesvirus G protein-coupled receptor signals through multiple pathways in endothelial cells. *J Biol Chem.* 2001;276(36):33805–33811.
102. Dwyer J, Le Guelte A, Galan Moya EM, et al. Remodeling of VE-cadherin junctions by the human herpes virus 8 G-protein coupled receptor. *Oncogene.* 2011;30(2):190–200.
103. He M, Bakken T, Kassimova A, Boshoff C, Philpott N, Cannon ML. Focal adhesion kinase is required for KSHV vGPCR signaling. *Mol Carcinog.* 2012;51(4):339–351.
104. Ma Q, Cavallin LE, Leung HJ, Chiozzini C, Goldschmidt-Clermont PJ, Mesri EA. A role for virally induced reactive oxygen species in Kaposi's sarcoma herpesvirus tumorigenesis. *Antioxid Redox Signal.* 2013;18(1):80–90.

105. Martin D, Galisteo R, Molinolo AA, Wetzker R, Hirsch E, Gutkind JS. PI3Kgamma mediates Kaposi's sarcoma-associated herpesvirus vGPCR-induced sarcomagenesis. *Cancer Cell*. 2011;19(6):805–813.
106. Shepard LW, Yang M, Xie P, et al. Constitutive activation of NF-kappa B and secretion of interleukin-8 induced by the G protein-coupled receptor of Kaposi's sarcoma-associated herpesvirus involve G alpha(13) and RhoA. *J Biol Chem*. 2001;276(49): 45979–45987.
107. Smit MJ, Verzijl D, Casarosa P, Navis M, Timmerman H, Leurs R. Kaposi's sarcoma-associated herpesvirus-encoded G protein-coupled receptor ORF74 constitutively activates p44/p42 MAPK and Akt via G(i) and phospholipase C-dependent signaling pathways. *J Virol*. 2002;76(4):1744–1752.
108. Doniger J, Muralidhar S, Rosenthal LJ. Human cytomegalovirus and human herpesvirus 6 genes that transform and transactivate. *Clin Microbiol Rev*. 1999;12(3):367–382.
109. Michaelis M, Baumgarten P, Mittelbronn M, Driever PH, Doerr HW, Cinatl Jr J. Oncomodulation by human cytomegalovirus: novel clinical findings open new roads. *Med Microbiol Immunol*. 2011;200(1):1–5.
110. Michaelis M, Doerr HW, Cinatl J. The story of human cytomegalovirus and cancer: increasing evidence and open questions. *Neoplasia*. 2009;11(1):1–9.
111. Ogata M. Human herpesvirus 6 in hematological malignancies. *J Clin Exp Hematop*. 2009;49(2):57–67.
112. Dziurzynski K, Chang SM, Heimberger AB, et al. Consensus on the role of human cytomegalovirus in glioblastoma. *Neuro-Oncology*. 2012;14(3):246–255.
113. Kofman A, Marcinkiewicz L, Dupart E, et al. The roles of viruses in brain tumor initiation and oncomodulation. *J Neuro-Oncol*. 2011;105(3):451–466.
114. Soroceanu L, Cobbs CS. Is HCMV a tumor promoter? *Virus Res*. 2011;157(2): 193–203.
115. Cobbs CS, Harkins L, Samanta M, et al. Human cytomegalovirus infection and expression in human malignant glioma. *Cancer Res*. 2002;62(12):3347–3350.
116. Harkins L, Volk AL, Samanta M, et al. Specific localisation of human cytomegalovirus nucleic acids and proteins in human colorectal cancer. *Lancet*. 2002;360(9345):1557–1563.
117. Harkins LE, Matlaf LA, Soroceanu L, et al. Detection of human cytomegalovirus in normal and neoplastic breast epithelium. *Herpesviridae*. 2010;1(1):8.
118. Samanta M, Harkins L, Klemm K, Britt WJ, Cobbs CS. High prevalence of human cytomegalovirus in prostatic intraepithelial neoplasia and prostatic carcinoma. *J Urol*. 2003;170(3):998–1002.
119. Mitchell DA, Xie W, Schmittling R, et al. Sensitive detection of human cytomegalovirus in tumors and peripheral blood of patients diagnosed with glioblastoma. *Neuro-Oncology*. 2008;10(1):10–18.
120. Ranganathan P, Clark PA, Kuo JS, Salamat MS, Kalejta RF. Significant association of multiple human cytomegalovirus genomic Loci with glioblastoma multiforme samples. *J Virol*. 2012;86(2):854–864.
121. Scheurer ME, Bondy ML, Aldape KD, Albrecht T, El-Zein R. Detection of human cytomegalovirus in different histological types of gliomas. *Acta Neuropathol*. 2008;116(1):79–86.
122. Bhattacharjee B, Renzette N, Kowalik TF. Genetic analysis of cytomegalovirus in malignant gliomas. *J Virol*. 2012;86(12):6815–6824.
123. Cimino PJ, Zhao G, Wang D, Sehn JK, Lewis Jr JS, Duncavage EJ. Detection of viral pathogens in high grade gliomas from unmapped next-generation sequencing data. *Exp Mol Pathol*. 2014;96(3):310–315.
124. Khoury JD, Tannir NM, Williams MD, et al. Landscape of DNA virus associations across human malignant cancers: analysis of 3,775 cases using RNA-Seq. *J Virol*. 2013;87(16):8916–8926.

125. Lau SK, Chen YY, Chen WG, et al. Lack of association of cytomegalovirus with human brain tumors. *Mod Pathol.* 2005;18(6):838–843.
126. Poltermann S, Schlehofer B, Steindorf K, Schnitzler P, Geletneky K, Schlehofer JR. Lack of association of herpesviruses with brain tumors. *J Neurovirol.* 2006;12(2):90–99.
127. Tang KW, Hellstrand K, Larsson E. Absence of cytomegalovirus in high-coverage DNA sequencing of human glioblastoma multiforme. *Int J Cancer.* 2014. http://dx.doi.org/10.1002/ijc.29042. [EPub ahead of print].
128. Yamashita Y, Ito Y, Isomura H, et al. Lack of presence of the human cytomegalovirus in human glioblastoma. *Mod Pathol.* 2014;27(7):922–929.
129. Schuessler A, Smith C, Beagley L, et al. Autologous T-cell therapy for cytomegalovirus as a consolidative treatment for recurrent glioblastoma. *Cancer Res.* 2014;74(13):3466–3476.
130. Stragliotto G, Rahbar A, Solberg NW, et al. Effects of valganciclovir as an add-on therapy in patients with cytomegalovirus-positive glioblastoma: a randomized, double-blind, hypothesis-generating study. *Int J Cancer.* 2013;133(5):1204–1213.
131. Dey M, Ahmed AU, Lesniak MS. Cytomegalovirus and glioma: putting the cart before the horse. *J Neurol Neurosurg Psychiatry.* 2014. http://dx.doi.org/10.1136/jnnp-2014-307727. [EPub ahead of print].
132. Soderberg-Naucler C, Rahbar A, Stragliotto G. Survival in patients with glioblastoma receiving valganciclovir. *N Engl J Med.* 2013;369(10):985–986.
133. Wick W, Wick A, Platten M. Challenging cytomegalovirus data in glioblastoma. *Neuro-Oncology.* 2014;16(1):165.
134. Pleskoff O, Casarosa P, Verneuil L, et al. The human cytomegalovirus-encoded chemokine receptor US28 induces caspase-dependent apoptosis. *FEBS J.* 2005; 272(16):4163–4177.
135. Maussang D, Langemeijer E, Fitzsimons CP, et al. The human cytomegalovirus-encoded chemokine receptor US28 promotes angiogenesis and tumor formation via cyclooxygenase-2. *Cancer Res.* 2009;69(7):2861–2869.
136. Soroceanu L, Matlaf L, Bezrookove V, et al. Human cytomegalovirus US28 found in glioblastoma promotes an invasive and angiogenic phenotype. *Cancer Res.* 2011; 71(21):6643–6653.
137. Baryawno N, Rahbar A, Wolmer-Solberg N, et al. Detection of human cytomegalovirus in medulloblastomas reveals a potential therapeutic target. *J Clin Invest.* 2011;121(10):4043–4055.
138. Frostegard J. Immunity, atherosclerosis and cardiovascular disease. *BMC Med.* 2013;11:117.
139. Rosenfeld ME, Campbell LA. Pathogens and atherosclerosis: update on the potential contribution of multiple infectious organisms to the pathogenesis of atherosclerosis. *Thromb Haemost.* 2011;106(5):858–867.
140. Lemstrom K, Tikkanen J, Koskinen P, Hayry P. Effect of cytomegalovirus on cardiac allograft arteriosclerosis—indirect or direct? *Am J Transplant.* 2005;5(3):421–422.
141. Streblow DN, Dumortier J, Moses AV, Orloff SL, Nelson JA. Mechanisms of cytomegalovirus-accelerated vascular disease: induction of paracrine factors that promote angiogenesis and wound healing. *Curr Top Microbiol Immunol.* 2008;325:397–415.
142. Nicholas J, Cameron KR, Honess RW. Herpesvirus saimiri encodes homologues of G protein-coupled receptors and cyclins. *Nature.* 1992;355(6358):362–365.
143. Fickenscher H, Fleckenstein B. Herpesvirus saimiri. *Philos Trans R Soc Lond B Biol Sci.* 2001;356(1408):545–567.
144. Chung YH, Means RE, Choi JK, Lee BS, Jung JU. Kaposi's sarcoma-associated herpesvirus OX2 glycoprotein activates myeloid-lineage cells to induce inflammatory cytokine production. *J Virol.* 2002;76(10):4688–4698.

145. Rezaee SA, Cunningham C, Davison AJ, Blackbourn DJ. Kaposi's sarcoma-associated herpesvirus immune modulation: an overview. *J Gen Virol.* 2006;87(Pt 7): 1781–1804.
146. Bais C, Santomasso B, Coso O, et al. G-protein-coupled receptor of Kaposi's sarcoma-associated herpesvirus is a viral oncogene and angiogenesis activator. *Nature.* 1998;391(6662):86–89.
147. Cobo F, Hernandez S, Hernandez L, et al. Expression of potentially oncogenic HHV-8 genes in an EBV-negative primary effusion lymphoma occurring in an HIV-seronegative patient. *J Pathol.* 1999;189(2):288–293.
148. Guo HG, Browning P, Nicholas J, et al. Characterization of a chemokine receptor-related gene in human herpesvirus 8 and its expression in Kaposi's sarcoma. *Virology.* 1997;228(2):371–378.
149. Kirshner JR, Staskus K, Haase A, Lagunoff M, Ganem D. Expression of the open reading frame 74 (G-protein-coupled receptor) gene of Kaposi's sarcoma (KS)-associated herpesvirus: implications for KS pathogenesis. *J Virol.* 1999;73(7): 6006–6014.
150. Nador RG, Milligan LL, Flore O, et al. Expression of Kaposi's sarcoma-associated herpesvirus G protein-coupled receptor monocistronic and bicistronic transcripts in primary effusion lymphomas. *Virology.* 2001;287(1):62–70.
151. Montaner S, Sodhi A, Pece S, Mesri EA, Gutkind JS. The Kaposi's sarcoma-associated herpesvirus G protein-coupled receptor promotes endothelial cell survival through the activation of Akt/protein kinase B. *Cancer Res.* 2001;61(6):2641–2648.
152. Pati S, Cavrois M, Guo HG, et al. Activation of NF-kappaB by the human herpesvirus 8 chemokine receptor ORF74: evidence for a paracrine model of Kaposi's sarcoma pathogenesis. *J Virol.* 2001;75(18):8660–8673.
153. Cesarman E, Mesri EA, Gershengorn MC. Viral G protein-coupled receptor and Kaposi's sarcoma: a model of paracrine neoplasia? *J Exp Med.* 2000;191(3):417–422.
154. Guo HG, Pati S, Sadowska M, Charurat M, Reitz M. Tumorigenesis by human herpesvirus 8 vGPCR is accelerated by human immunodeficiency virus type 1 Tat. *J Virol.* 2004;78(17):9336–9342.
155. Holst PJ, Rosenkilde MM, Manfra D, et al. Tumorigenesis induced by the HHV8-encoded chemokine receptor requires ligand modulation of high constitutive activity. *J Clin Invest.* 2001;108(12):1789–1796.
156. Yang TY, Chen SC, Leach MW, et al. Transgenic expression of the chemokine receptor encoded by human herpesvirus 8 induces an angioproliferative disease resembling Kaposi's sarcoma. *J Exp Med.* 2000;191(3):445–454.
157. Wang Y, Lu X, Zhu L, et al. IKK epsilon kinase is crucial for viral G protein-coupled receptor tumorigenesis. *Proc Natl Acad Sci USA.* 2013;110(27):11139–11144.
158. Bottero V, Kerur N, Sadagopan S, Patel K, Sharma-Walia N, Chandran B. Phosphorylation and polyubiquitination of transforming growth factor beta-activated kinase 1 are necessary for activation of NF-kappaB by the Kaposi's sarcoma-associated herpesvirus G protein-coupled receptor. *J Virol.* 2011;85(5):1980–1993.
159. Feng H, Sun Z, Farzan MR, Feng P. Sulfotyrosines of the Kaposi's sarcoma-associated herpesvirus G protein-coupled receptor promote tumorigenesis through autocrine activation. *J Virol.* 2010;84(7):3351–3361.
160. Bais C, Van Geelen A, Eroles P, et al. Kaposi's sarcoma associated herpesvirus G protein-coupled receptor immortalizes human endothelial cells by activation of the VEGF receptor-2/KDR. *Cancer Cell.* 2003;3(2):131–143.
161. Montaner S, Sodhi A, Ramsdell AK, et al. The Kaposi's sarcoma-associated herpesvirus G protein-coupled receptor as a therapeutic target for the treatment of Kaposi's sarcoma. *Cancer Res.* 2006;66(1):168–174.

162. Boudry C, Markine-Goriaynoff N, Delforge C, et al. The A5 gene of alcelaphine herpesvirus 1 encodes a constitutively active G-protein-coupled receptor that is non-essential for the induction of malignant catarrhal fever in rabbits. *J Gen Virol.* 2007;88(Pt 12):3224–3233.
163. Lyngaa R, Norregaard K, Kristensen M, Kubale V, Rosenkilde MM, Kledal TN. Cell transformation mediated by the Epstein-Barr virus G protein-coupled receptor BILF1 is dependent on constitutive signaling. *Oncogene.* 2010;29(31):4388–4398.
164. Palmeira L, Sorel O, Van Campe W, et al. An essential role for gamma-herpesvirus latency-associated nuclear antigen homolog in an acute lymphoproliferative disease of cattle. *Proc Natl Acad Sci USA.* 2013;110(21):E1933–E1942.
165. Boomker JM, The TH, de Leij LF, Harmsen MC. The human cytomegalovirus-encoded receptor US28 increases the activity of the major immediate-early promoter/enhancer. *Virus Res.* 2006;118(1–2):196–200.
166. Wen DQ, Zhang YY, Lv LP, et al. Human cytomegalovirus-encoded chemokine receptor homolog US28 stimulates the major immediate early gene promoter/enhancer via the induction of CREB. *J Recept Signal Transduct Res.* 2009;29(5):266–273.
167. Hertel L, Mocarski ES. Global analysis of host cell gene expression late during cytomegalovirus infection reveals extensive dysregulation of cell cycle gene expression and induction of Pseudomitosis independent of US28 function. *J Virol.* 2004;78(21):11988–12011.
168. Pleskoff O, Treboute C, Alizon M. The cytomegalovirus-encoded chemokine receptor US28 can enhance cell-cell fusion mediated by different viral proteins. *J Virol.* 1998;72(8):6389–6397.
169. O'Connor CM, Shenk T. Human cytomegalovirus pUS27 G protein-coupled receptor homologue is required for efficient spread by the extracellular route but not for direct cell-to-cell spread. *J Virol.* 2011;85(8):3700–3707.
170. Cheung AK, Abendroth A, Cunningham AL, Slobedman B. Viral gene expression during the establishment of human cytomegalovirus latent infection in myeloid progenitor cells. *Blood.* 2006;108(12):3691–3699.
171. Goodrum FD, Jordan CT, High K, Shenk T. Human cytomegalovirus gene expression during infection of primary hematopoietic progenitor cells: a model for latency. *Proc Natl Acad Sci USA.* 2002;99(25):16255–16260.
172. Patterson BK, Landay A, Andersson J, et al. Repertoire of chemokine receptor expression in the female genital tract: implications for human immunodeficiency virus transmission. *Am J Pathol.* 1998;153(2):481–490.
173. Beisser PS, Vink C, Van Dam JG, Grauls G, Vanherle SJ, Bruggeman CA. The R33 G protein-coupled receptor gene of rat cytomegalovirus plays an essential role in the pathogenesis of viral infection. *J Virol.* 1998;72(3):2352–2363.
174. Davis-Poynter NJ, Lynch DM, Vally H, et al. Identification and characterization of a G protein-coupled receptor homolog encoded by murine cytomegalovirus. *J Virol.* 1997;71(2):1521–1529.
175. Margulies BJ, Browne H, Gibson W. Identification of the human cytomegalovirus G protein-coupled receptor homologue encoded by UL33 in infected cells and enveloped virus particles. *Virology.* 1996;225(1):111–125.
176. Cardin RD, Schaefer GC, Allen JR, Davis-Poynter NJ, Farrell HE. The M33 chemokine receptor homolog of murine cytomegalovirus exhibits a differential tissue-specific role during in vivo replication and latency. *J Virol.* 2009;83(15):7590–7601.
177. Oliveira SA, Shenk TE. Murine cytomegalovirus M78 protein, a G protein-coupled receptor homologue, is a constituent of the virion and facilitates accumulation of immediate-early viral mRNA. *Proc Natl Acad Sci USA.* 2001;98(6):3237–3242.

178. Michel D, Milotic I, Wagner M, et al. The human cytomegalovirus UL78 gene is highly conserved among clinical isolates, but is dispensable for replication in fibroblasts and a renal artery organ-culture system. *J Gen Virol.* 2005;86(Pt 2):297–306.
179. Compton T, Nepomuceno RR, Nowlin DM. Human cytomegalovirus penetrates host cells by pH-independent fusion at the cell surface. *Virology.* 1992;191(1):387–395.
180. Ryckman BJ, Jarvis MA, Drummond DD, Nelson JA, Johnson DC. Human cytomegalovirus entry into epithelial and endothelial cells depends on genes UL128 to UL150 and occurs by endocytosis and low-pH fusion. *J Virol.* 2006;80(2):710–722.
181. Wang D, Yu QC, Schroer J, Murphy E, Shenk T. Human cytomegalovirus uses two distinct pathways to enter retinal pigmented epithelial cells. *Proc Natl Acad Sci USA.* 2007;104(50):20037–20042.
182. Kaptein SJ, Beisser PS, Gruijthuijsen YK, et al. The rat cytomegalovirus R78 G protein-coupled receptor gene is required for production of infectious virus in the spleen. *J Gen Virol.* 2003;84(Pt 9):2517–2530.
183. Chiou CJ, Poole LJ, Kim PS, et al. Patterns of gene expression and a transactivation function exhibited by the vGCR (ORF74) chemokine receptor protein of Kaposi's sarcoma-associated herpesvirus. *J Virol.* 2002;76(7):3421–3439.
184. Sandford G, Choi YB, Nicholas J. Role of ORF74-encoded viral G protein-coupled receptor in human herpesvirus 8 lytic replication. *J Virol.* 2009;83(24):13009–13014.
185. Chen J, Ye F, Xie J, Kuhne K, Gao SJ. Genome-wide identification of binding sites for Kaposi's sarcoma-associated herpesvirus lytic switch protein, RTA. *Virology.* 2009;386(2):290–302.
186. Zhang J, Wang J, Wood C, Xu D, Zhang L. Kaposi's sarcoma-associated herpesvirus/human herpesvirus 8 replication and transcription activator regulates viral and cellular genes via interferon-stimulated response elements. *J Virol.* 2005;79(9):5640–5652.
187. Bottero V, Sharma-Walia N, Kerur N, et al. Kaposi sarcoma-associated herpes virus (KSHV) G protein-coupled receptor (vGPCR) activates the ORF50 lytic switch promoter: a potential positive feedback loop for sustained ORF50 gene expression. *Virology.* 2009;392(1):34–51.
188. Lee BJ, Koszinowski UH, Sarawar SR, Adler H. A gammaherpesvirus G protein-coupled receptor homologue is required for increased viral replication in response to chemokines and efficient reactivation from latency. *J Immunol.* 2003;170(1):243–251.
189. Moorman NJ, Virgin HWt, Speck SH. Disruption of the gene encoding the gammaHV68 v-GPCR leads to decreased efficiency of reactivation from latency. *Virology.* 2003;307(2):179–190.

CHAPTER THIRTEEN

EBV, the Human Host, and the 7TM Receptors: Defense or Offense?

Kristine Niss Arfelt, Suzan Fares, Mette M. Rosenkilde[1]
Faculty of Health and Medical Sciences, Department of Neuroscience and Pharmacology, Laboratory for Molecular Pharmacology, University of Copenhagen, Copenhagen, Denmark
[1]Corresponding author: e-mail address: rosenkilde@sund.ku.dk

Contents

Abstract

Being present in around 90% of the worldwide population, Epstein–Barr virus (EBV) is an exceptionally prevalent virus. This highly successful virus establishes a latent infection in resting memory B cells and is maintained in a balance between viral homeostasis on one side and antiviral defense of the immune system on the other side. The life cycle of EBV is dependent on many viral proteins, but EBV also regulates a number of endogenous proteins. 7TM receptors and ligands of viral and host origin are examples of such

Progress in Molecular Biology and Translational Science, Volume 129
ISSN 1877-1173
http://dx.doi.org/10.1016/bs.pmbts.2014.10.011

proteins. 7TM receptors are highly druggable and they are among the most popular class of investigational drug targets. The 7TM receptor encoded by EBV-BILF1, is known to downregulate cell surface MHC class I expression as part of the immune evasion strategy of EBV. However, the functional impact of the relationship between EBV and the regulated endogenous 7TM receptors and ligands is still unclear. This is for instance the case for the most upregulated 7TM receptor EBI2 (EBV-induced gene 2 or GPR183). Whereas some regulated genes have been suggested to be involved in the EBV life cycle, others could also be important for the antiviral immune defense. As many of these 7TM receptors and ligands have been shown to be modulated in EBV-associated diseases, targeting these could provide an efficient and specific way to inhibit EBV-associated disease progression. Here, we will review current knowledge on EBV infection, the immune defense against EBV and 7TM receptors and ligands being either encoded or manipulated by EBV.

1. EBV INFECTION

Epstein–Barr virus (EBV, human herpesvirus 4) was discovered 50 years ago, when Epstein, Achong, and Barr used electron microscopy to identify viral particles in Burkitt's lymphoma cells.[1] It belongs to the lymphocryptovirus (LCV) genus of the gammaherpesvirus subfamily[2] (Fig. 1A). The EBV genome, which consists of a linear, double-stranded DNA molecule that encodes close to 100 viral genes, is enclosed in a nuclear capsid surrounded by a protein tegument, which in turn, is surrounded by a glycoprotein-coated viral envelope.[6] The glycoproteins are important for virus tropism, host selectivity, and infection.

1.1. Viral infection, entry, and tropism

It is generally believed that primary infection with EBV occurs virus in saliva from already infected persons, which infects B cells and epithelial cells in the Waldeyers tonsillar ring in the oropharynx. From Waldeyers ring, the virus replicates and further infects resting naïve B cells trafficking through the oropharynx or resting naïve B cells resident in the tonsillar crypts.[6] The mechanisms of virus entry into epithelial cells and B cells differ, but both depend on the attachment and membrane fusion mediated by the envelope glycoproteins. For B cell entry, the virion attaches to the B cell surface by binding of the viral gp350/220 to the complement component receptor CD21 (CR2); binding is associated with triggering of signaling and endocytosis.[7] After attachment and endocytosis, the glycoprotein gp42, which forms a tripartite complex with gH/gL, interacts with the major histocompatibility

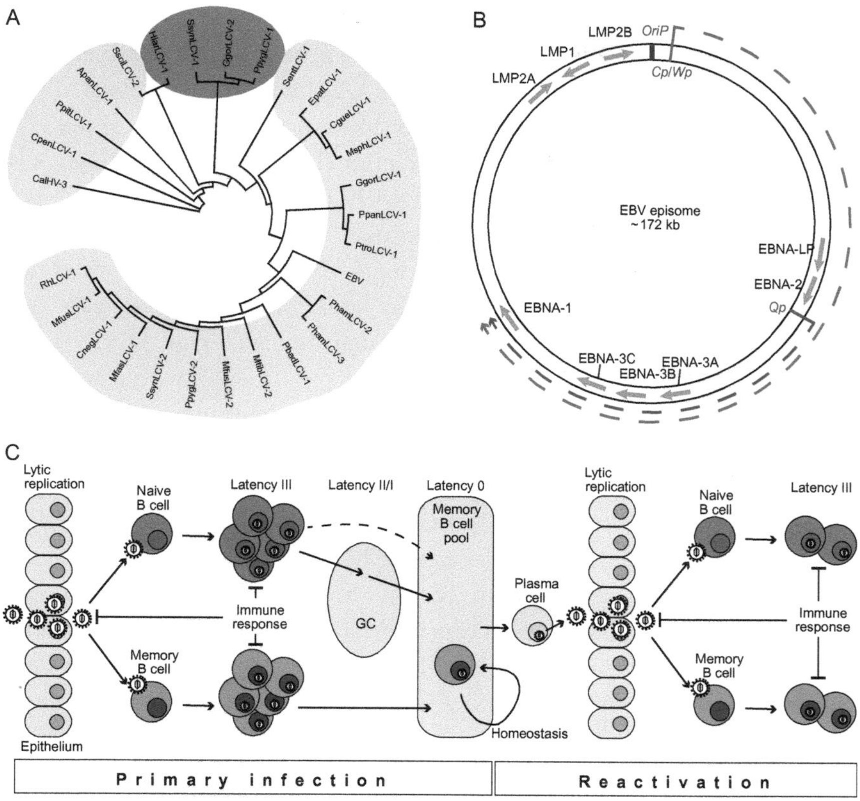

Figure 1 Overview of lymphocryptovirus, genome organization, and life cycle of EBV. (A) Dendrogram of the lymphocryptovirus (LCV) genus of the gammaherpes virus subfamily. The phylogenetic analysis was based on the amino acid sequence for the DPOL gene sequences from amino acid no 1–693 (for EBV). Amino acid sequence alignments were made using MAFFT.[3] The dendrogram was built using the maximum likelihood method in the PhyML 2.2.0 plug-in.[4] The LG nucleotide substitution model was applied. The major clades A (LCV hosted by New World monkeys), B and C (LCV hosted by Old World monkeys) are shown in blue, green, and purple, respectively. (B) Schematic overview of the EBV episome with indication of latent protein-coding genes (black) and promoters (pink). The latent genes are marked by green arrows. The latent origin of replication, OriP, is shown in dark blue. The dotted, long, and outer blue arrow shows the genes transcribed from the Cp and Wp promoter (all the EBNAs), whereas the dotted, inner, and pink arrow shows the transcription promoted from Qp (EBNA-1 only). (C) Proposed life cycle of EBV: the virus establishes a lytic infection in epithelial cells and/or B cells. The B cells express the growth-promoting latency III program and enter a state of highly proliferating lymphoblasts. Many of these are detected and removed by the immune system, but some express the latency II and I programs and secure entry into the memory B cell pool, where the virus can reside without immune detection. The

(Continued)

complex (MHC) class II and triggers membrane fusion mediated by the glycoproteins gH/gL and gB.[8] Attachment of virions to epithelial cells may occur through binding of gp350/220 to CD21, which is expressed at low levels on tonsillar epithelial cells, but also through an interaction of the viral BMRF-2 protein with integrins on the epithelial surface.[9] Unlike B cells, epithelial cells do not express MHC class II. The fusion with these cells is triggered by the interaction between cellular integrins and the gH/gL glycoproteins. The gB glycoprotein mediates the fusion, which does not require endocytosis.[10] Following fusion with both epithelial cells and B cells, virion capsids are transported to deliver linear viral DNA to the nucleus.[6]

The expression of gp42 on the viral envelope determines the viral tropism for either B cells or epithelial cells. The interaction of gH with an epithelial receptor is inhibited by gp42, meaning that gp42 expressing virions preferentially infect B cells.[11] Interestingly, virions produced by B cells do not express gp42 as gp42 molecules bind to MHC class II in the endoplasmic reticulum (ER) and are degraded. Thus, virions from B cells show tropism for epithelial cells. Conversely, virions produced in epithelial cells express gp42 and show tropism for B cells. The virions shed in saliva have been shown to be B cell tropic, suggesting that in the persisting infection the epithelial cells are used for B cell tropic virion amplification.[12] This observation also suggests that B cells, rather than epithelial cells, may be the initial target of virus infection when transmitted via saliva.

1.2. Lytic replication

EBV is well known for establishing an immediate latent infection in B cells, which provide the reservoir for the virus in the persisting infection. The lytic replication phase thus only occurs in B cells upon reactivation and in

Figure 1—Cont'd persisting virus infection is continuously maintained via homeostasis of the infected memory B cells. Infected memory B cells differentiating into plasmacells may activate lytic replication and virus shedding via epithelial cell infection. The reactivation is efficiently repressed by the memory immune response. Establishment of the latency in memory B cells has been proposed to occur via (1) direct infection and growth promotion of memory B cells. (2) Infection of naïve B cells and virus-driven transformation into memory B cells in a process, which mimics the antigen-driven memory cell differentiation of noninfected naïve B cells. This virus-driven transformation can either occur through the GCs, which represents a hostile environment or through GC-like differentiation outside the actual GCs (dotted arrow). *Adapted from Ref. 5.* (See the color plate.)

epithelial cells upon virus entry. The lytic replication is initiated by expression of the immediate-early genes, which act as transactivators for the early and late lytic genes. The most important immediate-early gene BZLF1 acts as a transactivator of late lytic genes and is critical to the initiation of lytic replication.[13] The early gene products have functions such as replication, metabolism, and blockade of antigen processing, whereas late gene products have more structural functions such as viral capsid antigens and immune evasion.[14]

1.3. Latent infection

When the linear viral DNA reaches the nucleus, it forms a circular DNA and initiates a short-prelatent phase, which is characterized by coexpression of both latent and some lytic genes. The latent genes activate the resting B cells to become highly proliferating lymphoblasts. No virions are produced in this phase and the lytic genes expressed interact with miRNAs to ensure cell cycle progression, protection from apoptosis and immune evasion.[6,15,16] The transformation efficiency of resting naïve B cells into highly proliferating lymphoblasts is only 1–10%.[17,18] This can be partly explained by a cellular DNA damage response initiated by the increased expression of the latent genes and resulting in apoptosis of virus-infected cells.[19]

An intriguing aspect of EBV infection is that infection of B cells differs dramatically between *in vivo* and *in vitro* settings. *In vivo*, EBV has the ability to establish a persisting latent infection in memory B cells, whereas infection of B cells *in vitro* results in an immortalized lymphoblastoid cell line (LCLs).[6] So, how can a growth-promoting virus end up in a quiescent state in resting memory B cells? The difference between these two outcomes of infection is the expression pattern of latent viral genes. Although the viral genome encodes nearly 100 protein-coding genes, only a small subset is expressed during latency. LCLs have been used as a platform to study these genes as all of them are expressed in the lymphoblastoid cells. Full latent gene expression is also present in newly infected B cells *in vivo*,[6] but in the *in vivo* settings EBV establishes several different patterns of gene expression, which are recognized in the various cancers associated with the virus. The latent protein-coding genes, along with their expression pattern and functions, are listed in Table 1 and shown in the EBV episome in Fig. 1B. In addition, abundant noncoding RNAs (EBERs) and viral miRNAs are also expressed during latency, but will not be discussed further.[26]

Table 1 EBV latency genes with latency program and function

Latent gene	Latency program	Function
EBNA-1	I, II, and III	Gene regulation, replication, and maintenance of the viral genome through regulation of viral promoters[20]
EBNA-2	III	Transactivator of viral and cellular genes, virus-induced growth transformation[21]
EBNA-LP	III	Coactivation with EBNA-2 and virus-induced growth transformation[21]
EBNA-3A	III	Virus-induced growth transformation and survival[6]
EBNA-3B	III	Viral tumor suppressor[22]
EBNA-3C	III	Virus-induced growth transformation and survival[6]
LMP1	II and III	CD40 mimic promoting B cell growth and survival[23]
LMP2A	II and III	B cell receptor mimic promoting B cell development and survival[24]
LMP2B	II and III	Modulator of LMP2A activation[25]

The fact that EBV is known to infect resting naïve B cells, but is maintained in the differentiated memory B cell pool, has led Thorley-Lawson and colleagues to postulate the germinal center (GC) model for EBV infection (Fig. 1C). In this model, the EBV infection is suggested to mimic the antigen-driven differentiation from naïve B cells to memory B cells via class switch recombination and somatic hypermutations in the GCs.[27] The GC model suggests that expression of latency program III induces B cell proliferation and formation of GCs. In GCs, the gene expression switches to latency program II with expression of LMP1 and LMP2A. LMP1 constitutively mimics the CD40 signal,[28] which induces survival and proliferation of infected B cells *in vitro*.[29] LMP2A, on the other hand, mimics the B cell receptor signaling properties and thereby bypasses induction of anergy and induces survival of the infected cells.[24] *In vitro* studies and studies with transgenic animals suggest that these two genes provide the necessary signals for the infected B cells to survive the critical selection in the GCs.[23] In addition, LMP1 induces class switching and somatic hypermutations by upregulation of activation-induced cytidine deaminase,[30] which is the most important regulator of secondary antibody maturation. After exit from the GCs, the EBV-infected cells differentiate into memory B cells and express latency program 0 (expression of noncoding RNAs but no viral proteins) or latency program I (expression of noncoding RNAs and EBNA-1 only). Expression of EBNA-1 ensures the viral homeostasis by replication and hence maintenance of the viral genome when infected memory B cells enter cell division.[31]

There is some controversy concerning the GC model. An argument in support of the model is the fact that the persisting virus infection is restricted to memory B cells. Also, the existence of the latency III program, which turns resting B cells into highly proliferating lymphoblastoids and thereby puts both the host and virus at risk for cancer development, suggests that successful latent infection cannot be achieved by simply infecting resting memory B cells.[27] Opposing the GC model, studies have shown that EBV infects both naïve and memory B cell populations.[32] The lack of somatic hypermutations in memory B cells sampled during infectious mononucleosis also indicates that the GC reaction is not required for the differentiation into memory B cells.[33] Also, very few EBV-infected cells are present in the GCs of both infectious mononucleosis and normal EBV-infected lymph tissue.[34,35] In summary, studies suggest that the GC model plays a role in the persisting EBV infection and that it is complemented by either direct infection of memory B cells or by differentiation of memory B cells located outside the GCs. It is important to state that in the absence of a good *in vivo* model for EBV, all studies on EBV-encoded or -regulated genes are made on samples from patients with EBV-associated diseases, *in vitro* experiments, or transgenic animals expressing one or two of the encoded or regulated genes. Thus, these studies cannot encompass the complex *in vivo* interactions between viral genes, or between virus and host. To fully understand these, there is a requirement for further studies in a reliable EBV *in vivo* model.

1.4. Regulation of latency, replication, and virus reactivation

As shown in Fig. 1B, the viral genome encodes three different promoter regions: Wp, Cp, and Qp. The virus uses Wp to ensure expression of EBNA-1–3C after initial infection. The use of Wp is transient and EBNA-1 expression induces promoter switch to Cp, which is used for full latent gene expression (program III). In latency I, the genome switches promoter to Qp resulting in the expression of EBNA-1 only. EBNA-1 also binds and activates the latent origin, OriP, and thereby ensures viral homeostasis during latent infection.[6] The regulatory mechanism behind the promoter switches determining virus life cycle through prelatent and latent phases followed by reactivation and lytic virus replication has long been unknown. In particular, the switch from a quiescent latent state to lytic replication has been a subject of attention for EBV researchers. Recent studies suggest that the virus takes advantage of the host epigenetic mechanisms in control of its promoter usage and different phases.[36] The viral genome in newly EBV-infected cells is, as is also the case for other herpesviruses and retroviruses, epigenetically naïve (has no epigenetic modifications).[37–39] Using the host cell epigenetic machinery, the virus genome acquires methylations and histone modifications during latency.[36] These modifications vary among the different latency programs. Thus, the Cp promoter has been found to be hypermethylated and silent in tumors expressing the latency I (Burkitt's lymphoma) or II (Hodgkin lymphoma), whereas it is unmethylated in LCLs and posttransplant lymphoproliferative disorders (PTLDs) expressing latency III.[40,41]

A solution to the question of how the virus switch from latency to lytic mode has also emerged within recent years: the immediate-early lytic transactivator BZLF1 has the capacity to bind to highly methylated DNA only. Thus, although this gene is expressed in the prelatent phase, it cannot bind to its binding motif in the unmethylated DNA. During latency, the DNA becomes increasingly methylated forming the preferred binding motif for BZLF1 and can then overcome epigenetic repression.[42]

2. IMMUNE RESPONSE AND IMMUNE EVASION

Up to 10% of the host B cells are infected with EBV during acute infectious mononucleosis.[43] Most of these cells are effectively cleared by the immune system, but some downregulate viral gene expression and differentiate into safety in the resting memory B cell pool. In the persisting infection, the virus and host coexist, so that homeostasis of the infected memory B cell pool is maintained by continuous low-level virus shedding while the immune system ensures that no full blown lytic replication is initiated. The number of infected B cells in the persisting infection is around 1–50 per 10^6 B cell.[43]

The host uses both the innate and adaptive immune response to eliminate infected B cells. The innate immune response relies heavily on virus recognition by Toll-like receptors (TLR) followed by interferon (IFN)-γ secretion and activation of natural killer (NK) cells. In particular, TLR3 expressed in classical dendritic cells (DCs) recognizes double-stranded

RNA such as EBV EBERs and induces secretion of IFN-γ, antiviral cytokines, and NK cell activation.[44] TLR9 expressed in plasmacytoid DCs recognizes unmethylated CpG motifs in the epigenetically naïve viral DNA[45] and induces IFN-γ secretion and NK cell activation.[46] The importance of NK cells is suggested by the finding that the levels of NK cells are significantly increased in patients with infectious mononucleosis.[47] In addition, recent *in vivo* studies of infectious mononucleosis in mice with reconstituted human immune components showed that depletion of NK cells led to loss of immune control and enhanced symptoms.[48]

The innate immune system ensures virus inhibition until the adaptive immune system is ready to launch a much more comprehensive antiviral response, which mainly consists of cytotoxic CD8 T cells and antibody production induced by CD4+ T cells to a minor degree.[43] Initially, cytotoxic T cells (CTL) are directed against lytic epitopes such as the immediate-early gene BZLF1. Later in the infection, the CTL are directed against latent epitopes, in particular.[43] In contrast to the CD8+ T cell response, the CD4+ T cell response is very modest and seems to be directed against a different set of early and late lytic epitopes as compared to the CD8+ T cell response.[49] The important role of the immune response and especially the adaptive immune response in the inhibition of EBV infection can be illustrated by the fact that a large number of patients with EBV-associated diseases have immunodeficiencies related to either genetic, iatrogenic (PTLD), or other diseases (like HIV).[6]

Like other herpesviruses, EBV has adapted to the immune system for instance by expressing many genes with immune evasive properties. Among these, LMP1 and BGL5 downregulate TLR9 in the latent and lytic phases, respectively.[50,51] Also, the late lytic function as a deubiquitinase in the TLR signaling pathway and thereby inhibits TLR signaling.[52] The CD8+ T cell cytotoxic response is inhibited by the downregulation of viral epitope presentation on MHC class I molecule on the cell surface by LMP1, BILF1, BGL5, viral IL-10 homolog BCRF1, and BNLF2a.[15,53–56] BCRF1 has also been shown to have broad inhibitory effects on the immune system by impairing NK cell-mediated killing of infected B cells, interference with CD4+ T cell activity and modulation of cytokine responses.[15] Finally, the CD4+ T cell response is inhibited by binding of the viral glycoprotein gp42 to MHC class II and the ensuing lowered recognition of CD4+ cells.[57]

The host-virus coexistence is completely dependent on the balance between a strong immune response inhibiting virus replication and immune evasion and survival of the virus with periodic shedding to enable transmission to new hosts. In the process of ensuring this balance, the virus manipulates 7TM receptors and their ligands.[58–60] Thus, like many other herpesviruses, EBV encodes a 7TM receptor and in addition regulates the expression of both 7TM receptors and ligands in the host. In the following parts of the review, we will give an overview of the 7TM receptor exploitation and regulation of EBV along with the potential druggability of these receptors.

3. EBV-BILF1—A VIRUS-ENCODED 7TM RECEPTOR WITH IMMUNE EVASIVE FUNCTIONS

The EBV-encoded BILF1 receptor (EBV-BILF1) is thought to be implicated in the immune evasion strategy of EBV.[56,61,62] This orphan 7TM receptor is expressed at significant levels during the early lytic phase of the virus infection, intermediate levels in LCLs and nasopharyngeal carcinoma derived C666-1 line and at low levels during the viral latent phase.[63,64] Several studies have shown that the pivotal role in immune evasion played by EBV-BILF1 occurs during the lytic replication rather than the latent phase.[56,61–64] In addition to EBV, the BILF1 gene is present in the two so far characterized nonhuman primate LCVs, Old World rhesus LCV-1 (RhLCV-1), and New World callitrichine herpesvirus-3 (CalHV-3).[62,65] Furthermore, ungulate gammaherpesviruses belonging to the genera *Macavirus* (Porcine lymphotropic herpesviruses 1–3, Alcelaphine herpesvirus 1, and Ovine herpesvirus 2) and *Percavirus* (Equine herpesvirus 2) encode putative 7TM receptors at the homologous genomic position. Although these BILF1 receptor homologs remain uncharacterized, the strong conservation reflects a significance of this gene in the viral pathogenesis.

3.1. Immune evasion strategy of EBV-BILF1

Unlike most of the virus-encoded 7TM receptors[59] BILF1 is not a chemokine receptor and therefore does not sequester chemokines in order to circumvent the host immune system. BILF1, however, uses another strategy, which is the downregulation of MHC class I cell surface expression (Fig. 2A) and inhibition of the cytotoxic $CD8^+$ T cell recognition of the virus-infected cells.[56,61,62] The exact mechanism by which BILF1 targets the MHC class I molecules is far from being understood; however, a study by Zuo *et al.*[61] reported that BILF1 is involved in the MHC class I downregulation by targeting the molecules on both the endocytic and the exocytic pathways. The mechanism of the MHC class I targeting on

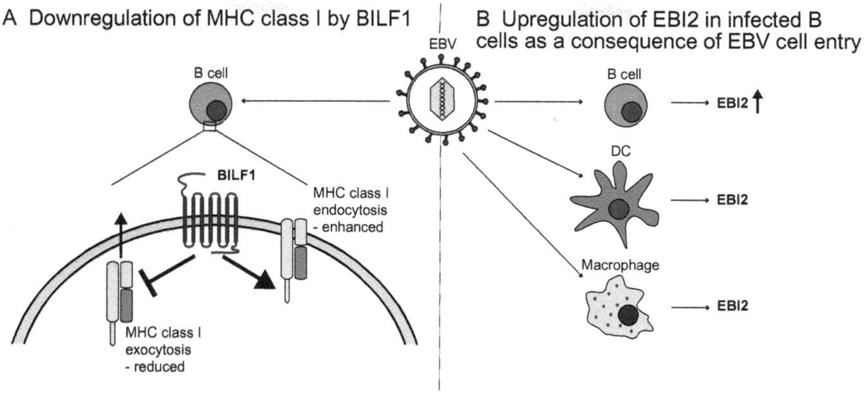

Figure 2 Suggested roles of the viral EBV-BILF1 and the human EBI2 receptor. (A) The EBV-encoded 7TM receptor BILF1 has immune evasive properties by reducing the cell surface levels of MHC class I molecule. This is achieved by BILF1-mediated induction of endocytosis of MHC class I molecules from the cell surface or by inhibition of the normal exocytic pathway transferring newly synthesized MHC class I molecules to the cell surface. (B) EBV infection induces expression of EBI2 on the B cell surface. Whether EBI2 plays a role in the virus life cycle or in the antiviral immune defense is yet unknown. Under normal physiological conditions, EBI2 is also expressed in DCs and macrophages. These cells could also express EBI2 upon EBV infection and thereby assist the antiviral immune response.

the endocytic pathway may involve a putative physical association of BILF1 with the MHC class I molecule and further enhancement of the MHC class I lysosomal degradation.[56,61] In contrast, the exocytic pathway targeting of the MHC class I molecules by BILF1 is not fully understood, albeit a physical association of BILF1 with the MHC class I molecules in the ER has been reported.[56] It is noteworthy that this physical association in the ER does not seem to hinder MHC class I trafficking to Golgi suggesting that the EBV association with MHC class I in the ER may instead interfere with the transporter associated with antigen processing (TAP)-associated glycoprotein or the TAP-complex binding.[56,65]

The wild-type EBV-BILF1 receptor was shown to potentiate internalization[61,62] and degradation of the MHC class I molecules resulting in a reduction of MHC class I on the cell surface.[61] Conversely, mutational studies of the DRY-like EKT motif in the bottom of TM3 of EBV-BILF1 showed that the K122A mutation impaired the ability of BILF1 to induce MHC class I endocytosis[61] and lead to abolished signaling properties of the receptor.[61,66] This motif is partially conserved among virus-encoded 7TM receptors[67] and has also been shown to impact the signaling of other virus-encoded receptors like the KSHV-ORF74,[68,69] HCMV-US28,[70,71] and MCMV-M33.[72] Interestingly, it was shown to be dispensable for proper G-protein coupling in the CXC-chemokine receptor ORF74 encoded by Equine herpesvirus 2, as this receptor lacks the most important positive charge in the middle of this motif.[73,74]

Another region with impact on MHC class I surface expression has been identified in the C-terminus of EBV-BILF1, as truncation of this (ΔC) impairs the ability of the receptor to target the MHC class I molecule for lysosomal degradation and downregulation.[61,62]

Nevertheless, as mentioned above, BILF1 is not the only player and it acts in accord with three other (BGLF5, BNLF2a, and vIL-10) to downregulate the MHC class I molecule cell surface expression.[15,53,54,56] BILF1 can be detected as soon as 4 h postinfection.[62] The initial expression of BILF1 occurs during the host cell protein synthesis suppression, which usually happens under the effect of BGLF5 as well as the TAP inhibition mediated by BNLF2a.[62] The transient expression of BNLF2a as well as possible detection of BILF1 16 h postinfection makes it compelling to suggest that BILF1 also is capable of targeting the MHC class I molecules in the late lytic cycle.[62] Furthermore, a marked impairment of CD8 + T cell recognition of EBV-infected cells was reported by Zuo *et al.*[61] and this clearly reflects that the effect of BILF1 on MHC class I molecule is significant. It is noteworthy that BILF1 targets multiple forms of MHC class I molecules including HLA-A, B, and E with a minimal effect on HLA-C. This broad inhibition of multiple subsets of the MHC class I molecule may help the virus escape the immune system and persist in the host suggesting that BILF1 is important for the immune evasion strategy of EBV.[62,65] Downregulation of MHC class I is also exhibited by the RhLCV-1 BILF1 homolog, suggesting evolutionary conservation of function at least for Old World primates.[62]

3.2. Signaling and tumorigenesis of EBV BILF1

In addition to its immune evasive effects, EBV-BILF1 is highly tumorigenic both *in vitro* and *in vivo*.[66] It signals in a constitutive ligand-independent manner through Gαi as it inhibits forskolin-induced adenylate cyclase activity in a gene–dose-dependent manner.[63,66] This constitutive signaling is shared with other virus-encoded receptor families such as US28 from HCMV[70,75–77] and the ORF74 receptors from human herpesvirus 8 (KSHV-ORF74),[69,78–80] Herpesvirus Saimirii,[81] Equine herpesvirus

2,[73] and murine gammaherpesvirus 68 (MHV68).[82] Most of these viral receptors also bind multiple chemokines in a broad spectrum manner, and scavenge these from the surroundings of virus-infected cells as part of an immune evasion strategy.[78,79,81,83–85] Lyngaa *et al.*[66] reported that BILF1 induces NIH 3T3 cell transformation and vascular endothelial growth factor secretion and *in vivo* tumor formation in nude mice. Importantly, these properties were linked to the constitutive activity of EBV-BILF1, as they were ameliorated by an activity-silencing point-mutation in the conserved DRY-like EKT motif in the bottom of TM3. Similar tumorigenesis, also with linkage to constitutive activity, has been shown for US28 and KSHV-ORF74.[78,80,86] This suggests that the constitutive activity of EBV-BILF1 could be implicated in EBV-associated cancer pathogenesis.[63–66]

4. EBI2: AN ENDOGENOUS 7TM RECEPTOR MANIPULATED BY EBV

EBI2 was identified in 1993, when Kieff and colleagues used subtractive hybridization of DNA from Burkitt's lymphoma cells to screen for upregulated genes in EBV-infected cells.[87] They found, being more than 200-fold upregulated (hence the name: EBV-induced gene 2). Since 1993, this finding has been confirmed in four different studies describing EBI2 upregulation in both lytic and latent settings[29,88–90] (see Table 2).

The name of EBI2 suggests that EBV directly induces expression of the receptor, but the fact is that we, after more than two decades, still do not know if EBI2 is *induced by EBV* or if it *is induced as part of the immune response* to the virus infection. As described in detail below, EBI2 has moved from being an orphan receptor with no known biological function to an oxysterol-induced receptor with important functions in the immune system within a few years; in addition, the molecular basis for activation and inhibition of EBI2 has been studied extensively.

4.1. A family of oxysterols acts as ligands for the EBI2 receptor

When first identified, EBI2 was proposed to have a polypeptide cytokine ligand.[87] It was thus surprising, when two groups simultaneously identified oxysterols as the endogenous ligands for EBI2.[91,92] Importantly, although 7α25-dihydroxycholesterol (7α,25-OHC) is the most potent ligand, the two groups discovered a family of oxysterols, all able to activate the EBI2

Table 2 Expression pattern of EBI2 in various cell types in both lytic replication and latency

Phase of infection/ gene		EBI2 expression	Assay system	References
Lytic		Upregulated	Akata and B95.8 cell RT-qPCR	90
Latency	0			
	I	>200-fold	Subtractive hybridization of mRNA in BL cells	87
		Upregulated	Akata cell RT-qPCR	90
	II			
	III	>4-fold	LCLs	89
		>2-fold	Gene expression profiling of PTLD samples	88
		Upregulated	B95.8 cell RT-qPCR	90
LMP1-induced		Rapid upregulation	Run-on RNA assays with conditional LMP1 expressing cells	29

receptor. Further studies on the EBI2–oxysterol interaction has revealed that EBI2 is strongly active through Gαi and that it signals through both G-protein-dependent and -independent pathways, although the receptor seems to be biased toward the former.[90–93] A defined set of anchor residues in EBI2 consisting of an arginine in the extracellular end of TM helix 2 and tyrosine residues in TM-3 and -6 are highly important as alanine substitutions of these residues significantly decreases oxysterol binding and EBI2 activation.[93,94] In addition, *ex vivo* proliferation studies on mouse B cells expressing human EBI2 showed that EBI2 induces proliferation, which can be inhibited by treatment with a small-molecule EBI2 antagonist.[95] This proliferation was also observed when comparing the growth of B cells from wt mice to that of B cells from mice with targeted deletion of EBI2.[95]

Oxysterols are oxidized derivatives of cholesterol, which regulate many cellular functions and have also been recognized for having functions within the innate and adaptive immune response. Hence, the oxysterol activated liver X receptors (LXRs) can rescue macrophages from pathogen-induced apoptosis, induce pathogen clearance and inhibit an aberrant inflammatory response.[96–98] Also, LXRs have an important checkpoint function on B and T cell proliferation as activated B and T cells induce oxysterol metabolism and inhibition of LXR leading to proliferation.[99] Accordingly, antigen-challenged mice lacking LXRs develop lymphoid hyperplasia.[99] Finally, LXRs are negative regulators of Th17 helper cells, which are implicated in autoimmune diseases and expression of LXRs in an *in vivo* model of multiple sclerosis thus improved disease.[100]

4.2. Roles of the EBI2–oxysterol axis in the immune response

EBI2 was initially shown to be expressed primarily in B cells and later studies confirmed this and added expression in DCs, macrophages, and to a lesser degree T cells and NK cells.[87,90–92,101] Interestingly, abundant expression of EBI2 in activated B cells is opposed by downregulation in GCs, leading researchers to the biological role of EBI2.[89,102] In 2009, two groups simultaneously published EBI2 as an important regulator of B cell localization in the lymphoid follicle.[103,104] Together with chemokine receptors CXCR4, CXCR5, and CCR7 and a highly controlled expression of these and their ligands (CXCL12, CXCL13, and CCL21, respectively), EBI2 mediates the segregation of B cells between the inner and outer follicle during antibody responses (for a recent review, see Ref. 105). Shortly, during the primary early antibody response activated B cells express high levels of EBI2 and migrate toward the inner- and outer follicular areas where oxysterol accumulation is generated by stromal cells.[103,104,106,107] Here, the B cells commit to one of two paths: plasmablast differentiation and antibody production by continuous expression of EBI2 or GC formation by downregulation of EBI2 by BCL-6.[102–104] The importance of EBI2 in the immune response is illustrated by a greatly reduced early antibody response in EBI2-deficient mice.[104]

Recent studies have expanded the role of EBI2 in the immune response to include directed localization of DCs in the marginal zone (MZ) bridging channels.[108,109] CD4 expressing DCs reside in the MZ bridging channels and survey the blood for antigens. Upon detection of antigen, DCs migrate toward the T cell zone and thereby initiate T helper cell activation and immune response. Two subsequent studies found that EBI2 expression and accumulation of 7α,25-OHC in the MZ bridging channel were required for generation of CD4+ DCs and localization of these to the MZ bridging channel.[108,109] EBI2 deficiency in DCs resulted in reduced activation of CD4+ T cells as well as reduced antibody response.[109] Also, EBI2 functions as a negative regulator of interferon responses in plasmacytoid and myeloid DCs, which could possibly balance aberrant TLR-mediated interferon responses to foreign and self-nucleic acids.[110]

In addition to its role in the adaptive immune response, EBI2 may also play a role in the innate immune response through actions on macrophages. Recently, macrophages were shown to express both EBI2 and the enzymes required for 7α,25-OHC production. Lipopolysaccharide (LPS) stimulation of a macrophage cell line leads to activation of EBI2, calcium mobilization and directed cell migration, suggesting that EBI2 indeed does play a role in macrophage-mediated immune responses.[101]

4.3. A potential role for the EBI2–oxysterol axis in EBV infection

Although very little is known of the role of EBI2 in EBV infections, knowledge on the role of EBI2 and in particular oxysterols in the general antiviral immune defense is steadily emerging. Production of oxysterols by macrophages might thus represent an antiviral host strategy as treatment of macrophages with TLR ligand LPS or type I interferons leads to an increase of 25-OHC oxysterol and its synthesizing enzyme cholesterol-25-hydroxylase (CH25H) both *in vitro* and *in vivo*.[111–113] *In vitro* studies with expression of CH25H or treatment with 25-OHC broadly inhibits enveloped viruses, among them herpes simplex, HIV, and MHV68.[114] *In vivo* studies with CH25H-deficient mice showed an increased susceptibility to MHV68 and also, 25-OHC treatment of humanized mice reduced HIV replication.[114] These effects of CH25H and 25-OHC are probably due to antiviral effects of the molecules themselves and not mediated by altered EBI2 activity.[114] On the contrary, studies on time course-dependent changes in the macrophage transcriptome after stimulation with *Salmonella* strains revealed that EBI2 messenger RNA (mRNA) levels increased up to 20-fold[115] directly linking EBI2 to the anti-*Salmonella* immune response.

The high expression of EBI2 during EBV infection suggests that it is required for either a successful primary and persisting EBV infection or in the immune response toward EBV infection. It has previously been suggested that EBV-mediated upregulation of EBI2 could ensure positioning of EBV-infected cells in a survival-promoting niche.[103] Thus, upregulation of EBI2 would keep infected B cells from participating in GC formation and risk GC-induced apoptosis. This is supported by the fact that EBI2-mediated migration is promoted by CD40 engagement[116] and that the CD40-mimic LMP1 has been shown to directly induce EBI2 in *in vitro* settings.[29] However, the high expression of EBI2 during latency program I,[87,90] when most of the viral genome is shutdown and only EBNA-1 is expressed, hints toward a role of EBI2 in the immune surveillance instead.

This is supported by the previously described finding that also *Salmonella* strains induce EBI2 upregulation in macrophages.[115]

Although the biological function of EBI2 has primarily been described for the antibody production in the adaptive immune response, antibodies (as described above) play a minor role in the immune anti-EBV response. The expression of EBI2 during EBV infection has for now focused on expression levels in B cells, but considering that EBI2 is also expressed and has biologic functions in macrophages and DCs, it would be interesting to measure expression of EBI2 in these cell compartments during EBV infection. Thus, it could be hypothesized that EBV infection also results in EBI2 upregulation in macrophages and DCs and that this could be part of an antiviral response mediated by these cells (Fig. 2B). As such, it has been shown that virus-infected macrophages can directly activate NK cells and thereby kill EBV-infected cells.[117] Also, DCs in the MZ bridging zone, which are activated by virus, activate CD4+ T cells, which in turn mediate antibody response as well as CD8+ CTL activation.[109] In addition, CD4+ T cells can acquire cytotoxic activity themselves and *in vivo* cytotoxicity assays in mice persistently infected with MHV68 show CD4 T cell-dependent killing of MHV68-infected cells.[118]

5. MANIPULATION OF THE HOST IMMUNE SYSTEM 7TM RECEPTORS AND LIGANDS BY EBV—THE CHEMOKINE SYSTEM

Chemokines induce chemotaxis of leukocytes through interactions with 7TM chemokine receptors. Approximately, 40 chemokines and 18 functional chemokine receptors have been identified in humans. Based on the pattern of conserved cysteine residues in the N-terminal region they are classified into four subfamilies: CC, CXC, C, and CX3C.[119] By controlling leukocyte migration, the chemokine system has important functions in coordinating the immune system, leukocyte homeostasis, lymphocyte activation, and the host immune response to infectious pathogens.[119] It is therefore not surprising that many viruses have been shown to modulate the chemokine system, several of them by encoding chemokine receptors or ligands.[60,120] EBV has not yet been shown to encode either chemokine receptors or ligands, but the system is being markedly manipulated during EBV infection, as some chemokines and chemokine receptors are upregulated, while others are downregulated (Fig. 3). Hence, the chemokine system may play an important role in the tissue localization

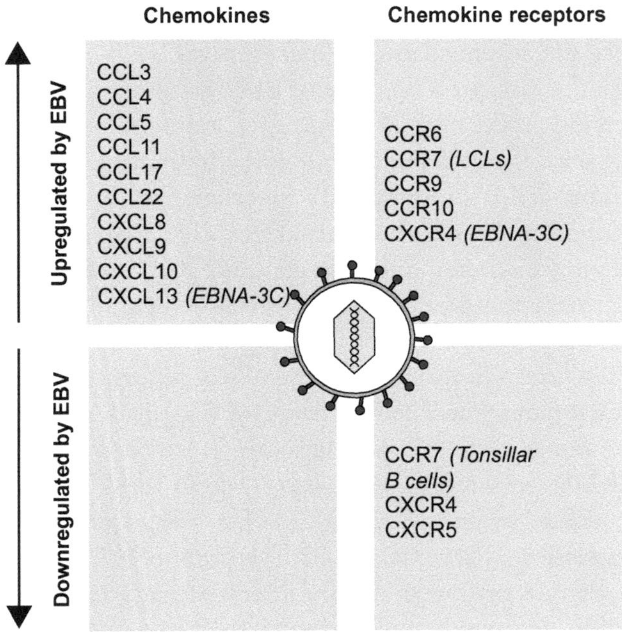

Figure 3 Regulation of the human chemokine system during EBV infections. EBV upregulates and downregulates both chemokine receptors and chemokine ligands. Notably, CCR7 is upregulated in Burkitt's lymphoma cells and LCLs, but time-dependent studies in tonsillary B cells showed that CCR7 was downregulated 7 days after infection suggesting that CCR7 is differentially expressed by EBV depending on B cell origin.[121] Also, CXCR4 expression is low in LCL with full latency III program expression, whereas LCLs expressing only EBNA-3C showed high expression of CXCR4 as well as its ligand CXCL13.

of EBV-infected B cells and thereby contribute to the pathogenesis of EBV-associated diseases.[121]

A study profiling chemokine expression in LCLs revealed high expression levels of CCR6, CCR7, and CCR10 and low expression of CXCR4 and CXCR5 in EBV immortalized cells.[122] Accordingly, LCLs showed a markedly increase in migration in response to the ligands of CCR6 (CCL20), CC7 (CCL21), and CCR10 (CCL28), but showed only weak migration in response to the ligands of CXCR4 (CXCL12) and CXCR5 (CXCL13).[122] Selective expression ofEBNA-2, LMP1 or both in an EBV negative cell line showed that EBNA-2 directly induces expression of CCR6, while both EBNA-2 and LMP1 downregulate CXCR4. In contrast neither CCR10 or CXCR5 was directly induced by EBNA-2 or LMP1 and the regulation of these two chemokine receptors was more likely a consequence of the plasmablast state of the

LCLs.[122] CXCR4 has also been found to be upregulated in EBNA-3B-deficient LCLs suggesting that this viral protein is a negative regulator of CXCR4.[123] In addition, BILF1 has been found to heterodimerize with CXCR4 and thereby inhibit CXCR4-mediated signaling.[124] The importance of the chemokine pattern described above can be deduced from the fact that also human herpesviruses 6A, 6B, and 7 has been shown to induce CCR7 expression and reduce CXCR4 expression.[125,126] Interestingly, a study examining the lymphoproliferation in EBV-infected huSCID mice showed that the axis was important for lymphoma development as CXCR4 was highly expressed on the tumors and inhibition of the CXCR4/CXCL12 axis reduced tumor development.[127] Consistently, the gene, which is expressed in latency III program and is important for growth transformation of infected B cells, upregulated the levels of both CXCL12 and CXCR4.[128] The diverging expression of CXCR4 by the viral tumor suppressor EBNA-3B (downregulation)[123] and the growth transforming EBNA-3C (upregulation)[128] taken together with increased expression of CXCR4 in EBV-mediated huSCID lymphoma development[127] clearly suggests that CXCR4 is important for the growth transformation of infected B cells. Another study on the effect of EBV on CXCR5 and CCR7 expression in tonsillary B cells showed that 2 days after EBV infection there were minor changes in the expression levels of CXCR5. By day 7, however, the expression levels of both CXCR5 and CCR7 go down and both of the aforementioned receptors were no longer expressed at the cell surface by day 14. Also, the chemotactic response to CXCL13 and CCL21 was reduced by day 2, when CXCR5 and CCR7 was still expressed, suggesting that the virus impairs chemokine-directed migration even in the presence of the receptors.[121]

The histologic finding that EBV-infected B cells in infectious mononucleosis tend to avoid the GCs and instead accumulate under the epithelium of certain mucosal tissues[34,129] could be explained by the above-mentioned expression patterns of chemokine receptors on EBV-infected B cells. CXCR4 and CXCR5 respond to their ligands, which are highly expressed in the center of lymphoid follicles and thereby direct B cells to GC formation.[105] CCL20 and CCL28, on the other hand, are normally expressed by the epithelial cells of mucosal tissues such as salivary glands and the tonsils.[130–132] As such, downregulation of CXCR4 and CXCR5 and upregulation of CCR6 and CCR10 could ensure migration toward the tonsillary epithelial cells.[122]

EBV also induces the expression of the mRNA levels of CCR9, which is responsible for B cell homing to mucosal tissues, as well as C5AR1, the receptor for the complement factor C5a. Alterations of the expression level of the aforementioned molecules may cause the EBV-infected cells to be retained in the interfollicular region of the tonsils.[121]

The manipulation of the chemokine system by EBV seems to be in favor of the virus infection, but it is likely that some of the regulation also reflects cell protection against the intruding virus and antiviral immune responses. As such, it was found that EBV-infected cells as well as EBV negative cells expressing LMP1 upregulate the expression of CCL17 and CCL22, known to preferentially attract Th2 cells and regulatory T cells via CCR4. This expression pattern would benefit the virus and skew the immune response away from a Th1 and CTL response. In contrast, EBV-infected cells were found to upregulate CCL3, CCL4, and CCL5, which are known to attract Th1 cells and activated CTL via CCR5.[133] Expression of CCL5 was also found to be elevated in LMP1 expressing cells as well as Burkitt's lymphoma cell lines.[134] Thus, the chemokine system may work in favor of both the virus and the immune system.

Regardless of its effect on EBV infection, the chemokine system has been found to be regulated in both nasopharyngeal cancer and Hodgkin lymphoma. As such, Hodgkin and Reed–Sternberg (HRS) cells from Hodgkin lymphoma and nasopharyngeal carcinoma cells have been shown to express CXCL8 and CCL17.[135] Furthermore, Hodgkin lymphoma cells express high levels of CXCL9, CXCL10, CCL3, CCL5, and CCL11.[136]

6. EBV-ASSOCIATED DISEASES

Primary EBV infection often occurs in young immunocompetent children and is asymptomatic or present with nonspecific mild symptoms.[6] However, given that EBV was originally identified in Burkitt's lymphoma cells and has the unique ability to transform resting B cells into highly proliferating lymphoblasts, it is not surprising that this virus is associated with a number of cancers in both immunocompromised and immunocompetent patients. In addition, EBV has been associated with autoimmune diseases involving infected B cells (see Table 3 for an overview of confirmed EBV-associated diseases).

6.1. Infectious mononucleosis

In adolescent and young people, primary infection with EBV may lead to infectious mononucleosis, which is usually a self-limiting disease with symptoms ranging from mild to severe. More than 50% of patients with

Table 3 Confirmed EBV-associated malignant and nonmalignant diseases

Disease	EBV association	Cell of origin	Latency program
Nonmalignant			
Infectious mononucleosis			
Malignant			
Immunocompetent			
Burkitt's lymphoma			
– Endemic	95–100%	GC or post-GC B cell	I
– Sporadic	15–85%	GC or post-GC B cell	I
Hodgkin lymphoma (classical)			
– Nodular sclerosing	20–40%	Post-GC B cell	II
– Lymphocyte-depleted	80–90%	Post-GC B cell	II
– Lymphocyte-rich	<10%	Post-GC B cell	II
– Mixed cellularity	60–75%	Post-GC B cell	II
Extranodal NK/T cell lymphoma	100%	$CD3^{-}$, $CD56^{+}$ T cell (derived from NK cell)	II
Nasopharyngeal carcinoma			
– Nonkeratinizing	100%	Undifferentiated epithelial cell	I/II
– Keratinizing	30–100%	Undifferentiated epithelial cell	I/II
Gastric carcinoma			
– Lymphoepithelioma like	90–100%	Undifferentiated gastric epithelial cell	I
– Conventional	5–15%	Undifferentiated gastric epithelial cell	I
Immunocompromised			
Lymphoproliferative disease in congenital immune deficiency	100%	GC or post-GC B cell	III
Posttransplant lymphoproliferative diseases (PTLDs)			

Continued

Table 3 Confirmed EBV-associated malignant and nonmalignant diseases—cont'd

Disease	EBV association	Cell of origin	Latency program
– Early lesions and polymorphic <1 year after transplant	90–100%	GC or post-GC B cell	II/III
– Monomorphic >1 year after transplant	50%	GC or post-GC B cell	II/III
Leiomyosarcomas	100%	Smooth muscle cell	III
HIV malignancies			
– Burkitt's lymphoma	55%	GC B cell	I
– Hodgkin lymphoma	100%	Post-GC B cell	II
– Diffuse large B cell lymphoma (DLBCL)—centroblast	30%	GC or post-GC B cell	I
– DLBCL—immunoblast	90%	GC or post-GC B cell	III
– Primary CNS lymphoma	100%	GC or post-GC B cell	III
– Primary effusion lymphoma	90–100%	GC or post-GC B cell	I

The diseases are shown along with association, cell of origin, and latency program expressed in the implicated cells.

infectious mononucleosis present with the triad of fever, lymphadenopathy, and pharyngitis.[137] The peripheral blood shows leukocytosis with an increase in mononuclear cells, CD8+ T cells, NK cells, and a decrease in CD4+ T cells.[6] The symptoms observed in infectious mononucleosis are due to the increased level of CD8+ cells and the cytokines they secrete.[138] Infected B cells in the throat of infectious mononucleosis patients vary in morphology as well as expression of viral genes. Thus, the infected cells express variations of latency programs III, II, and I.[33,139] In the periphery, more than 50% of the memory B cells are infected and these memory cells are resting and express the latency 0 program.[140] Interestingly, even in the early massively infectious phase of infectious mononucleosis, the restriction of virus infection to the memory B cell compartment is intact.[140] The risk of EBV positive Hodgkin lymphoma is increased after infectious mononucleosis,[141] which has been proposed to be due to MHC class I-restricted EBV-specific CTL responses in the early anti-EBV immune response.[142]

6.2. Diseases in immunocompetent patients

In the immunocompetent host, EBV is primarily associated with Burkitt's lymphoma and Hodgkin lymphoma originating in B cells; and nasopharyngeal carcinoma and gastric carcinoma originating in epithelial cells. Burkitt's lymphoma is a malignancy most commonly located in the abdomen and jaw. Three distinct clinical forms of Burkitt's lymphoma are recognized: endemic (African), sporadic (nonendemic), and immunodeficiency-associated. In central parts of Africa and New Guinea, Burkitt's lymphoma occurs holoendemic with malaria. The strong association of EBV with Burkitt's lymphoma is based on the high frequency of tumors that carry the virus in endemic areas (97%)[143] and the presence of clonal EBV in all tumor cells.[144] How EBV participates in the pathogenesis of Burkitt's lymphoma is still unclear. The tumor originates from GC B cells and expresses the latency program I.[145] All Burkitt's lymphoma cells carry one of three characteristic chromosomal translocations (from chromosome 8 to chromosome 14) that place the cellular oncogene c-myc under the control of the Ig heavy chain or one of the light-chain loci.[145] Classical Hodgkin lymphoma is characterized by the presence of large multinucleated B-lineage cells called HRS cells. There has also been found in some cases of infectious mononucleosis, thereby supporting the increased risk of Hodgkin lymphoma following a severe infectious mononucleosis.[129] However, EBV is only present in 40–60% of cases of Hodgkin lymphoma.[146] EBV-positive Hodgkin lymphoma cells origin in post-GC B cells and express latency program II.[6]

EBV infection of epithelial cells is associated with several carcinomas, among them nasopharyngeal carcinoma and gastric carcinoma. Nonkeratinizing nasopharyngeal carcinoma is extremely common in certain high-risk groups such as the Southern Chinese and Inuits and is almost always EBV positive.[147,148] The cancers originate in epithelial cells and express latency program 0.[149] Other cofactors are involved in the pathogenesis and both environmental and genetic susceptibilities have been implicated.[23] Almost 10% of gastric carcinomas worldwide origin in EBV-infected cells[150] making it the most common EBV-associated cancer.[6] All EBV-associated gastric carcinoma cells express latency program I and EBNA-1, which has been associated with promoted cancer cell survival.[151]

6.3. Diseases in immunocompromised patients

Impaired T cell immune surveillance of EBV-infected B cells caused by HIV-associated genetic or iatrogenic (transplantations) immunosuppression is associated with lymphomas and to a lesser extent leiomyosarcomas. PTLDs- and HIV-associated lymphomas constitute the most abundant lymphomas in immunocompromised patients. PTLDs include a heterogeneous group of lymphomas, of which almost all carry the virus and express the latency program III.[144] The incidence varies from 1% to 20%, depending upon the transplantation type and level of the immunosuppression.[139] The lymphomas mostly originate in GC B cells or post-GC B cells. PTLDs arising early after transplantation are almost always associated with EBV, whereas PTLDs arising more than 1 year after transplantation are monomorphic displaying mutations in genes such as BCL-6, p53, and c-myc and are less often associated with EBV.[152]

Similar to PTLDs, HIV-associated lymphomas are a highly heterogeneous group, with a varying degree of association to EBV. EBV positive HIV-cancers mostly originate in GC or post-GC B cells and express different latency programs.[153]

Besides cancers, EBV has been associated with several autoimmune diseases such as multiple sclerosis, rheumatoid arthritis, and systemic lupus.[6] It is very difficult to demonstrate the causative role of EBV in these diseases as almost all (90%) are infected with EBV. The linkage between EBV and multiple sclerosis is, however, growing stronger as 99% of multiple sclerosis patients are EBV positive[154] and infectious mononucleosis increases the risk of multiple sclerosis by 2.3-fold.[155] In addition, EBV-infected B cells have been discovered in the CNS of multiple sclerosis patients.[156]

7. DRUG-TARGET POTENTIAL

7TM receptors and in particular class A receptors to which the 7TM receptors mentioned in this review belong, are highly druggable molecules. Approximately, 35% of all currently marketed drugs target class A 7TM receptors.[157] There is no specific antiviral treatment for EBV. Current treatment strategies for EBV-associated lymphomas include B cell antibodies (rituximab, chimeric, and monoclonal antibody against CD20), reducing immunosuppression, EBV-specific CTL infusion, and chemotherapy for PTLDs; and anti-HIV treatment for HIV-associated lymphomas.[158,159] The use of rituximab in particular has dramatically changed the overall survival of PTLD patients, but a strategy directed specifically against EBV-infected cells and genes, which play a role in the EBV-mediated pathogenesis, could likely reduce the incidence of malignancies. As such, BILF1 constitutes an obvious target as it plays a role for EBV immune evasion[56,61,62] and has oncogenic properties both *in vitro* and *in vivo*.[66] EBI2 has been shown to have proliferative effects *ex vivo*[95] and in addition, the receptor is upregulated in PTLDs.[88] Though the role of EBI2 in EBV life cycle is still uncertain, the aforementioned knowledge suggests that EBI2 could play a role in EBV-mediated lymphoma development and therefore EBI2 constitutes another potential drug target. Finally, the chemokine expression pattern is skewed in both Hodgkin lymphoma and nasopharyngeal carcinoma[135] meaning that inhibition of some chemokine receptors could prove beneficial in the antiviral treatment strategy.

8. CONCLUSIONS

In summary, EBV encodes one 7TM receptor and manipulates endogenous 7TM receptors, some of which are involved in the virus life cycle and some of which may be involved in the antiviral host defense. In the case of EBI2, we are still unclear on whether the upregulation is virus mediated or immune system mediated. Nevertheless, EBI2 is upregulated in EBV-associated PTLD and, together with BILF1 and the chemokine receptors constitute potential drug targets in EBV-mediated diseases. Further studies exploring the relation between these receptors and EBV are highly warranted.

ACKNOWLEDGMENT

The authors thank Nick Davis-Poynter and Katja Spiess for critical reading of this manuscript and for helpful comments.

REFERENCES

1. Epstein MA, Achong BG, Barr YM. Virus particles in cultured lymphoblasts from Burkitt's lymphoma. *Lancet*. 1964;1(7335):702–703.
2. Wang F, Rivailler P, Rao P, Cho Y. Simian homologues of Epstein-Barr virus. *Philos Trans R Soc Lond B Biol Sci*. 2001;356(1408):489–497.
3. Katoh K, Misawa K, Kuma K-I, Miyata T. MAFFT: a novel method for rapid multiple sequence alignment based on fast Fourier transform. *Nucleic Acids Res*. 2002;30(14):3059–3066.
4. Guindon S, Dufayard J-F, Lefort V, Anisimova M, Hordijk W, Gascuel O. New algorithms and methods to estimate maximum-likelihood phylogenies: assessing the performance of PhyML 3.0. *Syst Biol*. 2010;59(3):307–321.
5. Young LS, Rickinson AB. Epstein-Barr virus: 40 years on. *Nat Rev Cancer*. 2004;4(10):757–768.
6. Knipe DM, Howley PM. *Fields Virology*. Philadelphia, PA: Wolters Kluwer/Lippincott Williams & Wilkins Health; 2013.
7. Tanner J, Weis J, Fearon D, Whang Y, Kieff E. Epstein-Barr virus gp350/220 binding to the B lymphocyte C3d receptor mediates adsorption, capping, and endocytosis. *Cell*. 1987;50(2):203–213.

8. Hutt-Fletcher LM. Epstein-Barr virus entry. *J Virol.* 2007;81(15):7825–7832.
9. Xiao J, Palefsky JM, Herrera R, Tugizov SM. Characterization of the Epstein-Barr virus glycoprotein BMRF-2. *Virology.* 2007;359(2):382–396.
10. Miller N, Hutt-Fletcher LM. Epstein-Barr virus enters B cells and epithelial cells by different routes. *J Virol.* 1992;66(6):3409–3414.
11. Borza CM, Hutt-Fletcher LM. Alternate replication in B cells and epithelial cells switches tropism of Epstein-Barr virus. *Nat Med.* 2002;8(6):594–599.
12. Hadinoto V, Shapiro M, Sun CC, Thorley-Lawson DA. The dynamics of EBV shedding implicate a central role for epithelial cells in amplifying viral output. *PLoS Pathog.* 2009;5(7):e1000496.
13. Wen W, Iwakiri D, Yamamoto K, Maruo S, Kanda T, Takada K. Epstein-Barr virus BZLF1 gene, a switch from latency to lytic infection, is expressed as an immediate-early gene after primary infection of B lymphocytes. *J Virol.* 2007;81(2):1037–1042.
14. Odumade OA, Hogquist KA, Balfour HH. Progress and problems in understanding and managing primary Epstein-Barr virus infections. *Clin Microbiol Rev.* 2011;24(1):193–209.
15. Jochum S, Moosmann A, Lang S, Hammerschmidt W, Zeidler R. The EBV immunoevasins vIL-10 and BNLF2a protect newly infected B cells from immune recognition and elimination. *PLoS Pathog.* 2012;8(5):e1002704.
16. Altmann M, Hammerschmidt W. Epstein-Barr virus provides a new paradigm: a requirement for the immediate inhibition of apoptosis. *PLoS Biol.* 2005;3(12):e404.
17. Sugden B, Mark W. Clonal transformation of adult human leukocytes by Epstein-Barr virus. *J Virol.* 1977;23(3):503–508.
18. Henderson E, Miller G, Robinson J, Heston L. Efficiency of transformation of lymphocytes by Epstein-Barr virus. *Virology.* 1977;76(1):152–163.
19. Nikitin PA, Yan CM, Forte E, et al. An ATM/Chk2-mediated DNA damage-responsive signaling pathway suppresses Epstein-Barr virus transformation of primary human B cells. *Cell Host Microbe.* 2010;8(6):510–522.
20. Westhoff Smith D, Sugden B. Potential cellular functions of Epstein-Barr nuclear antigen 1 (EBNA1) of Epstein-Barr virus. *Viruses.* 2013;5(1):226–240.
21. Portal D, Zhou H, Zhao B, et al. Epstein-Barr virus nuclear antigen leader protein localizes to promoters and enhancers with cell transcription factors and EBNA2. *Proc Natl Acad Sci U S A.* 2013;110(46):18537–18542.
22. White RE, Rämer PC, Naresh KN, et al. EBNA3B-deficient EBV promotes B cell lymphomagenesis in humanized mice and is found in human tumors. *J Clin Invest.* 2012;122(4):1487–1502.
23. Thorley-Lawson DA. Epstein-Barr virus: exploiting the immune system. *Nat Rev Immunol.* 2001;1(1):75–82.
24. Caldwell RG, Wilson JB, Anderson SJ, Longnecker R. Epstein-Barr virus LMP2A drives B cell development and survival in the absence of normal B cell receptor signals. *Immunity.* 1998;9(3):405–411.
25. Rovedo M, Longnecker R. Epstein-Barr virus latent membrane protein 2B (LMP2B) modulates LMP2A activity. *J Virol.* 2007;81(1):84–94.
26. Swaminathan S. Noncoding RNAs produced by oncogenic human herpesviruses. *J Cell Physiol.* 2008;216(2):321–326.
27. Babcock GJ, Hochberg D, Thorley-Lawson DA. The expression pattern of Epstein-Barr virus latent genes in vivo is dependent upon the differentiation stage of the infected B cell. *Immunity.* 2000;13(4):497–506.
28. Gires O, Zimber-Strobl U, Gonnella R, et al. Latent membrane protein 1 of Epstein-Barr virus mimics a constitutively active receptor molecule. *EMBO J.* 1997;16(20):6131–6140.

29. Dirmeier U, Hoffmann R, Kilger E, et al. Latent membrane protein 1 of Epstein-Barr virus coordinately regulates proliferation with control of apoptosis. *Oncogene.* 2005;24(10):1711–1717.
30. He B, Raab-Traub N, Casali P, Cerutti A. EBV-encoded latent membrane protein 1 cooperates with BAFF/BLyS and APRIL to induce T cell-independent Ig heavy chain class switching. *J Immunol.* 2003;171(10):5215–5224.
31. Thorley-Lawson DA, Hawkins JB, Tracy SI, Shapiro M. The pathogenesis of Epstein-Barr virus persistent infection. *Curr Opin Virol.* 2013;3(3):227–232.
32. Ehlin-Henriksson B, Gordon J, Klein G. B-lymphocyte subpopulations are equally susceptible to Epstein-Barr virus infection, irrespective of immunoglobulin isotype expression. *Immunology.* 2003;108(4):427–430.
33. Kurth J, Spieker T, Wustrow J, et al. EBV-infected B cells in infectious mononucleosis: viral strategies for spreading in the B cell compartment and establishing latency. *Immunity.* 2000;13(4):485–495.
34. Niedobitek G, Herbst H, Young LS, et al. Patterns of Epstein-Barr virus infection in non-neoplastic lymphoid tissue. *Blood.* 1992;79(10):2520–2526.
35. Roughan JE, Torgbor C, Thorley-Lawson DA. Germinal center B cells latently infected with Epstein-Barr virus proliferate extensively but do not increase in number. *J Virol.* 2010;84(2):1158–1168.
36. Woellmer A, Hammerschmidt W. Epstein-Barr virus and host cell methylation: regulation of latency, replication and virus reactivation. *Curr Opin Virol.* 2013;3(3):260–265.
37. Shalginskikh N, Poleshko A, Skalka AM, Katz RA. Retroviral DNA methylation and epigenetic repression are mediated by the antiviral host protein Daxx. *J Virol.* 2013;87(4):2137–2150.
38. Bechtel JT, Winant RC, Ganem D. Host and viral proteins in the virion of Kaposi's sarcoma-associated herpesvirus. *J Virol.* 2005;79(8):4952–4964.
39. Kalla M, Schmeinck A, Bergbauer M, Pich D, Hammerschmidt W. AP-1 homolog BZLF1 of Epstein-Barr virus has two essential functions dependent on the epigenetic state of the viral genome. *Proc Natl Acad Sci USA.* 2010;107(2):850–855.
40. Takacs M, Segesdi J, Banati F, et al. The importance of epigenetic alterations in the development of Epstein-Barr virus-related lymphomas. *Mediterr J Hematol Infect Dis.* 2009;1(2):e2009012.
41. Robertson KD, Manns A, Swinnen LJ, Zong JC, Gulley ML, Ambinder RF. CpG methylation of the major Epstein-Barr virus latency promoter in Burkitt's lymphoma and Hodgkin's disease. *Blood.* 1996;88(8):3129–3136.
42. Woellmer A, Arteaga-Salas JM, Hammerschmidt W. BZLF1 governs CpG-methylated chromatin of Epstein-Barr Virus reversing epigenetic repression. *PLoS Pathog.* 2012;8(9):e1002902.
43. Hislop AD, Taylor GS, Sauce D, Rickinson AB. Cellular responses to viral infection in humans: lessons from Epstein-Barr virus. *Annu Rev Immunol.* 2007;25:587–617.
44. Iwakiri D, Zhou L, Samanta M, et al. Epstein-Barr virus (EBV)-encoded small RNA is released from EBV-infected cells and activates signaling from Toll-like receptor 3. *J Exp Med.* 2009;206(10):2091–2099.
45. Fiola S, Gosselin D, Takada K, Gosselin J. TLR9 contributes to the recognition of EBV by primary monocytes and plasmacytoid dendritic cells. *J Immunol.* 2010;185(6):3620–3631.
46. Lim WH, Kireta S, Russ GR, Coates PTH. Human plasmacytoid dendritic cells regulate immune responses to Epstein-Barr virus (EBV) infection and delay EBV-related mortality in humanized NOD-SCID mice. *Blood.* 2007;109(3):1043–1050.
47. Williams H, McAulay K, Macsween KF, et al. The immune response to primary EBV infection: a role for natural killer cells. *Br J Haematol.* 2005;129(2):266–274.

48. Chijioke O, Müller A, Feederle R, et al. Human natural killer cells prevent infectious mononucleosis features by targeting lytic Epstein-Barr virus infection. *Cell Rep.* 2013;5(6):1489–1498.
49. Adhikary D, Behrends U, Boerschmann H, et al. Immunodominance of lytic cycle antigens in Epstein-Barr virus-specific CD4+ T cell preparations for therapy. *PLoS One.* 2007;2(7):e583.
50. Fathallah I, Parroche P, Gruffat H, et al. EBV latent membrane protein 1 is a negative regulator of TLR9. *J Immunol.* 2010;185(11):6439–6447.
51. van Gent M, Griffin BD, Berkhoff EG, et al. EBV lytic-phase protein BGLF5 contributes to TLR9 downregulation during productive infection. *J Immunol.* 2011;186(3): 1694–1702.
52. van Gent M, Braem SGE, de Jong A, et al. Epstein-Barr virus large tegument protein BPLF1 contributes to innate immune evasion through interference with Toll-like receptor signaling. *PLoS Pathog.* 2014;10(2):e1003960.
53. Zeidler R, Eissner G, Meissner P, et al. Downregulation of TAP1 in B lymphocytes by cellular and Epstein-Barr virus-encoded interleukin-10. *Blood.* 1997;90(6): 2390–2397.
54. Zuo J, Thomas W, van Leeuwen D, et al. The DNase of gammaherpesviruses impairs recognition by virus-specific CD8+ T cells through an additional host shutoff function. *J Virol.* 2008;82(5):2385–2393.
55. Smith C, Wakisaka N, Crough T, et al. Discerning regulation of cis- and trans-presentation of CD8+ T-cell epitopes by EBV-encoded oncogene LMP-1 through self-aggregation. *Blood.* 2009;113(24):6148–6152.
56. Zuo J, Currin A, Griffin BD, et al. The Epstein-Barr virus G-protein-coupled receptor contributes to immune evasion by targeting MHC class I molecules for degradation. *PLoS Pathog.* 2009;5(1):e1000255.
57. Ressing ME, van Leeuwen D, Verreck FAW, et al. Epstein-Barr virus gp42 is post-translationally modified to produce soluble gp42 that mediates HLA class II immune evasion. *J Virol.* 2005;79(2):841–852.
58. Rosenkilde MM, Waldhoer M, Lüttichau HR, Schwartz TW. Virally encoded 7TM receptors. *Oncogene.* 2001;20(13):1582–1593.
59. Rosenkilde MM, Smit MJ, Waldhoer M. Structure, function and physiological consequences of virally encoded chemokine seven transmembrane receptors. *Br J Pharmacol.* 2008;153(Suppl. 1):S154–S166.
60. Vischer HF, Siderius M, Leurs R, Smit MJ. Herpesvirus-encoded GPCRs: neglected players in inflammatory and proliferative diseases? *Nat Rev Drug Discov.* 2014;13(2): 123–139.
61. Zuo J, Quinn LL, Tamblyn J, et al. The Epstein-Barr virus-encoded BILF1 protein modulates immune recognition of endogenously processed antigen by targeting major histocompatibility complex class I molecules trafficking on both the exocytic and endocytic pathways. *J Virol.* 2011;85(4):1604–1614.
62. Griffin BD, Gram AM, Mulder A, et al. EBV BILF1 evolved to downregulate cell surface display of a wide range of HLA class I molecules through their cytoplasmic tail. *J Immunol.* 2013;190(4):1672–1684.
63. Paulsen SJ, Rosenkilde MM, Eugen-Olsen J, Kledal TN. Epstein-Barr virus-encoded BILF1 is a constitutively active G protein-coupled receptor. *J Virol.* 2005;79(1): 536–546.
64. Beisser PS, Verzijl D, Gruijthuijsen YK, et al. The Epstein-Barr virus BILF1 gene encodes a G protein-coupled receptor that inhibits phosphorylation of RNA-dependent protein kinase. *J Virol.* 2005;79(1):441–449.
65. Spiess K, Rosenkilde MM. *G Protein-Coupled Receptor Genetics.* 2014;45–65.

66. Lyngaa R, Nørregaard K, Kristensen M, Kubale V, Rosenkilde MM, Kledal TN. Cell transformation mediated by the Epstein-Barr virus G protein-coupled receptor BILF1 is dependent on constitutive signaling. *Oncogene*. 2010;29(31):4388–4398.
67. Jensen A-SM, Sparre-Ulrich AH, Davis-Poynter N, Rosenkilde MM. Structural diversity in conserved regions like the DRY-Motif among viral 7TM receptors—a consequence of evolutionary pressure? *Adv Virol*. 2012;2012:231813.
68. Burger M, Burger JA, Hoch RC, Oades Z, Takamori H, Schraufstatter IU. Point mutation causing constitutive signaling of CXCR2 leads to transforming activity similar to Kaposi's sarcoma herpesvirus-G protein-coupled receptor. *J Immunol*. 1999;163(4):2017–2022.
69. Rosenkilde MM, Kledal TN, Holst PJ, Schwartz TW. Selective elimination of high constitutive activity or chemokine binding in the human herpesvirus 8 encoded seven transmembrane oncogene ORF74. *J Biol Chem*. 2000;275(34):26309–26315.
70. Waldhoer M, Casarosa P, Rosenkilde MM, et al. The carboxyl terminus of human cytomegalovirus-encoded 7 transmembrane receptor US28 camouflages agonism by mediating constitutive endocytosis. *J Biol Chem*. 2003;278(21):19473–19482.
71. Pleskoff O, Casarosa P, Verneuil L, et al. The human cytomegalovirus-encoded chemokine receptor US28 induces caspase-dependent apoptosis. *FEBS J*. 2005;272(16):4163–4177.
72. Case R, Sharp E, Benned-Jensen T, Rosenkilde MM, Davis-Poynter N, Farrell HE. Functional analysis of the murine cytomegalovirus chemokine receptor homologue M33: ablation of constitutive signaling is associated with an attenuated phenotype in vivo. *J Virol*. 2008;82(4):1884–1898.
73. Rosenkilde MM, Kledal TN, Schwartz TW. High constitutive activity of a virus-encoded seven transmembrane receptor in the absence of the conserved DRY motif (Asp-Arg-Tyr) in transmembrane helix 3. *Mol Pharmacol*. 2005;68(1):11–19.
74. Flanagan CA. A GPCR that is not "DRY" *Mol Pharmacol*. 2005;68(1):1–3.
75. Casarosa P, Bakker RA, Verzijl D, et al. Constitutive signaling of the human cytomegalovirus-encoded chemokine receptor US28. *J Biol Chem*. 2001;276(2): 1133–1137.
76. Waldhoer M, Kledal TN, Farrell H, Schwartz TW. Murine cytomegalovirus (CMV) M33 and human CMV US28 receptors exhibit similar constitutive signaling activities. *J Virol*. 2002;76(16):8161–8168.
77. McLean KA, Holst PJ, Martini L, Schwartz TW, Rosenkilde MM. Similar activation of signal transduction pathways by the herpesvirus-encoded chemokine receptors US28 and ORF74. *Virology*. 2004;325(2):241–251.
78. Bais C, Santomasso B, Coso O, et al. G-protein-coupled receptor of Kaposi's sarcoma-associated herpesvirus is a viral oncogene and angiogenesis activator. *Nature*. 1998;391(6662):86–89.
79. Rosenkilde MM, Kledal TN, Bräuner-Osborne H, Schwartz TW. Agonists and inverse agonists for the herpesvirus 8-encoded constitutively active seven-transmembrane oncogene product, ORF-74. *J Biol Chem*. 1999;274(2):956–961.
80. Holst PJ, Rosenkilde MM, Manfra D, et al. Tumorigenesis induced by the HHV8-encoded chemokine receptor requires ligand modulation of high constitutive activity. *J Clin Invest*. 2001;108(12):1789–1796.
81. Rosenkilde MM, McLean KA, Holst PJ, Schwartz TW. The CXC chemokine receptor encoded by herpesvirus saimiri, ECRF3, shows ligand-regulated signaling through Gi, Gq, and G12/13 proteins but constitutive signaling only through Gi and G12/13 proteins. *J Biol Chem*. 2004;279(31):32524–32533.
82. Verzijl D, Fitzsimons CP, Van Dijk M, et al. Differential activation of murine herpesvirus 68- and Kaposi's sarcoma-associated herpesvirus-encoded ORF74

G protein-coupled receptors by human and murine chemokines. *J Virol.* 2004;78(7):3343–3351.

83. Ahuja SK, Murphy PM. Molecular piracy of mammalian interleukin-8 receptor type B by herpesvirus saimiri. *J Biol Chem.* 1993;268(28):20691–20694.
84. Kledal TN, Rosenkilde MM, Schwartz TW. Selective recognition of the membrane-bound CX3C chemokine, fractalkine, by the human cytomegalovirus-encoded broad-spectrum receptor US28. *FEBS Lett.* 1998;441(2):209–214.
85. Rosenkilde MM, Schwartz TW. Potency of ligands correlates with affinity measured against agonist and inverse agonists but not against neutral ligand in constitutively active chemokine receptor. *Mol Pharmacol.* 2000;57(3):602–609.
86. Maussang D, Verzijl D, van Walsum M, et al. Human cytomegalovirus-encoded chemokine receptor US28 promotes tumorigenesis. *Proc Natl Acad Sci USA.* 2006;103(35):13068–13073.
87. Birkenbach M, Josefsen K, Yalamanchili R, Lenoir G, Kieff E. Epstein-Barr virus-induced genes: first lymphocyte-specific G protein-coupled peptide receptors. *J Virol.* 1993;67(4):2209–2220.
88. Craig FE, Johnson LR, Harvey SAK, et al. Gene expression profiling of Epstein-Barr virus-positive and -negative monomorphic B-cell posttransplant lymphoproliferative disorders. *Diagn Mol Pathol.* 2007;16(3):158–168.
89. Cahir-McFarland ED, Carter K, Rosenwald A, et al. Role of NF-kappa B in cell survival and transcription of latent membrane protein 1-expressing or Epstein-Barr virus latency III-infected cells. *J Virol.* 2004;78(8):4108–4119.
90. Rosenkilde MM, Benned-Jensen T, Andersen H, et al. Molecular pharmacological phenotyping of EBI2. An orphan seven-transmembrane receptor with constitutive activity. *J Biol Chem.* 2006;281(19):13199–13208.
91. Hannedouche S, Zhang J, Yi T, et al. Oxysterols direct immune cell migration via EBI2. *Nature.* 2011;475(7357):524–527.
92. Liu C, Yang XV, Wu J, et al. Oxysterols direct B-cell migration through EBI2. *Nature.* 2011;475(7357):519–523.
93. Benned-Jensen T, Norn C, Laurent S, et al. Molecular characterization of oxysterol binding to the Epstein-Barr virus-induced gene 2 (GPR183). *J Biol Chem.* 2012;287(42):35470–35483.
94. Zhang L, Shih AY, Yang XV, et al. Identification of structural motifs critical for Epstein-Barr virus-induced molecule 2 function and homology modeling of the ligand docking site. *Mol Pharmacol.* 2012;82(6):1094–1103.
95. Benned-Jensen T, Smethurst C, Holst PJ, et al. Ligand modulation of the Epstein-Barr virus-induced seven-transmembrane receptor EBI2: identification of a potent and efficacious inverse agonist. *J Biol Chem.* 2011;286(33):29292–29302.
96. A-Gonzalez N, Bensinger SJ, Hong C, et al. Apoptotic cells promote their own clearance and immune tolerance through activation of the nuclear receptor LXR. *Immunity.* 2009;31(2):245–258.
97. Valledor AF, Hsu L-C, Ogawa S, Sawka-Verhelle D, Karin M, Glass CK. Activation of liver X receptors and retinoid X receptors prevents bacterial-induced macrophage apoptosis. *Proc Natl Acad Sci USA.* 2004;101(51):17813–17818.
98. Joseph SB, Bradley MN, Castrillo A, et al. LXR-dependent gene expression is important for macrophage survival and the innate immune response. *Cell.* 2004;119(2):299–309.
99. Bensinger SJ, Bradley MN, Joseph SB, et al. LXR signaling couples sterol metabolism to proliferation in the acquired immune response. *Cell.* 2008;134(1):97–111.
100. Cui G, Qin X, Wu L, et al. Liver X receptor (LXR) mediates negative regulation of mouse and human Th17 differentiation. *J Clin Invest.* 2011;121(2):658–670.

101. Preuss I, Ludwig M-G, Baumgarten B, et al. Transcriptional regulation and functional characterization of the oxysterol/EBI2 system in primary human macrophages. *Biochem Biophys Res Commun*. 2014;446(3):663–668.
102. Shaffer AL, Rosenwald A, Hurt EM, et al. Signatures of the immune response. *Immunity*. 2001;15(3):375–385.
103. Pereira JP, Kelly LM, Xu Y, Cyster JG. EBI2 mediates B cell segregation between the outer and centre follicle. *Nature*. 2009;460(7259):1122–1126.
104. Gatto D, Paus D, Basten A, Mackay CR, Brink R. Guidance of B cells by the orphan G protein-coupled receptor EBI2 shapes humoral immune responses. *Immunity*. 2009;31(2):259–269.
105. Gatto D, Brink R. B cell localization: regulation by EBI2 and its oxysterol ligand. *Trends Immunol*. 2013;34:336–341.
106. Pereira JP, Kelly LM, Cyster JG. Finding the right niche: B-cell migration in the early phases of T-dependent antibody responses. *Int Immunol*. 2010;22(6):413–419.
107. Yi T, Wang X, Kelly LM, et al. Oxysterol gradient generation by lymphoid stromal cells guides activated B cell movement during humoral responses. *Immunity*. 2012;37(3):535–548.
108. Yi T, Cyster JG. EBI2-mediated bridging channel positioning supports splenic dendritic cell homeostasis and particulate antigen capture. *Elife*. 2013;2:e00757.
109. Gatto D, Wood K, Caminschi I, et al. The chemotactic receptor EBI2 regulates the homeostasis, localization and immunological function of splenic dendritic cells. *Nat Immunol*. 2013;14(8):876.
110. Chiang EY, Johnston RJ, Grogan JL. EBI2 is a negative regulator of type I interferons in plasmacytoid and myeloid dendritic cells. *PLoS One*. 2013;8(12):e83457.
111. Diczfalusy U, Olofsson KE, Carlsson A-M, et al. Marked upregulation of cholesterol 25--hydroxylase expression by lipopolysaccharide. *J Lipid Res*. 2009;50(11):2258–2264.
112. Park K, Scott AL. Cholesterol 25-hydroxylase production by dendritic cells and macrophages is regulated by type I interferons. *J Leukoc Biol*. 2010;88(6):1081–1087.
113. Bauman DR, Bitmansour AD, McDonald JG, Thompson BM, Liang G, Russell DW. 25-Hydroxycholesterol secreted by macrophages in response to Toll-like receptor activation suppresses immunoglobulin A production. *Proc Natl Acad Sci USA*. 2009;106(39):16764–16769.
114. Liu S-Y, Aliyari R, Chikere K, et al. Interferon-inducible cholesterol-25-hydroxylase broadly inhibits viral entry by production of 25-hydroxycholesterol. *Immunity*. 2013;38(1):92–105.
115. Nau GJ, Richmond JFL, Schlesinger A, Jennings EG, Lander ES, Young RA. Human macrophage activation programs induced by bacterial pathogens. *Proc Natl Acad Sci USA*. 2002;99(3):1503–1508.
116. Kelly LM, Pereira JP, Yi T, Xu Y, Cyster JG. EBI2 guides serial movements of activated B cells and ligand activity is detectable in lymphoid and nonlymphoid tissues. *J Immunol*. 2011;187(6):3026–3032.
117. Sirén J, Sareneva T, Pirhonen J, et al. Cytokine and contact-dependent activation of natural killer cells by influenza A or Sendai virus-infected macrophages. *J Gen Virol*. 2004;85(Pt. 8):2357–2364.
118. Stuller KA, Flaño E. CD4 T cells mediate killing during persistent gammaherpesvirus 68 infection. *J Virol*. 2009;83(9):4700–4703.
119. Viola A, Luster AD. Chemokines and their receptors: drug targets in immunity and inflammation. *Annu Rev Pharmacol Toxicol*. 2008;48:171–197.
120. Rosenkilde MM, Kledal TN. Targeting herpesvirus reliance of the chemokine system. *Curr Drug Targets*. 2006;7(1):103–118.

121. Ehlin-Henriksson B, Liang W, Cagigi A, Mowafi F, Klein G, Nilsson A. Changes in chemokines and chemokine receptor expression on tonsillar B cells upon Epstein-Barr virus infection. *Immunology*. 2009;127(4):549–557.
122. Nakayama T, Fujisawa R, Izawa D, Hieshima K, Takada K, Yoshie O. Human B cells immortalized with Epstein-Barr virus upregulate CCR6 and CCR10 and down-regulate CXCR4 and CXCR5. *J Virol*. 2002;76(6):3072–3077.
123. Chen A, Zhao B, Kieff E, Aster JC, Wang F. EBNA-3B- and EBNA-3C-regulated cellular genes in Epstein-Barr virus-immortalized lymphoblastoid cell lines. *J Virol*. 2006;80(20):10139–10150.
124. Nijmeijer S, Leurs R, Smit MJ, Vischer HF. The Epstein-Barr virus-encoded G protein-coupled receptor BILF1 hetero-oligomerizes with human CXCR4, scavenges Gαi proteins, and constitutively impairs CXCR4 functioning. *J Biol Chem*. 2010;285(38):29632–29641.
125. Hasegawa H, Utsunomiya Y, Yasukawa M, Yanagisawa K, Fujita S. Induction of G protein-coupled peptide receptor EBI 1 by human herpesvirus 6 and 7 infection in CD4+ T cells. *J Virol*. 1994;68(8):5326–5329.
126. Yasukawa M, Hasegawa A, Sakai I, et al. Down-regulation of CXCR4 by human herpesvirus 6 (HHV-6) and HHV-7. *J Immunol*. 1999;162(9):5417–5422.
127. Piovan E, Tosello V, Indraccolo S, et al. Chemokine receptor expression in EBV-associated lymphoproliferation in hu/SCID mice: implications for CXCL12/CXCR4 axis in lymphoma generation. *Blood*. 2005;105(3):931–939.
128. Zhao B, Mar JC, Maruo S, et al. Epstein-Barr virus nuclear antigen 3C regulated genes in lymphoblastoid cell lines. *Proc Natl Acad Sci USA*. 2011;108(1):337–342.
129. Anagnostopoulos I, Hummel M, Kreschel C, Stein H. Morphology, immunophenotype, and distribution of latently and/or productively Epstein-Barr virus-infected cells in acute infectious mononucleosis: implications for the inter-individual infection route of Epstein-Barr virus. *Blood*. 1995;85(3):744–750.
130. Tanaka Y, Imai T, Baba M, et al. Selective expression of liver and activation-regulated chemokine (LARC) in intestinal epithelium in mice and humans. *Eur J Immunol*. 1999;29(2):633–642.
131. Pan J, Kunkel EJ, Gosslar U, et al. A novel chemokine ligand for CCR10 and CCR3 expressed by epithelial cells in mucosal tissues. *J Immunol*. 2000;165(6):2943–2949.
132. Wang W, Soto H, Oldham ER, et al. Identification of a novel chemokine (CCL28), which binds CCR10 (GPR2). *J Biol Chem*. 2000;275(29):22313–22323.
133. Nakayama T, Hieshima K, Nagakubo D, et al. Selective induction of Th2-attracting chemokines CCL17 and CCL22 in human B cells by latent membrane protein 1 of Epstein-Barr virus. *J Virol*. 2004;78(4):1665–1674.
134. Uchihara J-N, Krensky AM, Matsuda T, et al. Transactivation of the CCL5/RANTES gene by Epstein-Barr virus latent membrane protein 1. *Int J Cancer*. 2005;114(5): 747–755.
135. Beck A, Päzolt D, Grabenbauer GG, et al. Expression of cytokine and chemokine genes in Epstein-Barr virus-associated nasopharyngeal carcinoma: comparison with Hodgkin's disease. *J Pathol*. 2001;194(2):145–151.
136. Teruya-Feldstein J, Jaffe ES, Burd PR, Kingma DW, Setsuda JE, Tosato G. Differential chemokine expression in tissues involved by Hodgkin's disease: direct correlation of eotaxin expression and tissue eosinophilia. *Blood*. 1999;93(8):2463–2470.
137. Vouloumanou EK, Rafailidis PI, Falagas ME. Current diagnosis and management of infectious mononucleosis. *Curr Opin Hematol*. 2012;19(1):14–20.
138. Cohen JI. Epstein-Barr virus infection. *N Engl J Med*. 2000;343(7):481–492.
139. Kutok JL, Wang F. Spectrum of Epstein-Barr virus-associated diseases. *Annu Rev Pathol*. 2006;1:375–404.

140. Hochberg D, Souza T, Catalina M, Sullivan JL, Luzuriaga K, Thorley-Lawson DA. Acute infection with Epstein-Barr virus targets and overwhelms the peripheral memory B-cell compartment with resting, latently infected cells. *J Virol.* 2004;78(10): 5194–5204.
141. Hjalgrim H, Askling J, Rostgaard K, et al. Characteristics of Hodgkin's lymphoma after infectious mononucleosis. *N Engl J Med.* 2003;349(14):1324–1332.
142. Hjalgrim H, Rostgaard K, Johnson PCD, et al. HLA-A alleles and infectious mononucleosis suggest a critical role for cytotoxic T-cell response in EBV-related Hodgkin lymphoma. *Proc Natl Acad Sci USA.* 2010;107(14):6400–6405.
143. de-Thé G. Epstein-Barr virus and Burkitt's lymphoma worldwide: the causal relationship revisited. *IARC Sci Publ.* 1985;60:165–176.
144. Thorley-Lawson DA, Gross A. Persistence of the Epstein-Barr virus and the origins of associated lymphomas. *N Engl J Med.* 2004;350(13):1328–1337.
145. Küppers R. B cells under influence: transformation of B cells by Epstein-Barr virus. *Nat Rev Immunol.* 2003;3(10):801–812.
146. Glaser SL, Lin RJ, Stewart SL, et al. Epstein-Barr virus-associated Hodgkin's disease: epidemiologic characteristics in international data. *Int J Cancer.* 1997;70(4):375–382.
147. Muir CS. Cancer of the head and neck. Nasopharyngeal cancer. Epidemiology and etiology. *JAMA.* 1972;220(3):393–394.
148. Friborg JT, Melbye M. Cancer patterns in Inuit populations. *Lancet Oncol.* 2008;9(9):892–900.
149. Saha A, Robertson ES. Epstein-Barr virus-associated B-cell lymphomas: pathogenesis and clinical outcomes. *Clin Cancer Res.* 2011;17(10):3056–3063.
150. Shibata D, Weiss LM. Epstein-Barr virus-associated gastric adenocarcinoma. *Am J Pathol.* 1992;140(4):769–774.
151. Sivachandran N, Dawson CW, Young LS, Liu F-F, Middeldorp J, Frappier L. Contributions of the Epstein-Barr virus EBNA1 protein to gastric carcinoma. *J Virol.* 2012;86(1):60–68.
152. Capello D, Rossi D, Gaidano G. Post-transplant lymphoproliferative disorders: molecular basis of disease histogenesis and pathogenesis. *Hematol Oncol.* 2005;23(2):61–67.
153. Bibas M, Antinori A. EBV and HIV-related lymphoma. *Mediterr J Hematol Infect Dis.* 2009;1(2):e2009032.
154. Pender MP. Preventing and curing multiple sclerosis by controlling Epstein-Barr virus infection. *Autoimmun Rev.* 2009;8(7):563–568.
155. Thacker EL, Mirzaei F, Ascherio A. Infectious mononucleosis and risk for multiple sclerosis: a meta-analysis. *Ann Neurol.* 2006;59(3):499–503.
156. Serafini B, Rosicarelli B, Franciotta D, et al. Dysregulated Epstein-Barr virus infection in the multiple sclerosis brain. *J Exp Med.* 2007;204(12):2899–2912.
157. Hopkins AL, Groom CR. The druggable genome. *Nat Rev Drug Discov.* 2002;1(9):727–730.
158. Heslop HE. How I, treat EBV lymphoproliferation. *Blood.* 2009;114(19):4002–4008.
159. Little RF, Dunleavy K. Update on the treatment of HIV-associated hematologic malignancies. *Hematol Am Soc Hematol Educ Program.* 2013;2013:382–388.

INDEX

Note: Page numbers followed by "*f*" indicate figures and "*t*" indicate tables.

W

X

Y

COLOR PLATES

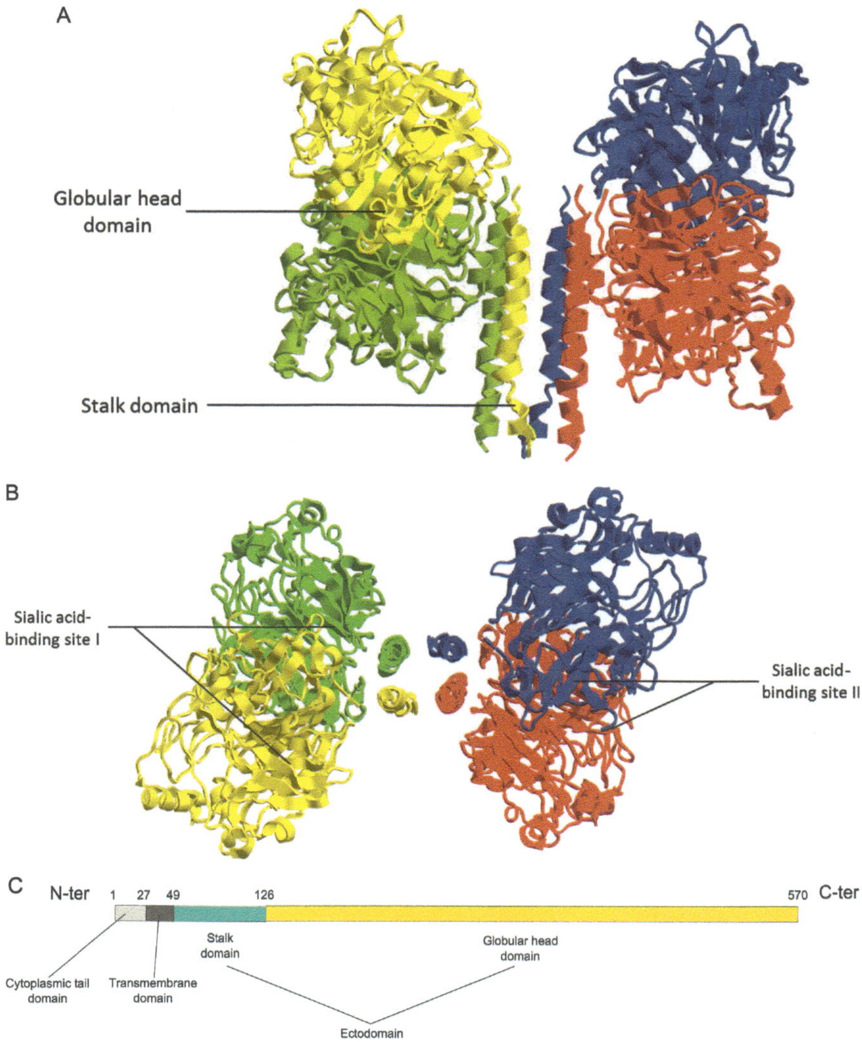

PLATE 1 ***Structure of the Newcastle disease virus hemagglutinin–neuraminidase protein.*** (A) Side view of the crystal structure of the tetramerized NDV-HN ectodomain showing the stalk and the globular domains of each monomer. Each color represents one monomer of the receptor-binding protein. One dimer is composed of green and yellow monomers, and the other of red and blue monomers. (PDB ID: 3T1E; Yuan *et al.*[48]). (B) Top view of the crystal structure of the tetramerized NDV-HN ectodomain, showing sialic acid-binding sites I and II. Each monomer bears a site I and a site II (PDB ID: 3T1E; Yuan *et al.*[48]). (C) Schematic representation of the domains of NDV-HN. (Figure 3 on page 9 of this Volume)

A

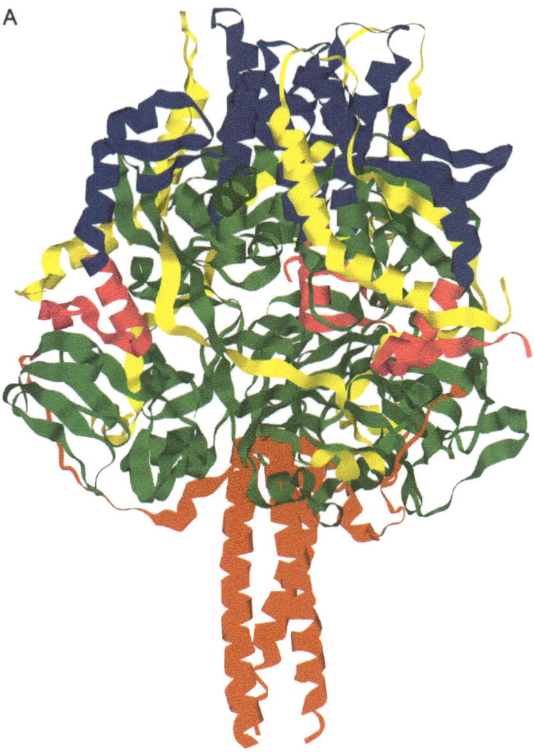

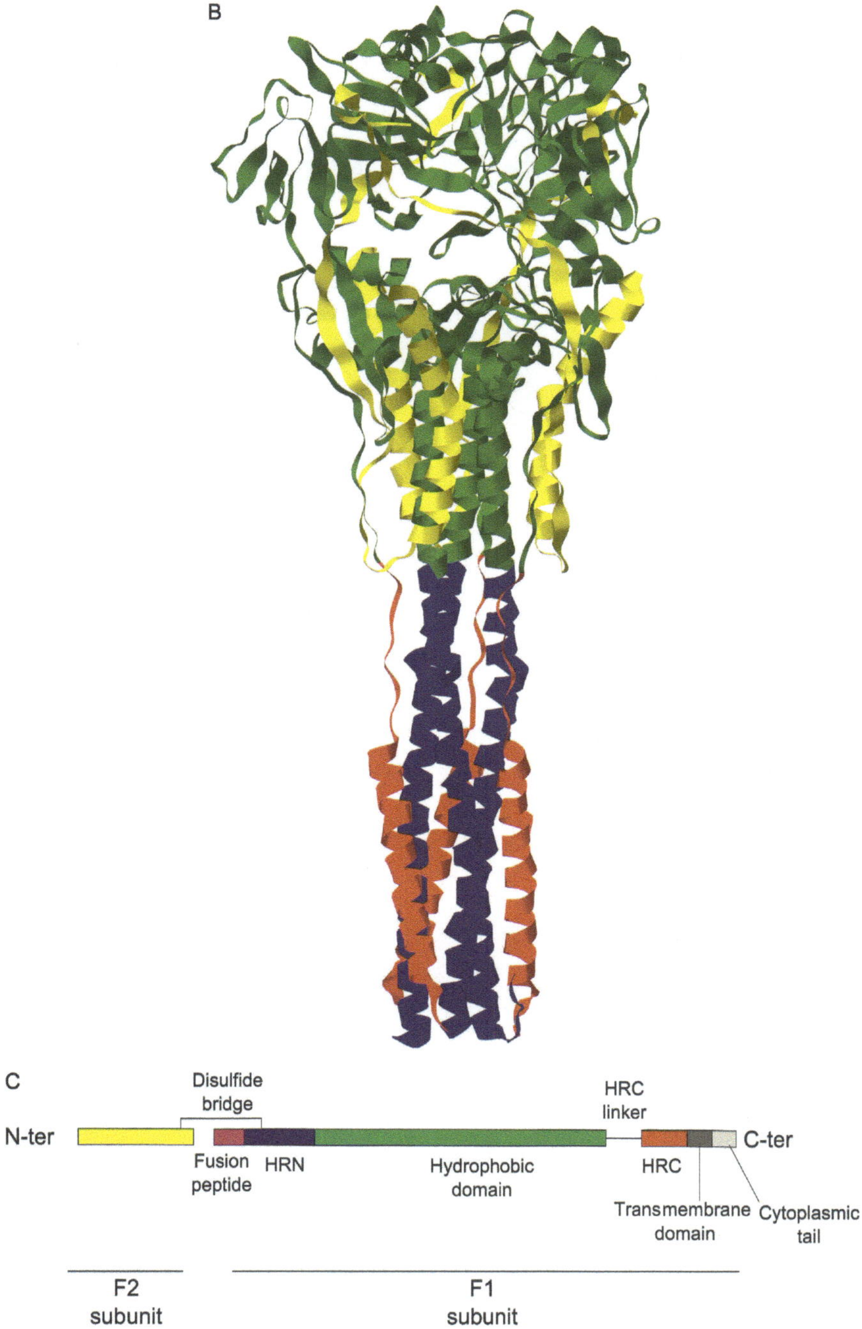

PLATE 2 ***Structure of the paramyxovirus fusion protein.*** (A) Crystal structure of the prefusion state of the trimeric fusion protein of PIV5 showing the fusion peptide (purple) in the hydrophobic pocket formed by a hydrophobic domain (deep green), the HRN domain (deep blue), and the F_2 subunit (yellow) (PDB ID: 4GIP; Welch *et al.*[69]). (B) Crystal structure of the postfusion state of the fusion protein of HPIV3 showing the 6-helix bundle structure formed by the HRN (deep blue) and HRC (red) domains interacting together (PDB ID: 1ZTM; Yin *et al.*[70]). (C) Schematic representation of the main domains of a monomer of the cleaved paramyxovirus fusion protein. HRC/HRN: heptad repeat C-/N-terminal domain. (Figure 4 on page 11 of this Volume)

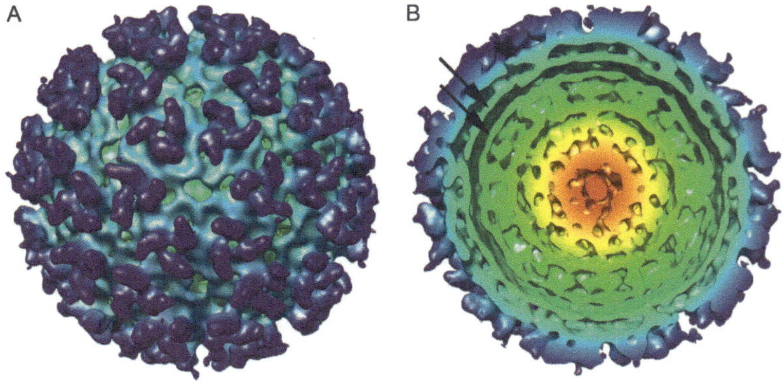

PLATE 3 ***Sindbis virus structure.*** (A) Cryo-EM reconstruction of Sindbis virus, view from the fivefold axis. The outer most envelope of E1/E2 trimers is shown in blue. (B) Cutaway view of the structure, the inner most nucleocapsid shell of the virus particle, is shown in green/yellow/red. The host-derived lipid bilayer is sandwiched between the layers (arrows). (Figure 1 on page 36 of this Volume)

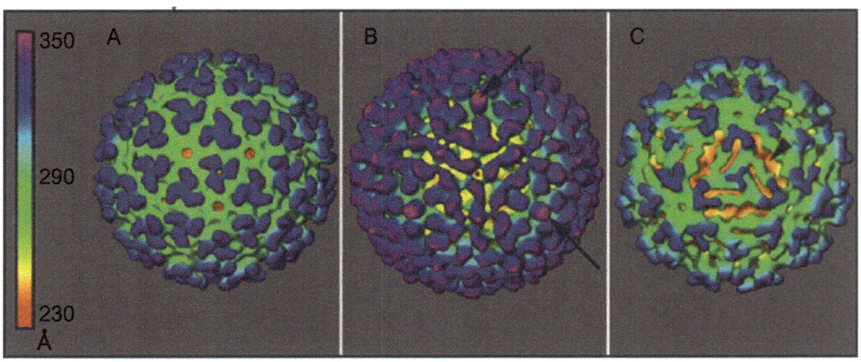

PLATE 4 ***The three-dimensional structures of Sindbis virus surface at 28 Å resolution viewed along icosahedral threefold axis.*** (A) Sindbis virus structure at pH 7.2. The conformational changes in Sindbis virus after exposure to pH 5.3 (B) and back to neutral pH (C) are shown. The arrows point to the protruding structure at the fivefold vertices. Arrowhead shows the fissure that appears at the twofold axis upon return to neutral pH. *Reprinted with permission from Elsevier from Ref. 32.* (Figure 3 on page 40 of this Volume)

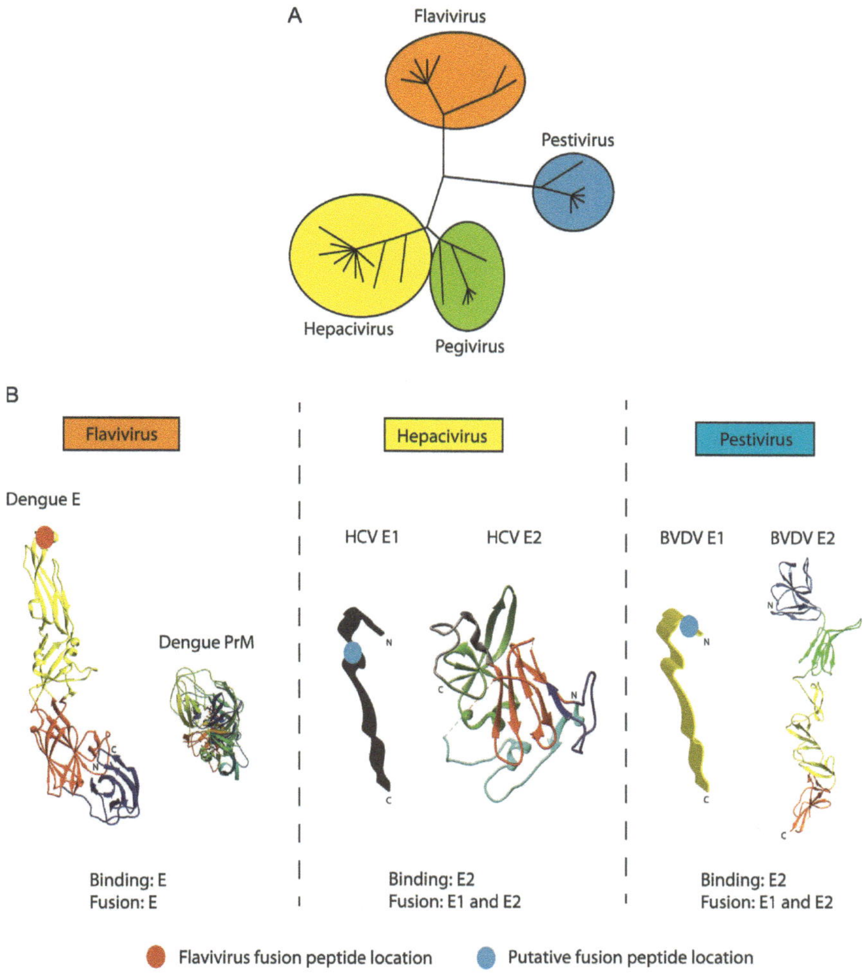

PLATE 5 ***Flaviviridae envelope glycoproteins.*** (A) Schematic representation of the phylogenetic tree of the Flaviviridae family. Each genus is labeled with a particular color. Flavivirus, orange; pestivirus, blue; pegivirus, green; hepacivirus, yellow. (B) Envelope glycoproteins of flavivirus (Dengue E and PrM), hepacivirus (HCV E1 and E2), and pestivirus (BDVD-1 E1 and E2) are compared. Flavivirus fusion peptide location is highlighted as a large red dot within E domain II. HCV and BVDV E1 putative structures are suggested as well as the putative location of their fusion peptide (large blue dot). Known or putative functions for each protein are indicated below the protein structures. HCV, Hepatitis C Virus; BVDV, Bovine viral diarrhea virus. Dengue E (PDB 1K4R), Dengue PrM (PDB 3C6E), HCV E2 core (PDB 4MWF), and BVDV-1 E2 (PDB 4JNT). (Figure 3 on page 92 of this Volume)

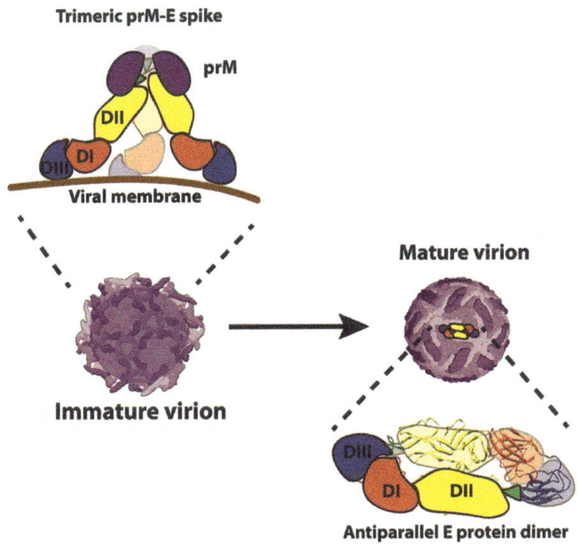

PLATE 6 ***The structure of the flavivirus envelope proteins and their organization on the virus particle.*** Flaviviruses are small spherical virions that incorporate a dense array of prM and E proteins that function to promote virus assembly, budding, and entry. The E protein is an elongated three domain molecule that is the principal target of neutralizing antibodies. Domain III (DIII, shown in blue) is thought to interact with receptors on target cells. The conserved 13-amino acid fusion loop (shown in green) is located at the distal end of domain II (DII, shown in yellow). DIII and DII are connected by the central domain I (DI, shown in red). In this cartoon, the prM protein is depicted as a purple spherical oval. Flaviviruses assemble as immature virus particles on which the E protein exists as 60 trimers of prM-E dimers that project as spikes away from the virion surface. Virus maturation is mediated by cleavage of prM by a cellular furin protease. On mature virions, the E protein exists as antiparallel homodimers that lie flat against the surface on the virion. The arrangement of E proteins on the mature DENV virion is depicted. Each virus particle is composed of 30 rafts of three antiparallel dimers in a herringbone pattern. (Figure 1 on page 145 of this Volume)

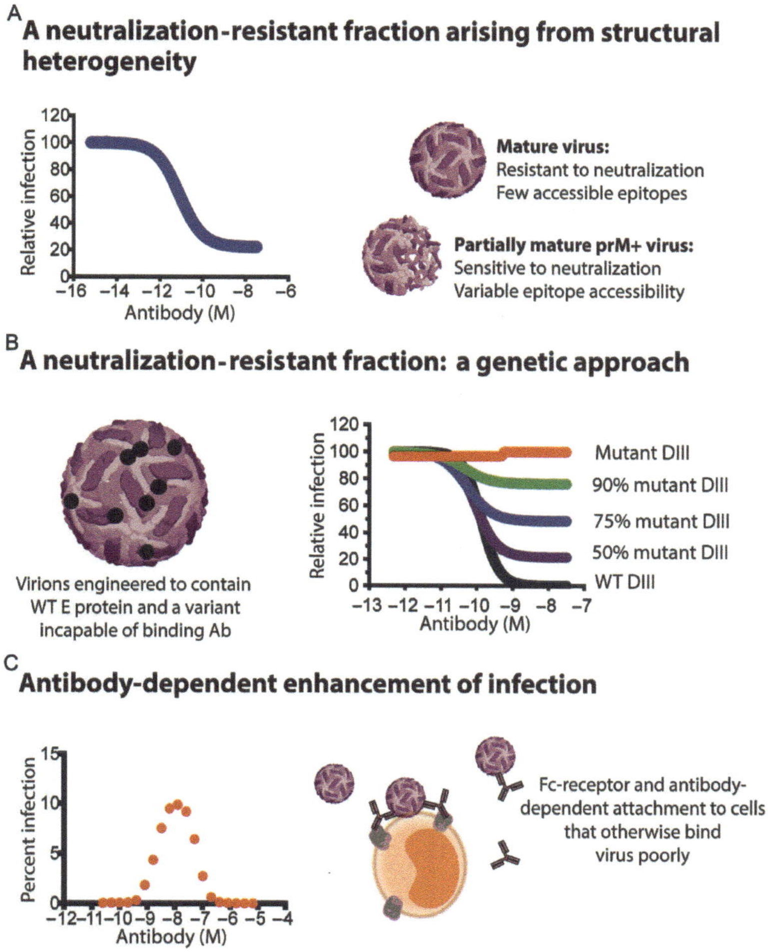

PLATE 7 ***Evidence in support of a multiple-hit requirement for neutralization.*** Several types of experimental studies support a multiple-hit model of antibody neutralization of flaviviruses. (A) In some instances, neutralization assays reveal a plateau effect at high concentrations of antibody. For flaviviruses, the neutralization-resistant fraction is a consequence of the significant structural heterogeneity of viruses released from infected cells. Mature virions in the population that do not display enough epitopes for a given antibody with a stoichiometry sufficient for neutralization are resistant to inhibition. Thus, infectious virions can be decorated by antibody with a stoichiometry insufficient for neutralization even under conditions of antibody excess or saturation. (B) A fraction of WNV virions resistant to neutralization can be created experimentally by reducing the number of epitopes on the average virion using genetic approaches in which E proteins with mutations (in this instance on DIII) that do not bind a given antibody are increasingly expressed on virions. (C) ADE describes a phenomenon in which antibody binds at an occupancy that is insufficient for neutralization yet capable of enhancing the efficiency of virus attachment and entry into cells via interactions with Fcγ-receptors expressed on cells. Because incubation with subneutralizing concentrations of antibody can enhance the infectivity via ADE, docking of a single antibody to the virion is insufficient for neutralization. (Figure 2 on page 148 of this Volume)

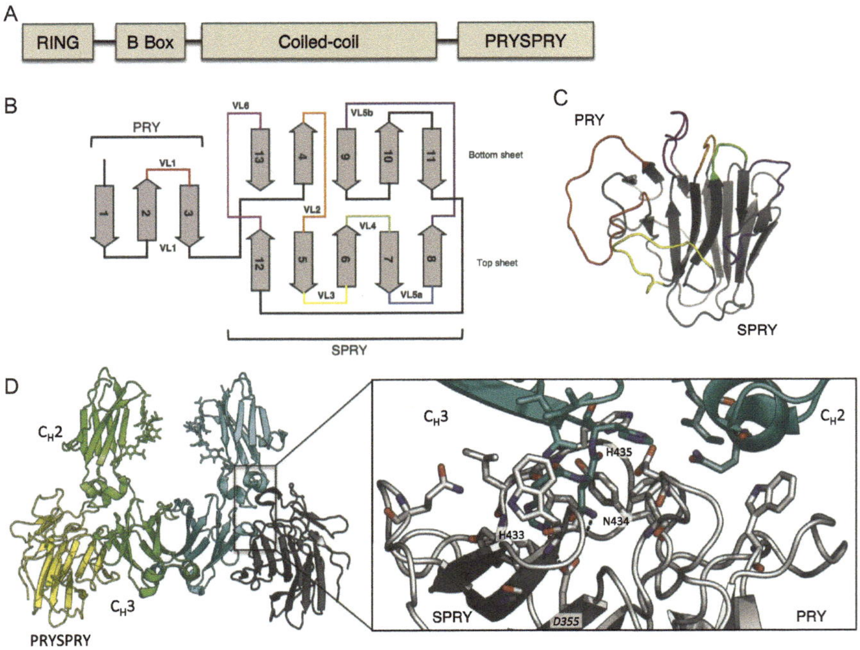

PLATE 8 ***TRIM21 structure and antibody binding mechanism.*** (A) TRIM21 domain structure. (B) Schematic diagram of TRIM21 PRYSPRY secondary structure in which the β-sheets and flexible binding site loops (VLs) are numbered. (C) Structure of TRIM21 PRYSPRY domain with VLs depicted in the same color scheme as in (B). (D) Structure of TRIM21 PRYSPRY:IgG Fc based on PDB 2IWG. The Fc is indicated in green and cyan and the PRYSPRY in yellow and gray. The boxed region is a close-up view of the interface. The "HNH" motif from Fc is marked. (Figure 1 on page 169 of this Volume)

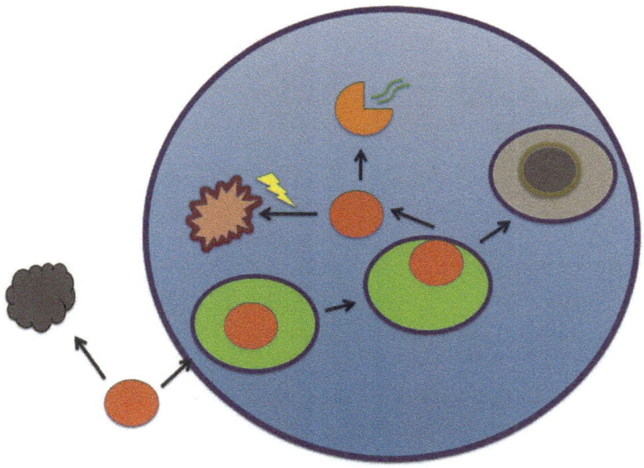

PLATE 9 ***Infectivity decay in suspension and abortive events during attachment and after entry.*** While virions are diffusing at random, they may lose their infectivity and become inert particles (gray) because of the exponential decay of their components; for enveloped viruses the decay of the envelope glycoproteins may often determine the infectivity half-life. The virion (red) is depicted as getting endocytosed (the endosomal lumen is bright green) and as it later either fuses with or penetrates from the endosome. In case of excessive delay before cytoplasmic entry, the endosome continues to the lysosomal compartment and the virion gets degraded (gray). In case of successful entry of the capsid into the cytoplasm, a productive uncoating (orange) may lead to genome (green curved double lines) release for transcription or, in competition with this, degradation of the whole core by the proteasome (yellow flash). (Figure 3 on page 310 of this Volume)

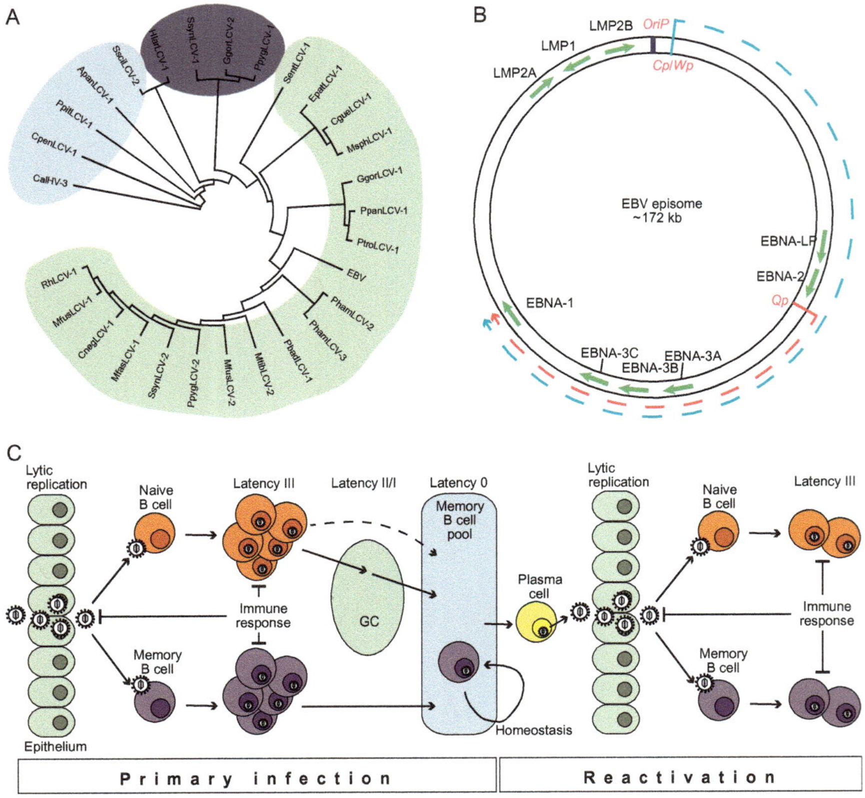

PLATE 10 ***Overview of lymphocryptovirus, genome organization, and life cycle of EBV.*** (A) Dendrogram of the lymphocryptovirus (LCV) genus of the gammaherpes virus subfamily. The phylogenetic analysis was based on the amino acid sequence for the DPOL gene sequences from amino acid no 1–693 (for EBV). Amino acid sequence alignments were made using MAFFT.[3] The dendrogram was built using the maximum likelihood method in the PhyML 2.2.0 plug-in.[4] The LG nucleotide substitution model was applied. The major clades A (LCV hosted by New World monkeys), B and C (LCV hosted by Old World monkeys) are shown in blue, green, and purple, respectively. (B) Schematic overview of the EBV episome with indication of latent protein-coding genes (black) and promoters (pink). The latent genes are marked by green arrows. The latent origin of replication, OriP, is shown in dark blue. The dotted, long, and outer blue arrow shows the genes transcribed from the Cp and Wp promoter (all the EBNAs), whereas the dotted, inner, and pink arrow shows the transcription promoted from Qp (EBNA-1 only). (C) Proposed life cycle of EBV: the virus establishes a lytic infection in epithelial cells and/or B cells. The B cells express the growth-promoting latency III program and enter a state of highly proliferating lymphoblasts. Many of these are detected and removed by the immune system, but some express the latency II and I programs and secure entry into the memory B cell pool, where the virus can reside without immune detection. The persisting virus infection is continuously maintained via homeostasis of the infected memory B cells. Infected memory B cells differentiating into plasmacells may activate lytic replication and virus shedding via epithelial cell infection. The reactivation is efficiently repressed by the memory immune response. Establishment of the latency in memory B cells has been proposed to occur via (1) direct infection and growth promotion of memory B cells. (2) Infection of naïve B cells and virus-driven transformation into memory B cells in a process, which mimics the antigen-driven memory cell differentiation of noninfected naïve B cells. This virus-driven transformation can either occur through the GCs, which represents a hostile environment or through GC-like differentiation outside the actual GCs (dotted arrow). *Adapted from Ref. 5.* (Figure 1 on page 397 of this Volume)

Kolibri 360 GmbH
Ferdinand-Jühlke-Straße 7
99095 Erfurt, Deutschland
produktsicherheit@kolibri360.de